A Molecular Approach to Primary Metabolism in Higher Plants

Edited by

CHRISTINE H. FOYER

Institute of Grassland and Environmental Research, Aberystwyth

and

W. PAUL QUICK

Department of Animal and Plant Sciences, University of Sheffield

Taylor & Francis
Publishers since 1798

UK	Taylor & Francis Ltd, 1 Gunpowder Square, London, EC4A 3DE
USA	Taylor & Francis Inc., 1900 Frost Road, Suite 101, Bristol, PA 19007

Copyright © Taylor & Francis Ltd 1997

All rights reserved. No part of this publication may be reproduced, stored in a retrieval system, or transmitted, in any form or by any means, electronic, electrostatic, magnetic tape, mechanical, photocopying, recording or otherwise, without the prior permission of the copyright owner.

British Library Cataloguing in Publication Data
A catalogue record for this book is available from the British Library.
ISBN 0-7484-0418-X (cased)
ISBN 0-7484-0419-8 (paperback)

Library of Congress Cataloging in Publication Data are available

Cover design by Jim Wilkie
Typeset in Times 10/12 pt by Acorn Bookwork, Salisbury, UK
Printed in Great Britain by T. J. International Ltd, Padstow

DEDICATION

This book is dedicated to Professor Tom ap Rees as a mark of his enthusiasm for plant biochemistry in the UK through research teaching and increasing public awareness.

...If I have seen further it is because I stood on the shoulders of giants.

Contents

Preface ix

List of Contributors xi

List of Abbreviations xv

SECTION ONE   Primary Nitrogen Assimilation, Carbon Assimilation and Carbon Partitioning 1

1   The photosynthetic electron transport system: efficiency and control 3
    Christine H. Foyer and Jeremy Harbinson

2   The regulation and control of photosynthetic carbon assimilation 41
    W. Paul Quick and H. Ekkehard Neuhaus

3   Modulation of sucrose metabolism 63
    Uwe Sonnewald

4   Modulation of starch synthesis 81
    Jack Preiss

5   Molecular crosstalk and the regulation of C- and N-responsive genes 105
    Karen E. Koch

6   Manipulation of the pathways of sucrose biosynthesis and nitrogen assimilation in transformed plants to improve photosynthesis and productivity 125
    Sylvie Ferrario-Méry, Erik Murchie, Bertrand Hirel, Nathalie Galtier, W. Paul Quick and Christine H. Foyer

7   Manipulating amino acid biosynthesis 155
    Stephen J. Temple and Champa Sengupta-Gopalan

Contents

8   Regulation of C/N metabolism by reversible protein phosphorylation   179
    Carol MacKintosh

SECTION TWO   Compartmentation, Transport and Whole Plant
Interactions   203

9   Compartmentation of C/N metabolism   205
    Dieter Heineke, Gertrud Lohaus and Heike Winter

10 Plasmodesmal-mediated plant communication network: implications for
    controlling carbon metabolism and resource allocation   219
    Shmuel Wolf and William J. Lucas

11 Nitrogen uptake and assimilation in roots and root nodules   237
    Alain Ourry, Anthony J. Gordon and James H. Macduff

12 Probing the carbon and nitrogen interaction: a whole plant perspective   255
    Thomas W. Rufty

13 The role of mycorrhiza   275
    Rüdiger Hampp and Astrid Wingler

SECTION THREE   Related Metabolism   293

14 Respiration and the alternative oxidase   295
    Hans Lambers

15 Manipulation of oil biosynthesis and lipid composition   311
    Matthew J. Hills and Stephen Rawsthorne

16 Resource use efficiency and crop performance: what is the link?   327
    Christopher J. Pollock

Preface

The recent developments in molecular biology that have led to a dramatic increase in our understanding of carbon and nitrogen assimilation, partitioning and interaction, prompted us to compile this volume. In the modern world students and young research workers have to be conversant with more than one discipline and we therefore perceived a need for a comprehensive up-to-date multidisciplinary volume covering all the major facets of the carbon–nitrogen relationship in higher plants.

In the past few years the application of molecular genetic techniques has led to a substantial increase in the available information on the regulation of plant primary metabolism, particularly with regard to compartmentation, co-ordination, transport, sensing, signal transduction and gene transcription. Strategies for manipulating plant metabolism involve the specific targeting of individual enzymes or pathways. This approach has been remarkably successful and has not only helped elucidate the roles of specific enzymes, but has also demonstrated the plasticity of plant metabolism which leads to compensatory changes as the plant attempts to counteract the metabolic effects of human intervention.

We have also become increasingly aware of the mechanisms whereby plants sense changes in their carbon and nitrogen status and relay this information to the nucleus where changes in gene expression are brought about. We have attempted to provide relevant information on the molecular co-ordination at the level of the cell and between different tissues that are separated both spatially and developmentally. Superimposed on this is the co-ordination of resource allocation in response to environmental stimuli as well as endogenous controls.

Crop productivity depends primarily on how efficiently light energy is used to drive carbon and nitrogen assimilation and how effectively the products of primary assimilation are allocated and utilized during plant growth and development. We believe that increasing plant yield potential is still a realistic goal for plant breeding. In the past, the most successful approach has been either to improve the environmental conditions in which crops grow by irrigation or by application of fertilizers, or to identify genotypes that are successful in a given environment. To realize the potential benefit with respect to sustained agriculture,

Preface

the former is costly and often damaging to the environment. The use of transformation technology to customize plants for improved vigour and yield is, hence, becoming increasingly attractive. The need for innovative agricultural technology is urgent since most of the productive sustainable farmland on Earth is already under cultivation.

In this volume the reader is introduced to the concepts of carbon and nitrogen metabolism in plants, brought up-to-date with current opinion and how these interactive processes are co-ordinated, and is led towards a realization of the immense potential that this new understanding has to offer for agriculture. We aim to cover what we perceive to be all the relevant aspects of the carbon–nitrogen interactions, from primary reactions to molecular genetics, and the development of strategies for engineering increased plant productivity by optimizing resource allocation.

We have encouraged our authors to provide personal perspectives of their topics while discussing them in depth. Furthermore, we asked that some chapters include relatively simple introductions and conclusions so that they would be accessible to the non-specialist as well as the specialist. We hope that molecular biologists will find that the information on biochemistry and physiology is accessible and vice versa.

Finally, we hope that this work provides a forum for new ideas, concepts and approaches. We acknowledge the patience of our contributors who responded rapidly and with enthusiasm to our call and we thank them for following our exacting guidelines in the preparation of this volume.

Christine H. Foyer
W. Paul Quick

List of contributors

Sylvie Ferrario-Méry
Laboratoire du Metabolisme, INRA, Route de Saint Cyr, 78026 Versailles Cedex, France

Christine H. Foyer
Environmental Biology Department, Institute of Grassland and Environmental Research, Plas Gogerddan, Aberystwyth, Ceredigion SY23 3EB, UK

Nathalie Galtier
Laboratoire du Metabolisme, INRA, Route de Saint Cyr, 78026 Versailles Cedex, France

Anthony J. Gordon
Environmental Biology Department, Institute of Grassland and Environmental Research, Plas Gogerddan, Aberystwyth, Ceredigion SY23 3EB, UK

Rüdiger Hampp
Physiologische Ökologie der Pflanzen, Universität Tübingen, Auf der Morgenstelle 1, D-72076 Tübingen, Germany

Jeremy Harbinson
Agrotechnological Research Institute (ATO-DLO), Bornsesteeg 59, PO Box 17, NL-6700 AA Wageningen, The Netherlands

Dieter Heineke
Institut für Biochemie der Pflanze der Georg-August-Universität, Untere Karspüle, D-37073 Göttingen, Germany

Matthew J. Hills
Department of Brassica and Oilseeds Research, John Innes Centre, Norwich Research Park, Colney, Norwich NR4 7UH, UK

List of contributors

Bertrand Hirel
Laboratoire du Biologie Cellulaire, INRA, Route de Saint Cyr, 78026 Versailles Cedex, France

Karen E. Koch
Graduate Program in Plant Molecular and Cellular Biology, University of Florida, Gainesville, Florida 32611, USA

Hans Lambers
Plant Ecology and Evolutionary Biology Program, Horticultural Sciences Department, Utrecht University, Sorbonnelaan 16, 3584 CA Utrecht, The Netherlands

Gertrud Lohaus
Institut für Biochemie der Pflanze der Georg-August-Universität, Göttingen, Germany

William J. Lucas
Section of Plant Biology, Division of Biological Sciences, University of California, Davis, California 95616, USA

James H. Macduff
Environmental Biology Department, Institute of Grassland and Environmental Research, Plas Gogerddan, Aberystwyth, Ceredigion SY23 3EB, UK

Carol MacKintosh
MRC Protein Phosphorylation Unit, Department of Biochemistry, Medical Sciences Institute, University of Dundee, Dundee DD1 4HN, UK

Erik Murchie
Department of Molecular Biology and Biotechnology, The University of Sheffield, Sheffield S10 2TN, UK

H. Ekkehard Neuhaus
Lehrstuhl für Pflanzenphysiologie, Universität Osnabruek, D-49069 Osnabruek, Germany

Alain Ourry
Laboratoire de Physiologie et de Biochimie Végétales, IRBA Université, 14032 Caen Cedex, France

Christopher J. Pollock
Research Director, Institute of Grassland and Environmental Research, Plas Gogerddan, Aberystwyth, Ceredigion SY23 3EB, UK

Jack Preiss
Biochemistry Department, Michigan State University, East Lansing, Michigan 48824-1319, USA

List of contributors

W. Paul Quick
Department of Animal and Plant Sciences, University of Sheffield, Box 601 Western Bank, Sheffield S10 2UQ, UK

Stephen Rawsthorne
Department of Brassica and Oilseeds Research, John Innes Centre, Norwich Research Park, Colney, Norwich NR4 7UH, UK

Thomas W. Rufty
Department of Crop Science, College of Agriculture and Life Sciences, North Carolina State University, Box 7620, Raleigh, North Carolina 27695-7620, USA

Champa Sengupta-Gopalan
Department of Agronomy and Horticulture, New Mexico State University, Las Cruces, New Mexico 88003, USA

Uwe Sonnewald
Institute für Pflanzengenetik und Kulturpflanzenforschung, Corrensstrasse 3, D-06466 Gatersleben, Germany

Stephen J. Temple
Department of Agronomy and Horticulture, New Mexico State University, Las Cruces, New Mexico 88003, USA

Astrid Wingler
Robert Hill Institute, Department of Animal and Plant Sciences, University of Sheffield, Sheffield S10 2TN, UK

Heike Winter
Institut für Biochemie der Pflanze der Georg-August-Universität, Untere Karspüle 2, D-37073 Göttingen, Germany

Shmuel Wolf
Department of Field and Vegetable Crops, Faculty of Agriculture, The Hebrew University of Jerusalem, Rehovot 76-100, Israel

List of abbreviations

γ-ECS	γ-glutamyl cysteine synthetase
$\Delta\mu_{H+}$	transthylakoid proton electrochemical potential gradient
$\Delta\Psi$	transthylakoid electrical potential gradient
ΔpH	transthylakoid pH gradient
Φ_{CO_2}	quantum yield for CO_2 fixation (the rate of CO_2 fixation divided by the irradiance)
Φ_{PSI}	quantum yield of PSI electron transport
Φ_{PSII}	quantum yield of PSII electron transport
A_0	primary electron acceptor chlorophyll of PSI
A_1	the primary quinone electron acceptor of rc_I
AAT	aspartate aminotransferase
ABA	abscisic acid
ACCase	acetyl-CoA carboxylase
ACP	acyl carrier protein
ADPGlc	ADP-glucose
ADPGlcPPase	ADP-glucose pyrophosphorylase
AGC	the cyclic nucleotide-regulated and lipid-activated protein kinases
AHAS	acetohydroxacid synthase
AK	aspartate kinase
Ala	Alanine
AOS	active oxygen species
APX	ascorbate peroxidase
A_{max}	maximum rate of photosynthesis
Arg	Arginine
AS	asparagine synthetase
AsA	ascorbic acid
ASN	asparagine
ASP	aspartate
Aspase	Aspartase
b_6 (outer)	high potential cytochrome b_6 of the cytochrome b_6/f complex, also called b_n or b_h

List of abbreviations

b_6 (inner)	low potential cytochrome b_6 of the cytochrome b_6/f complex, also called b_p or b_l
BS	bundle sheath
BS/PP	bundle sheath/phloem parenchyma
BSP	bark storage protein
C_3	plants where the major primary product of photosynthetic CO_2 assimilation is a 3-carbon compound
C_4	plants where the major primary product of photosynthetic CO_2 assimilation is a 4-carbon organic acid
C	carbon
CA1P	carboxy-arabinitol-1-phosphate
CAM	Crassulacean acid metabolism
CaM	calmodulin
CaMK	Ca^{2+}-dependent protein kinases
CaMV	cauliflower mosaic virus
CAT	catalase
CC	companion cell
CC-SE	companion cell-sieve element
CDPK	calmodulin-domain protein kinase
CF_0–CF_1	chloroplast ATP synthetase, also called the chloroplast coupling factor
CHATS	constitutive high affinity transport system
chl	chlorophyll
C^J_E	flux control coefficient
CMGC	cyclin-dependent protein kinases and relatives
CMV	cucumber mosaic virus
CoA	coenzyme A
CPT	choline phosphotransferase
cyt f	cytochrome f
DAG	diacylglycerol
DAHP	3-deoxy-D-arabino-heptulosonate-7-phosphate
DGAT	diacylglycerol acyltransferase
DHAP	dihydroxyacetone phosphate
DHAR	dehydroascorbate reductase
DHPS	3-aspartyl phosphate, 2-dihydrodipicolinate synthase
DP	degree of polymerization
DPGA	glycerate 1,3-bisphosphate
DRAG	dinitrogenase reductase activating glycohydrolase
DRAT	dinitrogenase reductase ADP-ribosyl transferase
DW	dry weight
E4P	erythrose 4-phosphate
E	enzyme
ECM	ectomycorrhiza
EPSPS	5-enolpyruvyl-shikimate-3-phosphate synthase
ER	entoplasmic reticulum
F_A, F_B	the primary and secondary electron accepting iron–sulphur centres of rc_I

F_{AB}	iron-sulphur acceptor to PSI
FAS	fatty acid synthetase
Fd	ferredoxin
FNR	feredoxin-NADP oxidoreductase
FQR	ferredoxin-plastoquinol oxidoreductase
Fru	fructose
$Fru1,6P_2$	fructose 1,6-bisphosphate
$Fru1,6P_2$ase	fructose 1,6-bisphosphase
$Fru2,6P_2$	fructose 2,6-bisphosphate
$Fru2,6P_2$ase	fructose 2,6-bisphosphase
Fru6P	fructose 6-phosphate
F'_v/F'_m	ratio of the relative fluorescence yield increase produced as the Q_A pool is reduced by a saturating light pulse and the maximum relative fluorescence yield achieved when the Q_A is completely reduced
F_x	iron-sulphur acceptor to PSI
G3P	glyceraldehyde phosphate
GAPDH	glyceraldehyde 3-phosphate dehydrogenase
GBSS	granule-bound starch synthase
GDH	glutamate dehydrogenase
Glc1P	glucose 1-phosphate
Glc6P	glucose 6-phosphate
Glc6PDH	glucose 6-phosphate dehydrogenase
Gln	glutamine
Glu	glutamate
GOGAT	glutamate synthase
gor	bacterial gene for glutathione reductase
GPAT	glycerol-3-phosphate acyltransferase
GR	glutathione reductase
GS	glutamine synthase
GSH	reduced glutathione
GSH-S	glutathione synthetase
GSSG	oxidized glutathione
HMGR	3-hydroxy-3-methyl glutaryl CoA reductase
HK	hexokinase
HPT	hygromycin phosphotransferase
HSD	homoserine dehydrogenase
htm	high melting temperature
IHATS	inducible high affinity transport system
Ivr	invertase
J	flux
α-keto	α-ketoglutarate
KAPP	kinase-associated protein phosphatase
LATS	low affinity transport system
LHCII	the light-harvesting chlorophyll *a/b* binding protein of photosystem II
LPAT	lysophosphatidic acid acyltransferase
LPC	lysophosphatidyl choline
MAP	mitogen-activated protein

List of abbreviations

MDAR	monodehydroascorbate reductase
MDH	malate dehydrogenase
MDHA	monodehydroascorbate radical
MP	movement protein
MV	methyl viologen
MW	molecular weight
ΔNR	NR carrying a N-terminal deletion
N	nitrogen
Nase	nitrogenase
ndh	genes for the NAD(P)H dehydrogenase complex
NIP	nitrate reductase inhibitor protein
NiR	nitrite reductase
NR	nitrate reductase
NR-P	the phosphorylated form of nitrate reductase
OAA	oxaloacetic acid
6PF2K	6-phosphofructo-2-kinase
6PF2K/Fru2,6P$_2$ase	6-phosphofructo-2-kinase/fructose-2,6-bisphosphatase
2PG	2-phosphoglycolate
P5C	Δ-pyrroline-5-carboxylase
P5CR	Δ-pyrroline-5-carboxylate reductase
P680	the primary electron donor chlorophyll of PSII
P680*	the excited state of P680
P700	the primary electron donor chlorophyll of PSI
P700*	the excited state of P700
P700$^+$	the oxidized form of P700
PAI	phosphoribosylanthranilate isomerase
PAR	photosynthetically active radiation
Pc	plastocyanin
PC	phosphatidyl choline
PDC	pyruvate dehydrogenase multienzyme complex
PEP	phosphoenolpyruvate
PEPCase	phosphoenolpyruvate carboxylase
PEPCK	phospho*enol*pyruvate carboxykinase
PFK	phosphofructokinase
PFP	pyrophosphate:fructose-6-phosphate-1-phosphotransferase
PG	phosphatidylglycerol
PGA	3-phosphoglycerate
PGI	phosphoglucose isomerase
PGK	phosphoglycerate kinase
PGM	phosphoglucomutase
phaeo	phaeophytin
PHB	polyhydroxybutyric acid
Pi	inorganic phosphate
PK	protein kinase
PKA	cAMP-dependent protein kinase (protein kinase A)
PKG	cGMP-dependent protein kinase (protein kinase G)
PKK	protein kinase kinase
PLP	pyridoxal phosphate
PP	phloem parenchyma

PPase	protein phosphatase
PPdK	pyruvate Pi dikinase
PPi	inorganic pyrophosphate
PPRP	pyruvate Pi dikinase regulatory protein
PPT	L-phosphinothricin
PQH_2	the mobile pool of plastoquinol
PRK	phosphoribulokinase
PS	photosystem
PSK	protein (serine/theronine) kinase
PTK	protein tyrosine kinase
PX	peroxidase
PYR	pyruvate
Q	ubiquinone
Q_A	the primary quinone acceptor of PSII
Q_B	the secondary quinone acceptor of PSII
q_N	non-photochemical quenching of chlorophyll *a* fluorescence
q_P	photochemical quenching of chlorophyll *a* fluorescence
Q_r	reduced ubiquinone
Q_t	total ubiquinone pool
R5P	ribose 5-phosphate
rbcL	large sub-unit of Rubisco
rbcS	small sub-unit of Rubisco
rc_I	reaction centre of PSI
rc_{II}	reaction centre of PSII
Reiske FeS	the Reiske iron–sulphur centre of the cytochrome b_6/f complex
RGR	relative growth rate
RKIN1	protein of 502 amino acids from rye endosperm similar to the SNF1 gene product of yeast
Rubisco	ribulose 1,5-bisphosphate carboxylase-oxygenase
$Ru1,5P_2$	ribulose 1,5-bisphosphate
Ru5P	ribulose 5-phosphate
35S	the 35S promoter from CaMV
35S-NR	a *nia2* gene construct where expression of the nitrate reductase protein is under the control of the 35S promoter from cauliflower mosaic virus
S_0–S_3	S-states of the water splitting complex
$S1,7P_2$	sedoheptulose 1,7-bisphosphate
$S1,7P_2$ase	sedoheptulose 1,7-bisphosphatase
S7P	sedoheptulose 7-phosphate
SAM synthase	S-adenosylmethionine synthetase
SBE	starch branching enzyme
SE	sieve element
SEL	size exclusion limit
SHAM	salicylhydroxamic acid
SM	signal molecule
SNF	(sucrose non-fermenting), family sub-group of calcium-dependent protein kinases

List of abbreviations

Suc6P	sucrose-6-phosphate
SOD	superoxide dismutase
SPP	sucrose-phosphate phosphatase
SPS	sucrose-phosphate synthase
S/R	shoot to root growth ratio
SSS	soluble starch synthase
STP	starch pyrophosphorylase
TAG	triacyl glycerol
TCA	tricarboxylic acid
TMV	tobacco mosaic virus
TP	triose phosphate
Tyr_z	a special tyrosine that functions as an electron transfer component in rc_{II}
UDPGlc	UDPGlucose
UDPGlcPPase	UDPGlc pyrophosphorylase
VA	vesicular-arbuscular
VAM	vesicular-arbuscular mycorrhiza
VSP	vegetative storage protein
Xu5P	xylose 5-phosphate

SECTION ONE

Primary Nitrogen Assimilation, Carbon Assimilation and Carbon Partitioning

1

The photosynthetic electron transport system: efficiency and control

CHRISTINE H. FOYER AND JEREMY HARBINSON

1.1 Introduction

The physiology and biochemistry of photosynthesis have been the subject of intense investigation for well over 100 years. A range of physiological tools have been developed which have allowed characterization of photosynthetic light-harvesting and electron transport both *in vitro* and *in vivo*. The functions of most of the essential components of the electron transport system are well characterized, and their interrelationships are for the most part understood, at least qualitatively. Many of the uncertainties and hypotheses that are currently proving difficult to analyze or test by conventional physiological techniques could be more elegantly resolved using molecular physiology, for example, the presence and role of cyclic electron transport (Scheller, 1996). Although molecular techniques have advanced our understanding of the synthesis, assembly and functional organization of proteins and protein complexes involved in photosynthesis (Andersson and Styring, 1991; Chitnis *et al.*, 1995; Cramer *et al.*, 1996; Hanson and Wydrzynski, 1990) they have made little impact on our understanding of photosynthesis as an integrated, regulated and adaptive physiological process.

In theory, the ability to enhance or decrease activity of a single enzyme in an otherwise isogenic background is a powerful approach with which to determine the relevance of any component of the photosynthetic system. Even when function appears to have been resolved, major advances can be made by alterations in the activity of specific components with a view to determining how this influences photosynthetic metabolism and physiology. The mRNA antisense technique has provided a particularly useful tool for the production of transformed plants with decreased levels of key components of photosynthetic metabolism. The first target of this kind in plants was an enzyme of carbon assimilation, ribulose-1,5-bisphosphate carboxylase-oxygenase (Rubisco; Quick *et al.*, 1991, 1992; Stitt *et al.*, 1991). Subsequently manipulations of Rubisco activase, glyceraldehyde-3-phosphate dehydrogenase, fructose-1,6-bisphosphatase and phosphoribulokinase were reported (Jiang *et al.*, 1994; Kassmann *et al.*,

1994; Mate *et al.*, 1993; Paul *et al.*, 1995; Price *et al.*, 1995a; Zrenner *et al.*, 1996). In comparison, relatively little information is available on the components that regulate carbon assimilation, electron transport and light-harvesting in plants exposed to suboptimal environmental conditions. The regulatory mechanisms that maintain the balance between the light and dark reactions of photosynthesis are largely unresolved (Chow *et al.*, 1988; Rock *et al.*, 1992; Sims and Pearcy, 1994; Thayer and Björkman, 1990). However, transformants in protein complexes involving the electron transport processes and ATP synthesis are becoming available (Price *et al.*, 1995b) and these may allow the regulation of photosynthetic electron transport to be critically analysed.

The control of photosynthesis in stress situations is a complex manifestation of physical and chemical switches and barriers that prevent over-excitation and over-reduction of photosystem II and over-reduction of the electron acceptors of photo-system I. Environmental stress causes considerable losses in yield and deterioration in the quality of products. In such situations primary metabolism is frequently modified or inhibited. Unless efficient, rapid acclimation mechanisms to the stress are established, damage ensues. Transfer of electrons or energy to oxygen can lead to cellular damage, but active oxygen species (AOS) also signal that a change in environmental conditions has occurred. In addition, oxygen activation maybe an integral part of the processes that protect thylakoid function when the enzymes of metabolism are impaired by stress. In contrast to photosynthetic light-harvesting and electron transport, molecular biology has made a considerable contribution to our understanding of the roles of AOS and the antioxidative enzymes.

1.2 Photosynthesis: general principles

Photosynthesis, which couples the energy of absorbed light with endothermic metabolic processes, is found in some bacteria, and in most algae and higher plants. Though the detailed operation of the photosynthetic processes in these different groups can be very different both in terms of its mechanism and physiological role, the general principles are the same in all cases. Light is absorbed and used to drive electron transport. This is coupled in various ways to metabolism and physiology. Light absorption and electron transport occur in selectively permeable lipid bilayer membranes in which various protein complexes are embedded. These are principally involved in light-harvesting, electron transport and ATP synthesis.

Light is absorbed by specialized light-harvesting pigments (chlorophylls, bacteriochlorophylls, phycobilins, carotenes and xanthophylls) associated with specific proteins. Light absorption by a pigment molecule results in the formation of an excited state. This excitation energy (commonly referred to as an 'exciton') is then used to produce transmembrane charge separation in specialized structures called 'reaction centres'. Though the light-trapping pigment–protein complexes are either wholly or partially embedded in a membrane, the reaction centre complexes always span the membrane. Essentially, charge separation temporarily stores some of the energy present in the otherwise short-lived exciton and makes it available for coupling to other processes. This stabilization occurs via two routes: first, the transmembrane charge separation produces an electrical

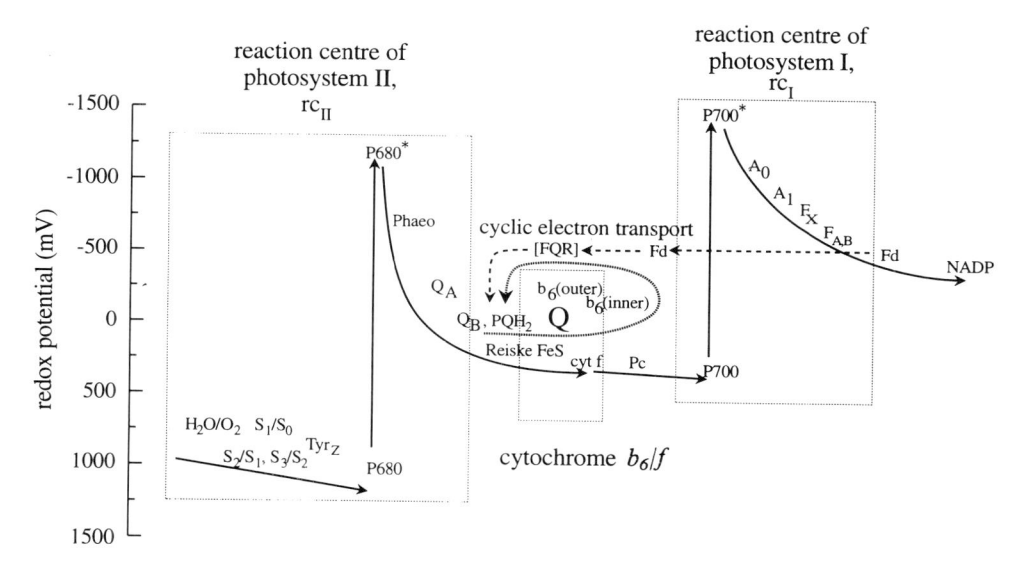

Figure 1.1 A schematic diagram of the photosynthetic electron transport chain showing the non-cyclic electron transport path (unbroken arrows). One possible cyclic electron path is shown (broken arrows), though at least two such paths exist (Bendall and Manasse, 1995; Scheller, 1996), and the Q-cycle path is shown (dotted arrow) (Cramer et al., 1996). For further information on the reaction centre of photosystem II see Andersson and Styring (1991), for the cytochrome b_6/f complex see Cramer et al. (1996), and for the reaction centre of photosystem I see Golbeck and Bryant (1991) or Chitnis et al. (1995)

potential difference across the membrane, and second the charge separation disrupts the redox equilibrium of the reaction centre and surrounding molecules by producing oxidizing and reducing molecules that cause further electron transfer events.

Photosynthetic pigments, reaction centres, and the use of the energy stored in the transmembrane potential and redox changes vary between different types of photosynthetic system. In higher plants the pigment–protein complexes are embedded in membranes called thylakoids, and they are associated in two groups that comprise photosystems I and II (Figure 1.1). Each photosystem has its own type of reaction centre, rc_I and rc_{II}, of which each have their own special reaction centre chlorophyll, P700 for rc_I (Armbrust et al., 1996; Chitnis et al., 1995; Golbeck and Bryant, 1991) and P680 for rc_{II} (Andersson and Styring, 1991; Renger, 1992). These reaction centres are usually portrayed as discrete protein complexes, but within the thylakoid they are associated with other light absorbing but non-photochemical pigment/protein complexes (Jansson, 1994) and possibly with other reaction centre complexes of the same type as dimers or trimers (Boekema et al., 1995; Kruip et al., 1994). The other major protein complexes in the thylakoid membrane are the cytochrome b_6/f complex (Cramer et al., 1996) and the ATPase complex. Two electron transport proteins associate only temporarily with the thylakoid membrane. These are plastocyanin (Gross, 1993; Haehnel, 1982, 1984) in the intra-thylakoid space (the lumen), and ferredoxin in chloroplast stroma surrounding the thylakoids. These carriers transport electrons between the cytochrome b_6/f complex and rc_I and various components in the stroma such as ferredoxin–NADP oxidoreductase (FNR) respectively. Reduced ferredoxin can also feed electrons back to the ferredoxin-quinone reductase in the cyclic electron path around PSI (Bendall and Manasse, 1995). Plastocyanin and ferredoxin reversibly bind to proteins on the membrane which possesses 'docking' sites for these carriers (Chitnis et al., 1995; Cramer et al., 1996; Gross, 1993). Plastoquinol (or its oxidized form, plastoquinone) is a low molecular weight hydrophobic electron transport component that diffuses within the plane of the membrane transferring electrons from rc_{II} and the cytochrome b_6/f complex. It has specific sites of interaction on both rc_{II} and the cytochrome b_6/f complex (Cramer et al., 1991, 1996).

1.3 The Z-scheme, non-cyclic electron transport and photosynthetic energy conversion

The hypothesis that the two reaction centre types in higher plant photosynthesis work in series to perform non-cyclic photosynthetic electron transport is widely accepted (Cramer et al., 1991; Haehnel, 1984). This hypothesis is generally depicted as the 'Z-scheme' (Figure 1.1). In the Z-scheme a strong oxidant ($P680^+$) and a strong reductant (phaeophytin$^-$) are formed by charge separation in rc_{II} (Renger, 1992). The oxidant is used to oxidize water, resulting in the release of O_2 and four protons into the lumen, and the reductant is used to reduce the part of the electron transport chain that links rc_{II} with rc_I. In rc_I, photochemistry generates a weak oxidant ($P700^+$) and a strong reductant (A_0^-; Golbeck and Bryant, 1991). $P700^+$ is used to reoxidize the electron transport chain between the photosystems. A_0^- is used to reduce another electron transport chain leading to the reduction of ferredoxin (Chitnis et al., 1995; Golbeck and Bryant, 1991). In higher plants these two types of reaction centre co-operate to transfer electrons from a weak reductant (water) to a strong reductant (reduced ferredoxin). The energy required to do this is provided by the light absorbed by two photosystems.

The process of electron transport results in the release of protons into the lumenal space and an uptake of protons from the stroma (Schlodder et al., 1982). In addition electrons can be circulated from rc_I back to reduce $P700^+$ along the cyclic electron transport path (Hosler and Yocum, 1985). In the latter case, proton release into the lumen occurs but there is no net redox transfer across the thylakoid membrane. Both non-cyclic and cyclic electron transport produce a proton concentration difference (ΔpH) between the acidified thylakoid space and the alkalinized stroma. In addition, an electrical potential (voltage) difference ($\Delta\Psi$) is formed by charge transfer across the membrane (Schlodder et al., 1982; Witt, 1979). Adding to these is another proton-pumping device involving the cytochrome b_6/f complex. This is called the 'Q-cycle' and it is probably constitutive (Kramer and Crofts, 1993). The transmembrane electrical potential is positive on the inside of the thylakoid membrane, increasing the potential energy of cations (for example H^+) in the lumen relative to those in the stroma. The electrical potential is however largely neutralized under steady state condition by the movement of chloride ions into the intra-thylakoid space so the $\Delta\Psi$ is usually very small in chloroplasts (Remis et al., 1986; Schlodder et al., 1982; Witt, 1979). Together the ΔpH and the Δ_Ψ form the trans-thylakoid electrochemical potential difference ($\Delta\mu_H^+$). Under optimal conditions the dissipation of this disequilibrium state is largely coupled to ATP synthesis by the thylakoid ATPase (Schlodder et al., 1982; Witt, 1979) though some is also coupled to other processes such as protein import into the thylakoid membrane (Mould and Robinson, 1991). Uncoupled leakage across the membrane or through the ATPase can also cause dissipation *in vitro* (Davenport and McCarty, 1984; Evron and Avron, 1990) but the physiological role of this *in vivo* remains to be established. In summary, light is absorbed, charge separation and electron transport occur, and assimilatory power (reduced ferredoxin, NADPH and ATP) is formed. Photosynthetic electron transport and energy transduction form part of a cycle since reduced ferredoxin must be re-oxidized and the substrates for ATP synthesis by the ATPase (ADP and Pi) must be regenerated. The full cycle requires the participation of CO_2 assimilation and related assimilatory processes. The operation of light-harvesting and electron transport must therefore ultimately be viewed in the context of metabolic demands of the plant as a whole.

1.4 Carbon dioxide fixation and photosynthetic metabolism

Studies in the 1940s by Calvin and Benson showed that a compound containing three carbon atoms, 3-phosphoglycerate (PGA), was the major initial product of photosynthesis in *Chlorella*, *Scenedesmus* and several higher plant species (Edwards and Walker, 1983). Hence, plants where primary carboxylation of atmospheric CO_2 resulted in the production of PGA were denoted as 'C_3' species. In C_3 plants carboxylation occurs via the enzyme ribulose-1,5-bisphosphate carboxylase-oxygenase (Rubisco). A small percentage of the CO_2 fixed in C_3 plants (less than 5%) is incorporated into acids that contain four carbon atoms, principally malate and aspartate (Edwards and Walker, 1983). In the 1960s, Kortschack, Hatch and Slack showed that in some plants a large percentage of the CO_2 (70–80%) is incorporated into malate or asparate and these C_4 acids are the primary products of CO_2 assimilation. They called these 'C_4' plants. In C_4 plants CO_2 is initially incorporated into phosphoenolpyruvate (PEP) by the action of PEP carboxylase (producing a 4-carbon product) and is later transferred into 3-carbon and 6-carbon compounds. In C_4 leaves the C_4 cycle acts as an efficient CO_2 concentrating mechanism. A high CO_2 concentration in the vicinity of Rubisco serves to decrease the process of photorespiration. This process, which occurs in C_3 leaves in air as a result of competition between CO_2 and O_2 for $Ru1,5P_2$ at the active site of Rubisco, reduces the quantum efficiency of CO_2 fixation (Edwards and Walker, 1983; Ehleringer and Björkman, 1977; Farquhar and von Caemmerer, 1982). Many other metabolic processes such as nitrate reduction, sulphate reduction, and shikimic acid synthesis, are driven by photosynthetic assimilatory power (ap Rees, 1987; Brunold, 1993; Hanning and Heldt, 1993; Robinson, 1988). Molecular oxygen can be reduced directly by photosystem (PS) I in the process of pseudo-cyclic electron transport. Oxygen reduction, which was first observed by Mehler and is hence called the Mehler reaction (Badger, 1985; Robinson, 1988), produces superoxide (O_2^-). This is destroyed by the action of superoxide dismutase (SOD). SOD produces hydrogen peroxide (H_2O_2) which is, in turn, destroyed by the action of ascorbate peroxidase (APX). H_2O_2 oxidation produces monodehydroascorbate (MDHA) from ascorbate. MDHA is a powerful PSI electron acceptor. Together, O_2^- production, H_2O_2 production and destruction and the subsequent re-reduction of MDHA to ascorbate via the PSI electron transport chain are called the 'Mehler-peroxidase' reaction sequence. The Mehler-peroxidase sequence can act as a potent sink of electrons *in vitro* and it results in ΔpH formation and ATP synthesis (Forti and Elli, 1995). Oxygen is an alternative energy sink whenever electron transport is in excess of the capacity for CO_2 assimilation. In the absence of convincing evidence to the contrary, flux through the Mehler-peroxidase reaction must be considered to be relatively low (less than 10% that of the combined activities of CO_2 fixation and photorespiration).

Photosynthesis is commonly quantified either in terms of flux or the efficiency with which light is used to produce flux (Figure 1.1). It can be measured by the rate of CO_2 fixation per unit leaf area, or as the quantum yield of CO_2 fixation. The quantum yield is very different to the efficiency of energy conversion. Under light-limiting conditions when both photosystems have a maximal quantum yield of close to 1, the efficiency of energy conversion from light to carbohydrate is only about 34% (Hill and Rich, 1983). The rate of light-harvesting or the rate or efficiency of electron transport are not often used as measurements of photosynthesis *in vivo* because they are difficult to measure rigorously. The rate or efficiency of O_2 evolution is often used as a simple measure of non-cyclic electron transport though it cannot be used to measure pseudo-cyclic electron transport. CO_2 fixation *per se* is commonly measured since it is perceived to be the most relevant parameter for plant or crop productivity determinations. Given certain assumptions concerning the relative rates of carboxylation and oxygenation of $Ru1,5P_2$ by Rubisco and the flux of reducing equivalents and ATP required through the photorespiratory pathway, it is possible to calculate the flux of reducing equivalents that support a measured rate of CO_2 fixation (Farquhar and von Caemmerer, 1982). Such calculations ignore the use of reducing equivalents or ATP by other metabolic processes. Non-cyclic electron transport rates can, however, frequently account for the calculated requirements of photorespiration and CO_2 fixation alone (Cornic and Briantais, 1991; Kent *et al.*, 1992; Nespoulos *et al.*, 1989; Peterson, 1989).

Photosynthesis couples the flux of absorbed light to metabolic fluxes. Conventionally, changes in photosynthetic rate are measured in response to changing environmental factors, such as changes in irradiance (PAR), or temperature. Alternatively, the quantum efficiency of any particular process, such as CO_2 fixation or electron transport by PSII, can be measured and its dependency on the environment of the leaf and its physiological condition described. Both approaches are useful because *the maximum quantum efficiency of photosynthesis and the maximum flux through the photosynthetic system are basic, independent limiting factors for operation of photosynthesis*. It is important to understand the relationship between efficiency and flux. As irradiance increases, the rate of CO_2 fixation increases but the incremental rate of increase decreases as the irradiance increases, leading to the light saturation of CO_2 fixation (Figure 1.2). This relationship can be expressed in terms of efficiency. With increasing irradiance the quantum yield of CO_2 fixation (Φ_{CO2}) decreases progressively from a maximum value at the lowest irradiance (for example, see Genty *et al.*, 1990a; Oberhuber

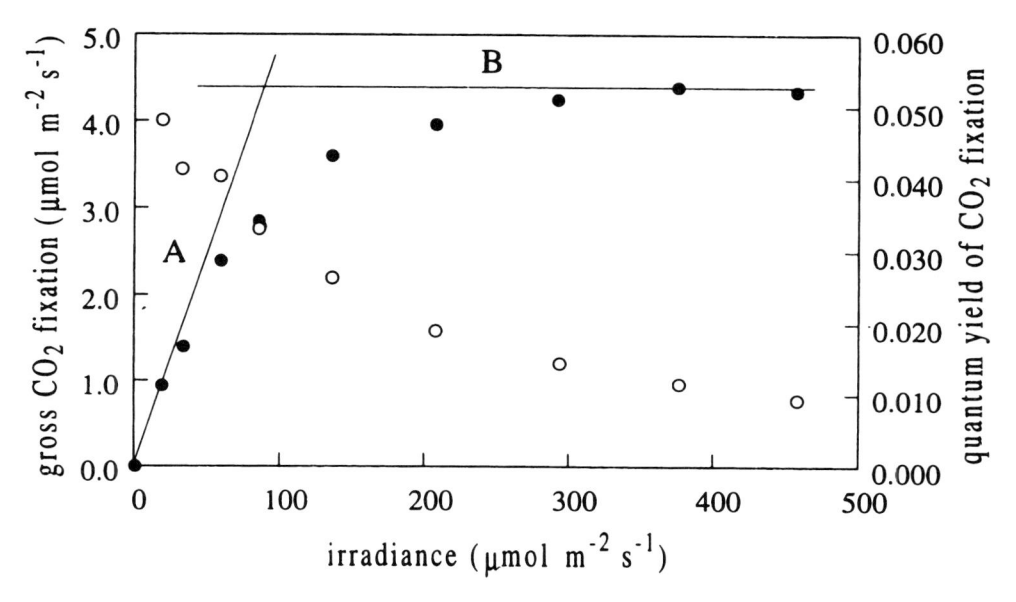

Figure 1.2 A light saturation curve for photosynthetic CO_2 fixation, measured in a leaf of the Mexican epiphytic shrub, *Juanulloa aurantiaca* (filled circles). Measurements were made at 350 ppm CO_2, 20% O_2, and N_2. Line A is the slope corresponding to the maximum quantum yield for CO_2 fixation on an incident irradiance basis while line B is the maximum rate of CO_2 fixation. Together lines A and B define the limits of photosynthetic CO_2 fixation. The light saturation of the quantum yield for CO_2 fixation on an incident irradiance basis (open circles) shows that the quantum yield declines as the rate of CO_2 fixation increases, and vice versa.

et al., 1993). Flux is maximum where efficiency is low, whereas efficiency is maximum where flux is minimum. Improvements to photosynthesis could be directed either towards increasing maximum efficiency or flux. It is important to note that quantum yields can be calculated either on the basis of absorbed or incident light intensities. In the former, yield is determined solely by the efficiency with which absorbed light is used to drive photosynthesis, whereas in the latter, the efficiency of light absorption is also important.

1.5 The maximum, limiting efficiency of photosynthesis: basic concepts

In higher plants estimates of the maximum efficiency of photosynthesis are based on the Z-scheme model where the two photosystems (PSII and PSI) act together in series to transfer electrons from water to ferredoxin and thence to NADPH, with O_2 and protons released from the water-splitting complex (Figure 1.1). If all absorbed light is used to drive photochemistry in either PSII or PSI, and if both photosystems are equally excited and act in series to produce non-cyclic electron transport, then eight quanta are needed for each oxygen evolved (four for each photosystem) and the quantum yield for O_2 evolution is 0.125. Four reducing equivalents are required for CO_2 fixation by the Benson–Calvin Cycle. In the absence of any competing process the maximum quantum yield of CO_2 fixation is also 0.125. Within the Z-scheme paradigm 0.125 is the best that can be achieved. A higher efficiency would require a different light-harvesting and electron transport mechanism.

The efficiency of light-harvesting and electron transport (on an absorbed light basis) is determined by the efficiency of photochemistry and the physiology of the electron transport system and non-photosynthetic pigments. Surprisingly, the internal quantum efficiencies for charge separation and electron transport within the reaction centres of both PSI and PSII are poorly characterized. For PSI the value is probably > 0.95 (Trissl and Wilhelm, 1993) while that of PSII is probably 0.85–0.95 (Kramer and Mathis, 1980; Mauzerall and Greenbaum, 1989; Schatz et al., 1988; Thielen and van Gorkom, 1981; Trissl and Wilhelm, 1993). Some PSII centres are incapable of sustaining electron transport; for example, they may be unable to transport electrons from Q_A to Q_B, or their access to plastoquinol may be blocked (Chylla and Whitmarsh, 1989; Lavergne and Leci, 1993). As a yield of 0.125 is dependent on 100% quantum efficiency for both photosystems working optimally in series; inherent inefficiencies mean that non-cyclic electron transport cannot have a quantum efficiency of 0.125 based on absorbed light. Furthermore, non-cyclic electron transport (within the Z-scheme) requires that the total rates of electron transport through both photosystems must be equal. If one or other photosystem has a potential rate of electron transport greater than the other, then inefficiency will result. PSI and PSII have different absorbance spectra. For a given irradiance spectrum it is probable that one photosystem will absorb more light than the other (Evans, 1986; Melis et al., 1987). This will result in an excitation imbalance and a loss of efficiency. Mechanisms for balancing the total rates of electron transport through the photosystems exist. These can involve adaptive changes in the relative numbers of reaction centres and their associated light-harvesting complexes (Melis, 1991). Short-term regulation of excitation distribution between the photosystems is also observed. Imbalances are sensed at the level of the plastoquinol pool. Increased reduction of the plastoquinol pool and, hence, the cytochrome b_6/f complex, activates a protein kinase that phosphorylates a component of the light-harvesting chlorophyll a/b-binding protein population, LHCII. LHCII is normally associated with PSII and this is called 'state I'. When LHCII becomes phosphorylated it dissociates from PSII. This is called 'state II'. This transition results in a decreased excitation of PSII, oxidizing the plastoquinol pool (Anderson, 1992; Bennett, 1991; Canaani and Malkin, 1984; Malkin and Canaani, 1994). Although these changes have been described in detail, their quantitative effect on the efficiency of photosynthesis has not been determined. In the cyanobacterium *Synechocystis* sp. strain PCC 6803 the inactivation of a gene which is homologous to open reading frame 184 (tobacco and *Marchantia*) or 185 (rice) of the chloroplast genome, produces an increased amount of PSII relative to PSI (Wilde et al., 1995). This gene product may therefore be implicated in regulating thylakoid development as occurs when plants are grown under different radiation environments (Melis, 1991).

The occurrence of cyclic electron transport (Figure 1.1) *in vivo* is debatable. The quantum efficiencies of PSI and PSII change in parallel (Genty and Harbinson, 1996), consistent with the operation of non-cyclic electron transport alone, or with the operation of cyclic electron transport at a constant proportion of the non-cyclic flux. Such analyses, based on comparisons of Φ_{PSI} and Φ_{PSII}, would also not reveal a cyclic flux of 10% or less than that of non-cyclic flux. Cyclic electron flux is estimated to be only about 15% (or less) of the non-cyclic flux (Bendall and Manasse, 1995; Fork and Herbert, 1993). The operation of a cyclic pathway will decrease the efficiency of non-cyclic electron transport.

The Mehler-peroxidase reaction will also serve to decrease the yield of non-cyclic electron transport. Note, however that if either cyclic electron flow or the Mehler-peroxidase reaction is necessary for regulatory or metabolic reasons it would not constitute an inefficiency *per se*. The yield of photosynthetic electron transport, based on the Z-scheme, must therefore be less than 0.125 because of inevitable inefficiencies in the operation of electron transport.

1.6 Measurements of the maximum quantum efficiency for higher plant photosynthesis

Photosynthetic electron transport drives metabolism. It is therefore more important to produce the correct ratio of reducing power and ATP in the stroma with highest overall quantum yield than it is to operate the non-cyclic electron transport pathway *per se*. In C_4 plants, electron transport in the bundle sheath cells is confined largely to cyclic electron flow. C_4 leaves, which have only negligible photorespiration, have a low quantum yield for O_2 evolution compared to C_3 leaves under non-photorespiratory conditions (Björkman and Demmig, 1987), but the absorbed light is used as efficiently as in C_3 leaves to produce electron transport.

Measurements of the quantum yield, on an absorbed light basis and in the absence of photorespiration, for net CO_2 fixation or net O_2 evolution, give values that are similar in all C_3 leaves. Using broad-band irradiance (white light), the yield for CO_2 fixation can be as high as 0.098 (Long *et al.*, 1993). The yields of O_2 evolution are higher with values of up to 0.1135 reported (Björkman and Demmig, 1987). Yields of CO_2 fixation are expected to be less than those of O_2 evolution because, even under non-photorespiratory conditions, many other light-driven biosynthetic processes such as NO_2^- reduction exist in leaves. However, when CO_2 assimilation and O_2 evolution were measured simultaneously at low irradiances they were found to be equal (Bloom *et al.*, 1989).

Using narrow-band excitation the yield of O_2 evolution on an absorbed light basis has been shown to be wavelength dependent as would be expected given the different absorbance spectra of the two photosystems and the need to have balanced excitation (Evans, 1987a). This implies that the short-term regulatory, physiological mechanisms for balancing the excitation of the photosystems do not fully compensate for spectral differences.

The measured yields of O_2 evolution have been considered to be incompatible with the Z-scheme model (Myers, 1980; Osborne, 1994; Osborne and Geider, 1987). Unfortunately, the quantitative impact of known inefficiencies on measurements of higher plant quantum efficiency is not known. Selecting a wavelength range which is expected to excite both photosystems equally, Osborne (1994) reported quantum yields for O_2 evolution of 0.130. Such high yields would suggest that a non-Z-scheme path for electron transport was operative. An example of an alternative scheme would involve electrons from PSII reducing ferredoxin or another terminal acceptor directly without passing through PSI. This kind of reaction has been observed *in vitro* (Arnon, 1995). However, for technical reasons these estimates of quantum yield may be in error (Genty and Harbinson, 1996). The situation in the algae may be somewhat different. Quantum yields measured in algae can be higher than 0.125 (Geider *et al.*, 1985; Osborne and Geider, 1987) although others are within the 0.125 limit (Ley, 1986; Senger, 1982). The mechanisms underlying these very high yields are unknown.

In conclusion, quantum yields for O_2 evolution in higher plants (expressed on an absorbed light basis) are consistent with the operation of the Z-scheme. Only when they are adjusted to account for inefficiencies (of an unknown magnitude) do they become incompatible with the Z-scheme. In algae there are some reports of yields higher than 0.125, implying that there may be potential for improving the photosynthetic quantum efficiency in higher plants.

1.7 Non-cyclic electron transport and photosynthetic metabolism

The operation of non-cyclic electron transport has implications for the supply of ATP and reducing power for photosynthetic CO_2 assimilation and other chloroplast metabolism. A balance must be maintained over the rates of ATP and NADPH formation. In C_3 plants for either CO_2 fixation or O_2 assimilation in photorespiration, the ATP/NADPH demands are 1.5 and 1.75 respectively (Farquhar *et al.*, 1980). Carboxylation and photorespiration occurring together in air require an ATP/NADPH ratio of 1.56 (Edwards and Walker, 1983). Whether non-cyclic electron transport alone can provide this required stoichiometry of ATP and NADPH production is unclear. It depends on the stoichiometry of protons (translocated into the lumen) for each electron transferred from water to ferredoxin, and the number of protons gated through the ATPase for each ATP synthesized (H^+/ATP). Until recently, it was widely accepted that the H^+/ATP ratio is 3. In the absence of a Q-cycle in the cytochrome b_6/f complex, two protons would be translocated for every electron moving from water to ferredoxin. This would result in four protons translocated for every NADPH synthesized, giving an ATP/NADPH ratio of 1.3. On the basis of this stoichiometry, non-cyclic electron transport from water to NADPH alone could not produce sufficient ATP to meet the demands of CO_2 fixation. There would consequently be a need for extra, parallel proton pumping potential which could be provided by cyclic electron transport or the Mehler reaction. If a Q-cycle (Figure 1.1) operates in the thylakoid cytochrome b_6/f complex, as is believed to occur (Kramer and Crofts, 1993), an extra proton will be pumped for every electron transported. The H^+/NADPH ratio would therefore be 6 and the ATP/NADPH would be 2. This is more than enough to satisfy the requirements of CO_2 assimilations and is in fact too much. It has been reported, however, that the H^+/ATP is 4 (Gräber *et al.*, 1987; Rumberg *et al.*, 1990). If this is the case then the ATP/NADPH ratio is 1.5 with a Q-cycle, which is just enough. While it has not been demonstrated that the exclusive operation of non-cyclic electron transport within the framework of the Z-scheme is consistent with the appropriate supply of NADP and ATP for CO_2 fixation and photorespiration in C_3 higher plants, this model is, nevertheless, feasible and the use of the Z-scheme model to define the maximum efficiency of photosynthesis is reasonable.

In algae, which have more diverse photosynthetic systems than higher plants, the exclusive operation of non-cyclic electron transport alone may not provide an appropriate ATP/NADPH ratio and the parallel operation of cyclic and non-cyclic electron transport may be required. Cyclic electron transport has been demonstrated in algae. In this case the use of the Z-scheme/non-cyclic electron transport model to define the maximum efficiency for photosynthetic electron transport is erroneous since it would essentially ignore the contribution of light-driven cyclic electron transport to the overall quantum efficiency of photosynthesis.

When manipulations of C_3 photosynthesis are directed towards increasing the maximum rate of photosynthesis (A_{max}) without altering the required ATP/NADPH ratio (for example, by increasing the total flux through the Benson–Calvin cycle), analyzing photosynthetic efficiency in terms of the dominant operation of non-cyclic electron transport is valid. If, on the other hand, additional metabolism requiring altered ATP/NADPH demands (that is different to those of CO_2 fixation and photorespiration, for example lipid synthesis), or in which cellular or chloroplast specialization is modified, for example converting C_3 plants to C_4 plants, then an assessment of the resultant efficiency of photosynthesis will have to account for the total quantum yield of photosynthetic energy conversion and not just that linked to non-cyclic electron flow.

Higher plant quantum yields expressed on an absorbed light basis are relatively constant (Ehleringer and Björkman, 1977; Ehleringer and Pearcy, 1983; Evans, 1987a; Long et al., 1993; Monson et al., 1982). In contrast, the quantum yield measured on an incident light basis is much more variable (Björkman and Demmig, 1987; Long et al., 1993). As far as leaf and plant productivity is concerned it is the quantum yield measured on an incident light basis that is most significant as it is this that will determine the response of electron transport to irradiance and determine the actual rate of CO_2 assimilation under light limiting conditions. The variation in quantum yield measured on an incident light basis is due to variations in leaf absorbance (Björkman and Demmig, 1987; Long et al., 1993). Leaf absorbance is determined partly by photosynthetic pigment concentration and partly by leaf structural organization (Osborne and Raven, 1986).

1.8 The light-saturated rate of photosynthesis and the regulation of photochemistry and electron transport

The light-saturated rate of gross CO_2 fixation (the net rate of CO_2 fixation corrected for the respiratory CO_2 loss) in air is extremely variable. In C_3 leaves it may be as low as 1 µmol m^{-2} s^{-1} for shade-adapted plants or as high as 60 µmol m^{-2} s^{-1} for sun-plants such as *Camissonia claviformis*, a Death Valley winter annual. Crop plants such as pea or wheat generally have CO_2 fixation rates of 20–30 µmol m^{-2} s^{-1}. Maximum photosynthetic capacity is determined by the environment and the history of the leaf. In C_3 leaves, CO_2 fixation is also dependent on the concentrations of CO_2 and O_2 in the air inside the leaf (Cornic and Briantais, 1991; Peterson, 1991, 1994).

The physiological, biochemical and structural differences between chloroplasts and leaves with high and low rates of CO_2 fixation have been widely studied

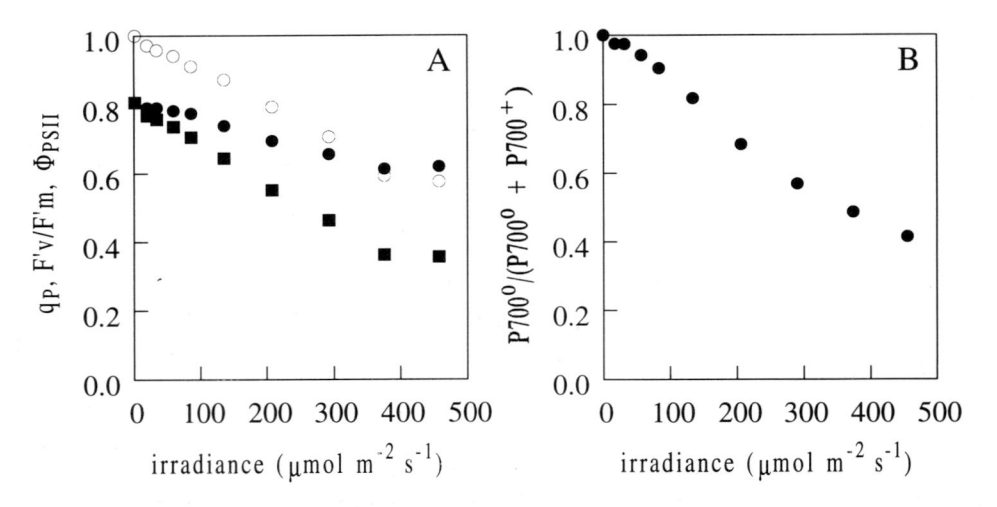

Figure 1.3 A: The light saturation of the components which determine the quantum yield for PSII photochemistry (Φ_{PSII}), q_P (○) and F'v/F'm (●). F'v/F'm is sometimes called 'excitation transfer efficiency' (Φ_{exc}). The product of q_P and F'v/F'm gives Φ_{PSII} (■). The maximum Φ_{PSII} for a leaf in optimal conditions is approximately 0.80–0.82 when measured with normal commercial fluorimeters. This optimal value is caused by the low intrinsic quantum efficiency of PSII photochemistry coupled to an error in the measuring system due to the presence of the weak chlorophyll fluorescence from PSI. B: The effect of increased irradiance on the ratio of oxidized P700 to the total amount of P700 (oxidized and unoxidized). If the PSI reaction centre is not limited by a shortage of electron acceptors (as is nearly always the case; Genty and Harbinson, 1996) then this ratio can be used as an index of the quantum yield for PSI photochemistry. To give an exact value for PSI, the quantum yield ratio must be multiplied by the actual maximum yield of PSI photochemistry (> 0.95)

(Anderson, 1986, 1992; Anderson and Osmond, 1987; Björkman, 1981; Evans, 1987b, 1988, 1989). Such differences arise not only from variations in the capacity for photosynthetic energy transduction but also the capacities of enzymes in the chloroplasts and the cytosol, and the translocation of assimilate to the sink tissues of the plant (van Bel and Gamalei, 1992). When photosynthesis is light-saturated the limitation by definition resides in the capacity to use absorbed light for electron transport coupled to metabolism. A decrease in the energy use efficiency results in a decrease in Φ_{PSI} and Φ_{PSII} (Figure 1.3) and a loss of photosynthetic capacity (Genty and Harbinson, 1996). High rates of photosynthesis are associated with high rates of electron transport and high concentrations of electron transport components (Anderson, 1986, 1992; Boardman, 1977; Evans, 1987b, 1988, 1989; Terashima and Evans, 1988).

In most cases the correlations between the concentrations of electron transport components and photosynthetic capacity have been obtained using plants grown under a range of irradiances. In this situation, changes at the molecular level modify and optimize the organization of the photosynthetic machinery, a process known as acclimation. Plants grown at high irradiances typically have higher rates of photosynthesis, measured as CO_2 fixation or O_2 evolution, than those grown at low irradiances. The concentration of the cytochrome b_6/f complex

(Björkman, 1981; Evans, 1987b, 1988; Terashima and Evans, 1988) and the ATPase (Anderson, 1986; Chow and Hope, 1987; Chow et al., 1991; Evans, 1987b; Leong and Anderson, 1984a) are particularly well correlated with photosynthetic capacity, but plastoquinol (Anderson, 1986; Björkman, 1981), rc_{II} (Anderson, 1986; Anderson and Osmond, 1987; Chow and Hope, 1987; Evans, 1987b; Wild et al., 1986) and plastocyanin (Burkey, 1993, 1994) are also positively correlated with increasing rates of photosynthesis. Little is known about the relationships with other components, such as ferredoxin. These responses are frequently discussed in terms of increasing the light use efficiency under conditions of increasing irradiance but they also optimize the allocation of nitrogen amongst the components of the photosynthetic apparatus (Evans, 1987b, 1988, 1989).

Antisense Reiske FeS tobacco transformants with a range of cytochrome b_6/f concentrations, have been produced (Price et al., 1995b). The rate of CO_2 fixation measured in air at an irradiance of 1000 μmol m^{-2} s^{-1} was linearly related to the cytochrome b_6/f concentration in these plants, implying that the control coefficient of the cytochrome b_6/f complex for photosynthesis was 1 at a high irradiance. Similar results have been obtained using isolated thylakoids treated with selective inhibitors of the cytochrome b_6/f complex (Heber et al., 1988).

When plants were grown under equal quantum fluxes of red and blue light, the blue light-grown plants had thylakoid membranes that resembled those of high-light grown plants, whereas thylakoids of the red light-grown plants resembled those of low-light grown plants (Leong and Anderson, 1984b; Lichtenthaler et al., 1980; Walters and Horton, 1995a,b; Wild and Holzapfel, 1980). This indicates an involvement of the blue-light signal transduction system (Kaufman, 1993). An alternative mechanism of developmental control, based on the redox state of the photosynthetic electron transport chain, has also been proposed (Allen, 1993; Melis, 1991; Melis et al., 1985). This mechanism has been demonstrated in the chlorophyte alga Dunaliella where the expression of the cab genes (which code for the apoprotein of LHCII) appear to be repressed as the plastoquinol pool becomes more reduced (Escoubas et al., 1995). Protein phosphorylation may play a role in this regulation. Increased phosphorylation appears to repress gene expression. This response appears to have much in common with the protein phosphorylation-controlled transition from state I to II as described previously. The tobacco transformants with decreased cytochrome b_6/f concentrations (Price et al., 1995b) have greater ATPase activities. In this case restricted intersystem electron transport capacity results in a rapid loss of Φ_{PSII} (and probably also of Φ_{PSI}) with increased irradiance. Consequently, it was suggested that the amounts of both ATPase and the cytochrome b_6/f are controlled by a common signal pathway that may involve the redox state of the electron transport chain. Adaptation to increased irradiance involves increased populations of both the cytochrome b_6/f and the ATPase. Other aspects of thylakoid organization in these transformants, however, were more like those of shade leaves; the chl a/b, the rc_I/chl, and the rc_{II}/chl decrease and the rc_{II}/rc_I increases. A regulatory role for cytochrome b_6/f in controlling the development of the light-harvesting and electron transport systems was inferred from these results. The thioredoxin system may also be important in controlling the translation of genes for thylakoid proteins (Danon and Mayfield, 1994).

Increases in the degree of reduction of the thioredoxin pool increase translation of the *psb*A gene.

The capacity for acclimation is determined by genetic factors. Plants which occur naturally in shaded or nutrient limited habitats in general cannot achieve high rates of photosynthesis, whatever their growth conditions. Similarly, plants which are capable of high rates of photosynthesis and normally grow in high light, nutrient rich habitats (such as many crop plants), cannot in general develop sustainable low rates of photosynthesis.

In addition to genetic and developmental factors, the rate of photosynthesis is controlled by a series of short-term physiological regulatory mechanisms (Foyer, 1993; Foyer and Harbinson, 1994; Foyer *et al*., 1990; Genty and Harbinson, 1996). These modulate the activity of the photosynthetic machinery in response to environmental factors (such as irradiance or changes in CO_2 supply) or the demands of the plant for photosynthate. To balance the supply of reducing power and ATP generated by electron transport to the demands of metabolism there are feedforward processes that serve to activate metabolism and feedback processes that act to down-regulate the efficiency of electron transport (Foyer, 1993; Foyer and Harbinson, 1994; Foyer *et al*., 1990; Genty and Harbinson, 1996).

The need for effective regulation arises because of the side reactions of photosynthesis. The operation of light-harvesting and electron transport requires the formation of reactive intermediates, such as singlet state chlorophyll, powerful oxidizing agents like $P680^+$, or powerful reducing agents such as reduced ferredoxin. These intermediates are used to drive electron transport to NADP, but they can also give rise to other damaging species and products (Foyer and Harbinson, 1994; Genty and Harbinson, 1996). Amongst the most damaging products of these side reactions are AOS such as superoxide formed by the Mehler reaction or singlet oxygen formed when normal ground state oxygen reacts with the triplet chlorophylls formed from excited singlet chlorophyll (Asada and Takahashi, 1987; Foyer and Harbinson, 1994). Powerful oxidants and reductants can cause damage, such as photoinhibition (Baker and Bowyer, 1994). These side reactions occur even under low light conditions (Aro *et al*., 1993; Mattoo *et al*., 1981), but net damage to the photosynthetic machinery only occurs when the quantum input exceeds metabolic demand (Aro *et al*., 1993; Sundby *et al*., 1993; van Wijk and van Hasselt, 1993). Regulation acts to reduce the rate of these damaging side reactions. In addition, effective detoxification pathways serve to eliminate the AOS as they are formed.

The mechanisms of the feedforward processes that activate the Benson–Calvin cycle are well understood. These are based on coarse control of enzyme activity by the increase of pH and Mg^{2+} concentration that occurs in the stroma following illumination, coupled to the fine control of certain enzymes, for example, by the reduction of specific dithiol residues by reduced thioredoxin (Buchanan, 1994; Foyer, 1993; Leegood *et al*., 1985; Scheibe, 1990). In contrast, the mechanisms of the feedback inhibition are not understood, nor is the means by which a balance is achieved between these up-regulation and down-regulation mechanisms (Foyer, 1993; Foyer *et al*., 1990; Genty and Harbinson, 1996). Regulation of light-harvesting and electron transport is observed in the response to increasing irradiance when the rates of CO_2 fixation or the O_2 evolution cease to be light-limited and become progressively light-saturated (Figure 1.2).

Increased irradiance is inevitably accompanied by a loss of quantum efficiency of CO_2 fixation, O_2 evolution, Φ_{PSI} and Φ_{PSII} (Figure 1.3; Genty and Harbinson, 1996; Genty et al., 1990a). Regulation occurs principally at two sites; PSII, and the plastoquinol/cytochrome b_6/f electron transfer step (Genty and Harbinson, 1996).

The loss of efficiency in PSII is caused by the formation of quenchers that can compete with reaction centre photochemistry (and chlorophyll fluorescence) to quench the exciton formed by light absorption (Genty and Harbinson, 1996; Horton et al., 1996). As a consequence of this competitive excitation energy quenching mechanism, the quantum efficiency of PSII is decreased and the phenomenon of non-photochemical quenching of chlorophyll fluorescence is produced (Genty et al., 1989, 1990b; Horton et al., 1996; Lavergne and Trissl, 1995). This mechanism and the fluorescence changes it produces are frequently referred to as non-photochemical quenching (q_N). The molecular basis of the competitive de-excitation process and its regulation are unknown. Decreased lumen pH values and zeaxanthin formation have both been implicated in this process (Genty and Harbinson, 1996; Horton et al., 1996). Experimentally, this regulation of PSII is measured as the decrease in the efficiency of excitation capture by open PSII reaction centres ($F'v/F'm$) and by an increase in q_N (Figure 1.3; Genty et al., 1989; Schreiber et al., 1986). Added to the loss of PSII efficiency due to the q_N is the decrease of quantum efficiency due to Q_A reduction (Genty et al., 1989). This results in a decrease in the probability that an exciton in the PSII pigment bed will encounter an open PSII reaction centre before it decays. Increasing Q_A reduction increases the exciton lifetime in PSII (Genty et al., 1992; Holzwarth, 1991; Schatz et al., 1988). The loss of Φ_{PSII} with increasing irradiance (Figure 1.3) is therefore due to both Q_A reduction and competitive de-excitation processes in the pigment bed (Genty et al., 1989; Harbinson et al., 1989). From the point of view of damaging side reactions, these mechanisms have different repercussions. The loss of Φ_{PSII} caused by a loss of photochemical quenching, q_P (via Q_A reduction) is accompanied by an increase in the lifetime of the exciton in PSII (Figure 1.3). This will increase the probability of chlorophyll triplet formation and the associated formation of singlet oxygen (Foyer and Harbinson, 1994). This problem is avoided if the loss of efficiency is mediated by an increase in q_N since this acts by decreasing the exciton lifetime.

The development of q_N is also associated with protection from irreversible or slowly reversible damage to PSII (photoinhibition). The mechanism of photoinhibition in vivo is still unclear. Current hypotheses are based on damage originating on both the acceptor and donor sides of P680 (Aro et al., 1993; Baker and Bowyer, 1994). On the acceptor side damage has been linked to the degree of reduction of Q_A, while on the donor side formation of $P680^+$ may be significant, possibly associated with the lowering of lumen pH. Whatever the mechanism, damage is related to photochemistry since this results in the formation of both Q_A^- and $P680^+$. Decreasing the rate of photochemistry by introducing a competitive de-excitation mechanism will, therefore, decrease the rate of damage.

Regulation of the rate of the plastoquinol/cytochrome b_6/f reaction is the principle physiological means by which the rate of electron transport is controlled to balance the supply and demand for reducing power. This phenomenon is known as 'photosynthetic control' of electron transport. The rate constant of the reaction between plastoquinol and the Reiske FeS centre of the cytochrome b_6/f complex decreases with decreasing lumen pH (Bendall, 1982; Nishio and Whitmarsh, 1993; Tikhonov et al., 1984). By controlling lumen pH, electron transport is controlled (Foyer et al., 1990; Horton, 1985; Weis et al., 1987). The mechanism by which metabolic processes modulate lumen pH is unclear (Foyer, 1993; Genty and Harbinson, 1996). Lumen pH is also involved in the control of the q_N (Horton et al., 1996) which acts to decrease electron flow from PSII to the plastoquinol pool. It may thus contribute to the control of electron transport by modifying the degree of reduction of the plastoquinol pool and thus the rate of the reaction between this pool and the cytochrome b_6/f pool (Bendall, 1982; Hope et al., 1988; Rich, 1982). Little is known of the redox state of the plastoquinol pool in illuminated thylakoids either in vivo or in vitro (Amesz et al., 1971, 1972). The effect of q_N on the redox state of the plastoquinol pool and thus electron transport from plastoquinol to the cytochrome b_6/f complex has not been investigated.

Regulation of electron transport at the plastoquinol/cytochrome b_6/f step has several consequences for the damaging side reactions associated with photosynthetic activity. First, PSI (like PSII) is sensitive to photoinhibition (Inoue et al., 1989; Sonoike, 1995, 1996; Terashima et al., 1994). The factors influencing the susceptibility of PSI to photoinhibition in vivo have not been fully characterized but illumination at chilling temperatures will cause damage to PSI (Sonoike, 1996; Sonoike and Terashima, 1994; Terashima et al., 1994). In vitro PSI is sensitive to damage in vitro when electron transport is blocked by over-reduction of its acceptor pool. Over-reduction can occur if electron transport is limited on the PSI acceptor side (Inoue et al., 1989). The sites at which damage to PSI is first apparent are on the acceptor side (Sonoike, 1996; Sonoike and Terashima, 1994). In contrast, whenever the acceptor side of PSI is oxidized and able to accept electrons (as is generally the case for leaves in air), decreases in Φ_{PSI} occur without injury. Limitation of electron transport on the donor side of PSI rather than on the acceptor side also decreases the rate of the Mehler reaction which is highly dependent on the redox state of electron transport carriers on the PSI acceptor side, such as ferredoxin and FNR (Goetze and Carpentier, 1994; Hosler and Yocum, 1985; Robinson, 1988).

1.9 The detoxification of active oxygen species

The hypothesis that AOS play a role in the destruction of the photosynthetic apparatus and the loss of membrane integrity in stress situations is widely accepted. The presence of AOS is associated with changes in gene expression that allow the plant to acclimate to environmental and biotic stresses. There is no evidence to suggest that amelioration of the antioxidant systems in transformed plants by increasing the activities of individual components of the antioxidant defence system prohibits signal transduction, although it increases the protection of the delicate cellular matrix against AOS-induced damage (Foyer et al., 1994). The apparent paradox between the universal production of AOS when the plant is challenged by any kind of stress and the potential threat that they pose to metabolism is rationalized by the requirement for a powerful but quickly destroyed alarm signal. This requirement for signalling explains the absence of a large and immediate increase in antioxidant capacity when stress conditions are imposed (Foyer *et al.*, 1994).

The molecular nature of the signal transduction sequences associated with the acquisition of stress tolerance are largely unresolved, but H_2O_2 has been implicated in the signalling pathways involved in the stimulation of the synthesis of antioxidant enzymes and other proteins (Prasad *et al.*, 1994, 1995). H_2O_2 is toxic to photosynthesis (Foyer and Harbinson, 1994). In the presence of transition metal ions, H_2O_2 and O_2 can react together to produce the hydroxyl radical ($\cdot OH$), one of the most reactive species known to biology (Scandalios, 1993). Generation of hydroxyl radicals leads to lipid peroxidation, DNA breakage and protein denaturation (Scandalios, 1993).

When nature harnessed the potential of oxygen reduction and water oxidation, the products and intermediates of such aerobic metabolism had also to be dealt with. The potential of the photosynthetic and mitochondrial electron transport chains to produce AOS is considerable and occurs under all conditions (Foyer and Harbinson, 1994). It is now generally accepted that plant cells monitor the endogenous levels of AOS, because H_2O_2 is a putative signalling vector in stress response in plants. The oxidative burst produced during the hypersensitive response has an important role in the induction of plant defence mechanisms (Chen *et al.*, 1993; Conruth *et al.*, 1995; Levine *et al.*, 1994; Matters and Scandalios, 1986; Mehdy, 1994). Superoxide and H_2O_2 production together with subsequent increases in the activity of antioxidant enzymes, are important features in the acquisition of systemic acquired resistance (SAR) and the hypersensitive response (HR) in plants. The plasmalemma-associated oxidative burst associated with HR produces H_2O_2. H_2O_2 production has a dual effect: first, it strengthens the physical barriers to prevent pathogen penetration into the cell (Bradley *et al.*, 1992; Brisson *et al.*, 1994), and second it is a signalling vector. An increase in cellular H_2O_2 is not only implicated in initiation of the defences against pathogens (Antoniw and White, 1980; Levine *et al.*, 1994) but has also been implicated in abiotic stress acclimation, for example chilling tolerance in maize (Prasad *et al.*, 1994). Interestingly, H_2O_2 is used in commercial products to protect plants against pathogens. Application of H_2O_2 to tobacco and *Arabidopsis* caused accumulation of benzoic acid and salicylic acid (Leon *et al.*, 1995; Summermatter *et al.*, 1995). Salicylic acid belongs to a diverse group of plant phenolic compounds which have both direct and indirect effects on stress physiology. Salicylic acid has been shown to inhibit catalase activity (Chen *et al.*, 1993; Conruth *et al.*, 1995; Sanchez-Casas and Klessig, 1994). Through its inhibitory effect on catalase, salicylic acid can increase the H_2O_2 concentration *in vivo*. While salicylic acid accumulation was induced by H_2O_2 in *Arabidopsis* leaves (Summermatter *et al.*, 1995), inhibition of catalase activity was not observed. Both H_2O_2 and salicylic acid hence have putative roles in plant defence reactions against pathogens and the induction of tolerance to environmental stresses.

The strategy of all aerobic organisms is to minimize AOS production under optimal conditions and to remove superoxide and H_2O_2 as they are formed by the concerted action of superoxide dismutases (SOD) and catalases (CAT) or peroxidases (PX). The antioxidant defence system is composed of non-enzymic as well as enzymic components, in particular, ascorbic acid, glutathione, α-tocopherol and β-carotene. The ascorbate/glutathione cycle (Foyer and Halliwell, 1976) involves successive oxidations and reductions of ascorbate, glutathione and NADPH by the enzymes ascorbate peroxidase (APX), monodehydroascorbate reductase (MDAR), dehydroascorbate reductase (DHAR), and glutathione reductase (GR). SOD, which essentially converts one AOS to another, i.e. O_2^- to H_2O_2, determines the concentration of superoxide available for the metal catalyzed Haber–Weiss and Fenton reactions and consequently controls the rate of hydroxyl formation. H_2O_2 is destroyed either by the action of CAT or by radical APX (Dalton *et al.*, 1993; Gillham and Dodge, 1986; Klapheck *et al.*, 1990; Nakano and Asada, 1981). In the chloroplast the re-reduction of monodehydroascorbate to ascorbate is largely non-enzymatic and is facilitated by reduced ferredoxin produced by non-cyclic electron flow. The production and destruction of H_2O_2 in this system occurs on the surface of the thylakoid membranes (Miyake and Asada, 1992). In the stroma and other cellular compartments the scavenging of O_2^- and H_2O_2 involves MDAR, DHAR and GR (Dalton *et al.*, 1993; Foyer and Harbinson, 1994; Hossain *et al.*, 1984). The relative importance of the latter in the regeneration of ascorbate in the chloroplast *per se* remains debatable. However, it is clear that transformed poplar leaves over-expressing GR maintain greater levels of ascorbate, as well as glutathione, in their leaves than do untransformed controls (Foyer *et al.*, 1995a,b).

Over-production of antioxidant enzymes provides an elegant approach with which to study mechanisms of tolerance to environmental stress. Similarly, mutants altered in antioxidant enzymes can play a pivotal role in testing the importance of specific components of the acclimatory and defence systems associated with environmental stress tolerance. In the following discussion we have tried to pinpoint some of the essential elements of the regulatory systems involved in the thylakoid function which serve to protect the photosynthetic machinery from light-induced damage and to discuss the current state of understanding of the role of the antioxidant enzymes in these processes.

1.10 Manipulation of SOD activity

SODs occur in three distinct molecular forms in plants having either Mn, Fe or Cu/Zn as metal cofactors (Fridovich, 1986). These enzymes respond to oxidative stress to varying degrees and are found in most, if not all, cellular compartments (Matters and Scandalios, 1986; Tsang *et al.*, 1991). All of these SOD types have been expressed in transformed plants and targeted to either chloroplasts, mitochondria or cytosol. These transformants have been tested for tolerance to paraquat, a herbicide which transfers electrons from PSI to oxygen and is hence an active pro-oxidant (Bowler *et al.*, 1991; Perl *et al.*, 1993; Sen Gupta *et al.*, 1993a; Tepperman and Dunsmuir, 1990), to ozone (Pitcher *et al.*, 1991; Van Camp *et al.*, 1994) or to photoinhibition (Sen Gupta *et al.*, 1993a,b; Tepperman and Dunsmuir, 1990) and to freezing (McKersie *et al.*, 1993). Taken together, the experimental evidence discussed below largely supports the view that constitutive SOD expression tends to increase tolerance to oxidative stress. However, Tepperman and Dunsmuir (1990) obtained 30–50-fold increases in foliar SOD activity yet the transformed tobacco plants showed no additional resistance to oxidative stress. Transformed tomato plants with SOD activities 2–4-fold above those of the untransformed controls showed no additional tolerance to photoinhibition (Tepperman and Dunsmuir, 1990). Since SOD only converts one form of active oxygen (O_2^-) to another (H_2O_2), high levels of SOD expression alone may not be advantageous. Subsequent reports on transformed tobacco demonstrated varying degrees of enhanced tolerance in lines containing 2–3-fold increases in SOD activity (Bowler *et al.*, 1991; Foyer *et al.*, 1994; Sen Gupta *et al.*, 1993a,b; Van Camp *et al.*, 1994). The discrepancy in the results obtained in the different studies may be related to: differences in assay procedures and levels of exposure to the applied stress, e.g. differing levels and regimes of ozone fumigation (Pitcher *et al.*, 1991; Van Camp *et al.*, 1994); induction of other components of the antioxidant defences via the SOD-induced elevation of tissue H_2O_2 levels; and the ability of the introduced SOD form to function efficiently within the environment to which it is directed.

Modifying plants for enhanced SOD activity has revealed mechanisms of coordinate regulation of the antioxidant systems (Prasad *et al.*, 1994). Suppression of the activities of native SOD isoforms (Perl *et al.*, 1993) and induction of higher levels of APX activities in the cytosol and chloroplasts accompanied the increase in steady state levels of cytosolic APX transcript (Sen Gupta *et al.*, 1993b). Over-production of SOD in tobacco chloroplasts resulted in increased oxidative stress tolerance in tobacco (Bowler *et al.*, 1991; Foyer *et al.*, 1994; Sen Gupta *et al.*, 1993a; Slooten *et al.*, 1995; Van Camp *et al.*, 1994), alfalfa (McKersie *et al.*, 1993), potato (Perl *et al.*, 1993) and cotton (Allen, 1995). Similar results were obtained with overproduction of SOD in the mitochondria of tobacco and alfalfa (Bowler *et al.*, 1991; McKersie *et al.*, 1993) and in the cytosol of tobacco and potato (Perl *et al.*, 1993; Pitcher and Zilinskas, 1996). In tobacco (*Nicotiana tabacum* PB06), overproduction of Mn SOD had a clear protective effect on paraquat-induced ion leakage and also on paraquat-induced inactivation of PSII. In *N. tabacum* SR1, protection of membrane integrity related to decreased ion leakage, but no protection of the PSII reaction centre *per se* was demonstrated (Slooten *et al.*, 1995), suggesting an improvement in the protection of the plasmalemma but not PSII, in spite of the fact that the over-produced SOD was localized in the chloroplasts. One possibility for the lack of protection of PSII might reside in the nature of the SOD that was overexpressed. In this case the mitochondrial Mn SOD was used in contrast to previous reports of over-production of Cu/Zn SOD in potato chloroplasts (Perl *et al.*, 1993) and of *E. coli* Mn SOD in tobacco (Foyer *et al.*, 1994), where increased resistance to the inhibition of photosynthesis by paraquat was observed. Over-production of Fe SOD in the chloroplasts of *N. tabacum* cv. SR1 protected both the plasmalemma and PSII from paraquat-induced damage. This may suggest that Fe SOD gives better protection in chloroplasts than mitochondrial Mn SOD. Functional differences between the *E. coli* Fe SOD and Mn SOD isoforms have been reported with regard to protection (Hopkin *et al.*, 1992; Kunert *et al.*, 1990). Transformed tobacco constitutively over-expressing a pea cytosolic Cu/Zn SOD gene in the cytosol were less sensitive to ozone than untransformed controls (Pitcher and Zilinskas, 1996). Transformed alfalfa plants expressing the same Mn SOD cDNA from *Nicotiana plumbaginifolia* have been shown to be more resistant to drought stress. These transformants showed increased survival and vigour after exposure to sub-lethal freezing temperatures

(McKersie *et al.*, 1993). A three-year field trial with these plants indicated that yield and survival were significantly improved (McKersie *et al.*, 1996). Despite the variations in response observed in different studies, preliminary results indicate that such manipulations may aid the genetic improvement of stress tolerance. Furthermore, they support the hypothesis that tolerance to oxidative stress is important in adaptation to field environments.

1.11 Manipulation of the enzymes involved in glutathione synthesis and turnover

Constitutive expression of the *Escherichia coli GR (gor)* gene in the cytosol of transformed tobacco resulted in increases of between 1.4 and 10-fold in the foliar GR activity (Aono *et al.*, 1991; Foyer *et al.*, 1991). Expression in the chloroplast led to a threefold increase in the GR activity (Aono *et al.*, 1993). Using SO_2 or paraquat-induced visual damage to leaves or leaf discs as a screen for tolerance, Aono *et al.* (1991, 1993) showed enhanced tolerance in these plants. CO_2-dependent O_2 evolution was equally sensitive to paraquat-induced inhibition in transformed plants with elevated cytosolic GR activity as in the untransformed controls (Foyer *et al.*, 1991). The transformed leaves were, however, able to maintain higher levels of reduced ascorbate than the controls in the presence of paraquat (Foyer *et al.*, 1991). The elevated GR level in the chloroplast, however, did not protect the pigments from destruction by oxidative processes. Decreased pigment loss has been reported in plants expressing bacterial GR in the chloroplasts following paraquat treatment (Aono *et al.*, 1993), but this was not observed in poplars expressing the same DNA.

Using bacterial genes, transformed poplar lines expressing γ-glutamyl cysteine synthetase (γ-ECS) and glutathione synthetase (GSH-S) in the cytosol, GR in either the chloroplasts or the cytosol has been produced (Broadbent *et al.*, 1995; Foyer *et al.*, 1995a,b; Noctor *et al.*, 1996; Strohm *et al.*, 1995). Expression of the respective genes coding for these enzymes yielded plants with significantly increased enzyme activities (Foyer *et al.*, 1995a,b; Noctor *et al.*, 1996; Strohm *et al.*, 1995). Expression of the *gor* gene in the chloroplast resulted in extractable foliar GR activities up to 1000-fold those of the untransformed controls (Foyer *et al.*, 1995a,b). In contrast, cytosolic expression resulted in only 2–10-fold elevation of GR activity. This difference was attributed to greater stability of the *gor* gene product in the chloroplast. Expression of GSH-S in the cytosol led to up to 500-fold increases in GSH-S activity (Foyer *et al.*, 1995a,b). Increased GSH-S in the cytosol did not result in increases in the foliar glutathione levels but labelling experiments pointed to the availability of cysteine and the first step in the glutathione biosynthetic pathway (catalyzed by γ-ECS) as the limiting steps of glutathione biosynthesis (Strohm *et al.*, 1995). This was confirmed in transformed poplar plants over-expressing γ-ECS where the leaves contained approximately ten times the amount of γ-glutamyl cysteine compared to the untransformed controls (Noctor *et al.*, 1996). The foliar glutathione and ascorbate contents were approximately doubled in those plants that dramatically over-produced bacterial GR in the chloroplast (Foyer *et al.*, 1995a,b) but glutathione was increased when γ-ECS was over-expressed in the cytosol (Noctor *et al.*, 1996). The transformed poplar leaves over-expressing γ-ECS had three times the level of glutathione in their leaves but there was no change in the ascorbate pool.

In *Nicotiana tabacum* over-expression of *Pisum sativum* GR in the chloroplast, but not the cytosol, led to an increase in foliar glutathione by up to 50 per cent (Broadbent *et al.*, 1995). These results confirm that GR maintains the glutathione pool. An *E. coli gor* deletion mutant with little detectable total glutathione (Kunert *et al.*, 1990) was restored by over-expression of GR, suggesting that an absence of GR activity or cycling of glutathione in some way controls the size of the cellular glutathione pool. There is a clear relationship between GR activity and the size of the foliar glutathione pool.

Plants over-expressing GR in the chloroplast were found to be much less sensitive to the stress imposed by low temperatures combined with high light (Foyer *et al.*, 1995a,b). CO_2 assimilation was much less affected by this photoinhibitory treatment in plants over-expressing GR in the chloroplast than in untransformed controls (Foyer *et al.*, 1995a,b). Poplar plants over-expressing bacterial GR either in the cytosol (fivefold higher than the untransformed controls) or in the chloroplast (150–200 times that in the leaves of the untransformed controls) showed equal sensitivity to ozone as the untransformed controls.

Transformed tobacco plants constitutively expressing the GR coding sequence from pea (Broadbent *et al.*, 1995; Creissen *et al.*, 1992, 1995) exhibited a range of enhanced GR activities from 2–12-fold those of the untransformed controls. In these plants GR activities were enhanced twofold in the cytosol (GR32 lines), 2–3-fold in the chloroplast (GR36 lines) and 4.5–12-fold in the chloroplast and mitochondrion combined (GR46 lines). Enhanced tolerance to paraquat was observed in three out of seven of the GR32 and GR36 lines but the GR46 lines were equally sensitive. Conversely, two out of four of the GR46 lines displayed tolerance to a single eight-hour ozone fumigation (200 nl l^{-1}) whereas none of the GR36/GR32 lines displayed increased tolerance. The nature of the differences in these observed effects is unknown but different degrees of GR enhancement may confer tolerance to different stresses or they may reflect the different modes of damage induced by paraquat and ozone.

1.12 Other systems associated with stress protection

The use of the antioxidative system as a target for modulation of photosynthesis and for enhanced tolerance to oxidative processes is attractive. So far these studies have largely been limited to the enzymes of the antioxidative defence system discussed above. The nature of these has largely been dictated by the availability of relevant genes. Prasad *et al.* (1994) considered that resistance to photoinhibition (Sen Gupta *et al.*, 1993a,b) and tolerance to paraquat (Bowler *et al.*, 1991) produced by SOD over-expression was actually the result of a peroxide-induced signal. They point out that none of the studies on plants modified in SOD activity measured the level of peroxide or explored the effects of the transformation on H_2O_2 scavenging enzymes. Hence, it is feasible that H_2O_2 accumulated as a result of increased SOD activity and that this may itself have been a signal for the increased expression of enzymes involved in antioxidant defence. This H_2O_2-inducible defence system will function continuously even in the absence of environmental stress and will be sensitive to even very small changes in H_2O_2 levels. Therefore the mechanisms conferring stress tolerance may not be those that are immediately evident in these studies. Increased tolerance to oxidation is a multifactorial function and not solely dependent on the increase in the activity of one enzyme. Nevertheless, the acquisition of tolerance involves increased levels of glutathione and ascorbate as well as optimization of cellular thiol-disulphide exchange reactions and stabilization of enzymes in addition to the increase in the potential for protein synthesis and gene expression. All of these parameters must be studied in order to understand fully the nature of stress tolerance and its relationship to the regulation of photosynthesis. Many key components of the defence system associated with photosynthesis respond to environmental stimuli. In most cases precise information on the molecular regulation of the genes coding for structural or protective proteins is largely unavailable. The following provides a brief description of some of the systems which will be increasingly exploited via molecular genetic manipulations in the future.

1.12.1 Chlororespiration

The role of chlororespiration in mediating regulation of thylakoid electron flow remains speculative. It has long been postulated that a chloroplast NAD(P)H dehydrogenase could be involved in the regulation of electron flow (Bennoun, 1982). In plants, at least twelve open reading frames corresponding to *ndh* genes have been described. These are homologous to the genes encoding subunits of the mitochondrial NADH dehydrogenase complex 1 but bear no relationship to the equivalent cyanobacterial complex. The function of these genes has been difficult to study because of the relatively low levels of the *ndh* transcripts and proteins found in plants (Berger *et al.*, 1993; Martin *et al.*, 1996; Nixon *et al.*, 1989; Wu *et al.*, 1989). The expression of the *ndh* genes has been associated with greening and with conditions that produce oxidative stress (Berger *et al.*, 1993; Schantz and Bogorad, 1988).

1.12.2 Catalases and peroxidases

Catalases are largely localized in peroxisomes and glyoxysomes but are also found in the cytosol. There are three major catalase isoforms (CAT 1, 2, 3) which are distinguished by the type of activity they catalyze and their differential expression and tissue localization (Willekens *et al.*, 1994, 1995). Catalase functions in both peroxidatic and catalatic modes. In the latter, H_2O_2 is directly decomposed to O_2 and water, while in the former it catalyzes the reduction of substrates such as methanol, formaldehyde and nitrite. Catalase photoinactivation accompanies photoinhibition (Feierabend *et al.*, 1992) and is sensitive to inhibition by the putative signal metabolite salicylic acid (Chen *et al.*, 1993). Interestingly Durner and Klessig (1995) demonstrated that APX was also sensitive to inhibition by salicylic acid. Regulation of catalase expression in response to environmental and metabolic stimuli is an intriguing feature of the plants' responses to sub-optimal or threatening situations. Non-specific peroxidases may be important in H_2O_2 scavenging in stress conditions (Prasad *et al.*, 1995).

1.13 Ascorbic acid

Ascorbate (vitamin C) is an essential antioxidant and is a powerful regulator of major functions in plant cells, particularly photosynthesis. The pathway of ascorbate biosynthesis in plants has not been resolved. An *Arabidopsis thaliana* mutant selected via its sensitivity to the anthropogenic oxidizing air pollutant O_3 (*soz* 1, sensitive to ozone), is deficient in ascorbic acid and shows increased sensitivity to other AOS-generating compounds such as SO_2 and UV-B light (Conklin *et al.*, 1996). The nature of the lesion in ascorbate biosynthesis in this mutant remains to be resolved. L-ascorbic acid is a product of hexose metabolism (Nishikimi and Yagi, 1996). The involvement of phosphorylated intermediates in the pathway of ascorbate synthesis has been demonstrated in algae, but not in higher plants. Two discrete biosynthetic pathways for the conversion of D-glucose to L-ascorbic acid are possible: a direct pathway which conserves the carbon chain in the same order and carbon sequence as that found in D-glucose; and an inversion pathway in which D-glucose is converted to uronic acids and their derivatives by oxidation of C-6, a reduction of C-1 and an enediol-generating oxidation. Until the pathway of ascorbate biosynthesis has been elucidated and the component enzymes and their respective genes discovered, manipulation of ascorbate levels in a similar manner to that currently being undertaken for glutathione is not feasible.

1.14 Conclusions and perspectives

The major determinants of plant productivity are light, water, CO_2 and nitrogen. All of these have direct effects on photosynthetic efficiency. Photosynthesis drives plant growth, development and biomass production. While much of the fundamental framework and operation of the photosynthetic machinery can be described in detail, our current understanding of the integral processes of regulation, acclimation and adaptation is fragmentary and far from complete. Physiological and metabolic approaches have allowed a detailed description of the former. We envisage that molecular biology will have its greatest impact in elucidating the latter. Through the use of transformation technology we have the capacity to engineer plants genetically for specific purposes including improved efficiency and yield under non-optimal conditions. In contrast to carbon metabolism, the application of molecular approaches to the improvement of photosynthetic electron transport efficiency is a largely unexplored area. In the present discussion we have therefore only been able to outline the concepts that are pivotal to the understanding of the operation of the photosynthetic electron transport system and the nature of its relationships to plant productivity in optimal and stress situations. In addition, the associated elaborate defence mechanisms that serve to scavenge reactive molecules have been considered. Of necessity this has been limited to the enzyme systems where most information is available, notably the SODs and the enzymes of the ascorbate–glutathione cycle, but there are many other antioxidants and quenchers with their associated enzymes which remain to be fully characterized. In conclusion, the following summarizes the principle points of our arguments:

1. Higher plant photosynthesis uses the energy in electromagnetic radiation (in the range 400–700 nm) as a driving force for metabolism, especially the fixation of CO_2 into an organic form. The energy of the absorbed radiation is used to produce reducing power and a transmembrane proton electrochemical potential difference that can be used to drive the formation of ATP (together these are called assimilatory power).

2. The photosynthetic machinery associated with the absorption of light and the stabilization of light energy as a driving force is comprised of proteins that are embedded in, or adjacent to, lipid bilayer membranes which enclose an aqueous space. The proteins contain specific co-factors such as chlorophylls, auxiliary light trapping pigments and inorganic, organic or organometallic redox components. The distribution of the proteins within and around the bilayer is highly organized. Proteins are asymmetrically inserted into the membrane and proteins adjacent to the membrane are associated with one or other of its faces.

3. Following the absorption of a light quantum by a photosynthetic pigment an excited molecule is produced. Excitation energy migrates through a population of protein-bound chlorophyll molecules to a reaction centre containing a specific chlorophyll molecule (or pair of molecules). When excited they are oxidized by an adjacent electron acceptor, producing a chlorophyll radical cation that is then reduced by an electron donor. In oxygenic photosynthesis there are two types of reaction centre, each associated with specific pigment-binding proteins.

4. The activity of photosynthesis is limited at low light by the maximum quantum efficiency of light utilization, and at high light by the maximum rate of turnover of photosynthetic carbon metabolism.

5. In oxygenic photosynthesis the maximum light use efficiency of electron transport on a quantum basis is less than 0.125, consistent with the Z-scheme for photosynthetic electron transport where the two types of reaction centre operate in series.

6. The photosynthetic machinery is highly regulated and adaptive. Its operation is subject to short-term physiological modification and regulation, and to longer-term developmental changes that affect the total and relative sizes of protein populations and of other non-protein low molecular weight components.

7. The regulation and adaptation of photosynthesis serves partly to protect the system from damage that can result as a consequence of side reactions associated with reactive intermediates formed during the operation of electron transport. They also serve to increase the light and nitrogen use efficiency of photosynthesis at limiting light and to adjust the light-saturated rate of photosynthesis so that it matches the ranges of irradiances that the leaf experiences.

8. Photosynthesis both produces and consumes oxygen. Oxygen is consumed principally via photorespiration and by direct transfer of electrons to O_2 by the electron transport system producing superoxide. These events are intimately involved with the regulation of photosynthetic electron transport and protect the photosynthetic apparatus from damage in times of stress.

9. Down-regulation of photosynthetic efficiencies at times when light energy is in excess of the capacity to use available energy effectively, is essential to reduce the risk of singlet oxygen formation. Photorespiration and the Mehler-peroxidase reaction are also sinks for excess assimilatory power.

10. Amelioration of the antioxidative defences associated with the photosynthetic apparatus as a means of preventing stress-induced damage is currently being tested in several higher plant species. It has not yet been unequivocally demonstrated that increasing single components of the antioxidant system increases stress tolerance but initial results are sufficiently encouraging to stimulate increasing interest in this approach.

References

ALLEN, J.F. (1993) Redox control of gene expression and the function of chloroplast genomes – an hypothesis. *Photosyn. Res.* 36, 95–102.

ALLEN, R.D. (1995) Dissection of oxidative stress tolerance using transgenic plants. *Plant Physiol.* 107, 1047–1054.

AMESZ, J., VAN DEN ENGH, G.J. and VISSER, J.W.M. (1971) Reactions of plastoquinone and other photosynthetic intermediates in intact algae and chloroplasts. In: *2nd International Congress of Photosynthesis* (FORTI, G., AVRON, M. and MELANDRI, A., eds), pp. 419–430. Dr W. Junk, The Hague.

AMESZ, J., VISSER, J.W.M., VAN DEN ENGH, G.J. and DIRKS, G.J. (1972) Reaction kinetics of the intermediates of the photosynthetic chain between the two photosystems. *Biochim. Biophys. Acta* 256, 370–380.

ANDERSON, J.M. (1986) Photoregulation of the composition, function, and structure of thylakoid membranes. *Ann. Rev. Plant Physiol. Plant Mol. Biol.* 37, 93–136.

ANDERSON, J.M. (1992) Cytochrome-b6f complex – dynamic molecular organization, function and acclimation. *Photosyn. Res.* 34, 341–357.

ANDERSON, J.M. and OSMOND, C.B. (1987) Sub-shade responses: compromises between acclimation and photoinhibition. In: *Photoinhibition* (KYLE, D.J., OSMOND, C.B. and ARNTZEN, C.J., eds), pp. 1–37. Elsevier Science Publishers, Amsterdam.

ANDERSSON, B. and STYRING, S. (1991) Photosystem-II – molecular-organization, function, and acclimation. *Curr. Topics Bioenerg.* 16, 1–81.

ANTONIW, J.F., and WHITE, R.F. (1980) The effects of aspirin and polyacrilic acid on soluble leaf proteins and resistance to virus infection in five cultivars of tobacco. *Phytopath. Z.* 98, 331–341.

AONO, M., KUBO, A., SAJI, H., NATORI, T., TANAKA, K. and KONDO, N. (1991) Resistance to active oxygen toxicity of transgenic *Nicotiana tabacum* that expresses the gene for glutathione reductase from *Escherichia coli*. *Plant Cell Physiol.* 32, 691–697.

AONO, M., KUBO, A., SAJI, H., TANAKA, K. and KONDO, N. (1993) Enhanced tolerance to photooxidative stress of transgenic *Nicotiana tabacum* with high chloroplastic glutathione reductase activity. *Plant Cell Physiol.* 34, 129–

135.
AP REES, T. (1987) Compartmentation of plant metabolism. In: *The Biochemistry of Plants*, Vol. 12 (DAVIES, D.D., ed.), pp. 87–115. Academic Press, San Diego.

ARMBRUST, T.S., CHITNIS, P.R. and GULKEMA, J.A. (1996) Organization of photosystem I polypeptides examined by chemical cross-linking. *Plant Physiol.* **111**, 1307–1312.

ARNON, D.I. (1995) Divergent pathways of photosynthetic electron transfer – the autonomous oxygenic and anoxygenic photosystems. *Photosyn. Res.* **46**, 47–71.

ARO, E.M., VIRGIN, I. and ANDERSSON, B. (1993) Photoinhibition of photosystem II. Inactivation, protein damage and turnover. *Biochim. Biophys. Acta* **1143**, 113–134.

ASADA, K. and TAKAHASHI, M. (1987) Production and scavenging of active oxygen in photosynthesis. In: *Photoinhibition* (KYLE, D.J., OSMOND, C.B. and ARNTZEN, C.J., eds), pp. 227–287. Elsevier Science Publishers, Amsterdam.

BADGER, M.R. (1985) Photosynthetic oxygen exchange. *Ann. Rev. Plant Physiol.* **36**, 27–53.

BAKER, N.R. and BOWYER, J.R. (eds) (1994) *Photoinhibition of Photosynthesis. From Molecular Mechanisms to the Field.* Bios, Oxford.

BENDALL, D.S. (1982) Photosynthetic cytochromes of oxygenic organisms. *Biochim. Biophys. Acta* **683**, 119–152.

BENDALL, D.S. and MANASSE, R.S. (1995) Cyclic photophosphorylation and electron transport. *Biochim. Biophys. Acta* **1229**, 23–38.

BENNETT, J. (1991) Protein-phosphorylation in green plant chloroplasts. *Annu. Rev. Plant Physiol. Plant Mol. Biol.* **42**, 281–311.

BENNOUN, P. (1982) Evidence for a respiratory chain in the chloroplast. *Proc. Natl Acad. Sci. USA* **79**, 4352–4356.

BERGER, S., ELLERSILK, U., WESTHOFF, P. and STEINMULLER, K. (1993) Studies on the expression of NDH-H, a subunit of the NAD(P)H-plastoquinone oxidoreductase of higher plant chloroplasts. *Planta* **190**, 25–31.

BJÖRKMAN, O. (1981) Responses to different quantum flux densities. In: *Physiological Plant Ecology* (LANGE, O., NOBEL, P.S., OSMOND, C.B. and ZIEGLER, H., eds), pp. 57–107. Springer Verlag, Berlin.

BJÖRKMAN, O. and DEMMIG, B: (1987). Photon yield of O_2 evolution and chlorophyll fluorescence characteristics at 77 K among vascular plants of diverse origins. *Planta* **170**, 489–504.

BLOOM, A.J., CALDWELL, R.M., FINAZZO, J., WARNER, R.L. and WEISSBART, J. (1989) Oxygen and carbon dioxide fluxes from barley shoots depend on nitrate assimilation. *Plant Physiol.* **91**, 352–356.

BOARDMAN, N.K. (1977) Comparative photosynthesis of sun and shade plants. *Annu. Rev. Plant Physiol. Plant Mol. Biol.* **28**, 355–377.

BOEKEMA, E.J., HANKAMER, B., BALD, D., KRUIP, J., NIELD, J., BOONSTRA, A.F., BARBER, J. and RÖGNER, M. (1995) Supramolecular structure of the photosystem II complex from green plants and cyanobacteria. *Proc. Natl Acad. Sci. USA* **92**, 175–179.

BOWLER, C., SLOOTEN, L., VANDENBRANDEN, S., DE RYCKE, R., BOTTERMAN, J., SYBESMA, C., VAN MONTAGU, M. and INZÉ, D. (1991) Manganese superoxide dismutase can reduce cellular damage mediated by oxygen radicals in transgenic plants. *EMBO J.* **10**, 1723–1732.

BRADLEY, D.J., KJELLHORN, P. and LAMB, C.J. (1992) Elicitor- and wound-induced oxidative cross-linking of a proline-rich plant cell wall structural protein: a novel, rapid plant defence response. *Cell* **70**, 21–30.

BRISSON, L.F., TENBAKEN, R. and LAMB, C.J. (1994) Function of oxidative cross-linking of cell wall structural proteins in plant disease resistance. *Plant*

Cell **6**, 1703–1712.
BROADBENT, P., CREISSEN, G.P., KULAR, B., WELLBURN, A.R. and MULLINEAUX, P.M. (1995) Oxidative stress responses in transgenic tobacco containing altered levels of glutathione reductase activity. *Plant J.* **8**, 247–255.
BRUNOLD, C. (1993) Regulatory interactions between sulfate and nitrate assimilation. In: *Sulfur Nutrition and Assimilation in Higher Plants* (DE KOK, L.J., STULEN, I., RENNENBERG, H., BRUNOLD, C. and RAUSER, W.E., eds), pp. 61–75. SPB Academic Publishing, The Hague.
BUCHANAN, B.B. (1994) The ferredoxin-thioredoxin system: update on its role in the regulation of oxygenic photosynthesis. *Adv. Mol. Cell Biol.* **10**, 337–354.
BURKEY, K.O. (1993) Effect of growth irradiance on plastocyanin levels in barley. *Photosyn. Res.* **36**, 103–110.
BURKEY, K.O. (1994) Genetic variation of photosynthetic electron transport in barley – identification of plastocyanin as a potential limiting factor. *Plant Sci.* **97**, 177–187.
CANAANI, O. and MALKIN, S. (1984) Distribution of light excitation in an intact leaf between the two photosystems of photosynthesis. Changes in absorption cross-sections following state 1–state 2 transitions. *Biochim. Biophys. Acta* **766**, 513–524.
CHEN, Z., SILVA, H. and KLESSIG, D.F. (1993) Active oxygen species in the induction of systemic acquired resistance by salicylic acid. *Science* **262**, 1883–1886.
CHITNIS, P.R., XU, Q., CHITNIS, V.P. and NECHUSHTAI, R. (1995) Function and organization of photosystem-I polypeptides. *Photosyn. Res.* **44**, 23–40.
CHOW, W.S. and HOPE, A.B. (1987) The stoichiometries of supramolecular complexes in thylakoid membranes from spinach chloroplasts. *Austr. J. Plant Physiol.* **14**, 21–28.
CHOW, W.S., LUPING, Q., GOODCHILD, D.J. and ANDERSON, J.M. (1988) Photosynthetic acclimation of *Alocisia macrorrhiza* (L) G. Don to growth irradiance: structure, function and composition in chloroplasts. *Austr. J. Plant Phys.* **15**, 107–122.
CHOW, W.S., ADAMSON, H.Y. and ANDERSON, J.M. (1991) Photosynthetic acclimation of *Tradescantia albiflora* to growth irradiance: lack of adjustment of light-harvesting components and its consequences. *Physiol. Plant.* **81**, 175–182.
CHYLLA, R.A. and WHITMARSH, J. (1989) Inactive photosystem II complexes in leaves. Turnover rate and quantitation. *Plant Physiol.* **90**, 765–772.
CONKLIN, P.L., WILLIAMS, E.H. and LAST R.L. (1996) Environmental stress sensitivity of an ascorbate deficient *Arabidopsis* mutant. *Proc. Natl Acad. Sci USA* **93**, 9970–9974.
CONRUTH, U., CHEN, Z.X., RICIGLIANO, J.R. and KLESSIG, D.F. (1995) 2-inducers of plant defence responses, 2,6-dichloroisionicotinic acid and salicylic acid inhibit catalase activity in tobacco. *Proc. Natl Acad. Sci USA* **92**, 7143–7147.
CORNIC, G. and BRIANTAIS, J.-M. (1991) Partitioning of photosynthetic electron flow between CO_2 and O_2 reduction in a C_3 leaf (*Phaseolus vulgaris* L.) at different CO_2 concentrations and during drought stress. *Planta* **183**, 178–184.
CRAMER, W.A., FURBACHER, P.N., SZCZEPANIAK, A. and TAE, G.S. (1991) Electron-transport between photosystem II and photosystem I. *Curr. Topics Bioenerg.* **16**, 179–222.
CRAMER, W.A., SORIANO, G.M., PONOMAREV, M., HUANG, D., ZHANG, H., MARTINEZ, S.E. and SMITH, J.L. (1996) Some new structural aspects and old controversies concerning the cytochrome *b6f* complex of oxygenic photosynthesis. *Annu. Rev. Plant Physiol. Plant Mol. Biol.* **47**, 477–508.

CREISSEN, G.P., EDWARDS, E.A., ENARD, C., WELLBURN, A. and MULLINEAUX, P. (1992) Molecular characterisation of glutathione reductase cDNAs from pea (*Pisum sativum* L.). *Plant J.* **2**, 129–131.

CREISSEN, G., REYNOLDS, H., XUE, Y. and MULLINEAUX, P. (1995) Simultaneous targeting of pea glutathione reductase and of a bacterial fusion protein to chloroplasts and mitochondria in transgenic tobacco. *Plant J.* **8**, 167–175.

DALTON, D.A., BAIRD, L.M., LANGEBERG, L., TAUGHER, C.Y., ANYAN, W.R., VANCE, C.P. and SARATH, G. (1993) Subcellular localization of oxygen defense enzymes in soybean (*Glycine max* L. Merr.) root nodules. *Plant Physiol.* **102**, 481–489.

DANON, A. and MAYFIELD, S.P. (1994) Light-regulated translation of chloroplast messenger RNAs through redox potential. *Science* **266**, 1717–1719.

DAVENPORT, J.W. and McCARTY, R.E. (1984) An analysis of proton fluxes coupled to electron transport and ATP synthesis in chloroplast thylakoids. *Biochim. Biophys. Acta* **766**, 363–374.

DURNER, J. and KLESSIG, D.F. (1995) Inhibition of ascorbate peroxidase by salicylic acid and 2,6-dichloroisonnicotinic acid 2 inducers of plant defence responses. *Proc. Natl Acad. Sci. USA* **92**, 11312–11316.

EDWARDS, G. and WALKER, D.A. (1983) C_3,C_4: *Mechanisms, and Cellular and Environmental Regulation, of Photosynthesis.* Blackwell Scientific Publications, Oxford.

EHLERINGER, J. and BJÖRKMAN, O. (1977) Quantum yields for CO_2 uptake in C_3 and C_4 plants. *Plant Physiol.* **59**, 86–90.

EHLERINGER, J. and PEARCY, R.W. (1983) Variation in quantum yield for CO_2 uptake among C_3 and C_4 plants. *Plant Physiol.* **73**, 555–559.

ESCOUBAS, J-M., LOMAS, M., LaROCHE, J. and FALKOWSKI, P.G. (1995) Light intensity regulation of CAB gene transcription is signalled by the redox state of the plastoquinone pool. *Proc. Natl Acad. Sci. USA* **92**, 10237–10241.

EVANS, J.R. (1986) A quantitative analysis of light distribution between the two photosystems, considering variation in both the relative amounts of the chlorophyll-protein complexes and the spectral quality of light. *Photobiochem. Photobiophys.* **10**, 135–147.

EVANS, J.R. (1987a) The dependence of quantum yield on wavelength and growth irradiance. *Austr. J. Plant Physiol.* **14**, 69–79.

EVANS, J.R. (1987b) The relationship between electron transport components and photosynthetic capacity in pea leaves grown at different irradiances. *Austr. J. Plant Physiol.* **14**, 157–170.

EVANS, J.R. (1988) Acclimation by the thylakoid membranes to growth irradiance and partitioning of nitrogen between soluble and thylakoid proteins. *Austr. J. Plant Physiol.* **15**, 93–106.

EVANS, J.R. (1989) Photosynthesis and nitrogen relationships in leaves of C_3 plants. *Oecologia* **78**, 9–19.

EVRON, Y. and AVRON, M. (1990) Characterization of an alkaline pH-dependent proton 'slip' in the ATP synthase of lettuce thylakoids. *Biochim. Biophys. Acta* **1019**, 115–120.

FARQUHAR, G.D. and VON CAEMMERER, S. (1982) Modelling the photosynthetic response to environmental conditions. In: *Encyclopaedia of Plant Physiology (New Series), Physiological Plant Ecology II*, Vol. 12B (NOBEL, P.S., OSMOND, C.B. and ZIEGLER, H., eds), pp. 549–587. Springer Verlag, Berlin.

FARQUHAR, G.D., VON CAEMERRER, S. and BERRY, J.A. (1980) A biochemical model of photosynthetic CO_2 assimilation in leaves of C_3 species. *Planta* **149**, 78–90.

FEIERABEND, J., SCHAAN, C. and HERTWIG, B. (1992) Photoinactivation of catalase occurs under high- and low-temperature stress conditions and accompanies photoinhibition of photosystem II. *Plant Physiol.* **100**, 1554–1561.

FORK, D.C. and HERBERT, S.K. (1993) Electron transport and photophosphorylation by photosystem I in vivo in plants and cyanobacteria. *Photosyn. Res.* **36**, 149–168.

FORTI, G. and ELLI, G. (1995) The function of ascorbic acid in photosynthetic phosphorylation. *Plant Physiol.* **109**, 1027–1211.

FOYER, C.H. (1993) Interactions between electron transport and carbon assimilation in leaves: coordination of activities and control. In: *Photosynthesis: Photoreactions to Plant Productivity* (ABROL, Y.P., MOHANTY, P. and GOVINDJEE, eds), pp. 199–224. Oxford and IBH, New Delhi.

FOYER, C.H. and HALLIWELL, B. (1976) The presence of glutathione and glutathione reductase in chloroplasts: a proposed role in ascorbic acid metabolism. *Planta* **133**, 21–25.

FOYER, C.H. and HARBINSON, J. (1994) Oxygen metabolism and the regulation of photosynthesis electron transport. In: *Causes of Photooxidative Stresses and Amelioration of Defense Systems in Plants* (FOYER, C.H. and MULLINEAUX, P., eds), pp. 1–42. CRC Press, Boca Raton, FL.

FOYER, C.H., FURBANK, R., HARBINSON, J. and HORTON, P. (1990) The mechanisms contributing to photosynthetic control of electron transport by carbon assimilation in leaves. *Photosyn. Res.* **25**, 83–100.

FOYER, C.H., LELANDAIS, M., GALAP, C. and KUNERT, K.J. (1991) Effects of elevated cytosolic glutathione reductase activity on the cellular glutathione pool and photosynthesis in leaves under normal and stress conditions. *Plant Physiol.* **97**, 863–872.

FOYER, C.H., DESCOURVIÈRES, P. and KUNERT, K.J. (1994) Protection against oxygen radicals: an important defense mechanism studied in transgenic plants. *Plant, Cell and Environ.*, Special Issue **17**, 507–524.

FOYER, C.H., SOURIAU, N., PERRET, S., LELANDAIS, M., KUNERT, K.-J., PRUVOST, C. and JOUANIN, L. (1995a) Overexpression of glutathione reductase but not glutathione synthetase leads to an increase in antioxidant capacity and improved photosynthesis in poplar (*Populus tremula* x *P. alba*) trees. *Plant Physiol.* **109**, 1047–1057.

FOYER, C.H., JOUANIN, L., SOURIAU, N., PERRET, S., LELANDAIS, M., KUNERT, K.-J., NOCTOR, G., PRUVOST, C., STROHM, M., MEHLHORN, H., POLLE, A. and RENNENBERG, H. (1995b) The molecular, biochemical and physiological function of glutathione and its action in poplar. In: *Eurosilva Contribution to Forest Tree Physiology, Dourdan, France* (SANDERMANN, H. JR and BONNET-MASIMBERT, M., eds), pp. 141–170. INRA Editions, Paris (Les Colloques, No. 76).

FRIDOVICH, E. (1986) Superoxide dismutases. *Adv. Enzymol.* **58**, 61–97.

GEIDER, R.J., OSBORNE, B.A. and RAVEN, J.A. (1985) Light dependence of growth and photosynthesis in *Phaeodactylum tricornutum* (Bacillariophyceae). *J. Phycol.* **21**, 609–619.

GENTY, B. and HARBINSON, J. (1996) The regulation of light utilisation for photosynthetic electron transport. In: *Environmental Stress and Photosynthesis* (BAKER, N.R., ed). Kluwer Academic Press, Dordrecht pp. 67–99.

GENTY, B., BRIANTAIS, J.-M. and BAKER, N.R. (1989) The relationship between the quantum yield of photosynthetic electron transport and quenching of chlorophyll fluorescence. *Biochim. Biophys. Acta* **990**, 87–92.

GENTY, B., HARBINSON, J. and BAKER, N.R. (1990a) Relative quantum efficiencies of photosystems I and II of leaves in photorespiratory and non-photorespiratory conditions. *Plant Physiol. Biochem.* **28**, 1–10.

GENTY, B., HARBINSON, J., BRIANTAIS, J.M. and BAKER, N.R.

(1990b) The relationship between nonphotochemical quenching of chlorophyll fluorescence and the rate of photosystem II photochemistry in leaves. *Photosyn. Res.* **25**, 249–257.

GENTY, B., GOULAS, Y., DIMON, B., PELTIER, J.M. and MOYA, I. (1992) Modulation efficiency of primary conversion in leaves, mechanisms involved at PSII. In: *Research in Photosynthesis*, Vol. 4 (MURATA, N., ed.), pp. 603–610. Kluwer Academic Publishers, Dordrecht.

GILLHAM, D.J. and DODGE, A.D. (1986) Hydrogen peroxide scavenging systems within pea chloroplasts. *Planta* **167**, 246–251.

GOETZE, D.C. and CARPENTIER, R. (1994) Ferredoxin-NADP$^+$ reductase is the site of oxygen reduction in pseudocyclic electron transport. *Can. J. Bot.* **72**, 256–260.

GOLBECK, J.H. and BRYANT, D.A. (1991) Photosystem I. *Curr. Topics Bioenerg.* **16**, 83–177.

GRÄBER, P., JUNESCH, U. and THULKE, G. (1987) The chloroplast ATP-synthase: the rate of the catalytic reaction. In: *Progress in Photosynthesis Research*, Vol. 3 (BIGGINS, J., ed.), pp. 2.177–2.184. Martinus Nijhoff, Dordrecht.

GROSS, E.L. (1993) Plastocyanin: structure and function. *Photosyn. Res.* **37**, 103–116.

HAEHNEL, W. (1982) On the functional organization of electron transport from plastoquinone to photosystem I. *Biochim. Biophys. Acta* **682**, 245–257.

HAEHNEL, W. (1984) Photosynthetic electron transport in higher plants. *Annu. Rev. Plant Physiol. Plant Mol. Biol.* **35**, 659–683.

HANNING, I. and HELDT, H.W. (1993) On the function of mitochondrial metabolism during photosynthesis in spinach (*Spinacia oleracea* L.) leaves. Partitioning between respiration and export of redox equivalents and precursors for nitrate assimilation products. *Plant Physiol.* **103**, 1147–1154.

HANSON, O. and WYDRZYNSKI, T. (1990) Current perceptions of photosystem II. *Photosyn. Res.* **23**, 131–162.

HARBINSON, J., GENTY, B. and BAKER, N.R. (1989) Relationships between the quantum efficiencies of photosystems I and II in pea leaves. *Plant Physiol.* **90**, 1029–1034.

HEBER, U., NEIMANIS, S. and DIETZ, K.-J. (1988) Fractional control of photosynthesis by the Q_B protein, the cytochrome f/b_6 complex and other components of the photosynthetic apparatus. *Planta* **173**, 267–274.

HILL, R. and RICH, P.R. (1983) A physical interpretation for the natural photosynthetic process. *Proc. Natl Acad. Sci. USA* **80**, 978–982.

HOLZWARTH, A.R. (1991) Excited states kinetics in chlorophyll systems and its relationship to the functional organisation of the system. In: *The Chlorophylls* (SHEERS, H., ed.), pp. 1125–1152. CRC Press, Boca Raton, FL.

HOPE, A.B., LIGGINS, J. and MATTHEWS, D.B. (1988) The kinetics of reactions in and near the cytochrome *b/f* complex of chloroplast thylakoids. I. Proton deposition. *Austr. J. Plant Physiol.* **15**, 695–703.

HOPKIN, K.A., PAPAZIAN, M.A. and STEINMAN, H.M. (1992) Functional differences between manganese and iron superoxide dismutases in *Escherichia coli* K-12. *J. Biol. Chem.* **267**, 24253–24258.

HORTON, P. (1985) Interactions between electron transfer and carbon assimilation. In: *Photosynthetic Mechanisms and the Environment*, 1st edition, Vol. 1 (BARBER, J. and BAKER, N.R., eds), pp. 135–187. Elsevier Science Publishers, Amsterdam.

HORTON, P., RUBAN, A.V. and WALTERS, R.G. (1996) Regulation of light harvesting in green plants. *Annu. Rev. Plant Physiol. Plant Mol. Biol.* **47**, 655–684.

HOSLER, J.P. and YOCUM, C.F. (1985) Evidence for two photophosphorylation reactions concurrent with ferredoxin-catalyzed non-cyclic electron transport. *Biochim. Biophys. Acta* **808**, 21–31.

HOSSAIN, M.A., NAKANO, Y. and ASADA, Y. (1984) Monodehydroascorbate reductase in spinach chloroplasts and its participation in regeneration of ascorbate for scavenging hydrogen peroxide. *Plant Cell Physiol.* **25**, 385–395.

INOUE, K., FUJII, T., YOKOYAMA, E., MATSUURA, K., HIYAMA, T. and SAKURAI, H. (1989) The photoinhibition site of photosystem I in isolated chloroplasts under extremely reducing conditions. *Plant Cell Physiol.* **30**, 65–71.

JANSSON, S. (1994) The light-harvesting chlorophyll a/b-binding proteins. *Biochim. Biophys. Acta* **1184**, 1–19.

JIANG, C.Z., QUICK, W.P., ALRED, R., KLIEBENSTEIN, D. and RODERMEL, S.R. (1994) Antisense RNA inhibition of Rubisco activase expression. *Plant J.* **5**, 787–798.

KASSMANN, J., SONNEWALD, U. and WILLMITZER, L. (1994) Reduction of the chloroplastic fructose-1,6-bisphosphatase in transgenic potato plants impairs photosynthesis and plant growth. *Plant J.* **6**, 637–650.

KAUFMAN, L.S. (1993) Transduction of blue-light signals. *Plant Physiol.* **102**, 333–337.

KENT, S.S., COURNAC, L. and FARINEAU, J. (1992) An integrated model for the determination of the Rubisco specificity factor, respiration in the light and other photosynthetic parameters of C_3 plants. *Plant Physiol. Biochem.* **30**, 625–637.

KLAPHECK, S., ZIMMER, I. and COSSE, H. (1990) Scavenging of hydrogen peroxide in the endosperm of *Ricinus communis* by ascorbate peroxidase. *Plant Cell Physiol.* **31**, 1005–1013.

KRAMER, D.M. and CROFTS, A.R. (1993) The concerted reduction of the high and low potential chains of the bf complex by plastoquinol. *Biochim. Biophys. Acta* **1183**, 72–84.

KRAMER, H. and MATHIS, P. (1980) Quantum yield and rate of formation of the carotenoid triplet state in photosynthetic structures. *Biochim. Biophys. Acta* **593**, 319–329.

KRUIP, J., BALD, D., BOEKEMA, E. and ROGNER, M. (1994) Evidence for the existence of trimeric and monomeric photosystem-I complexes in thylakoid membranes from cyanobacteria. *Photosyn. Res.* **40**, 279–286.

KUNERT, K.J., CRESSWELL, C.F., SCHMIDT, A., MULLINEAUX, P.M. and FOYER, C.H. (1990) Variations in the activity of glutathione reductase and the cellular glutathione content in relation to sensitivity to methylviologen in *Escherichia coli*. *Arch. Biochem. Biophys.* **282**, 233–238.

LAVERGNE, J. and LECI, E. (1993). Properties of inactive photosystem-II centers. *Photosyn. Res.* **35**, 323–343.

LAVERGNE, J. and TRISSL, H.-W. (1995) Theory of fluorescence induction in photosystem II: derivation of analytical expressions in a model including exciton-radical-pair equilibrium and restricted energy transfer between photosynthetic units. *Biophys. J.* **68**, 2474–2492.

LEEGOOD, R.C., WALKER, D.A. and FOYER, C.H. (1985) Regulation of the Benson–Calvin cycle. In: *Photosynthetic Mechanisms and the Environment* (BARBER, J. and BAKER, N.R., eds), pp. 189–258. Elsevier Science Publishers, Amsterdam.

LEON, J., LANTON, M.A. and RASKIN, I. (1995) Hydrogen peroxide stimulates salicylic acid biosynthesis in tobacco. *Plant Physiol.* **103**, 1673–1678.

LEONG, T.Y. and ANDERSON, J.M. (1984a) Adaptation of the thylakoid membranes of pea chloroplast to light intensities. II. Regulation of electron transport capacities, electron carriers, coupling factor (CF1) activity and rates of photosynthesis. *Photosyn. Res.* **5**, 117–128.

LEONG, T.Y. and ANDERSON, J.M. (1984b) Effect of light quality on the composition and function of thylakoid membranes in *Atriplex triangularis*. *Biochim. Biophys. Acta* **766**, 533–541.

LEVINE, A., TENHAKEN, R., DIXON, R.A. and LAMB, C.J. (1994) H_2O_2 from the oxidative burst orchestrates the plant hypersensitive disease resistance response as a local trigger of programmed cell death and a diffusible inducer of cellular protectant genes. *Cell* **79**, 583–593.

LEY, A.C. (1986) Relationships among cell chlorophyll content, photosystem II light-harvesting and the quantum yield for oxygen production in *Chlorella*. *Photosyn. Res.* **10**, 189–196.

LICHTENTHALER, H.K., BUSHMANN, C. and RAHMSDORF, U. (1980) The importance of blue light for the development of sun-type chloroplasts. In: *The Blue Light Syndrome* (SENGER, H., ed.), pp. 485–494. Springer Verlag, Berlin.

LONG, S.P., POSTL, W.F. and BOLHAR-NORDENKAMPF, H.R. (1993) Quantum yields for uptake of carbon dioxide in C3 vascular plants of contrasting habitats and taxonomic groupings. *Planta* **189**, 226–234.

MALKIN, S. and CANAANI, O. (1994) The use and characteristics of the photoacoustic method in the study of photosynthesis. *Annu. Rev. Plant Physiol. Plant Mol. Biol.* **45**, 493–526.

MARTIN, M., CASANO, L.M. and SABATER, B. (1996) Identification of NDH-A, the product of *ndhA* gene, as a thylakoid protein synthesised in response to photooxidative treatment. *Plant Cell Physiol.* **37**(3), 293–298.

MATE, C.J., HUDSON, G.S., VON CAEMMERER, S., EVANS, J.R. and ANDREWS, T.J. (1993) Reduction of ribulose bisphosphate carboxylase activase levels in tobacco by antisense RNA reduces ribulose-bisphosphate carboxylase carbamylation and impairs photosynthesis. *Plant Physiol.* **102**, 1119–1128.

MATTERS, G.L. and SCANDALIOS, J.G. (1986) Effect of the free radical generating herbicide paraquat on the expression of the superoxide dismutase (SOD) genes in maize. *Biochim. Biophys. Acta* **882**, 29–38.

MATTOO, A.K., PICK, U., HOFFMAN-FALK, H. and EDELMAN, M. (1981) The rapidly metabolized 32,000 dalton polypeptide is the proteinaceous shield regulating photosystem II electron transfer and mediating diuron herbicide sensitivity in chloroplasts. *Proc. Natl Acad. Sci. USA* **78**, 1572–1576.

MAUZERALL, D. and GREENBAUM, N.L. (1989) The absolute size of a photosynthetic unit. *Biochim. Biophys. Acta* **974**, 119–140.

MCKERSIE, B.D., CHEN, Y., DE BEUS, M., BOWLEY, S.R., BOWLER, C., INZÉ, D., D'HALLUIN, K. and BOTTERMAN, J. (1993) Superoxide dismutase enhances tolerance of freezing stress in transgenic alfalfa (*Medicago sativa* L.). *Plant Physiol.* **103**, 1155–1163.

MCKERSIE, B.D., BOWLEY, S.R., HARJANTO, E. and LEPRINCE, O. (1996) Water deficit tolerance and field performance of transgenic alfalfa overexpressing superoxide dismutase. *Plant Physiol.* **111**, 1177–1181.

MEHDY, M. (1994) Active oxygen species in plant defence against pathogens. *Plant Physiol.* **105**, 467–472.

MELIS, A. (1991) Dynamics of photosynthetic membrane-composition and function. *Biochim. Biophys. Acta* **1058**, 87–106.

MELIS, A., MANODORI, A., GLICK, R.E., GHIRARDI, M.L., MCCAULEY, S.W. and NEALE, P.J. (1985) The mechanism of photosynthetic membrane adaptation to environmental stress conditions: a hypothesis on the role of electron transport and of ATP/NADPH pool in the regulation of thylakoid membrane organization and function. *Physiol. Vég.* **23**, 757–765.

MELIS, A., SPANGFORT, M. and ANDERSSON, B. (1987) Light-absorption and electron transport balance between photosystem II and photosystem I in spinach chloroplasts. *Photochem. Photobiol.* **45**, 129–136.

MIYAKE, C. and ASADA, K. (1992). Thylakoid-bound ascorbate peroxidase in spinach chloroplasts and photoreduction of its primary oxidation product mono-

dehydroascorbate radicals in thylakoids. *Plant Cell Physiol.* **33**, 541–553.
MONSON, R.K., LITTLEJOHN, R.O. and WILLIAMS, G.J. (1982) The quantum yield for CO_2 uptake in C_3 and C_4 grasses. *Photosyn. Res.* **3**, 153–159.
MOULD, R.M. and ROBINSON, C. (1991) A proton gradient is required for the transport of 2 lumenal oxygen-evolving proteins across the thylakoid membrane. *J. Biol. Chem.* **266**, 12189–12193.
MYERS, J. (1980) On the algae: thoughts about physiology and measurements of efficiency. In: *Primary Productivity in the Sea* (FALKOWSKI, P.G., ed.), pp. 1–16. Plenum, New York.
NAKANO, Y. and ASADA, K. (1981) Hydrogen peroxide is scavenged by an ascorbate-specific peroxidase in spinach chloroplasts. *Plant Cell Physiol.* **22**, 867–880.
NESPOULOS, C., PELTIER, G. and GANS, P. (1989) Photosynthetic and photorespiratory gas exchange in *Lemna minor*. *Plant Physiol. Biochem.* **27**, 863–871.
NISHIKIMI, M. and YAGI, K. (1996) Biochemistry and molecular biology of ascorbic acid biosynthesis. In: *Subcellular Biochemistry Vol. 25: Ascorbic acid: Biochemistry and Biomedical Cell Biology* (HARRIS, J.R., ed.). Plenum, New York.
NISHIO, J.N. and WHITMARSH, J. (1993) Dissipation of the proton electrochemical potential in intact chloroplasts. II. The pH gradient monitored by cytochrome f reduction kinetics. *Plant Physiol.* **101**, 89–96.
NIXON, P.J., GOUNARIS, K., COOMBER, S.A., HUNTER, C.N., DRYER, T.A. and BARBER, J. (1989) *psbG* is not a photosystem two gene but may be a *ndh* gene. *J. Biol. Chem.* **264**, 14129–14135.
NOCTOR, G., STROHM, M., JOUANIN, L., KUNERT, K.-J., FOYER, C.H. and RENNENBERG, H. (1996) Synthesis of glutathione in leaves of transgenic poplar (*Populus tremula* x *P. alba*) overexpressing γ-glutamylcysteine synthetase. *Plant Physiol.* **112**, 1071–1078.
OBERHUBER, W., DAI, Z.Y. and EDWARDS, G.E. (1993) Light dependence of quantum yields of photosystem-II and CO_2 fixation in C_3 and C_4 plants. *Photosyn. Res.* **35**, 265–274.
OSBORNE, B.A. (1994) Photon requirement for O_2-evolution in Red ($\lambda = 680$ nm) light for some C_3 and C_4 plants and a C_3-C_4 intermediate species. *Plant Cell Environ.* **17**, 143–152.
OSBORNE, B.A. and GEIDER, R.J. (1987) The minimum photon requirement for photosynthesis. An analysis of the data of Warburg and Burk (1950) and Yuan, Evans and Daniels (1955). *New Phytol.* **106**, 631–644.
OSBORNE, B.A. and RAVEN, J.A. (1986) Light absorption by plants and its implications for photosynthesis. *Biol. Rev.* **61**, 1–61.
PAUL, M.J., KNIGHT, J.S., HABASH, D., PARRY, M.A.J., LAWLOR, D.W., BARNES, S.A., LOYNES, A. and GRAY, J.C. (1995) Reduction in phosphoribulokinase activity by antisense RNA in transgenic tobacco: effect on CO_2 assimilation and plant growth in low irradiance. *Plant J.* **7**, 535–542.
PERL, A., PERL-TREVES, R., GALILI, S., AVIV, D., SHALGI, E., MALKIN, S. and GALUN, E. (1993) Enhanced oxidative stress defence in transgenic potato expressing tomato Cu/Zn superoxide dismutases. *Theor. Appl. Genet.* **85**, 568–576.
PETERSON, R.B. (1989) Partitioning of non-cyclic photosynthetic electron transport to O_2-dependent dissipative processes as probed by fluorescence and CO_2. *Plant Physiol.* **90**, 1322–1328.
PETERSON, R.B. (1991) Effects of O_2 and CO_2 concentrations on quantum yields of photosystem I and photosystem II in tobacco leaf tissue. *Plant Physiol.* **97**, 1388–

1394.

PETERSON, R.B. (1994) Regulation of electron transport in photosystems I and II in C_3, C_3-C_4, and C_4 species of *Panicum* in response to changing irradiance and O_2 levels. *Plant Physiol.* **105**, 349–356.

PITCHER, L.H. and ZILINSKAS, B.A. (1996) Over-expression of copper/zinc superoxide dismutase in the cytosol of transgenic tobacco confers partial resistance to ozone-induced foliar necrosis. *Plant Physiol.* **110**, 583–588.

PITCHER, L.H., BRENNAN, E., HURLEY, A., DUNSMUIR, P., TEPPERMAN, J.M. and ZILINSKAS, B.A. (1991) Overproduction of petunia chloroplastic copper/zinc superoxide dismutase does not confer ozone tolerance in transgenic tobacco. *Plant Physiol.* **97**, 452–455.

PRASAD, T.K., ANDERSON, M.D., MARTIN, B.A. and STEWART, C.R. (1994) Evidence for chilling-induced oxidative stress in maize seedlings and a regulatory role for hydrogen peroxide. *Plant Cell.* **6**, 65–74.

PRASAD, T.K., ANDERSON, M.D. and STEWART, C.R. (1995) Localisation and characterisation of peroxidases in the mitochondria of chilling acclimated maize seedlings. *Plant Physiol.* **108**, 1597–1605.

PRICE, G.D., EVANS, J.R., VON CAEMMERER, S., YU, J.W. and BADGER, M.R. (1995a) Specific reduction of chloroplast glyceraldehyde 3-phosphate dehydrogenase activity by antisense RNA reduces CO_2 assimilation via a reduction in ribulose bisphosphate regeneration in transgenic tobacco plants. *Planta* **195**, 369–378.

PRICE, G.D., YU, J.-W., VON CAEMMERER, S., EVANS, J.R., CHOW, W.S., ANDERSON, J.M., HURRY, V. and BADGER, M.R. (1995b) Chloroplast cytochrome b_6/f and ATP synthase complexes in tobacco: transformants with antisense RNA against nuclear-encoded transcripts for the Rieske FeS and ATPδ polypeptides. *Austr. J. Plant Physiol.* **22**, 285–297.

QUICK, W.P., SCHURR, U., SCHEIBE, R.R., SCHULZE, E.D., RODERMEL, S.R., BOGORAD, L. and STITT, M. (1991) Decreased ribulose-1,5-bisphosphate carboxylase oxygenase in transgenic tobacco with 'antisense' rbcS. Impact on photosynthesis in ambient growth conditions. *Planta* **183**, 542–554.

QUICK, W.P., FICHTNER, K., SCHULZE, E.D., WENDLER, R., LEEGOOD, R.C., MOONEY, H., RODERMEL, S.R., BOGORAD, L. and STITT, M. (1992) Decreased ribulose-1,5-bisphosphate carboxylase oxygenase in transgenic tobacco transformed with 'antisense' rbcS. IV. Impact on photosynthesis in conditions of altered nitrogen supply. *Planta* **188**, 522–531.

REMIS, D., BULYCHEV, A.A. and KURELLA, G.A. (1986) The electrical and chemical components of the protonmotive force in chloroplasts as measured with capillary and pH-sensitive microelectrodes. *Biochim. Biophys. Acta* **852**, 68–73.

RENGER, G. (1992) Energy transfer and trapping in photosystem II. In: *The Photosystems: Structure, Function and Molecular Biology* (BARBER, J., ed.), pp. 45–99. Elsevier, Amsterdam.

RICH, P. (1982) A physicochemical model of quinone-cytochrome bc complex electron transfers. In: *Function of Quinones in Energy Conserving Systems* (TRUMPOWER, B.L., ed.), pp. 73–86. Academic Press, New York.

ROBINSON, J.M. (1988) Does O_2 photoreduction occur within chloroplasts *in vivo*? *Physiol. Plant.* **72**, 666–680.

ROCK, C.D., BOWLBY, N.R., HOFFMANN-BENNING, S. and ZUVANT, J.A.D. (1992) The aba mutant of *Arabidopsis thaliana* (L) Heynh. has reduced chlorophyll fluorescence yields and reduced thylakoid stacking. *Plant Physiol.* **100**, 1796–1801.

RUMBERG, B., SCHUBERT, K., STRELOW, F. and TRAN-ANH, T.

(1990) The H$^+$-ATP-synthase of spinach chloroplasts is four. In: *Current Research in Photosynthesis*, Vol. 3 (BALTSCHEFFSKY, M., ed.), pp. 125–128. Kluwer Academic Publishers, Dordrecht.

SANCHEZ-CASAS, P. and KLESSIG, D.F. (1994) A salicylic acid-binding activity and a salicylic acid-inhibitole catalase activity are present in a variety of plant species. *Plant Physiol.* **106**, 1675–1679.

SCANDALIOS, J.G. (1993) Oxygen stress and superoxide dismutases. *Plant Physiol.* **101**, 7–12.

SCHANTZ, R. and BOGORAD, L. (1988) Maize chloroplast genes *ndhD*, *ndhE* and *psaC*. Sequences, transcripts and transcript pools. *Plant Mol. Biol.* **11**, 239–247.

SCHATZ, G.H., BROCK, H. and HOLZWARTH, A.R. (1988) Kinetic and energetic model for the primary processes in photosystem II. *Biophys. J.* **54**, 397–405.

SCHEIBE, R. (1990) Light/dark modulation: regulation of chloroplast metabolism in a new light. *Bot. Acta* **103**, 327–334.

SCHELLER, H.V. (1996) In vitro cyclic electron transport in barley thylakoids follows two independent pathways. *Plant Physiol.* **110**, 187–194.

SCHLODDER, E., GRÄBER, P. and WITT, H.T. (1982) Mechanism of phosphorylation in chloroplasts. In: *Electron Transport and Photophosphorylation*, 1st edition (BARBER, J., ed.), pp. 103–175. Elsevier Science Publishers, Amsterdam.

SCHREIBER, U., SCHLIWA, U. and BILGER, W. (1986) Continuous recording of photochemical and non-photochemical chlorophyll fluorescence quenching with a new type of modulation fluorometer. *Photosyn. Res.* **10**, 51–62.

SENGER, H. (1982) Efficiency of incident light utilisation and quantum requirements of microalgae. In: *CRC Handbook of Biosolar Resources*, Vol. 1 (MITSUI, A. and BLACK, C.C., eds), pp. 55–58. CRC Press, Boca Raton, FL.

SEN GUPTA, A., HEINEN, J.L., HOLADAY, A.S., BURKE, J.J. and ALLEN, R.D. (1993a) Increased resistance to oxidative stress in transgenic plants that overexpress chloroplastic Cu/Zn superoxide dismutase. *Proc. Natl Acad. Sci. USA* **90**, 1629–1633.

SEN GUPTA, A., WEBB, R.P., HOLADAY, A.S. and ALLEN, R.D. (1993b) Overexpression of superoxide dismutase protects plants from oxidative stress. *Plant Physiol.* **103**, 1067–1073.

SIMS, D.A. and PEARCY, R.W. (1994) Scaling sun and shade photosynthetic acclimation of *Alocasia macrorrhiza* to whole plant performance. I. Carbon balance and allocation at different daily photon flux densities. *Plant Cell Environ.* **17**, 881–887.

SLOOTEN, L., CAPIAU, K., VAN CAMP., W., VAN MONTAGU, M., SYBESMA, C. and INZÉ, D. (1995) Factors affecting the enhancement of oxidative stress tolerance in transgenic tobacco overexpressing manganese superoxide dismutase in the chloroplasts. *Plant Physiol.* **107**, 737–750.

SONOIKE, K. (1995) Selective photoinhibition of photosystem-I in isolated thylakoid membranes from cucumber and spinach. *Plant Cell Physiol.* **36**, 825–830.

SONOIKE, K. (1996) Photoinhibition of photosystem I – its physiological significance in the chilling sensitivity of plants. *Plant Cell Physiol.* **37**, 239–247.

SONOIKE, K. and TERASHIMA, I. (1994) Mechanism of photosystem I photoinhibition in leaves of *Cucumis Sativus* L. *Planta* **194**, 287–293.

STITT, M., QUICK, W.P., SCHURR, U., SCHULZE, E.D., RODERMEL, S.R. and BOGORAD, L. (1991) Decreased ribulose-1,5-bisphosphate carboxylase oxygenase in transgenic tobacco transformed with 'antisense' rbcS. II. Flux control coefficients for photosynthesis in varying light, CO_2 and air humidity. *Planta* **183**, 555–566.

STROHM, M., JOUANIN, L., KUNERT, K.-J., PRUVOST, C., POLLE, A., FOYER, C.H. and RENNENBERG, H. (1995) Regulation of glutathione synthesis in leaves of transgenic poplar (*Populus tremula* x *P. alba*) overexpressing glutathione synthetase. *Plant J.* **7**, 141–145.

SUMMERMATTER, K., STICHER, L. and METRAUX, J.P. (1995) Systemic responses in *Arabidopsis thaliana* infected and challenged with *Pseudomas syringae p.v. syringae*. *Plant Physiol.* **108**, 1379–1385.

SUNDBY, C., MCCAFFERY, S. and ANDERSON, J.M. (1993) Turnover of the photosystem II D1 protein in higher plants under photoinhibitory and nonphotoinhibitory irradiance. *J. Biol. Chem.* **268**, 25476–25482.

TEPPERMAN, J.M. and DUNSMUIR, P. (1990) Transformed plants with elevated levels of chloroplastic SOD are not more resistant to superoxide toxicity. *Plant Mol. Biol.* **14**, 501–511.

TERASHIMA, I. and EVANS, J.R. (1988) Effects of light and nitrogen nutrition on the organization of the photosynthetic apparatus in spinach. *Plant Cell Physiol.* **29**, 143–155.

TERASHIMA, I., FUNAYAMA, S. and SONOIKE, K. (1994) The site of photoinhibition in leaves of Cucumis sativus L at low temperatures is photosystem I, not photosystem II. *Planta* **193**, 300–306.

THAYER, S.S. and BJÖRKMAN, O. (1990) Leaf xanthophyll content and composition in sun and shade determined by HPLC. *Photosyn. Res.* **23**, 331–343.

THIELEN, A.P.G.M. and VAN GORKOM, H. (1981) Energy transfer and quantum yield in photosystem II. *Biochim. Biophys. Acta* **637**, 439–446.

TIKHONOV, A.N., KHOMUTOV, G.B. and RUUGE, E.K. (1984) Electron transport control in chloroplasts. Effects of magnesium ions on the electron flow between two photosystems. *Photobiochem. Photobiophys.* **8**, 261–269.

TRISSL, H.W. and WILHELM, C. (1993) Why do thylakoid membranes from higher-plants form grana stacks. *Trends Biochem. Sci.* **18**, 415–419.

TSANG, E.W.T., BOWLER, C., HÉROUART, D., VAN CAMP, W., VILLAROEL, R., GENETELLO, C., VAN MONTAGU, M. and INZÉ, D. (1991) Differential regulation of superoxide dismutase in plants exposed to environmental stress. *Plant Cell* **3**, 783–792.

VAN BEL, A.E. and GAMALEI, V.V. (1992) Ecophysiology of phloem loading in source leaves. *Plant Cell Environ.* **92**, 265–270.

VAN CAMP, W., WILLEKENS, H., BOWLER, C., VAN MONTAGU, M., INZÉ, D., REUPOLD-POPP, P., SANDERMANN JR, H. and LANGEBARTELS, C. (1994) Elevated levels of superoxide dismutase protect transgenic plants against ozone damage. *Bio/technology* **12**, 165–168.

VAN WIJK, K.J. and VAN HASSELT, P.R. (1993) Kinetic resolution of different recovery phases of photoinhibited photosystem II in cold-acclimated and non-acclimated spinach leaves. *Physiol. Plant.* **87**, 187–198.

WALTERS, R.G. and HORTON, P. (1995a) Acclimation of *Arabidopsis thaliana* to the light environment – changes in photosynthetic function. *Planta* **197**, 306–312.

WALTERS, R.G. and HORTON, P. (1995b) Acclimation of *Arabidopsis thaliana* to the light environment – regulation of chloroplast composition. *Planta* **197**, 475–481.

WEIS, E., BALL, J.T. and BERRY, J. (1987) Photosynthetic control of electron transport in leaves of *Phaseolus vulgaris*: evidence for regulation of photosystem II by the proton gradient. In: *Progress in Photosynthesis Research*, Vol. 2 (BIGGINS, J., ed.), pp. 553–556. Martinus Nijhoff, Dordrecht.

WILD, A. and HOLZAPFEL, A. (1980) The effect of blue and red light on the content of chlorophyll, cytochrome f, soluble reducing sugars, soluble proteins and the nitrate reductase activity during growth of the primary leaves of *Sinapsis alba*. In: *The Blue Light Syndrome* (SENGER, H., ed.), pp. 444–451. Springer Verlag, Berlin.

WILD, A., HÖPFNER, M., RÜHLE, W. and RICHTER, M. (1986) Changes in the stoichiometry of photosystem II components as an adaptive response to high light and low light conditions during growth. *Z. Naturforsch. Teil. C.* **41**, 597–

603.
WILDE, A., HÄRTEL, H., HÜBSCHMANN, T., HOFFMAN, P., SHESTAKOV, S.V. and BÖRNER, T. (1995) Inactivation of a *Synechocystis* sp strain PCC 6803 gene with homology to conserved chloroplast open reading frame 184 increases the photosystem II-to-photosystem I ratio. *Plant Cell.* **7**, 649–658.

WILLEKENS, H., LANGEBARTELS, C., TIRE, C., VAN MONTAGU, M., INZÉ, D. and VAN CAMP, W. (1994) Differential expression of catalase genes in *Nicotiana plumbaginifolia*. *Proc. Natl Acad. Sci. USA* **91**, 10450–10454.

WILLEKENS, H., INZÉ, D., VAN MONTAGU, M. and VAN CAMP, W. (1995) Catalases in plants. *Mol. Breed.* **1**, 207–228.

WITT, H.T. (1979) Energy conversion in the functional membrane of photosynthesis. Analysis by light pulse and electric pulse methods. The central role of the electric field. *Biochim. Biophys. Acta* **505**, 355–427.

WU, M., NIE, Z.Q. and YANG, J. (1989) The 18-KD protein that binds to the chloroplast DNA replicative origin is an iron-sulphur protein related to a subunit of NADH-dehydrogenase. *Plant Cell* **1**, 551–557.

ZRENNER, R., KRAUSE, K.-P., APEL, P. and SONNEWALD, U. (1996) Reduction of the cytosolic fructose-1,6-bisphosphatase in transgenic potato plants limits photosynthetic sucrose biosynthesis with no impact on plant growth and tuber yield. *Plant J.* **9**, 671–681.

2

The regulation and control of photosynthetic carbon assimilation

W. PAUL QUICK AND H. EKKEHARD NEUHAUS

2.1 Introduction

The initial fixation of CO_2 in plants occurs within green plastids (chloroplasts) found not only in leaves but also in a range of other tissues where their precise function has yet to be established. This chapter will discuss the carbohydrate metabolism of chloroplasts; the first part will focus on photosynthetic CO_2 fixation in leaves and the second part will discuss our current understanding of the carbohydrate metabolism of green plastids found in other tissues. This chapter is not intended to be a comprehensive review of the literature, rather a survey of recent developments in this field. Readers are referred to several excellent reviews that have appeared in the literature and which we, the authors, have used as the background of this work (Leegood, 1996; Macdonald and Buchanan, 1990; Walker and Edwards, 1983).

2.2 Photosynthetic CO_2 fixation in green leaves: the pathway

Figure 2.1 outlines the reductive pentose phosphate pathway identified by Calvin and coworkers in the 1950s and responsible for photosynthetic CO_2 fixation. The cycle has several important features that will be expanded further below.

First, a carboxylation reaction unique to photosynthetic organisms incorporates inorganic CO_2 into ribulose 1,5-bisphosphate ($Ru1,5P_2$). This is catalyzed by the enzyme ribulose 1,5-bisphosphate carboxylase-oxygenase, often referred to as Rubisco. Rubisco has a poor affinity for CO_2 ($K_m \approx 10$ µM), which at present day concentrations of atmospheric CO_2 means that Rubisco operates at best below or close to 50% of its maximum activity. It is also a particularly sluggish enzyme having a low specific activity (3–4 µmol min^{-1} mg^{-1} protein). As a result plants require large amounts of Rubisco protein to maximize rates of photosynthesis, and this protein can account for up to 50% of leaf protein in C_3 plants. As the name suggests, the enzyme catalyzes both a carboxylase and an oxygenase reaction. The latter results in oxygenation of $Ru1,5P_2$ and the

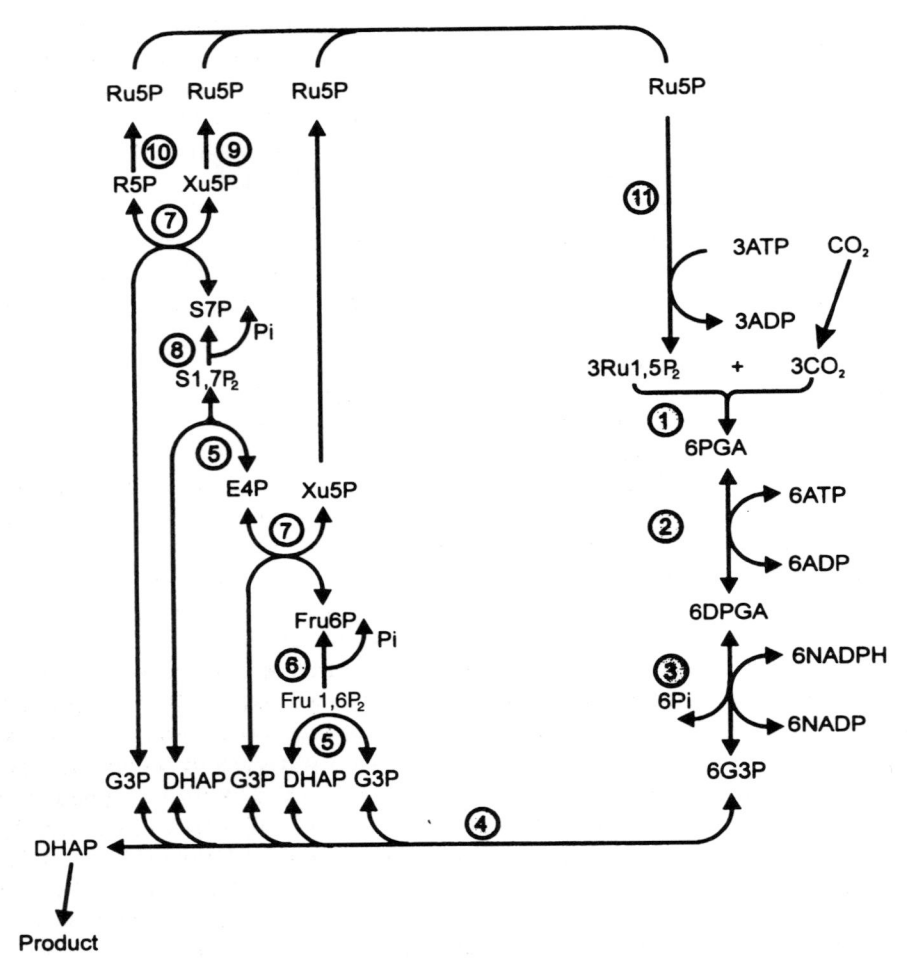

A molecular approach to primary metabolism in higher plants

Figure 2.1 Diagrammatic representation of the Calvin cycle. The pathway is portrayed assuming three molecules of CO_2 are fixed for each turn of the cycle. The subsequent fate of the six molecules of PGA produced is shown. The enzymes that catalyze the individual reactions are depicted as numbers enclosed in a circle. These refer to: (1) ribulose 1,5-bisphosphate carboxylase-oxygenase; (2) 3-phosphoglycerate kinase; (3) glyceraldehyde 3-phosphate dehydrogenase; (4) triose phosphate isomerase; (5) aldolase; (6) fructose 1,6-bisphosphatase; (7) transketolase; (8) sedoheptulose 1,7-bisphosphatase; (9) ribulose 5-phosphate 3-epimerase; (10) ribose 5-phosphate isomerase; (11) phosphoribulokinase.

production of a 2-carbon molecule of 2-phosphoglycolate (2PG) and a 3-carbon molecule of 3-phosphoglycerate (PGA); carboxylation results in the production of two molecules of PGA. The 2PG produced has then to be recycled via the photorespiratory pathway, resulting in a loss of CO_2 and a release of NH_3, which has to be effectively reassimilated if precious nitrogen reserves are not to be depleted; both of these processes require an input of energy in the form of ATP or reducing equivalents. The oxygenation reaction is therefore a wasteful process resulting in the loss of CO_2 and the expenditure of energy. CO_2 and O_2

are competitive substrates and therefore the rate of each reaction is dependent on the concentration of each of the gases dissolved in the chloroplast. The low CO_2 concentration of the atmosphere (and hence the low concentration dissolved within the aqueous phase of the leaf) requires that leaves maximize the availability of CO_2 from outside. Increased leaf porosity also results in large evaporative losses for the leaf and frequently a leaf must balance the requirements for CO_2 with the availability of water. Low water availability frequently leads to reduced CO_2 concentrations within the leaf and hence promotes the rate of the oxygenation reaction. It has been suggested that in these conditions of reduced CO_2 availability, the oxygenation reaction is useful in that it allows the continued dissipation of light energy which might otherwise damage the photosynthetic electron transport chain (see Chapter 1).

Second, PGA is also reduced to triose phosphate (TP). The two reactions required are perhaps better known for their function in cytoplasmic glycolysis where the reaction proceeds in the opposite direction, generating ATP and NADH. The reverse direction sustained in the chloroplast during photosynthesis is maintained by a high ATP/ADP and NADPH/NADP ratio, generated by the photosynthetic electron transport chain. This reductive phase of the cycle provides the link between the light driven membrane bound electron transport chain and the soluble enzymic components of the Calvin cycle (often referred to as the light and dark reactions, respectively).

Third, the 5-carbon metabolite, $Ru1,5P_2$, required for carboxylation, is regenerated from 3-carbon sugar phosphates; this is essential if continued CO_2 fixation is to be sustained. This regeneration occurs via a series of reactions involving both sugar condensation and carbon rearrangement reactions. Condensation of 3-carbon (triose phosphate) and 4-carbon (erythrose 4-phosphate E4P) sugar phosphates yields sedoheptulose 1,7-bisphosphate ($S1,7P_2$) and fructose 1,6-bisphosphate ($Fru1,6P_2$). These are subsequently dephosphorylated to produce their respective monophosphates. Two carbon unit transfers from fructose 6-phosphate (Fru6P) and sedoheptulose 7-phosphate (S7P) result in the production of three 5-carbon sugar phosphates and a 4-carbon compound that can be recondensed with TP. Epimerase and isomerase reactions convert the isomer (ribose 5P, R5P) and epimer (xylulose 5-P, Xu5P) to ribulose 5-phosphate (Ru5P). Finally, Ru5P is phosphorylated by Ru5P kinase, requiring further ATP generated by the electron transport chain.

The net effect of these reactions can be seen from Figure 2.1; this scheme shows how the pathway operates to form a net 3-carbon sugar phosphate (TP) product; three molecules of CO_2 are reacted with three molecules of $Ru1,5P_2$ to produce six molecules of the 3-carbon molecule, PGA; these are then phosphorylated and reduced to form six molecules of TP, requiring six molecules each of ATP and NADPH. Five TP molecules are then rearranged to regenerate the three molecules of $Ru1,5P_2$ in the regenerative phase of the Calvin cycle, consuming a further three molecules of ATP. Three ATP and two NADPH are thus required from the electron transport chain for each molecule of CO_2 that is fixed. The triose phosphate product is then primarily utilized to produce starch in the chloroplast or sucrose in the cytosol, but can also be used for a wide variety of other cellular functions. The rate of consumption of the product, TP, must be co-ordinated with the prevailing rate of photosynthesis in order that the correct proportion of TP is recycled to regenerate further acceptors for CO_2.

Rapid fluctuations can occur in the rate of photosynthesis (e.g. as a result of changing light intensity) and therefore, very strict control of the rate of the TP-consuming reactions is required. This regulation is discussed briefly below.

Starch and sucrose are the major end-products of photosynthesis. Carbon fixed during photosynthesis is either retained in the chloroplast and converted to the storage carbohydrate starch or it is transferred to the cytosol in the form of TP and converted to sucrose. The major substrates for starch and sucrose synthesis in photosynthetic tissues are three-carbon sugar phosphates. These can be exported from the chloroplast during photosynthesis, predominantly in the form of triose phosphates, via the phosphate translocator in a strict counter exchange for phosphate. Continued photosynthesis not only requires efficient regeneration of the CO_2 acceptor ($Ru1,5P_2$) but also that the phosphate incorporated into the TP-product is recycled. Work with isolated chloroplasts has demonstrated that the rate of photosynthesis is strictly dependent on the concentration of inorganic phosphate (Pi) in the external medium. Concentrations above or below an optimum (typically 0.5 mM Pi) inhibit the rate of photosynthesis: when Pi is too low, export of TP from the chloroplast is prevented, and a build-up of Calvin cycle intermediates occurs as all of the TP is fed back into the cycle. The concentration of Pi in the chloroplast falls which restricts the rate of ATP synthesis and hence further CO_2 fixation. Conversely, when Pi is too high, export of TP is increased beyond a rate that can be sustained by photosynthesis, the concentration of Calvin cycle intermediates falls and regeneration of $Ru1,5P_2$ necessary for further CO_2 fixation is reduced (Walker and Edwards, 1983). The maintenance of an optimal Pi concentration in the cytosol requires that Pi incorporated into triose phosphates is recycled efficiently at a rate that matches the prevailing rate of photosynthesis. This requires co-ordination of not only sucrose synthesis with photosynthesis, but also other competing pathways for photosynthate, for example starch, fructan and amino acid synthesis. These are the topics of subsequent chapters of this book to which the reader is referred.

2.3 Regulation of the Calvin cycle

2.3.1 Regulation of Rubisco

Rubisco catalyzes the initial carboxylation reaction and the first irreversible reaction of the Calvin cycle. The activity of the enzyme is highly regulated at several levels. First, enzyme activity requires that Mg^{2+} and CO_2 are bound to a lysine residue adjacent to the active site (carbamylation); this can be achieved spontaneously in the test-tube if the enzyme is incubated at alkaline pH in the presence of Mg^{2+} and CO_2 and in the absence of $Ru1,5P_2$. The CO_2 molecule is not the one used for catalysis, but is thought to play a role in the binding of Mg^{2+} required for catalysis. Besides the carboxylation and oxygenation reaction described above, Rubisco also catalyzes at least two further reactions resulting in the production of two novel products, xylulose 1,5-bisphosphate and 3-keto-arabinitol-1,5-bisphosphate. Both of these compounds bind very tightly to the catalytic site and hence render the enzyme effectively inactive until displaced. The substrate of the enzyme can also inhibit activity under certain conditions; $Ru1,5P_2$ binds very tightly to the non-carbamylated (inactive) form of the enzyme; this prevents carbamylation of the enzyme and renders it effectively inactive until displaced (see Bainbridge et al., 1995, for a recent review). Furthermore, a naturally occurring inhibitor of Rubisco has been identified (2-carboxy-arabinitol-1-phosphate, CA1P) that is found in abundance during the night and is degraded during the photoperiod (Portis, 1992). This nocturnal inhibitor of Rubisco also binds tightly to the active site. CA1P is thought to play an important role in the regulation of Rubisco activity although the mechanisms that control the amount of this metabolite are far from clear as are the reasons for the nocturnal inhibition of Rubisco activity. CA1P is not found at all in some plants and at concentrations too low to be effective in others. A ubiquitous role for this metabolite has yet to be demonstrated. More recently a further inhibitory metabolite of Rubisco that is present in the light has also been demonstrated but has not yet been chemically identified (Keys et al., 1995).

A common feature of all of these potential inhibitors is that they bind very tightly to the Rubisco active site and have to be displaced before catalytic competence is restored. This is brought about by the enzyme Rubisco activase which effectively removes these tight binding inhibitors from the active site (Portis, 1995). The process requires ATP and is inhibited by ADP, hence the rate of Rubisco activation is determined in part by the availability of ATP from the photosynthetic electron transport chain. This is one proposed mechanism whereby the activation state of Rubisco can be co-ordinated with the activity of the light-driven electron transport chain. Indeed, a close correlation is observed between the rate of photosynthesis and the activation state of Rubisco, independently of how the rate is varied (light or CO_2).

2.3.2 Thiol-reduction

Several enzymes of the Calvin cycle are subject to light/dark regulation. Light dependent activation of several enzymes occurs rapidly upon illumination through a process that is coupled to photosynthetic electron transport via a ferredoxin/thioredoxin soluble electron transport system (Buchanan, 1980). Activation of target enzymes occurs when specific disulphide bonds (between sulphide groups contained in cysteine residues of the polypeptide) are reduced and hence cleaved. This process can be accomplished chemically in the test-tube with a reduced di-thiol such as dithiothreitol. In vivo this is accomplished by reduced thioredoxin. Thioredoxin is reduced by reduced ferredoxin and the reaction is catalyzed by the enzyme ferredoxin/thioredoxin reductase. Ferredoxin itself is reduced by the electron transport chain, hence initiation of electron transport brings about the activation of several photosynthetic enzymes. These include FBPase, SBPase, GAPDH and R5P kinase from the Calvin cycle. Two other enzymes are also light activated: the chloroplast ATP synthetase, also called the chloroplast coupling factor (CF_0–CF_1) and malate dehydrogenase. The first is responsible for the light dependent synthesis of ATP and the latter for the transfer of reducing equivalents out of the chloroplast. One further enzyme has been shown to be regulated by thiol-reduction, glucose-6-phosphate dehydrogenase; this enzyme is the starting point for entry of hexose phosphates into the oxidative pentose phosphate pathway where sugars are degraded, CO_2 is released and NADPH is produced. This pathway is required in the dark for the generation of reducing equivalents in chloroplast. The enzyme Glc6PDH is actually deactivated by thiol/reduction and hence is only active in the dark.

Originally it was proposed that this type of thiol-regulation merely acted as a light/dark switch to prevent the operation of the Calvin cycle in the dark, which could result in the operation of futile metabolic cycles (simultaneous synthesis and degradation of metabolites, e.g. Fru1,6P$_2$). More recently, evidence has accumulated to suggest that this regulation is much more subtle (Faske *et al.*, 1995). Frequently the substrate of the regulated enzyme has been shown to interact with the thiol-reduction process such that in the presence of substrate a lower redox potential is required to activate the enzyme. This provides a mechanism whereby the enzyme is only activated in the light but the extent of activation is modulated by the availability of substrate, allowing the activation state of the enzyme to match the flux through the Calvin cycle (Faske *et al.*, 1995).

2.4 Analysis of photosynthetic control

Recent advances in plant molecular biology have allowed scientists to manipulate genetically a variety of plants. These techniques can be used to generate stable transgenic plants that can have altered activity (up or down) of specific enzymes or even the introduction of either novel enzymes or modified enzymes that have different kinetic or regulatory properties from the native protein. This has allowed a wholly new approach to solving problems in plant metabolism. The relevance/function of individual enzymes for specific aspects of metabolism can now be directly investigated. This approach has recently been applied to several enzymes of the Calvin cycle, the results of which are described briefly below.

2.5 Genetic modification of the Calvin cycle

2.5.1 *Carboxylation phase*

Rubisco

The first transgenic plants to be utilized for control analysis in plants contained an antisense Rubisco cDNA construct. A variety of plants were produced that not only contained reduced activity of Rubisco but also to varying degrees, the extent of reduction closely matching the copy number of the inserted gene (Rodermel *et al.*, 1988). These plants allowed for the first time the direct consequences for photosynthesis of a reduction in a specific Calvin cycle enzyme. Moreover, these 'consequences' could be formalized and quantified by application of control theory (Heinrich and Rapoport, 1974; Kascer and Burns, 1973). This approach allows the calculation of a flux control coefficient (C_E^J):

$$C_E^J = \frac{\delta J/J}{\delta E/E}$$

where $\delta E/E$ is the fractional change in the amount of enzyme (E) and $\delta J/J$ is the resultant change in flux (J) through the pathway. For linear metabolic pathways the value of C_E^J can range from 0 for an enzyme that exerts no control on the flux through the pathway to 1 for an enzyme that exerts total control. There are several precautions that must be taken when applying control theory: first, that the amount of enzyme (not activity unless it is established that this is related to the amount of enzyme) can be quantified; second, that the flux through the pathway can be accurately determined; and third, that alteration of the enzyme activity does not lead to other pleiotropic changes within the plant. It must also be remembered that experimental error is often in the region of $\pm 5\%$, hence calculation of true flux control coefficients is impossible and at best we can only obtain approximations. This is particularly true when control is evenly shared between two or three enzymes of the pathway; minor changes in activity of one of these enzymes could lead to substantial reallocation of control that does not reflect the normal condition and can lead to the assignment of erroneous values. This has recently been demonstrated in a mathematical model of the Calvin cycle, where the control exerted by the SBPase was shown to be over-estimated by greater than 100% (Pettersson, 1996).

A large volume of work has now been published that is concerned with the analysis of transgenic tobacco plants with reduced Rubisco activity (see Quick (1994) and Stitt and Schulze (1994) for recent reviews). This reflects both the central role of this enzyme in photosynthetic metabolism and the fact that these transgenic plants provided a 'test-case' where the value of transgenic technology for addressing biochemical and physiological problems could be assessed. Initial studies with these transformants revealed that Rubisco protein could be reduced by 40% before the rate of photosynthesis was reduced significantly, i.e. the control exerted by Rubisco on photosynthesis was low (0.15) (Quick *et al.*, 1991a). Reduction of Rubisco protein by 40% had very little effect on the activity of other enzymes and little effect on the growth rate of the plant (Quick *et al.*, 1991b). Plants with reduced Rubisco protein were able to compensate by fully activating the remaining enzyme; typically only 50–60% was activated in the wild-type. These initial studies were confined to experimental conditions that approximated to the plant growth conditions (300 µmol m^{-2} s^{-1} irradiance and a temperature of 20°C). However if the experimental conditions were altered, in the short-term, it was shown that the control exerted by Rubisco on photosynthesis was not constant. For example, in conditions of high light intensity, control increased to a value of 0.76, showing that in these conditions Rubisco was the major factor determining the rate of photosynthesis. Similarly, changing the atmospheric concentration of CO_2 altered the control exerted by Rubisco; reducing the CO_2 concentration from 450 to 250 µbar resulted in an increase in the control strength from 0.1 to 0.8 (Stitt *et al.*, 1991).

In the longer term, growth of plants in altered environments of irradiance or CO_2 concentration, or even altered availability of nutrients, can lead to many morphological and biochemical adjustments within the plant. Work with Rubisco antisense plants demonstrated that the control exerted by Rubisco was influenced by the environmental growth conditions. The control exerted by Rubisco increased with increasing growth irradiance as was observed by increasing the irradiance in the short term. However the actual values obtained in the long-term experiments were much lower: plants grown at 300 µmol m^{-2} s^{-1} irradiance and measured at 1000 µmol m^{-2} s^{-1} gave a control strength of 0.8, whereas plants grown at 1000 µmol m^{-2} s^{-1} gave a value of 0.3 (Quick, 1994). These tobacco plants clearly acclimate to growth in high light so that a one-sided limitation of photosynthesis by a single enzyme is avoided; this was achieved by an increased production of Rubisco protein as the growth irradiance increased.

The production of large amounts of Rubisco requires a considerable investment of nitrogen, which for most plants represents a scarce commodity. It is well established that plants have reduced protein content when grown with low nitrogen availability. Changing protein content could result in the redistribution of control within metabolic pathways. The control exerted by Rubisco was investigated using the tobacco antisense plants by Fichtner et al. (1993). They were able to show that growth in low nitrogen leads to a selective loss of Rubisco protein and resulted in an increase in control exerted by this enzyme; a control strength of almost 1.0 was measured even when plants were grown in moderate irradiance (300 µmol m^{-2} s^{-1}).

It is worth noting that given the appropriate choice of plant growth and experimental conditions, any value of control strength could be determined for Rubisco! When setting out to evaluate the metabolic control exerted by a particular enzyme careful consideration of the influence of the plant growth regime and the experimental protocol should first be given.

Rubisco activase

As discussed above, Rubisco activase plays an important role in the removal of tightly bound end-products from the active site of Rubisco. The maintenance of active Rubisco requires constant removal of these unwanted products. This phenomenon is readily observed when Rubisco activity is measured *in vitro*. Linear rates of catalysis are only maintained for 1 or 2 minutes before the rate starts to subside; this is often referred to as fall-over (Edmondson et al., 1990; Zhu and Jensen, 1991). However, the significance of this enzyme for maintaining active Rubisco and hence photosynthetic rates remained uncertain. An activase null mutant of *Arabidopsis* was identified that proved lethal in plants grown at atmospheric CO_2 concentrations (Somerville et al., 1992) and suggested an important role for this enzyme in the control of the Calvin cycle. Transgenic tobacco plants expressing an antisense construct to the rca mRNA encoding Rubisco activase were produced simultaneously in two laboratories in order to investigate the control that this regulatory enzyme exerts on photosynthesis (Jiang et al., 1994; Mate et al., 1993). Plants that contained less than 25% of wild-type activase activity required growth at elevated CO_2 for survival (Mate et al., 1993). Photosynthesis was inhibited due to a low activity of Rubisco as a result of reduced carbamylation of the active site and increased amounts of tightly bound CA1P; inhibition of photosynthesis could be reversed by increasing the CO_2 concentration. Interestingly, the transgenic plants contained double the amount of Rubisco protein compared to wild-type, suggesting that some form of compensatory regulation was in operation.

The rate of activation of Rubisco during a dark to light transition was also reduced to 25% of wild-type suggesting that activase may play an important role in the rapid adjustment of Rubisco activity in environments where photosynthetic rates fluctuate dramatically. For example, shade plants frequently experience large fluctuations in light intensity due to the penetration of light through canopies. These plants have a higher ratio of Rubisco activase to Rubisco protein suggesting that activase also plays an important role during non-steady-state (fluctuating) photosynthesis (Ian Woodrow, personal communication).

In a second study where Rubisco activase was reduced less severely (up to 50% of wild-type) it was demonstrated that Rubisco activase only became limiting for plant growth and photosynthesis (measured in the ambient growth conditions) when activity was reduced below 60% of wild-type. Hence the control exerted by Rubisco activase was low (approaching zero). However, the light saturated rate of photosynthesis was sensitive to minor reductions in Rubisco activase as full activation of available Rubisco was not achieved in transgenic plants in these conditions (Jiang et al., 1994). Taken together these results show that a regulatory enzyme not directly involved in the metabolic pathway can exert control on the rate of photosynthesis.

2.5.2 Reduction phase

Glyceraldehyde 3-phosphate dehydrogenase (GAPDH)

Phosphoglycerate kinase (PGK) and GAPDH both catalyze highly reversible reactions; measurement of metabolite concentrations during steady-state photosynthesis shows that these reactions are very close to equilibrium. This infers that the rate of TP production is largely dependent on the rate of ATP and NADPH production by the photosynthetic electron transport chain rather than the activity of these enzymes *per se*. To test this hypothesis, Price et al. (1995) recently produced a series of transgenic plants with reduced chloroplastic GAPDH. GAPDH is a complex enzyme comprising two similar but not identical protein subunits (GapA and GapB). The exact number of these subunits in the native enzyme varies between A_2B_2 and A_8B_8, depending on the activation state of the protein, and is thought to be associated with the thiol-regulation mechanism, the smaller form of the enzyme being the activated form. Expression of an antisense construct to the GapA subunit under the control of the 35S promoter was found to be sufficient to decrease the activity of the native enzyme substantially (Price et al., 1995). Surprisingly, GAPDH activity could be reduced to 35% of wild-type activity before any effect on the rate of photosynthesis was observed. This was despite an immediate drop in the size of the $Ru1,5P_2$ pool, suggesting that there was an immediate limitation on the regeneration reactions of the Calvin cycle. As discussed earlier, GAPDH is subject to thiol-regulation; the activation status of GAPDH was not measured in this study and therefore compensatory increased activation of the enzyme cannot be ruled out. There was also no apparent accumulation of PGA even in plants with severely reduced GAPDH which might be expected given the block in the reductive reactions of the Calvin cycle. The possibility remains that a proportion of the reduction of PGA occurred in the cytosol utilizing the cytosolic NAD-dependent GAPDH. This would require effective export of PGA in the light and also a supply of ATP and NADH. However the major conclusion of this work is that GAPDH exerts very little control on the rate of photosynthesis in tobacco.

2.5.3 Regeneration

Fructose 1,6-bisphosphatase (Fru1,6P$_2$ase)

Transgenic potato plants expressing reduced levels of the chloroplastic Fru1,6P$_2$ase were produced by expressing an antisense cDNA construct (Kossmann *et al.*, 1994). Plants were produced with between 12 and 36% wild-type Fru1,6P$_2$ase activity. Plants with 36% wild-type activity had no discernible difference in growth, tuber yield or phenotype from wild-type plants. Photosynthesis measured at 100 or 1000 µmol m^{-2} s^{-1} irradiance and saturating CO$_2$ showed only marginal reductions in photosynthesis. These data suggest that Fru1,6P$_2$ase exerts little control on the rate of photosynthesis in potato. However, plants with further reductions in Fru1,6P$_2$ase activity show reduced plant growth and tuber yield. Photosynthetic capacity is reduced and the substrate F1,6P$_2$ increased eightfold. The concentration of triose phosphates was increased, whereas hexose phosphates remained largely unchanged. Given that Fru6P is a product of the Fru1,6P$_2$ase reaction, this might suggest that the distribution of hexose phosphates within the cell was altered, i.e. Fru6P declined in the chloroplast but increased in the cytoplasm. This situation would be predicted to shift the partitioning of photosynthate towards cytosolic sucrose synthesis. This was observed in the transgenic lines; ^{14}CO$_2$ feeding demonstrated increased partitioning of photosynthate to sucrose and although carbohydrate measurements showed a general decline in the transgenic plants this was predominantly associated with a decline in the amount of starch.

Sedoheptulose 1,7-bisphosphatase (S1,7P$_2$ase)

S1,7P$_2$ase is thought to be unique to the Calvin cycle, catalyzes an essentially irreversible reaction *in vivo*, and is a dimeric protein of two identical subunits. As seen in Figure 2.1, S1,7P$_2$ase is located at an interesting point in the Calvin cycle; it is a committed step in the pathway for regeneration of the CO$_2$-acceptor Ru1,5P$_2$; prior to this point both TP and Fru6P can be used for carbon export or starch metabolism, respectively. Hence the effect of altering S1,7P$_2$ase activity is difficult to predict; reduced activity could lead to an accumulation of substrates, especially E4P, a potent inhibitor of the enzyme phosphoglucoseisomerase (PGI), required for the conversion of Fru6P to glucose 6-phosphate in the pathway of starch synthesis. Alternatively, the regeneration of Ru1,5P$_2$ may be more sensitive and hence a general drop in Calvin cycle metabolites might ensue. The expression of an antisense cDNA construct to the chloroplastic S1,7P$_2$ase in tobacco under the control of a 35S promoter resulted in the production of a range of plants with reduced S1,7P$_2$ase mRNA and protein (Harrison *et al.*, 1997). A 40% reduction in S1,7P$_2$ase protein was sufficient to generate a distinct chlorotic leaf phenotype (chlorosis occurring predominantly around the major veins) and an inhibition of plant growth. The carbohydrate content of the leaves was reduced, largely as a result of a reduced starch content, suggesting that starch rather than sucrose synthesis was inhibited in transgenic lines. Analysis of the control exerted by S1,7P$_2$ase revealed that the control strength varied as a function of photosynthetic rate (Alred, Leegood and Quick, unpublished data). At light saturated rates of photosynthesis, the control exerted by S1,7P$_2$ase was 0.6 and this value declined almost linearly as the rate of photosynthesis was reduced by decreasing the light intensity; S1,7P$_2$ase exerted a control of about 0.15 in conditions of the ambient growth environment. These values are similar in magnitude to those obtained for Rubisco and demonstrate the considerable control that can be exerted by an enzyme traditionally not considered as an important control point within this pathway. Mathematical models of the Calvin cycle which include S1,7P$_2$ase have also predicted that a high control is exerted by this enzyme (Fell, personal communication; Pettersson, 1996). This probably reflects the low maximum catalytic activity of S1,7P$_2$ase that can be measured in crude leaf extracts and which is generally only just sufficient to catalyze the measured maximum photosynthetic flux. The partitioning of recently fixed $^{14}CO_2$ into starch was also found to be reduced in transgenic plants with reduced S1,7P$_2$ase activity, suggesting the importance of this enzyme in the regulation of partitioning of photosynthate between starch and sucrose.

Phosphoribulokinase (PRK)

PRK catalyzes the synthesis of Ru1,5P$_2$ from Ru5P and ATP, and as mentioned above, is also subject to regulation by thiol-reduction. Paul *et al.* (1995) produced a series of transgenic tobacco plants expressing a PRK antisense cDNA construct. The activity of PRK was reduced in a number of tobacco lines to varying extents and down to 5% of the wild-type activity. Analysis of these plants showed that PRK could be reduced by more than 80% before any effects on photosynthetic rate or plant growth were observed. Similarly phosphorylated intermediates of the Calvin cycle, including the substrate of PRK (Ru5P), were unaltered until PRK activities were dramatically reduced. Hence the control exerted by PRK on photosynthesis in wild-type tobacco was negligible. The tobacco plants used in this experiment were grown at relatively low irradiance (300 µmol m^{-2} s^{-1}) and photosynthesis was measured in the growth environment. In these conditions Rubisco was also found to exert a low control; only when the growth or measurement irradiance was increased did the control increase. More recent experiments have demonstrated that increasing irradiance of measurement increases the control exerted by PRK on photosynthesis; the value for the control coefficient increased from 0 at 300 µmol m^{-2} s^{-1} to 0.25 at 800 µmol m^{-2} s^{-1} measurement irradiance. Moreover, when these plants were grown at elevated irradiance, the control was reduced back to a negligible value (M. Paul, personal communication). This further highlights the need for caution when interpreting flux control coefficients; the plant species, growth environment and measurement environment must also be taken into consideration.

Aldolase

There are two major forms of aldolase in higher plants: class I, which are characterized by a tetrameric protein structure (160 kD) and are located within the plastid; and class II, which are characterized by a dimeric protein structure (80 kD) and located in the cytosol (Pelzer-Reith *et al.*, 1994). Aldolase catalyzes a freely reversible reaction *in vivo* and, at present, is not known to be subject to regulation by metabolites or covalent modification. This enzyme would traditionally be predicted to exert little control on the rate of photosynthesis. To test this hypothesis the plastidic isoform of aldolase was cloned from potato and the cDNA sequence was used to produce transgenic potato plants that expressed an antisense RNA under the control of CaMV 35S constitutive promoter. Analysis of these plants revealed a series of lines containing between 5 and 40% wild-type activity. Rates of photosynthesis and plant growth were not affected until activity was reduced below 60% of the wild-type and autotrophic growth was not possible if the activity was reduced below 10% of wild-type activity (Sonnewald et al., 1994). However, a control strength of 0.25 was calculated for aldolase on photosynthesis in conditions of saturating irradiance and CO_2 (Haake and Zrenner, unpublished results). Analysis of carbohydrates revealed that starch synthesis was much more inhibited than sucrose synthesis; a 70% reduction of aldolase activity led to 74% reduction in the amount of starch and yet had no effect on the leaf sucrose concentration. All perturbations of the regeneration reactions of the Calvin cycle (reduced $Fru1,6P_2$ase, $S1,7P_2$ase and aldolase) have so far resulted in a similar preferential inhibition of starch rather than sucrose metabolism, despite the lesions occurring before and after the formation of Fru6P. This would suggest that the primary effect of these metabolic lesions is on the rate of $Ru1,5P_2$ regeneration which in turn leads to generally low levels of phosphorylated metabolites (including PGA) and conversely increased availability of free phosphate. A low PGA/Pi ratio is known to inhibit ADPGlcPPase in the pathway of starch synthesis (see Chapter 4).

Control analysis of the Calvin cycle has revealed several important features of this pathway. First, that control is shared among several enzymes of the pathway and that the distribution of control changes when either the plant growth conditions or the short-term experimental conditions are varied (e.g. Rubisco). Second, that control does not necessarily reside with highly regulated enzymes that catalyze essentially irreversible reactions (e.g. $Fru1,6P_2$ase). Third, that enzymes considered to catalyze freely reversible reactions and that can achieve very high unidirectional rates of catalysis can also exert control when their activity is reduced below 50% of wild-type (e.g. aldolase). Fourth, considerable control can be located on enzymes where it is least expected as revealed by the antisense $S1,7P_2$ase transgenic tobacco plants. Based on data published thus far, it is clear that transgenic plants have provided and will continue to provide, a very productive tool for the analysis of the control of photosynthetic carbon assimilation.

2.6 Starch degradation in chloroplasts

Chloroplastic starch is referred to as transitory starch since, in contrast to storage starch, the level of this carbohydrate in leaves exhibits a strong diurnal rhythm. The controlled degradation of chloroplastic starch during the night is important for plant metabolism. Disturbances in starch metabolism induce severe negative changes in plant biochemistry and growth. Interestingly, our knowledge about the processes leading to starch degradation in leaves is much less than our knowledge about the mechanisms of starch synthesis in chloroplasts. This discrepancy is mainly due to difficulties in the purification of high starch containing chloroplasts as they are a prerequisite to the study of starch degradation. However, a few studies of starch degradation in chloroplasts have been carried out (for a review see Ziegler and Beck, 1989). In principal there are two metabolic routes for the degradation of transitory starch: an amylolytic and a phosphorylytic pathway. The amylolytic degradation of starch occurs through a concerted action of α-amylases, D-enzymes, and debranching enzymes. As a result of these enzymes the neutral sugars maltose and glucose are released. Most plastids possess a starch phosphorylase enzyme that catalyzes the transfer of Pi to a terminal starch glucose moiety leading to the synthesis of glucose 1-phosphate (Glc1P). This reaction is strongly favoured during the night when the level of phosphorylated intermediates is low and the corresponding level of orthophosphate (released from these metabolites) is high.

By analyzing the composition of metabolites released from high starch containing leaf chloroplasts it is possible to demonstrate that both metabolic pathways for starch degradation, the amylolytic and the phosphorylytic pathway, are active. This conclusion is based on the finding that both the neutral sugars (glucose and maltose) and phosphorylated intermediates (PGA and DHAP) are exported from the chloroplast during starch degradation (Stitt and ap Rees, 1980). The export of glucose and maltose is mediated by the chloroplastic glucose transporter, whereas the phosphorylated intermediates are exported by the triose-phosphate transporter. In addition, some of the phosphorylated intermediates are utilized in the plastidic oxidative pentose phosphate pathway to fuel the chloroplast stroma with reducing equivalents and five-carbon sugars necessary for a variety of anabolic reactions.

The concentration of Pi could potentially have a dramatic effect on the rate of starch degradation for at least two reasons. First, Pi stimulates the reaction of starch phosphorylase leading to higher rates of Glc1P synthesis. Second, increased cytosolic concentrations of inorganic phosphate promote the export of phosphorylated intermediates from the stroma into the cytosol. The latter effect is due to a stimulation of the triose-phosphate transporter known to carry phosphorylated intermediates in a strict counter-exchange with Pi.

Recently, it was demonstrated that the degradation of transitory starch was stimulated by the uptake of ATP into isolated chloroplasts. Using chloroplasts from leaves of *Mesembryanthemum crystallinum* L. it was possible to show that the presence of ATP in the incubation medium stimulated starch degradation by a factor of 2.5 (Neuhaus and Schulte, 1996). The important function of ATP in starch degradation was also demonstrated in heterotrophic plastids. Isolated amyloplasts purified from cauliflower inflorescences were able to export carbohydrates liberated from starch. As demonstrated for chloroplasts (see above) the addition of ATP substantially increased the rate of carbohydrate export from this type of plastid (Neuhaus *et al.*, 1995).

2.6.1 Are there plastid type specific products of starch degradation?

When we consider the processes that lead to starch degradation we need to bear in mind that photosynthetically active plastids are located in a wide range of different tissues and that these plastids fulfil very different metabolic functions.

For example, the metabolism of chloroplasts in C_3, C_4, or CAM plants have very different features. Therefore, it is not surprising that the major products of starch degradation released by mesophyll chloroplasts appear to be quite different and are species or metabolism specific. Mesophyll chloroplasts from pea leaves mainly export phosphorylated intermediates and divert a significant amount of carbon into the oxidative pentose phosphate pathway. In contrast to this, spinach chloroplasts export, in addition to phosphorylated intermediates, substantial amounts of neutral sugars like glucose and maltose (Stitt and Heldt, 1981). An *Arabidopsis thaliana* L. mutant was identified that had extremely high levels of transitory starch at the end of the night period. It was shown that the inability to degrade starch was due to the absence of a chloroplastic glucose transporter (Trethewey and ap Rees, 1994). The reduced capacity of starch degradation in this mutant correlated with an increased concentration of glucose in the chloroplast indicating a major role for the amylolytic starch degradation pathway in this species.

Recently, the analysis of starch degradation in chloroplasts from the facultative CAM plant *Mesembryanthemum crystallinum* L., isolated when the plant was in the C_3 or CAM mode, indicated that there were important differences in the composition of metabolites released during starch degradation. In contrast to C_3 chloroplasts from this species, the chloroplasts isolated from CAM induced plants release Glc6P as the major export metabolite. The release of Glc6P is not due to specific transport properties of the envelope membranes of the plastids as both types of plastids were able to transport Glc6P and glucose at the same rate (Neuhaus and Schulte, 1996). The specific release of Glc6P in the cytosol may act as an important metabolic signal necessary for CAM. In CAM the nocturnal degradation of starch coincides with an acidification of the tissues due to carboxylation of PEP by the cytosolic PEP carboxylase. This enzyme is allosterically activated by Glc6P and therefore the appearance of Glc6P in the cytosol is a metabolic feed-forward signal contributing to high rates of nocturnal CO_2 fixation required in CAM plants.

2.7 Carbon metabolism and photosynthesis in fruit chloroplasts

In contrast to leaf-mesophyll chloroplasts, the metabolism of fruit chloroplasts has received far less attention. In this section we discuss the metabolism of chloroplasts in fruit tissues and tissues closely related to fruits (like ear leaves from cereals), since study of photosynthetic processes in these tissues could contribute substantially to our knowledge of photosynthesis and plant biochemistry (see below). For the purposes of this chapter we refer to these plastids as chloroplasts.

Many fruit tissues appear green in colour at some phase of their development due to the presence of chloroplasts. These chloroplasts are present either in early stages of growth and subsequently turn into chromoplasts during ripening or they are present throughout the entire process of development. Besides the difference in the location of fruit and leaf chloroplasts there is also another important difference between both types of organelle. In contrast to the latter, fruit chloroplasts are located in plant organs which exhibit a strong sink capacity for carbohydrates and nutrients. As photosynthesis in fruit tissues is important for gaining a maximum crop yield (Blanke and Lenz, 1989; Thorne, 1965), we have an urgent need to increase our knowledge about how the metabolic events that occur within fruit chloroplasts contribute to the growth of the organ.

2.7.1 Fruit photosynthesis supports crop yield

There are several reports independently demonstrating that fruit chloroplasts contribute substantially to carbon fixation and crop yield. In this context it is important to differentiate between net CO_2 fixation by fruit chloroplasts and other functions of these organelles leading to higher crop yield. For example, photosynthesis in cereal-ear leaves contributes between 15 and 75% of the final grain weight (Caley et al., 1990). However, on the basis of the total carbon found in cereal endosperm, less than 2% is attributable to photosynthesis in ear-leaf chloroplasts (Watson and Duffus, 1985). This discrepancy clearly indicates that we should focus our attention on several metabolic functions of fruit chloroplasts that are driven by photosynthesis.

2.7.2 CO_2 fixation by fruit chloroplasts

The ability of fruit chloroplasts to fix exogenously supplied $^{14}CO_2$ into starch has been demonstrated in a wide range of fruit tissues (Blanke and Lenz, 1989). As such a conversion depends upon a complete complement of Calvin cycle enzymes it is perhaps not surprising that all of the required enzymes and those required for starch synthesis have been identified in appreciable amounts (Camara et al., 1995). The observation that only a small fraction of fruit carbon is fixed in the corresponding chloroplasts (see above) does not necessarily mean that CO_2 fixation occurs at low rates. The comparably low abundance of Rubisco in fruit chloroplasts (about 1:50–100 of the content in leaf-mesophyll chloroplasts) is partly compensated by very high CO_2 partial pressures occurring in fruit tissues. For example, in apple fruit tissues, CO_2 partial pressures between 50 to 100 mbar have been demonstrated to occur (Blanke and Lenz, 1989), preventing any likely photorespiratory activity of Rubisco.

The low rate of net CO_2 fixation is more likely due to the specific morphological constraints connected with the location of the chloroplasts in fruits. In most cases fruits are optimized to minimise water loss which is realized by a low number of stomata, waxy cuticles, low surface-area-to-volume ratio and tight connections of chloroplast containing perivascular cells lacking gas-filled caverns (Skene, 1963). These characteristics reduce the diffusion velocity of gases between the inner fruit tissues and the ambient air. The high internal CO_2 partial pressure, in some fruits, exhibits a significant diurnal rhythm, high in the dark and lower in the light, indicating a major role of fruit leaf photosynthesis in the refixation of carbon lost through respiratory activity. Indeed, this has been demonstrated for several types of fruit chloroplasts where the major source of CO_2 utilized for fruit photosynthesis was derived from respiratory CO_2 (Blanke and Lenz, 1989; Wullschleger et al., 1991). Therefore, light driven refixation of respiratory CO_2 can contribute to a significant reduction of carbon loss.

2.8 Other metabolic pathways affected by photosynthesis

The primary processes of photosynthesis occur within the thylakoid membranes and lead to the synthesis of ATP and NADPH. This production of phosphorylation energy and reducing equivalents can partly be used for CO_2 fixation but can also be used to fuel other anabolic reactions. Some of these are discussed below.

It is well-known that fruit tissues contain relatively high activities of PEP carboxylase catalyzing the cytosolic fixation of CO_2 by carboxylation of phosphoenolpyruvate. However, the product of this reaction (oxaloacetate, OAA), is only transitory and is rapidly reduced to malate. Interestingly, the cellular site of OAA reduction still remains unknown. Recently, it was demonstrated that even darkened chloroplasts isolated from green pepper (*Capsicum annuum* L.) fruits were able to reduce exogenously supplied OAA to malate (Neuhaus and co-workers, unpublished). It appears reasonable to assume that such a conversion can also proceed in illuminated fruit chloroplasts, as under these conditions the NADPH/NADP ratio strongly favours the reduction of OAA to malate. This assumption is reinforced by studies with spinach-leaf mesophyll cells where it was demonstrated that the low cytosolic NADH/NAD measured would prevent OAA reduction (Heinecke *et al.*, 1991). Therefore, the plastidic stroma appears to be the major cellular site for OAA reduction, allowing photosynthesis in fruit chloroplasts to contribute to the refixation of respiratory CO_2 by PEP carboxylase.

Fatty acids and lipids are not transported via the phloem. Therefore, the plastidic synthesis of fatty acids is the sole source of these compounds which are essential for the development of membranes as well as for the production of storage lipids in these organs. Analyzing the lipid content in olive fruits harvested from different parts of a tree demonstrated that fruits grown under highest irradiances (upper canopy) are bigger and contain significantly more oil (percentage of fresh weight) than fruits grown in the shade (for review, see Sanchez, 1994). $^{14}CO_2$ feeding to olive-fruit tissue slices revealed that this compound was primarily being used as a precursor for glycerolipid synthesis, including triacyl glycerols as well as phospholipids. That $^{14}CO_2$-dependent lipid synthesis was strongly light dependent is striking evidence for the involvement of photosynthesis. Moreover, radioactively labelled acetate or pyruvate fed to olive-fruit tissues also induced lipid synthesis provided that light was present (Sanchez, 1994). The latter data indicate that incorporation of metabolites supplies to the fruit (for example by phloem transport) is stimulated by light. This is most likely due to the light-dependent synthesis of NADPH and ATP within the stromal compartment, both compounds being necessary for fatty-acid synthesis.

Nitrite reduction and ammonia fixation are processes exclusively located in the plastidic compartment. These anabolic reactions have been extensively analyzed in leaf-mesophyll chloroplasts, whereas there is a dearth of studies concerning the biochemistry of nitrogen metabolism in fruit chloroplasts. Recently, Thom and Neuhaus (1995) demonstrated that isolated pepper-fruit chloroplasts were able to use reducing equivalents (generated by the oxidative pentose phosphate pathway) for these processes. Again, as photosynthesis provides the stroma with comparably high amounts of reducing equivalents it would seem reasonable to assume that fruit chloroplasts can contribute to the production of amino acids needed for the synthesis of proteins and hence biomass.

Starch is one of the major constituents in fruits. It has been demonstrated for tomato fruit tissues that the light induced assimilation of endogenous respiratory CO_2 is not sufficient for the observed difference in the starch content of illuminated or darkened fruits (Guan and Janes, 1991). Obviously, processes linked to photosynthesis, other than CO_2 fixation, positively influence starch synthesis. Using isolated chloroplasts from green pepper fruits it was possible to identify light as a strong activator of Glc6P-dependent starch synthesis (Batz *et al.*, 1995). As demonstrated for a wide range of heterotrophic plastids (Neuhaus *et al.*, 1993), isolated fruit chloroplasts are able to import exogenously supplied Glc6P via a specific hexose-phosphate translocator. Since photosynthesis leads to ATP synthesis in the stroma, this energy may be responsible for driving the production of ADP glucose by ADPGlc PPase, an enzyme known to exert considerable control over the rate of starch biosynthesis (see Chapter 4). The rate of CO_2-dependent starch synthesis is much lower than the rate of Glc6P-dependent starch synthesis in isolated illuminated chloroplasts from green pepper fruits. This clearly demonstrates that the major effect of light is the photosynthetic production of ATP that is used to drive starch synthesis from exogenously supplied carbohydrates (Glc6P) rather than internally fixed CO_2. Why do fruit chloroplasts possess a Glc6P transporting protein in the envelope membrane in addition to the TP transporter used in leaf chloroplasts? Starch synthesis from cytosolic precursors could theoretically proceed via the uptake of triose phosphates. To answer this question we should remember that fruit tissues are supplied with carbohydrates by day and night. This means that starch synthesis also has to continue in conditions when photosynthesis does not occur (in the night). The conversion of triose phosphates to hexose phosphates needed for starch synthesis is dependent upon the presence of an active $Fru1,6P_2$ase. The nocturnal synthesis of hexose phosphates from imported triose phosphates via this light activated enzyme is impossible. The import of Glc6P allows the inactive fructose 1,6-bisphosphate phosphatase reaction to be bypassed and starch synthesis can hence occur in the night (Batz *et al.*, 1995). By this mechanism nocturnal starch synthesis in these cells can maintain a high sink strength during the night.

In conclusion, photosynthesis in fruit chloroplasts does not only allow for refixation of respiratory CO_2. The light driven synthesis of reducing equivalents and ATP also fuels other anabolic reactions such as ammonia, amino acid, fatty acid, and starch synthesis which contribute substantially to crop yield.

2.9 Modulation of transport properties across the plastid envelope

Above we have demonstrated how carbohydrate metabolism in chloroplasts is regulated and how this metabolism contributes to plant biochemistry. The contribution of plastids to overall plant metabolism ultimately depends upon the co-ordinated transport of several metabolites and other compounds across the chloroplastic envelope. Interestingly, hardly anything is known about the interaction between the specific metabolic pathways that operate in chloroplasts and the transport properties of the envelope membranes. That specific types of plastids possess specific transport proteins has now been demonstrated for several organelles. For example, chloroplasts purified from spinach mesophyll tissues exhibit high rates of triose phosphate transport but they are unable to transport Glc6P across the envelope (Fliege et al., 1978). In contrast to this, chloroplasts located in other tissues exhibit substantial transport capacity for Glc6P. Transport of Glc6P is shown to occur across the envelope of chloroplasts purified from the algae *Codium fragile* (Rutter and Cobb, 1983) and from the CAM plant *Sedum praealtum* (Piazza et al., 1982). In addition, chloroplasts isolated from guard cells of *Pisum sativum* have been demonstrated to transport Glc6P across the envelope membrane (Overlach et al., 1993). These observations indicate that specific developmental and/or metabolic signals can trigger changes in the transport properties of the plastidic envelope. Comparison of the sugar phosphate transport properties of envelope membranes from autotrophic and heterotrophic plastids revealed that in most cases the latter types of plastid were capable of the rapid transport of hexose phosphates (Neuhaus et al., 1993). This observation indicates that heterotrophy may be one important metabolic situation leading to Glc6P transport across the envelope.

Interestingly, heterotrophic situations can be induced even in leaves previously exhibiting autotrophic metabolism. Feeding glucose via the transpiration stream into detached spinach or potato leads to a decrease of enzyme activities, pigmentation and membrane structures associated with photosynthetic CO_2 fixation. This phenomenon is referred to as a shift from autotrophy to heterotrophy (Krapp et al., 1991). This stimulation of heterotrophic metabolism in leaf tissues was also shown to induce Glc6P transport across the envelope of mesophyll chloroplasts (Quick et al., 1995). However, there are further indications that heterotrophy *per se* does not always induce hexose phosphate transport across the plastidic envelope. Analysis of the precursor dependency of starch synthesis in isolated etioplasts from dark grown barley plants revealed that this type of heterotrophic plastid was not able to import Glc6P (Batz et al., 1992). The latter observation indicates that other developmental signals also influence the transport properties across the plastidic envelope.

All chloroplasts eventually age and convert into gerontoplasts, typically found in senescent tissues. In leaves this process correlates with a massive decrease of enzyme activities involved in photosynthesis and a loss of chlorophyll. As these phenomena also occur after glucose feeding to detached leaves it is likely that gerontoplasts also possess the ability to transport Glc6P across the envelope membrane. Indeed, Glc6P transport was shown indirectly by the analysis of metabolites that stimulate chlorophyll degradation in isolated barley gerontoplasts. As chlorophyll degradation is dependent upon the supply of the stroma with NADPH it was assumed that the plastidic oxidative pentose phosphate pathway was involved in the synthesis of reducing equivalents. Interestingly, this metabolic pathway and chlorophyll degradation are stimulated in isolated barley gerontoplasts by the exogenous supply of Glc6P (Matile et al., 1992).

The possibility that plastid envelopes possess more than one type of translocator has also received very little attention in the literature. The demonstration that Glc6P transport can be induced in leaves, by glucose feeding, within one day (Quick et al., 1995), suggests that a novel transporter was induced rather than that an existing transporter was replaced. This notion has been reinforced by observations made with chromoplasts isolated from red-pepper fruits. Using this heterotrophic plastid it was possible to demonstrate that both a triose phosphate specific and a hexose phosphate specific transporter were present simultaneously in the same membrane (Quick and Neuhaus, 1996).

The examples given above clearly demonstrate that plastids have great flexibility to adapt even the transport properties of their membranes. This flexibility in metabolism and communication properties with the cytoplasm is only recently becoming apparent; future work in this area is likely to change/improve considerably our current understanding of metabolism and its regulation in these unique organelles.

References and further reading

BAINBRIDGE, G., MADGWICK, P., PARMAR, S., MITCHELL, R., PAUL, M., PITTS, J., KEYS, A.J. and PARRY, M.A.J. (1995) Engineering Rubisco to change its catalytic properties. *J. Exp. Bot.* **46**, 1269–1276.

BATZ, O., SCHEIBE, R. and NEUHAUS, H.E. (1992) Transport processes and corresponding changes in metabolite levels in relation to starch synthesis in barley (Hordeum vulgare L.) etioplasts. *Plant Physiol.* **100**, 184–190.

BATZ, O., SCHEIBE, R. and NEUHAUS, H.E. (1995) Purification of chloroplasts from fruits of green-pepper (Capsicum annuum L.) and characterization of starch synthesis. Evidence for a functional hexose-phosphate translocator. *Planta* **196**, 50–57.

BLANKE, M.M. and LENZ, F. (1989) Fruit photosynthesis. *Plant, Cell Environ.* **12**, 31–46.

BUCHANAN, B.B. (1980) Role of light in the regulation of chloroplast enzymes. *Annu. Rev. Plant Physiol. Plant Mol. Biol.* **31**, 341–374.

CALEY, C.Y., DUFFUS, C.M. and JEFFCOAT, B. (1990) Photosynthesis in the pericarp of developing wheat grains. *J. Exp. Bot.* **41**, 303–307.

CAMARA, B., HUGUENEY, P., BOUVIER, F., KUNTZ, M. and MONEGER, R. (1995) Biochemistry and molecular biology of chromoplast development. *Int. Rev. Cytol.* **163**, 175–247.

EDMONDSON, D.L., BADGER, M.R. and ANDREWS, T.J. (1990) Slow inactivation of Rubisco during catalysis is caused by the accumulation of a slow tight-binding inhibitor at the catalytic site. *Plant Physiol.* **93**, 1390–1397.

FASKE, M., HOLTGREFE, S., OCHERETINA, O., MEISTER, M., BACKHAUSEN, J.E. and SCHEIBE, R. (1995) Redox equilibria between the regulatory thiols of light/dark-modulated chloroplast enzymes and dithiothreitol: fine-tuning by metabolites. *Biochim. Biophys. Acta* **1247**(1), 135–142.

FICHTNER, K., QUICK, W.P., MOONEY, S.R., SCHULZE, E.-D., RODERMEL, S.R., BOGORAD, L. and STITT, M. (1993) Decreased Rubisco in transgenic tobacco transformed with antisense Rbcs. V. Relationship between photosynthetic rate, storage capacity, biomass allocation and vegetative plant growth at three different nitrogen supplies. *Planta* **190**, 1–9.

FLIEGE, R., FLÜGGE, U.-I., WERDAN, K. and HELDT, H.W. (1978) Specific transport of inorganic phosphate, 3-phosphoglycerate and triose phosphates across the inner membrane of the envelope in spinach chloroplasts. *Biochim. Biophys. Acta* **502**, 232–247.

GUAN, H.P. and JANES, H.W. (1991) Light regulation of sink metabolism in tomato fruit. I. Growth and sugar accumulation. *Plant Physiol.* **96**, 916–921.

HARRISON, L., LLOYD, J. and RAINES, C. (1997) Transgenic tobacco plants with reduced SBPase activity show reduced growth and rates of photosynthesis. *Planta* (submitted).

HEINECKE, D., RIENS, B., GROSSE, H., HOFERICHTER, P., PETER, U., FLUEGGE, U.-I. and HELDT, H.W. (1991) Redox transfer across the inner chloroplast envelope membrane. *Plant Physiol.* **95**, 1131–1137.

HEINRICH, R. and RAPOPORT, T.A. (1974) A linear steady-state treatment of enzymatic chains. General properties, control and effector strength. *Eur. J. Biochem.* **42**, 89–95.

JIANG, C.Z., QUICK, W.P., ALRED, R., KLIEBENSTEIN, D. and RODERMEL, S.R. (1994) Antisense RNA inhibition of Rubisco activase expression. *Plant J.* **5**, 787–798.

KACSER, H. and BURNS, J.A. (1973) The control of flux. *SEB Symp.* **27**, 65–104.

KEYS, A.J., MAJOR, I. and PARRY, M.A.J. (1995) Is there another player in the game of Rubisco regulation? *J. Exp. Bot.* **46**, 1245–1252.

KOSSMANN, J., SONNEWALD, U. and WILLMITZER, L. (1994) Reduction of the chloroplastic FBPase in transgenic potato plants impairs photosynthesis and plant growth. *Plant J.* **6**, 637–650.

KRAPP, A., QUICK, W.P. and STITT, M. (1991) Ribulose-1,5-bisphosphate carboxylase-oxygenase, other photosynthetic enzymes and chlorophyll decrease when glucose is supplied to mature spinach leaves via the transpiration stream. *Planta* **186**, 58–69.

LAUERER, M., QUICK, W.P., SAFTIC, D., FICHTNER, K., SCHULZE, E.-D., RODERMEL, S.R., BOGORAD, L. and STITT, M. (1993) Decreased Rubisco in transgenic tobacco transformed with antisense Rbcs. VI. Effects on plants grown in different irradiance. *Planta* **189**, 174–181.

LEEGOOD, R.C. (1996) Primary photosynthate production: physiology and metabolism. In: *Photoassimilate Distribution in Plants and Crops* (ZAMSKI, E. and SCHAFFER, A.A., eds), pp. 21–41. Marcel Dekker, New York.

MACDONALD, F.D. and BUCHANAN, B.B. (1990) The reductive pentose phosphate pathway and its regulation. In: *Plant Physiology, Biochemistry and Molecular Biology* (DENNIS, D.D. and TURPIN, D.H., eds). Longman Scientific, Singapore.

MATE, C.J., HUDSON, G.S., VON CAEMMERER, S., EVANS, J.R. and ANDREWS, T.J. (1993) Reduction of Rubisco activase levels in tobacco by antisense RNA reduces Rubisco carbamylation and impairs photosynthesis. *Plant Physiol.* **102**, 1119–1128.

MATILE, P., SCHELLENBERG, M. and PEISKER, C. (1992) Production and release of a chlorophyll catabolite in isolated senescent chloroplasts. *Planta* **187**, 230–235.

NEUHAUS, H.E. and SCHULTE, N. (1996) Starch degradation in chloroplasts isolated from C3 or CAM induced *Mesembryanthemum crystallinum* L. *Biochem. J.* **318**, 945–953.

NEUHAUS, H.E., THOM, E., BATZ, O. and SCHEIBE, R. (1993) Purification of highly intact plastids from various heterotrophic plant tissues. Analysis of enzymic equipment and precursor dependency for starch biosynthesis. *Biochem. J.* **296**, 496–501.

NEUHAUS, H.E., HENRICHS, G. and SCHEIBE, R. (1995) Starch degradation in intact amyloplasts from cauliflower buds (*Brassica oleracea* L.). *Planta* **195**, 496–504.

OVERLACH, S., DICKMANN, W. and RASCHKE, K. (1993) Phosphate translocator of isolated guard-cell chloroplasts from *Pisum sativum* transports glucose-6-phosphate. *Plant Physiol.* **101**, 1201–1207.

PAUL, M.J., KNIGHT, J.S., HABASH, D., PARRY, M.A.J., LAWLOR, D.W., BARNES, S.A., LOYNES, A. and GRAY, J.C.

(1995) Reduction in phosphoribulokinase activity by antisense RNA in transgenic tobacco: effect on CO_2 assimilation and plant growth in low irradiance. *Plant J.* **7**, 535–542.

PELZER-REITH, B., WIEGAND, S. and SCHNARRENBERGER, C. (1994) Plastid class I and cytosol class II aldolase of *Euglena gracilis*. *Plant Physiol.* **106**, 1137–1144.

PETTERSSON, G. (1996) Error associated with experimental flux control determinations in the Calvin cycle. *Biochim. Biophys. Acta* **1289**, 169–174.

PETTERSSON, G. and RYDE-PETTERSSON, U. (1990) Dependence of the Calvin cycle activity on kinetic parameters for the interaction of non-equilibrium cycle enzymes with their substrates. *Eur. J. Biochem.* **186**, 683–688.

PIAZZA, G.J., SMITH, M.G. and GIBBS, M. (1982) Characterization of the formation and distribution of photosynthetic products by *Sedum praealtum* chloroplasts. *Plant Physiol.* **70**, 1748–1758.

PORTIS, A.R. (1992) Regulation of Rubisco activity. *Annu. Rev. Plant Physiol. Plant Mol. Biol.* **43**, 415–437.

PORTIS, A.R. (1995) The regulation of Rubisco by Rubisco activase. *J. Exp. Bot.* **46**, 1285–1292.

PRICE, G.D., EVANS, J.R., VON CAEMMERER, S., YU, J.W. and BADGER, M.R. (1995) Specific reduction of chloroplast GAPDH by antisense RNA reduces CO_2 assimilation via a reduction in RuBP regeneration in transgenic tobacco plants. *Planta* **195**, 369–378.

QUICK, W.P. (1994) Analysis of transgenic tobacco plants containing varying amounts of Rubisco. *Biochem. Soc. Trans.* **22**, 899–904.

QUICK, W.P. and NEUHAUS, H.E. (1996) Evidence for two types of phosphate translocators in sweet-pepper (capsicum annuum L.) fruit chromoplasts. *Biochem. J.* **320**, 7–10.

QUICK, W.P., SCHURR, U., SCHULZE, E.-D., RODERMEL, S.R., BOGORAD, L. and STITT, M. (1991a) Decreased Rubisco in transgenic tobacco transformed with antisense rbcS. I. Impact on photosynthesis in the ambient growth conditions. *Planta* **183**, 555–566.

QUICK, W.P., SCHURR, U., FICHTNER, K., SCHULZE, E.-D., RODERMEL, S.R., BOGORAD, L. and STITT, M. (1991b) The impact of decreased Rubisco on photosynthesis, growth and storage in tobacco plants which have been transformed with antisense rbcS. *Plant J.* **1**, 51–59.

QUICK, W.P., FICHTNER, K., SCHULZE, E.-D., WENDLER, R., LEEGOOD, R.C., MOONEY, H., RODERMEL, S.R., BOGORAD, L. and STITT, M. (1992) Decreased Rubisco in transgenic tobacco transformed with antisense Rbcs. III. Impact on photosynthesis and plant growth in conditions of altered nitrogen supply. *Planta* **188**, 522–531.

QUICK, W.P., SCHEIBE, R. and NEUHAUS, H.E. (1995) Induction of hexose-phosphate translocator activity in spinach chloroplasts. *Plant Physiol.* **109**, 113–121.

RODERMEL, S.R., ABBOTT, M.S. and BOGORAD, L. (1988) Nuclear-organelle interactions: nuclear antisense gene inhibits Rubisco enzyme levels in transformed tobacco plants. *Cell* **6**, 673–681.

RUTTER, C.J. and COBB, A.H. (1983) Translocation of orthophosphate and glucose-6-phosphate in Codium fragile chloroplasts. *New Phytol.* **95**, 559–568.

SANCHEZ, J. (1994) Lipid photosynthesis in olive fruit. *Prog. Lipid Res.* **33**, 97–104.

SKENE, D.S. (1963) The fine structure of apple, pear and plum fruit surfaces, their changes during ripening, and their response to polishing. *Ann. Bot.* **27**, 581–587.

SOMERVILLE, C.R., PORTIS, A.R. and OGREN, W.L. (1982) A mutant of *Arabidopsis thalianna* which lacks activation of Rubisco *in vivo*. *Plant Physiol.* **70**, 381–387.

SONNEWALD, U., LERCHL, J., ZRENNER, R. and FROMMER, W. (1994) Manipulation of source-sink relations in transgenic plants. *Plant Cell Environ.* **17**, 649–658.

STITT, M. and AP REES, T. (1980) Carbohydrate breakdown by chloroplasts of Pisum sativum. *Biochim. Biophys. Acta* **627**, 131–143.

STITT, M. and HELDT, H.W. (1981) Physiological rates of starch breakdown in isolated intact spinach chloroplasts. *Plant Physiol.* **68**, 755–761.

STITT, N.M. and SCHULZE, E.D. (1994) Does Rubisco control the rate of photosynthesis and plant growth? An exercise in molecular ecophysiology. *Plant Cell Environ.* **17**, 465–487.

STITT, M., QUICK, W.P., SCHURR, U., SCHULZE, E.-D., RODERMEL, S.R. and BOGORAD, L. (1991) Decreased Rubisco in transgenic tobacco transformed with antisense rbcS. II. Flux control coefficients for photosynthesis in varying light, CO_2 and air humidity. *Planta* **183**, 566–576.

THOM, E. and NEUHAUS, H.E. (1995) Oxidation of imported or endogenous carbohydrates by isolated chloroplasts from green-pepper fruits (Capsicum annuum L.). *Plant Physiol.* **109**, 1421–1426.

THORNE, G.N. (1965) Photosynthesis of ears and flag leaves of wheat and barley. *Ann. Bot.* **29**, 317–329.

TRETHEWEY, R.N. and AP REES, T. (1994) A mutant of Arabidopsis thaliana lacking the ability to transport glucose across the chloroplast envelope. *Biochem J.* **301**, 449–454.

WALKER, D.A. and EDWARDS, G. (1983) *C3,C4*. Blackwell Scientific Publications, London.

WATSON, P.A. and DUFFUS, C.M. (1985) Carbon dioxide fixation by detached cereal caryopses. *Plant Physiol.* **87**, 504–509.

WULLSCHLEGER, S.D., OISTERHUIS, D.M., HURREN, R.G. and HANSON, P.J. (1991) Evidence for light-dependent recycling of respired carbon dioxide by the cotton fruit. *Plant Physiol.* **97**, 574–579.

ZHU, G. and JENSEN, R.G. (1991) Fallover of Rubisco activity. *Plant Physiol.* **97**, 1354–1358.

ZIEGLER, P. and BECK, E. (1989) Biosynthesis and degradation of starch in higher plants. *Annu. Rev. Plant Physiol. Plant Mol. Biol.* **40**, 95–117.

3

Modulation of sucrose metabolism

UWE SONNEWALD

Since the first report on the successful transformation of plants (Hernalsteens *et al.*, 1980) molecular tools (e.g. suitable transformation systems, regulatory sequences and genes) have been developed allowing the genetic manipulation of metabolic pathways in transgenic plants potentially leading to the production of tailor made plant products (Stitt and Sonnewald, 1995). The introduction of novel functions or the repression of endogenous routes in transgenic plants enables the creation of metabolic configurations which may be beneficial as long as cellular constituents necessary for growth and maintenance are not seriously affected. Thus understanding the regulation of metabolic partitioning is a prerequisite for metabolic engineering. Metabolic pathways commonly contain several enzymes, possibly regulating the flux of metabolites through the pathway; furthermore, cross-reaction between different pathways must exist, adapting metabolite flux according to developmental and environmental requirements. In the past metabolic regulation has been studied by determining substrate concentrations and characterizing the biochemical properties of a given enzyme *in vitro*. Although detailed information about the properties of individual enzymes have been obtained, the approach has several disadvantages, e.g. enzymes may be regulated by several effectors and alternative pathways may exist (Stitt, 1995). The use of mutants or transgenic plants altered in respect of the activity of a single enzyme allows the study of the function of the target enzyme *in vivo*, thereby omitting the problems discussed above. Furthermore ectopic expression of foreign enzymes enables the introduction of new pathways, allowing the manipulation of metabolite concentrations and/or end-products.

Carbon partitioning is one of the best studied cases in which genetic engineering as well as mutants have been used to study its regulation. This chapter will summarize recent progress in the field of transgenic plants altered in their sucrose metabolism.

3.1 Photosynthetic sucrose biosynthesis in C_3 plants

Photosynthetic carbon metabolism in higher plants is thought to be one determining factor for plant growth and crop yield. As primary products of photosynthesis, triose-phosphates are formed which are either used to support the Calvin cycle and starch biosynthesis in the stroma of chloroplasts or are exported into the cytosol as precursors for sucrose synthesis. Counter-exchange of inorganic phosphate, triose-phosphates and PGA between the stroma and cytosol is catalyzed by the triose-phosphate translocator (Flügge and Heldt, 1991). As sucrose is the major transport form of fixed carbon in most higher plants, enzyme activities involved in photosynthetic sucrose biosynthesis are of particular interest with respect to crop yield. The key regulatory steps of sucrose biosynthesis are supposed to be the interconversion of fructose 1,6-bisphosphate (Fru1,6P$_2$) and fructose 6-phosphate (Fru6P$_2$), and the formation of sucrose 6-phosphate (Suc6P) from UDP-glucose (UDPGlc) and Fru6P (Daie, 1993; Huber et al., 1985; Stitt and Quick, 1989). The interconversion of Fru1,6P$_2$ and Fru6P involves three enzymatic activities (illustrated in Figure 3.1): the cytosolic fructose 1,6-bisphosphatase (cytFru1,6P$_2$ase) catalyzes the forward reaction; the backward reaction is catalyzed by phosphofructokinase (PFK); the third enzyme activity, pyrophosphate:fructose-6-phosphate-1-phosphotransferase (PFP) is able to catalyze the reaction in both directions. In contrast only one enzyme, sucrose-phosphate synthase (SPS), is responsible for the formation of Suc6P, which is converted to the final product sucrose via the action of the enzyme sucrose-phosphate phosphatase (SPP).

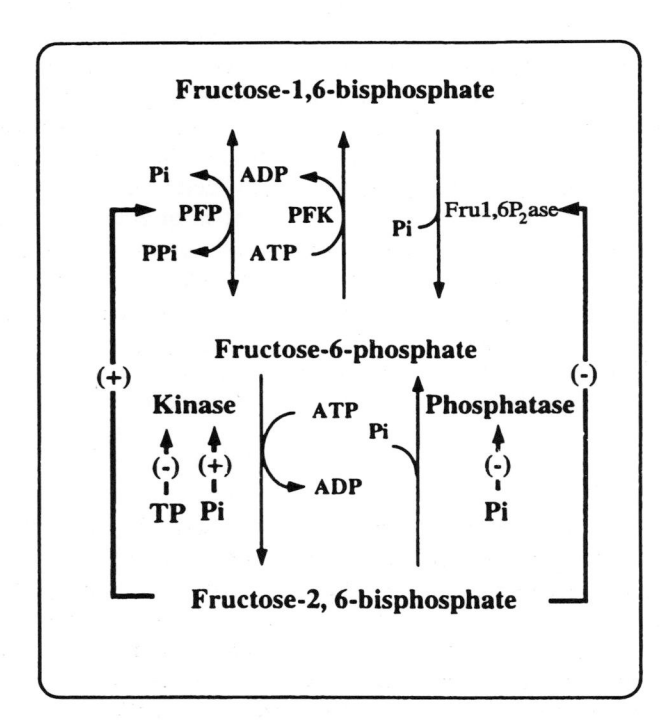

Figure 3.1 Schematic drawing of the regulatory mechanisms underlying the interconversion of Fru1,6P$_2$ and Fru6P

3.1.1 Molecular approaches to alter the interconversion of Fru1,6P$_2$ and Fru6P

Ectopic expression of alien enzymes

The interconversion of Fru1,6P$_2$ and Fru6P has been analyzed *in vivo* in transgenic tobacco and potato plants expressing either a pyrophosphatase from *E. coli* (Geigenberger *et al.*, 1996; Jellito *et al.*, 1992; Lerchl *et al.*, 1995; Sonnewald, 1992), a rat liver 6-phosphofructo-2-kinase (Scott *et al.*, 1995) or by antisense inhibition of PFP (Hajirezaei *et al.*, 1994; Paul *et al.*, 1995) and Fru1,6P$_2$ase (Zrenner *et al.*, 1996). The activities of the enzymes Fru1,6P$_2$ase and PFP are subject to allosteric control by the signal metabolite fructose 2,6-bisphosphate (Fru2,6P$_2$) (Stitt, 1990) which stimulates PFP and inhibits Fru1,6P$_2$ase (Figure 3.1). The level of Fru2,6P$_2$ *in vivo* is determined by the relative activities of 6-phosphofructo-2-kinase (PF2K) and fructose 2,6-bisphosphatase (Fru2,6P$_2$ase). To assess the extent to which Fru2,6P$_2$ regulates carbon metabolism, a modified rat liver enzyme possessing only PF2K activity was introduced into tobacco plants (Scott and Kruger, 1995; Scott *et al.*, 1995). The elevated PF2K activity in the transgenic plants resulted in levels of Fru1,6P$_2$ increased by up to 2.3-fold as determined at the end of the dark period. Analysis of the transgenic plants revealed that Fru2,6P$_2$ inhibits flux of assimilates towards sucrose and stimulates flux to starch during photosynthesis which is consistent with its proposed role.

As a result of the dephosphorylation of Fru1,6P$_2$ by PFP in the presence of inorganic phosphate (Pi), Fru6P and inorganic pyrophosphate (PPi) are formed (Figure 3.1). It might therefore be possible to shift the reaction into the direction of Fru6P by removing the PPi from the equilibrium and thus to overcome control exerted by Fru2,6P$_2$. In addition, the conversion of glucose-1-phosphate (Glu1P) to UDP-glucose (UDPGlc), catalyzed by the enzyme UDP-glucose pyrophosphorylase (UDPGlcPase), liberates PPi at a later step in sucrose synthesis. Thus by introducing an alien enzyme cleaving PPi in the cytosol, carbon flux in the direction of sucrose biosynthesis should be enhanced. As expected, the constitutive expression of a cytosolic *E. coli* pyrophosphatase resulted in an increased ratio between soluble sugars and starch by a factor of 3–4 in source leaves of transgenic tobacco and potato plants (Sonnewald, 1992). Unexpectedly, however the plants showed stunted growth and reduced root formation, suggesting that in addition to increased sucrose synthesis in leaf mesophyll cells the export of photoassimilates had been affected (see Section 3.2).

Analysis of transgenic plants with reduced pyrophosphate: fructose-6-phosphate-1-phosphotransferase activities

Based on data discussed above and on biochemical characterizations of the enzyme, PFP has been considered to play an essential role in plant metabolism. Several roles have been proposed, including a role in glycolysis, gluconeogenesis, regulation of cytosolic PPi concentrations, wound respiration and general stress adaptation (Paul *et al.*, 1995). To test these possibilities, antisense tobacco (Paul *et al.*, 1995) and potato (Hajirezaei *et al.*, 1994) plants with reduced PFP activities have been created. In both cases a more than 90% reduction of PFP activity was achieved. Growth performance of the transgenic plants was indistinguishable from wild-type plants. In source leaves of transgenic tobacco the rate of photosynthesis and partitioning between starch and sucrose was unaltered. To investigate the effect of decreased PFP activity on leaf glycolysis hexose-phosphates, glycolytic intermediates and the regulatory metabolite fructose 2,6-bisphosphate (Fru2,6P_2) were measured in sink leaves of transgenic plants at the end of the dark period. In agreement with its role in glycolysis, elevated hexose-phosphate levels and a decrease in 3PGA and PEP were found (Paul et al., 1995). However, reduced PFP activity could be compensated by: reduced inhibition of PFK due to lower PEP concentrations; and higher Fru2,6P_2 levels activating the residual PFP protein. Similar results have been obtained in case of transgenic potato plants (Hajirezaei et al., 1994).

Duff et al. (1989) reported that in response to phosphate starvation, PFP activity is strongly induced in *Brassica nigra* suspension cells. Therefore it was postulated that PFP could provide an ATP-independent bypass for glycolysis in case of phosphate limitation (Theodorou and Plaxton, 1993, 1994). In contradiction to this assumption, seedlings of transgenic tobacco with reduced PFP activity grown under limiting phosphate supply did not show any growth alterations when compared to wild-type controls. Furthermore growth under limiting nitrogen and low temperature conditions did not reveal an essential function of PFP under these suboptimal growth conditions. From these data Paul et al. (1995) concluded that PFP does not play an essential role in regulating normal vegetative plant growth, nor does it contribute to the adaptation to low nitrogen, phosphate and temperature.

Analysis of plants with reduced cytosolic fructose 1,6-bisphosphatase activity

The impact of reduced cytosolic Fru1,6P_2ase (cytFru1,6P_2ase) activity has been studied in a mutant or variant line of *Flaveria linearis* described to possess 75% reduction in enzyme activity (Sharkey et al., 1988) and in transgenic potato plants expressing an antisense cytosolic Fru1,6P_2ase gene (Zrenner et al., 1996).

Flaveria linearis, a C_3–C_4 intermediate, with reduced cytFru1,6P_2ase activity, showed loss of O_2 sensitivity of photosynthesis, indicating end-product limitation of photosynthesis (Sharkey, 1990), and changed carbohydrate partitioning in favour of starch synthesis (Sharkey et al., 1992). Decreased sucrose biosynthetic capacity lead to decreased growth rates and an increased shoot:root ratio in transgenic plants with low cytFru1,6P_2ase activity. Due to possible genetic uncertainties of the mutant line, additional mutations causing the observed changes could not unequivocally be ruled out. To investigate the genetic control of cytFru1,6P_2ase activity and to correlate the reduced capacity of sucrose biosynthesis with the loss of O_2 sensitivity of photosynthesis, Micallef and Sharkey (1996) studied the inheritance of the different traits. Analysis of F_1, F_2 and selfed lines generated from plants with low or high cytFru1,6P_2ase activity revealed that cytFru1,6P_2ase activity is controlled by one gene. Furthermore O_2 sensitivity of photosynthesis as well as the observed growth alterations, co-segregated with low cytFru1,6P_2ase activity.

cDNA clones encoding cytFru1,6P_2ase have been isolated from *Beta vulgaris*, *Brassica napus*, *Spinacia oleracea* and *Solanum tuberosum* (Harn and Daie, 1992; Hur et al., 1992; Laroche et al., 1995; Zrenner et al., 1996). To study the impact of reduced cytFru1,6P_2ase on plant growth and crop yield, Zrenner et al. (1996) generated transgenic potato plants with 9–55% of the wild-type cytFru1,6P_2ase activity using the antisense strategy. Analysis of the transgenic potato plants revealed that 45% reduction of the cytFru1,6P$_2$ase activity did not cause any measurable change in metabolite concentrations, growth behaviour or photosynthetic parameters of the transgenic plants. Inhibition of cytFru1,6P$_2$ase activity below 20% of the wild-type activity led to an accumulation of PGA, triose-phosphates and fructose 1,6-bisphosphate in source leaves. This resulted in a reduced light saturated rate of assimilation measured via gas exchange and a decreased photosynthetic rate under conditions of the leaf disc electrode with saturating light and CO_2. Measuring photosynthetic carbon fluxes by labelling leaf discs with $^{14}CO_2$ revealed a 53–65% reduction of sucrose synthesis whereas starch synthesis decreased by only 18–24%. Despite these changes steady-state sucrose concentrations were not affected in source leaves from transgenic plants. Starch accumulated by more than a factor of 3 compared to wild-type leaves and was degraded during the night. This provides strong evidence for the hypothesis that hexoses and/or hexose-phosphates are exported out of the chloroplasts, thereby circumventing the limitation of sucrose biosynthesis caused by the inhibition of cytosolic Fru1,6P$_2$ase in the dark. Accordingly, plant growth and potato tuber yield remained unaltered. From these data it can be concluded that a reduced photosynthetic sucrose biosynthetic capacity can be efficiently compensated without any reduction in crop yield under greenhouse or growth chamber conditions by changing carbon export strategy.

In favour of this hypothesis, Schäfer and Heber (1977) reported that a glucose transporter might be involved in carbohydrate export during the dark period. Furthermore, upon glucose feeding to spinach and potato leaves, Quick *et al.* (1995) could demonstrate that a hexose-phosphate translocator activity is inducible in the envelope membrane of chloroplasts. Additionally, an *Arabidopsis thaliana* mutant has been described (Trethewey and ap Rees, 1994), lacking the ability to transport glucose across the chloroplast envelope. This mutant is delayed in starch breakdown, which leads to the accumulation of starch in the leaf mesophyll. Moreover, in transgenic plants with antisense inhibition of the triose phosphate translocator (Heineke *et al.*, 1994; Riesmeier *et al.*, 1993), the ratio of assimilates translocated at night is much higher when compared to the translocation during the day, providing evidence that alternative translocator activities are responsible for assimilate export out of the chloroplast in the dark.

3.1.2 Molecular approaches to alter sucrose-phosphate synthase activity in transgenic plants

Several lines of evidence suggest that SPS might catalyze the rate-limiting step in sucrose biosynthesis. First, changes in SPS activity often correlate with changes in the rate of sucrose synthesis and export (Huber and Israel, 1982; Rocher *et al.*, 1989; Stitt *et al.*, 1988). Second, SPS can be modulated by a hierarchy of several interacting mechanisms (Figure 3.2), including: allosteric regulation via metabolites (Doehlert and Huber, 1983); post-translational modification via reversible protein phosphorylation (Huber *et al.*, 1989a); and developmental regulation of gene expression leading to changes in the amount of SPS protein (Walker and Huber, 1989). In spinach leaves (Doehlert and Huber, 1983) and potato tubers (Reimholz *et al.*, 1994), the enzyme is activated by glucose 6-

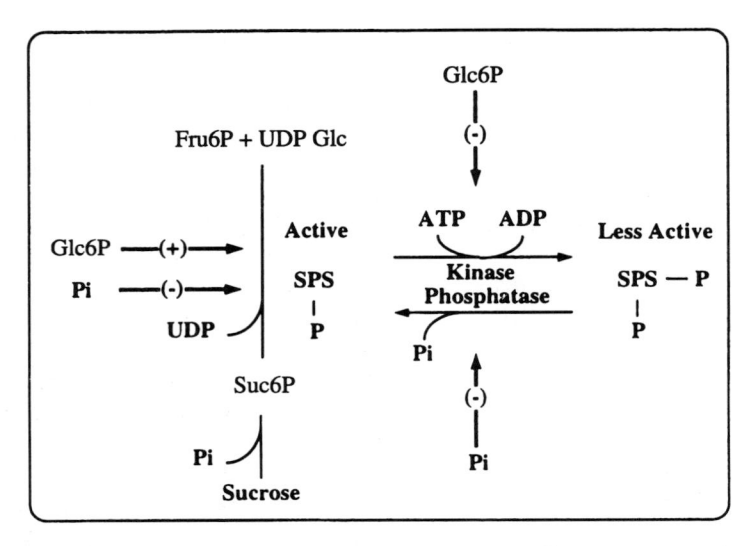

Figure 3.2 Simplified scheme of regulatory mechanisms controlling SPS activity (modified after Huber and Huber, 1992).

phosphate and inhibited by inorganic phosphate. Application of gibberellic acid to spinach and soybean leaves resulted in increased levels of SPS protein, indicating that gibberellin is one of the endogenous hormonal factors regulating the steady-state level of SPS protein (Cheikh and Brenner, 1992; Cheikh et al., 1992). In spinach, SPS exists in two kinetically distinct forms which differ in their degree of phosphorylation (Huber et al., 1989a; for review, see Huber and Huber, 1992). The highly phosphorylated form is characterized by a low substrate (fructose 6-phosphate) and effector (glucose 6-phosphate) affinity whereas the lower phosphorylated form has a high substrate and effector affinity. Sequencing of the major phosphopeptide showed that serine-158 was the major regulatory phosphorylation site (McMichael et al., 1993). Weiner (1995) produced antibodies directed against a synthetic peptide containing the phosphorylation site serine-158. The anti-peptide antibodies preferentially precipitated highly activated SPS, adding additional proof to the importance of serine-158 phosphorylation. Multiple protein kinases have been isolated which are able to phosphorylate spinach SPS as well as synthetic peptides derived from the major phosphorylation site in vitro (McMichael et al., 1995).

In addition to the biochemical investigations, the genetic variation of SPS activity has been studied and correlated with growth rates of maize seedlings. Rocher et al. (1989) investigated eight maize genotypes and found that SPS activity varied in the ratio of 1 to 3 from the less active to the more active genotype and this variation was highly correlated with growth rates. Similar experiments have been carried out by Causse et al. (1995). In two out of four field trials, SPS activity could be correlated to the forage dry matter yield of hybrids. In these studies other sucrose biosynthetic enzymes have not been considered and therefore do not provide direct proof that SPS regulates sucrose synthesis, nor do they allow assessment of the importance of SPS relative to other possible control sites. Transgenic plants with altered SPS activity would allow these questions to be answered.

cDNA clones encoding SPS have been isolated from various plants including maize, spinach, and sugar beet (Hesse *et al.*, 1995; Klein *et al.*, 1993; Sonnewald and Basner, EMBL Data Bank, accession number X73477; Sonnewald *et al.*, 1993; Worrell *et al.*, 1991). Two strategies, heterologous over-expression and antisense repression, have been followed to study the impact of altered SPS activity on plant growth and development. Worrell *et al.* (1991) created transgenic tomato plants carrying the maize SPS cDNA clone under control of the constitutive 35S CaMV or the leaf-specific Rubisco SSU promoter. In both cases transgenic plants with up to 12-fold higher SPS activities could be regenerated (Worrell *et al.*, 1991). However, in subsequent studies SPS activity was found to be only 3- to 6-fold higher (Galtier *et al.*, 1993, 1995). Elevated SPS activity led to a reduction of starch and an increase of sucrose in source leaves. Increased sucrose accumulation was due to a stimulation of carbon partitioning towards sucrose, thus it was concluded that SPS plays a pivotal role in carbon partitioning (Galtier *et al.*, 1993; Micallef *et al.*, 1995). Based on a classification of Huber *et al.* (1989a), maize SPS belongs to class I, i.e. exhibits changes in V_{max} as well as affinity constants upon activation. Determining the light regulation of maize SPS isolated from leaves of transgenic tomato plants, no diurnal regulation was found (Galtier *et al.*, 1993; Worrell *et al.*, 1991). The assay employed to determine the activation state was developed for spinach SPS. Since spinach SPS belongs to class II, i.e. only affinity constants are affected upon activation, it is questionable whether under these assay conditions the activation state of maize SPS is adequately described. To solve this question Galtier *et al.* (1995) determined the sensitivity of the introduced SPS to the effectors fructose 6-phosphate, glucose 6-phosphate and inorganic phosphate. Under these conditions a decreased activation state of the introduced SPS was found. However, SPS activity of transformed tomato plants remained higher (approximately 1.5-fold) as compared to control plants. Plants grown in growth chambers showed no detrimental effects and total dry matter production was unchanged (Galtier *et al.*, 1993). However the root:shoot ratio of transgenic plants was changed in favour of the shoot. The photosynthetic rate in air was not significantly increased as a result of elevated SPS activity. When grown at elevated (65 Pa) atmospheric carbon dioxide concentrations, SPS transformants had a 20% greater rate of photosynthesis as compared to untransformed controls, indicating a reduced limitation of photosynthesis imposed by end-product synthesis. Therefore it can be concluded that increasing the capacity of sucrose biosynthesis may affect acclimation to elevated carbon dioxide. Although increased photosynthetic rates have been observed when SPS transformants were grown at 65 Pa CO_2, fruit yield was not significantly increased. Similar to the results obtained with antisense cytFBPase potato plants (discussed above), a strict relationship between yield and leaf photosynthetic rate could not be established.

The effect of decreased SPS activity has been studied in transgenic potato plants carrying either a chimeric antisense or sense (co-suppression) SPS gene (Geigenberger *et al.*, 1995; Krause, 1994). In fully expanded source leaves a 60–70% decrease in SPS activity could be achieved. The reduced amount of SPS protein is partly compensated by increased activation state of the remaining polypeptide. Decreased SPS activity did not result in a reduced photosynthetic rate. However the ratio between starch and sucrose was altered in favour of starch. The decreased accumulation of soluble sugars was due to decreased carbon partitioning towards sucrose. Based on these results the control coefficient (for details see Small and Kacser, 1993) of SPS for sucrose synthesis in source leaves in saturating light and carbon dioxide was estimated to be approximately 0.5 (Geigenberger et al., 1995). If SPS had been the sole regulatory step in sucrose biosynthesis a control coefficient of 1 would have been expected. Therefore it can be concluded that SPS is important but is not the only factor regulating the rate of sucrose synthesis. Reduced SPS activity did not result in any visible phenotype nor did it influence tuber yield of the transgenic plants as compared to untransformed controls.

Taken together, up- and down-regulation of SPS leads to the expected changes with respect to carbon partitioning. However it remains questionable whether the actual rate of photosynthetic sucrose formation does determine final crop yield. Other factors such as phloem loading and/or sucrose utilization in sink tissues might be the limiting steps.

3.2 Manipulation of sucrose metabolism in phloem cells

Once sucrose has been synthesized in mesophyll cells of source leaves it has to be translocated to other tissues. The transport pathway of sucrose is the phloem system. Two hypotheses of how sucrose enters the phloem system have been postulated: symplastic loading via plasmodesmata; and apoplastic loading mediated by an active transport system (Frommer and Sonnewald, 1995). Via the ectopic expression of the *suc2* gene encoding a yeast-derived invertase targeted to the apoplast (Dickinson et al., 1991; Heineke et al., 1992; von Schaewen et al., 1990) or the antisense inhibition of a sucrose transport protein from potato (Riesmeier et al., 1994), it could unequivocally be demonstrated that phloem loading of sucrose involves an apoplastic step in tobacco, tomato, potato and *Arabidopsis* plants. The active loading is assumed to take place at the border between mesophyll and phloem cells. The proton gradient, which is needed for the proton sucrose cotransport (Delrot, 1981), is generated by an H^+-ATPase located in the plasma membrane of companion cells (Bouche-Pillon et al., 1994). To maintain an adequate carbon supply, a small proportion of incoming sucrose is cleaved either by invertase or sucrose synthase (Claussen et al., 1985). The role of invertases is still a matter of debate because tissue-specific abundance, compartmentation, and regulation of invertase activities are not well understood. Phloem cells contain a high sucrose synthase activity (Martin et al., 1993; Nolte and Koch, 1993; Yang and Russell, 1990) and metabolism during glycolysis within the phloem complex of *Ricinus* seedlings has been shown to involve sucrose synthase (Geigenberger et al., 1993). Breakdown of sucrose via sucrose synthase would lead to the formation of fructose and UDP-glucose. In the presence of inorganic pyrophosphate, UDP-glucose is converted to glucose 1-phosphate via the action of UDPGlcPPase. Thus removing cytosolic pyrophosphate would limit the utilization of UDP-glucose and thereby reduce the energy gain, resulting in the inhibition of sucrose loading into the phloem system.

To study the role of PPi in phloem cells, a chimeric gene was constructed using the phloem-specific *rolC* promoter of *Agrobacterium rhizogenes* to drive the expression of the *ppa* gene (Lerchl et al., 1995). Removal of cytosolic PPi in those cells resulted in photoassimilate accumulation in source leaves, chlorophyll loss, and reduced plant growth. In roots, a decrease of amino acids and sugars was found in the tip, whereas a relatively high level of sugars was found in the basal region of the root. This indicates that respiratory metabolism is required to maintain phloem function along the transport path and for initial phloem loading (Geigenberger et al., 1996). From these data it was postulated that sucrose hydrolysis via sucrose synthase is essential for assimilate partitioning. Supporting this hypothesis a decreased ATP/ADP ratio and a reduced rate of respiration were found in midribs of the transgenic plants (Geigenberger et al., 1996). To bypass the PPi-dependent sucrose synthase step, transgenic plants were created expressing the yeast *suc2* gene encoding cytosolic invertase in their phloem cells. Transgenic plants containing various levels of phloem-specific cytosolic invertase were generated. To combine both traits, crosses between invertase- and pyrophosphatase-containing transgenic plants were carried out. Analysis of their offspring revealed that the invertase can complement the phenotypic effects caused by the removal of PPi in phloem cells (Lerchl et al., 1995).

3.3 Sucrose utilization in sink tissues

Sink strength is defined as the ability of an organ to attract photoassimilates (Ho, 1988). It is dependent on the mechanism for the transfer of nutrients away from the transport system and the physical and biochemical isolation of the transported carbon in the sink tissue. Sink tissues are subdivided into metabolic sinks (i.e. meristems) and storage sinks (i.e. seeds, tubers etc.). In storage sinks imported assimilates are deposited in the form of soluble sugars, starch, proteins or fatty acids. Sink strength can be estimated by the rates of nutrient uptake as measured in incorporation studies with labelled photoassimilates or by dry matter accumulation during development as determined as yield after harvest. Respiratory activity can also contribute to sink strength. As sucrose is the major transport form of fixed photoassimilates into sink tissues, the mechanisms responsible for the immediate metabolism of sucrose is of particular interest. Two enzymes, invertase(s) and sucrose synthase, are able to metabolize imported sucrose in sink tissues. Sucrose synthase is confined to the cytosolic compartment whereas invertase(s) exist in several isoforms located in the cell wall, the vacuole or possibly the cytosol. In sink tissues different routes of sucrose utilization exist depending on the mechanism of unloading (Figure 3.3). In case of apoplastic unloading, sucrose may be hydrolyzed by an apoplastic acid invertase yielding glucose and fructose which are subsequently transported into the cytosol for further metabolism. Alternatively, sucrose delivered directly into the vacuole may be hydrolyzed via a vacuolar acid invertase. In case of symplastic unloading of sucrose into the cytosol it could either be utilized via the action of neutral invertase or sucrose synthase. The reaction catalyzed by all invertase isozymes is highly exothermic and irreversible (Avigad, 1982), whereas the reaction catalyzed by sucrose synthase is readily reversible *in vivo* in developing potato tubers (Geigenberger and Stitt, 1993). In most sink tissues both invertase(s) and sucrose synthase activities are found. However the relative activities of both enzymes may change during sink development. Based on genetic and molecular

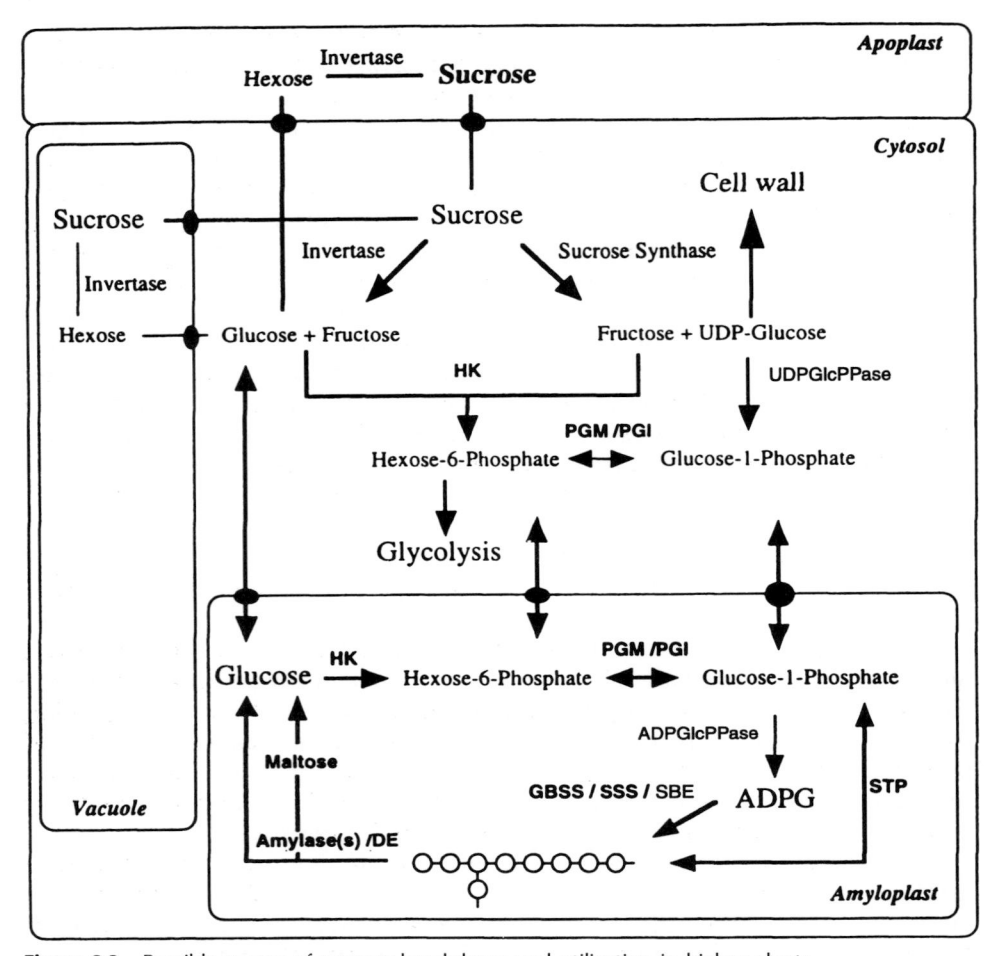

Figure 3.3 Possible routes of sucrose breakdown and utilization in higher plants

approaches, acid invertase has been shown to regulate the hexose to sucrose ratio in tomato fruits (Harada *et al.*, 1995; Ohyama *et al.*, 1995). In *Vicia faba*, a seed coat-associated invertase has been shown to be expressed during the prestorage phase of seeds (Weber *et al.*, 1995). Following the developmentally regulated degradation of the seed coat parenchyma cells and loss of invertase activity, sucrose synthase activity increases in cotyledons most likely initiating the storage phase. The importance of invertase activities during seed development has been established by characterizing the *miniature*-1 mutant of maize. Biochemical and physiological investigations of *miniature*-1 kernels revealed a loss of both soluble and wall-bound invertase activities (Miller and Chourey, 1992), and provided evidence that invertase-mediated maintenance of a photosynthate gradient between pedicel and endosperm contributes to normal seed development. In agreement with a postulated function of sucrose synthase during the storage phase of sink tissues a second maize mutant, *shrunken*-1, deficient in sucrose synthase activity, is characterized by a strong decrease of starch accumulation in the endosperm (Chourey and Nelson, 1976).

In potato the process of tuberization is characterized by an increase in starch synthesis and a decrease in reducing sugars (Davies, 1984). During early stages of tuber development a high hexose:sucrose ratio is found. During this stage invertase activity and transcript levels are high (Zrenner, 1993). Upon swelling of the stolon tip, invertase activity declines and sucrose synthase activity increases. The increase of sucrose synthase activity is paralled by an increase of starch synthesis and the accumulation of storage proteins. Measurements of maximum catalytic activities of invertase(s) and sucrose synthase in growing potato tubers revealed that only sucrose synthase activity can sustain a rate of sucrose breakdown comparable to that observed *in vivo* (Morrell and ap Rees, 1986). Thus it is likely that the first step in the conversion of sucrose to starch is largely catalyzed by sucrose synthase. Furthermore it is proposed that sucrose synthase activity could be taken as a marker for sink strength in potato tubers and other actively filling sinks (Sung *et al.*, 1989; Wang *et al.*, 1993; Xu *et al.*, 1989). In potato two differentially regulated classes of sucrose synthase genes, Sus3 and Sus4, have been identified (Fu and Park, 1995). Sus3 is mainly expressed in stems and roots and may provide the vascular function of sucrose synthase. In contrast Sus4 is primarily expressed in the storage and vascular tissue of growing tubers possibly providing the sink function (Fu and Park, 1995).

To investigate the unique role of sucrose synthase with respect to sucrose metabolism and sink strength in growing potato tubers, transgenic potato plants were created expressing sucrose synthase antisense RNA corresponding to the Sus4 isoform (Zrenner *et al.*, 1995). Although the constitutive 35S CaMV promotor was used to drive the expression of the antisense RNA, inhibition of sucrose synthase activity was tuber-specific. This is in agreement with the postulated functions of Sus3 and Sus4 isoforms. The inhibition of sucrose synthase led to no change in the sucrose content, but to a strong accumulation of reducing sugars and an inhibition of starch accumulation in developing potato tubers, which was accompanied by a decrease in total tuber dry weight and a reduction of soluble tuber proteins. The increase in hexoses was paralleled by a 40-fold increase in invertase activities but no considerable changes in hexokinase activities. The reduced protein content was mainly due to a decrease in the major storage proteins patatin, the 22 kDa proteins and the proteinase inhibitors. Altogether these data are in agreement with the assumption that sucrose synthase is the major determinant of potato tuber sink strength.

3.4 Conclusion

Genetic engineering has successfully been employed to alter photosynthetic sucrose biosynthesis. Transgenic plants with decreased or increased sucrose biosynthetic capacities enabled alteration of assimilate partitioning and study of its regulation. Calculating flux control coefficients for individual enzymes revealed that several enzymes rather than one enzyme control overall flux through a given pathway. A strict relationship between the rate of photosynthetic sucrose biosynthesis and crop yield could not be established. This may indicate that either the supply of assimilates is not a rate limiting step during sink development or that other factors such as long distance transport and/or assimilate utilization in sink tissues become limiting when supply is in excess.

References

AVIGAD, G. (1982) Sucrose and other disaccharides. In: *Encyclopedia of Plant Physiology*, Vol. 13A (LOEWUS, F.A. and TANNER, W., eds), pp. 217–347. Springer-Verlag, Berlin.

BOUCHE-PILLON, S., FLEURAT-LESSARD, P., FROMONT, J.C., SERRANO, R. and BONNEMAIN, J.L. (1994) Immunolocalization of the plasma membrane H+-ATPase in minor veins of *Vicia faba* in relation to phloem loading. *Plant Physiol.* **105**, 691–697.

CAUSSE, M., ROCHER, J.-P., PELLESCHI, S., BARRIÉRE, Y., DE VIENNE, D. and PRIOUL, J.-L. (1995) Sucrose phosphate synthase: an enzyme with heterotic activity correlated with maize growth. *Crop Sci.* **35**, 995–1001.

CHEIKH, N. and BRENNER, M.L. (1992) Regulation of key enzymes of sucrose biosynthesis in soybean leaves. *Plant Physiol.* **100**, 1230–1237.

CHEIKH, N., BRENNER, M.L., HUBER, J.L. and HUBER, S.C. (1992) Regulation of sucrose phosphate synthase by gibberellins in soybean and spinach plants. *Plant Physiol.* **100**, 1238–1242.

CHOUREY, P.S. and NELSON, O. (1976) The enzymatic deficiency conditioned by the shrunken-1 mutation in maize. *Biochem. Genet.* **14**, 1041–1055.

CLAUSSEN, W., LOVEYS, B.R. and HAWKER, J.S. (1985) Comparative investigations on the distribution of sucrose synthase activity and invertase activity within growing, mature and old leaves of some C3 and C4 plant species. *Physiol. Plant.* **65**, 275–280.

DAIE, J. (1993) Cytosolic fructose-1,6-bisphosphatase: a key enzyme in the sucrose biosynthetic pathway. *Photosyn. Res.* **38**, 5–14.

DAVIES, H.V. (1984) Sugar metabolism in stolon tips of potato during early tuberization. *Z. Pflanzenphysiol.* **113**, 377–381.

DELROT, S. (1981) Proton fluxes associated with sugar uptake in Vicia faba leaf tissue. *Plant Physiol.* **68**, 706–711.

DICKINSON, C.D., ALTABELLA, T. and CHRISPEELS, M.J. (1991) Slow-growth phenotype of transgenic tomato plants expressing apoplastic invertase. *Plant Physiol.* **95**, 420–425.

DOEHLERT, D.C. and HUBER, S.C. (1983) Regulation of spinach leaf sucrose-phosphate synthase by Glc6P, inorganic phosphate and pH. *Plant Physiol.* **73**, 989–994.

DUFF, S.M.G., MOOREHEAD, G.B.G., LEFEBVRE, D.D. and PLAXTON, W.C. (1989) Phosphate starvation inducible 'bypasses' of adenylate- and phosphate dependent glycolytic enzymes in Brassica nigra suspension cells. *Plant Physiol.* **90**, 1275–1278.

FLÜGGE, U.-I. and HELDT, H.W. (1991) Metabolite translocators of the chloroplast envelope. *Annu. Rev. Plant Physiol. Plant Mol. Biol.* **42**, 129–144.

FROMMER, W.B. and SONNEWALD, U. (1995) Molecular analysis of carbon partitioning in solanaceous species. *J. Exp. Bot.* **46**(287), 587–607.

FU H. and PARK, W.D. (1995) Sink- and vascular-associated sucrose synthase functions are encoded by different gene classes in potato. *Plant Cell* **7**, 1369–1385.

GALTIER, N., FOYER, C.H., HUBER, J., VOELKER, T.A. and HUBER, S.C. (1993) Effects of elevated sucrose-phosphate synthase activity on photosynthesis, assimilate partitioning, and growth in tomato (*Lycopersicon esculentum* var. UC82B). *Plant Physiol.* **101**, 535–543.

GALTIER, N., FOYER, C.H., MURCHIE, E., ALRED, R., QUICK, P., VOELKER, T.A., THÉPENIER, C., LASCÈVE, G. and BETSCHE, T. (1995) Effects of light and atmospheric carbon dioxide enrichment on photosynthesis and carbon partitioning in leaves of tomato (*Lycopersicon esculentum* L.) plants over-expressing sucrose phosphate synthase. *J. Exp. Bot.* **46**, 1335–1344.

GEIGENBERGER, P. and STITT, M. (1993) Sucrose synthase catalyses a readily reversible reaction in vivo in developing potato tubers and other plant tissues. *Planta* **189**, 329–339.

GEIGENBERGER, P., LANGENBERGER, S., WILKE, I., HEINEKE, D., HELDT, H.W. and STITT, M. (1993) Sucrose is metabolised by sucrose synthase and glycolysis within the phloem of Ricinus communis L. seedlings. *Planta* **190**, 446–452.

GEIGENBERGER, P., KRAUSE, K.-P., HILL, L.M., REIMHOLZ, R., MACRAE. E., QUICK, P., SONNEWALD, U. and STITT, M. (1995) The regulation of sucrose synthesis in leaves and tubers of potato plants. In: *Sucrose Metabolism, Biochemistry, Physiology and Molecular Biology* (PONTIS, H.G., SALERNO, G.L. and ECHEVERRIA, E.J., eds), pp. 14–24. American Society of Plant Physiologists, MD.

GEIGENBERGER, P., LERCHL, J., STITT, M. and SONNEWALD, U. (1996) Phloem-specific expression of pyrophosphatase inhibits long-distance transport of carbohydrates and amino acids in tobacco plants. *Plant Cell Environ.* **19**, 43–55.

HAJIREZAEI, M., SONNEWALD, U., VIOLA, R., CARLISLE, S., DENNIS, D. and STITT, M. (1994) Transgenic potato plants with strongly reduced expression of pyrophosphate:fructose-6-phosphate phosphotransferase show no visible phenotype and only minor changes in metabolic fluxes in their tubers. *Planta* **92**, 16–30.

HARADA, S., FUKUTA, S., TANAKA, H., ISHIGURO, Y. and SATO, T. (1995) Genetic analysis of the trait of sucrose accumulation in tomato fruit using molecular marker. *Breeding Sci.* **45**, 429–434.

HARN, C. and DAIE, J. (1992) Cloning and nucleotide sequence of a cDNA encoding the cytosolic fructose-1,6-bisphosphatase of sugar beet (Beta vulgaris L.). *Plant Physiol.* **98**, 790–791.

HEINEKE, D., SONNEWALD, U., BÜSSIS, D., GÜNTER, G., LEIDREITER, K., WILKE, I., RASCHKE, K, WILLMITZER, L. and HELDT, H.W. (1992) Apoplastic expression of yeast-derived invertase in potato: effect on photosynthesis, leaf solute composition, water relations, and tuber composition. *Plant Physiol.* **100**, 301–308.

HEINEKE, D., KRUSE, A., FLÜGGE, U.I., FROMMER, W.B., RIESMEIER, J.W., WILLMITZER, L. and HELDT, H.W. (1994) Effect of antisense repression of the chloroplast triose phosphate translocator on photosynthetic metabolism in transgenic potato plants. *Planta* **193**, 174–180.

HERNALSTEENS, J.-P., VAN VLIET, F., DE BEUCKELEER, M., DEPICKER, A., ENGLER, G., LEMMERS, M., HOLSTERS, M., VAN MONTAGUE, M. and SCHELL, J. (1980) The Agrobacterium tumefaciens Ti plasmid as a host vector system for introducing foreign DNA in plant cells. *Nature* **287**, 654–656.

HESSE, H, SONNEWALD, U. and WILLMITZER, L. (1995) Cloning and expression analysis of sucrose-phosphate synthase from sugar beet (*Beta vulgaris* L.). *Mol. Gen. Genet.* **247**, 515–520.

HO, L.C. (1988) Metabolism and compartmentation of imported sugars in sink organs in relation to sink strength. *Annu. Rev. Plant Physiol. Plant Mol. Biol.* **39**, 355–378.

HUBER, S.C. and HUBER, J.L. (1992) Role of sucrose-phosphate synthase in sucrose metabolism in leaves. *Plant Physiol.* **99**, 1275–1278.

HUBER, S.C. and ISRAEL, D.W. (1982) Biochemical basis for the partitioning of photosynthetically fixed carbon between starch and sucrose in soybean leaves. *Plant Physiol.* **69**, 691–696.

HUBER, S.C., KERR, P.S. and TORRES, W.K. (1985) Regulation of sucrose synthesis and movement. In: *Regulation of Carbon Partitioning in Photosynthetic Tissue* (HEATH, R.L. and PREISS, J. eds), pp. 199–214. Williams and Wilkins, Baltimore, MD.

HUBER, S.C., NIELSEN, T.H., HUBER, J.L.A. and PHARR, D.M. (1989a) Variation among species in light activation of sucrose-phosphate synthase. *Plant Cell Physiol.* **30**, 277–285.

HUBER, J.L.A., HUBER, S.C. and NIELSEN, T.H. (1989b) Protein phosphorylation as a mechanism for regulation of spinach leaf sucrose-phosphate synthase activity. *Arch. Biochem. Biophys.* **270**, 681–690.

HUR, Y., UNGER, E.A. and VASCONCELOS, A.C. (1992) Isolation and characterization of a cDNA encoding cytosolic fructose-1,6-bisphosphatase from spinach. *Plant Mol. Biol.* **18**, 788–802.

JELLITO, T., SONNEWALD, U., WILLMITZER, L., HAJIREZEAI, M. and STITT, M. (1992) Inorganic pyrophosphate content and metabolites in potato and tobacco plants expressing E. coli pyrophosphatase in their cytosol. *Planta* **188**, 238–244.

KLEIN, R.R., CRAFTS-BRANDNER, S.J. and SALVUCCI, M.E. (1993) Cloning and developmental expression of sucrose-phosphate-synthase gene from spinach. *Planta* **190**, 498–510.

KRAUSE, K.-P. (1994) Zur regulation von Saccharosephosphatsynthase. PhD diss., Univ. Bayreuth.

LAROCHE, A., FRICK, M.M., KAZALA, C., WESELAKE, R.J. and THOMAS, J.E. (1995) Isolation and characterization of an oilseed rape fructose-1,6-bisphosphatase cDNA. *Plant Physiol.* **108**, 1335–1336.

LERCHL, J., GEIGENBERGER, P., STITT, M. and SONNEWALD, U. (1995) Inhibition of long-distance sucrose transport by inorganic pyrophosphatase can be complemented by phloem-specific expression of cytosolic yeast-derived invertase in transgenic plants. *Plant Cell* **7**, 259–270.

MARTIN, T., FROMMER, W.B., SALANOUBAT, M. and WILLMITZER, L. (1993) Expression of an Arabidopsis sucrose synthase gene indicates a role in metabolization of sucrose both during phloem loading and in sink organs. *Plant Journal* **4** (2), 367–377.

MCMICHAEL, R.W., KLEIN, R.R., SALVUCCI, M.E. and HUBER, S.C. (1993) Identification of the major regulatory phosphorylation site in sucrose-phosphate synthase. *Arch. Biochem. Biophys.* **307**, 248–252.

MCMICHAEL, R.W., BACHMANN, M. and HUBER, S.C. (1995) Spinach leaf sucrose-phosphate synthase and nitrate reductase are phosphorylated/inactivated by multiple protein kinases in vitro. *Plant Physiol.* **108**, 1077–1082.

MICALLEF, B.J. and SHARKEY, T.D. (1996) Genetic and physiological characterization of Flaveria linearis plants having a reduced activity of cytosolic fructose-1,6-bisphosphatase. *Plant Cell Environ.* **19**, 1–9.

MICALLEF, B.J., HASKINS, K.A., VANDERVEER, P.J., ROH, K.-S., SHEWMAKER, C.K. and SHARKEY, T.D. (1995) Altered photosynthesis, flowering, and fruiting in transgenic tomato plants that have increased capacity for sucrose synthesis. *Planta* **196**, 327–334.

MILLER, M.E. and CHOUREY, P.S. (1992) The maize invertase-deficient miniature-1 seed mutation is associated with aberrant pedicel and endosperm development. *Plant Cell* **4**, 297–305.

MORRELL, S and AP REES, T. (1986) Sugar metabolism in developing tubers of Solanum tuberosum. *Phytochemistry* **25**, 1579–1585.

NOLTE, K.D. and KOCH, K.E. (1993) Companion cell specific localization of sucrose synthase in zones of phloem loading and unloading. *Physiol. Plant.* **101**, 899–905.

OHYAMA, A., ITO, H., SATO, T., NISHIMURA, S., IMAI, T. and HIRAI, M. (1995) Suppression of acid invertase activity by antisense RNA modifies the sugar composition of tomato fruit. *Plant Cell Physiol.* **36**, 369–376.

PAUL, M., SONNEWALD, U., HAJIREZAEI, M., DENNIS, D. and STITT, M. (1995) Transgenic tobacco plants with strongly decreased expression of pyrophosphate: fructose-6-phosphate 1-phosphotransferase do not differ significantly from wildtype in photosynthate partitioning, plant growth or their ability to cope with limiting phosphate, limiting nitrogen and suboptimal temperatures. *Planta* **196**, 277–283.

QUICK, W.P., SCHEIBE, R. and NEUHAUS, H.E. (1995) Induction of hexose-phosphate translocator activity in spinach chloroplasts. *Plant Physiol.* **109**, 113–121.

REIMHOLZ, R., GEIGENBERGER, P. and STITT, M. (1994) Sucrose-phosphate synthase is regulated via metabolites and protein phosphorylation in potato tubers, in a manner analogous to the enzyme in leaves. *Planta* **192**, 480–488.

RIESMEIER, J., FLÜGGE, U.I., SCHULZ, B., HEINEKE, D., HELDT, H.W., WILLMITZER, L. and FROMMER, W.B. (1993) Antisense repression of the chloroplast triose phosphate translocator affects carbon partitioning in transgenic potato plants. *Proc. Natl Acad. Sci. USA* **90**, 6160–6164.

RIESMEIER, J.W., FROMMER, W.B. and WILLMITZER, L. (1994) Evidence for an essential role of the sucrose transporter in phloem loading and assimilate partitioning. *EMBO J.* **13**(1), 1–7.

ROCHER, J.P., PRIOUL, S.L., LEEHAMY, A., REYSS, A. and JOUSSAUME, M. (1989) Genetic variability in carbon fixation, sucrose-P synthase and ADP-Glc pyrophosphorylase in maize plants of differing growth rate. *Plant Physiol.* **89**, 416–420.

SCHÄFER, G. and HEBER, U. (1977) Glucose transport into spinach chloroplasts. *Plant Physiol.* **60**, 286–289.

SCOTT, P. and KRUGER, N.J. (1995) Influence of elevated fructose-2,6-bisphosphate levels on starch mobilization in transgenic tobacco leaves in the dark. *Plant Physiol.* **108**, 1569–1577.

SCOTT, P., LANGE, A.J., PILKIS, S.J. and KRUGER, N.J. (1995) Carbon metabolism in leaves of transgenic tobacco (Nicotiana tabacum L.) containing elevated fructose 2,6-bisphosphate levels. *Plant J.* **7**, 461–469.

SHARKEY, T.D. (1990) Feedback limitation of photosynthesis and the physiological role of ribulose bisphosphate carboxylase carbamylation. *Botanical Magazine*, Tokyo Special Issue **2**, 87–105.

SHARKEY, T.D., KOBZA, J., SEEMAN, R. and BROWN, R.H. (1988) Reduced cytosolic fructose-1,6-bisphosphatase activity leads to loss of O_2 sensitivity in a Flaveria linearis mutant. *Plant Physiol.* **86**, 667–671.

SHARKEY, T.D., SAVITCH, L.V., VANDERVEER, P.J. and MICALLEF, B.J. (1992) Carbon partitioning in a Flaveria linearis mutant with reduced cytosolic fructose bisphosphatase. *Plant Physiol.* **100**, 210–215.

SMALL, J.R. and KACSER, H. (1993) Responses of metabolic systems to large changes in enzyme activities and effectors. *Eur. J. Biochem.* **213**, 613–624.

SONNEWALD, U. (1992) Expression of *E. coli.* inorganic pyrophosphatase in transgenic plants alters photoassimilate partitioning. *Plant J.* **2**, 571–581.

SONNEWALD, U., QUICK, W.P., MACRAE, E., KRAUSE, K.-P. and STITT, M. (1993) Purification, cloning and expression of spinach leaf sucrose-phosphate synthase in *E.coli*. *Planta* **189**, 174–181.

STITT, M. (1990) Fructose-2,6-bisphosphate as a regulatory molecule in plants. *Annu. Rev. Plant Physiol. Plant Mol. Biol.* **41**, 153–185.

STITT, M. (1995) The use of transgenic plants to study the regulation of plant carbohydrate metabolism. *Austr. J. Plant Physiol.* **22**, 635–646.

STITT, M. and QUICK, W.P. (1989) Photosynthetic carbon partitioning: its regulation and possibilities for manipulation. *Physiol. Plant.* **77**, 633–641.

STITT, M. and SONNEWALD, U. (1995) Regulation of metabolism in transgenic plants. *Annu. Rev. Plant Physiol. Plant Mol. Biol.* **46**, 341–368.

STITT, M., WILKE, I., FEIL, R. and HELDT, H.W. (1988) Coarse control of sucrose phosphate synthase in leaves: alterations of the kinetic properties in response to the rate of photosynthesis and the accumulation of sucrose. *Planta* **174**, 217–230.

SUNG, S.S., XU, W.P. and BLACK, C.C. (1989) Identification of actively filling sucrose sinks. *Plant Physiol.* **89**, 1117–1121.

THEODOROU, M.E. and PLAXTON, W.C. (1993) Metabolic adaptations of plant respiration to nutritional phosphate deprivation. *Plant Physiol.* **101**, 339–344.

THEODOROU, M.E. and PLAXTON, W.C. (1994) Induction of PPi-dependent phosphofructokinase by phosphate starvation in seedlings of Brassica nigra. *Plant Cell Environ.* **17**, 287–294.

TRETHEWEY, R.N. and AP REES, T. (1994) A mutant of *Arabidopsis thaliana* lacking the ability to transport glucose across the chloroplast envelope. *Biochem. J.* **301**, 449–454.

VON SCHAEWEN, A., STITT, M., SCHMIDT, R., SONNEWALD, U. and WILLMITZER, L. (1990) Expression of a yeast-derived invertase in the cell wall of tobacco and *Arabidopsis* plants leads to accumulation of carbohydrate and inhibition of photosynthesis and strongly influences growth and phenotype of transgenic tobacco plants. *EMBO J.* **9**, 3033–3044.

WALKER, J.L.A. and HUBER, S.C. (1989) Regulation of sucrose-phosphate synthase activity in spinach leaves by protein level and covalent modification. *Planta* **117**, 116–120.

WANG, F., SANZ, A., BRENNER, M.L. and SMITH, A. (1993) Sucrose synthase, starch accumulation, and tomato fruit sink strength. *Plant Physiol.* **101**, 321–327.

WEBER, H., BORISJUK, L, HEIM, U., BUCHNER, P. and WOBUS, U. (1995) Seed coat-associated invertases of fava bean control both unloading and storage functions: cloning of cDNAs and cell type-specific expression. *Plant Cell* **7**, 1835–1846.

WEINER, H. (1995) Antibodies that distinguish between the serine-158 phospho- and dephospho-form of spinach leaf sucrose-phosphate synthase. *Plant Physiol.* **108**, 219–225.

WORRELL, A.C., BRUNEAU, J.-M., SUMMERFELT, K., BOERSIG, M. and VOELKER, T.A. (1991) Expression of a maize sucrose phosphate synthase in tomato alters leaf carbohydrate partitioning. *Plant Cell* **3**, 1121–1130.

XU D.P., SUNG, S.S., LOBODA, T., KORMANIK, P.P. and BLACK, C.C. (1989) Characterization of sucrolysis via the uridine diphosphate and pyrophosphate-dependent sucrose synthase pathway. *Plant Physiol.* **90**, 635–642.

YANG, N.S. and RUSSELL, D. (1990) Maize sucrose synthase-1 promoter directs phloem cell specific expression of GUS gene in transgenic tobacco plants. *Proc. Natl Acad. Sci.* **87**, 4144–4148.

ZRENNER, R. (1993) Klonierung und funktionelle Analyse von Genen kodierend für am Saccharosestoffwechsel der Kartoffel beteiligte Proteine. PhD diss., Freie Univ. Berlin.

ZRENNER, R., SALANOUBAT, M., WILLMITZER, L. and SONNEWALD, U. (1995) Evidence of the crucial role of sucrose synthase for sink strength using transgenic potato plants (Solanum tuberosum L.). *Plant J.* **7**, 97–107.

ZRENNER, R., KRAUSE, K.-P., APEL, P. and SONNEWALD, U. (1996) Reduction of the cytosolic fructose-1,6-bisphosphatase in transgenic potato plants limits photosynthetic sucrose biosynthesis with no impact on plant growth and tuber yield. *Plant J.* **9**, 671–681.

4

Modulation of starch synthesis

JACK PREISS

4.1 Introduction

Starch is the most dominant storage polysaccharide in plants and is widely distributed, being present in all major organs of most higher plants and practically every type of tissue: leaves, pollen, grains, fruit, roots, shoots and stems. In storage tissues such as endosperm or tuber, starch can be 65–90% of the dry weight of the tissue. The starch is organized in granules and its size and shape varies with the species, organ and the stage of development.

In the leaf, starch is synthesized in the chloroplast during the day via CO_2 fixation during photosynthesis and degraded at night by respiration. At night, leaf starch is a major source of sucrose for the rest of the plant as suggested by the starch-deficient mutants of *Arabidopsis thaliana*. Caspar *et al.* (1985) showed that the growth rate of null-starch mutants was greatly reduced if grown in a day–night regimen, but normal growth was observed in continuous-light. In the light the mutant was able to synthesize sucrose which could then be transported from leaf to sink tissues while in the dark the mutant had no carbon reserve to synthesize sucrose. Thus, leaf starch is not just a passive overflow product of photosynthesis as it fulfils an important function in dark metabolism.

Starch in non-photosynthetic tissues is synthesized in the amyloplast during the developmental phase of the organ and usually synthesis ceases when the tissue reaches maturity. Mobilization of the storage starch occurs at a different physiological state. In sprouting of tubers or in seed germination starch is usually converted to sucrose for growth of the whole plant. In tomato fruit and banana, starch synthesis occurs during maturation of the fruit. When the fruit starts ripening the starch is mobilized to form sucrose and monosaccharides such as glucose and fructose ('soluble solids'). Thus, biosynthesis and degradation of starch in leaves is a more dynamic process compared to the similar processes in reserve tissue or fruit.

The starch granule is composed essentially of two types of polysaccharides, termed amylose and amylopectin. Amylose is usually composed of linear chains of about 840 to 22 000 α-D-glucopyranosyl residues linked by α-(1–>4) bonds (molecular weight 136 000 to 3.5×10^6). The number of anhydroglucose units varies quite widely with plant species and stage of development. A fraction of the amylose is branched to a small extent, about $1\to 6$ α-D-glucopyranose is found at one per 170 to 500 glucosyl units. In contrast, amylopectin, which usually comprises about 70% of the starch granule, is highly branched with about 4 to 5% of the glucosidic linkages being $\alpha\text{-}1\to 6$.

Amylopectin molecules are very large flattened disks consisting of α-(1,4)-glucan chains joined by frequent α-(1,6)-branch points. The adjacent branch structures in amylopectin may form double helices that are in an organized crystalline structure (Oostergetel and van Bruggen, 1993). Models of amylopectin structure which have been proposed by Robin et al. (1974), Manners and Matheson (1981) and by Hizukuri (1986, 1995) are those that best fit the experimental data available and are discussed in a thorough review by Morrison and Karkalas (1990).

The purpose of this chapter is to describe some properties of the enzymes involved in starch biosynthesis, its regulation and to present the present state of research aimed at understanding the details of starch biosynthesis. A working hypothesis of the specific functions of the starch biosynthetic enzymes involved in synthesis of both amylose and amylopectin is presented. The possibilities for modification of quantity and quality (i.e. modification of structure) of starch by genetic manipulation of plants are briefly discussed.

A number of reviews on starch biosynthesis have recently appeared which touch on many of the topics discussed in this chapter (Ball, 1995; Martin and Smith, 1995; Müller-Röber and Kossmann, 1994; Preiss, 1996; Preiss and Sivak, 1996; Sivak and Preiss, 1995; Smith et al., 1995).

4.2 Enzymes involved in the biosynthesis of starch

4.2.1 The accepted biosynthetic pathway

It is generally believed that the major route to starch biosynthesis involves three reactions. The first reaction, catalyzed by adenosine diphosphate-glucose pyrophosphorylase (ADPGlcPPase, EC 2.7.7.27), is the synthesis of the glucosyl donor, adenosine diphosphate-glucose (reaction 4.1). The second reaction, catalyzed by starch synthase (EC 2.4.1.21), transfers the glucosyl unit of ADPGlc to the non-reducing end of an α-1,4 glucan primer to form a new α-1,4 glucosidic bond. This reaction then continues, elongating further the glucan chain (reaction 4.2). The synthesis of the α-1,6 branch linkages found in amylopectin are catalyzed by the branching enzyme (EC 2.4.1.18; reaction 4.3). These three reactions are needed to synthesize all the glucosidic linkages found in the starch granule. However, it is interesting to note that the maize endosperm *sugary* 1 mutation, which does not affect the expression of the genes coding for the three activities, results in a significant reduction in starch granule formation (Pan and Nelson, 1984). Conversely, a water-soluble α-1,4 glucan, phytoglycogen, increases to the point that the total polysaccharide content approaches that of normal maize. Thus, another activity is required to complete formation of the starch granule. Pan and Nelson (1984) showed that the maize endosperm *Su* 1 mutation was defective in debranching enzyme activity. Recently, *Su* 1 was cloned (James *et al.*, 1995) and sequence analysis of its cDNA showed that it had strongest homology with a bacterial isoamylase (Yang *et al.*, 1996). It is believed therefore, that the *Su* 1 gene is the structural gene for debranching enzyme activity and it is needed for formation of the finished amylopectin product. Thus, a fourth enzyme would be needed for converting the branched product synthesized by the branching enzyme into amylopectin (reaction 4.4) which can then crystallize, trapping the amylose to form the starch granule.

$$\text{ATP} + \alpha\text{-glucose-1-P} \rightleftharpoons \text{ADP-glucose} + \text{PPi} \tag{4.1}$$

$$\text{ADP-glucose} + \text{glucosyl}_n \longrightarrow \text{ADP} + \text{glucosyl}_{n+1} \tag{4.2}$$

$$\text{Linear } \alpha\text{-1,4 glucan chain} \longrightarrow \text{Branched } \alpha\text{-1,4-}\alpha\text{-1,6 glucan chain} \tag{4.3}$$

$$\text{Branched } \alpha\text{-1,4-}\alpha\text{-1,6 glucan chain} \longrightarrow \text{Amylopectin} \tag{4.4}$$

4.2.2 Evidence for the pathway

A number of reports have suggested the possibility that plant UDP-glucose-specific starch synthases and starch phosphorylases may be involved in starch synthesis. However the high K_m values for their substrates (UDP-glucose and glucose-1-P respectively) as compared to their concentration in the relevant cellular compartments, argues against a significant role in starch biosynthesis. In addition, the synthesis of UDP-glucose, at least in some starch synthesizing plant tissues, only occurs in the cytosol and not in the amyloplast. No transport of UDP-glucose in the plastid has been reported. Moreover, phosphorylase catalyzes an equilibrium reaction in cells having Pi concentrations in excess of glucose-1-P. Thus, it is most likely that phosphorylase plays a degradative role in starch degradation rather than in its synthesis.

Data from a number of genetic and biochemical studies indicate that the ADP-glucose pathway, involving the reactions described above, is very important for starch synthesis. Mutants of maize endosperm *shrunken*-2 and *brittle*-2 (Dickinson and Preiss, 1969a,b; Tsai and Nelson, 1966), deficient in ADPGlcPPase activity are also deficient in starch. Smith *et al.* (1989) have shown that a pea line having recessive *rb* genes (the gene controlling the level of ADPGlcPPase activity in developing pea embryos) containing 3–5% of the ADPGlcPPase activity, had only 38–72% of the starch found in the normal pea line. In *Arabidopsis thaliana*, Lin *et al.* (1988a) isolated a mutant containing less than 2% of the starch seen in the normal strain and less than 2% of the ADPGlcPPase activity. Immunoblots indicated that the enzyme was absent from the *Arabidopsis* extracts. In potato tuber, Müller-Röber *et al.* (1992) expressed a chimeric gene encoding antisense RNA for the ADPGlcPPase small subunit which caused a reduction in enzymatic activity to 2–5% of the normal levels and this led to a reduction in starch content.

Thus, in four different plant systems, a reduction of ADPGlcPPase activity led to a reduction in starch accumulation. Alternatively, an increase in ADPGlcPPase activity was achieved by transformation of the potato tuber with a mutant *E. coli* ADPGlcPPase gene that is insensitive to the regulatory effectors of the plant enzyme (Stark *et al.*, 1992). This increased the potato tuber starch content by 30–60%. This suggested not only the important role of the

ADPGlcPPase in starch synthesis but also that the enzyme activity is normally rate-limiting. Other data showing a relationship between activity of the ADPGlcPPase and starch accumulation in other plant species have been previously reviewed (Okita, 1992; Preiss, 1988; Preiss and Levi, 1980; Preiss and Sivak, 1996; Sivak and Preiss, 1995). Thus, the ADPGlcPPase and the subsequent reactions utilizing ADP-glucose is the dominant route for starch synthesis in plants and ADP-glucose synthesis is perhaps rate-limiting.

4.3 Properties of ADP-glucose pyrophosphorylase

The properties of the plant and bacterial ADPGlcPPases such as subunit structure, substrate and allosteric kinetics, determination of the allosteric and substrate binding sites, have been extensively reviewed (e.g. Preiss, 1982, 1984, 1991; Preiss and Romeo, 1994; Preiss and Sivak, 1996; Sivak and Preiss, 1995) and the reader is referred to these reviews. Some of the data will be summarized below.

4.3.1 Subunit structure

Many bacterial ADPGlcPPases have been purified and in almost all cases their subunit structure has been determined. Invariably the native enzymes have been found to be tetrameric with only one subunit. The molecular masses of the bacterial ADPGlcPPase subunits range from 49 to 54 kDa. In contrast, the plant tissue ADPGlcPPases are tetrameric in subunit structure but are composed of two different subunits which are products of different genes. One subunit, designated the small subunit, has a molecular mass of about 50–54 kDa while the other subunit, labelled as the large subunit, has a molecular mass of 51–60 kDa. The molecular masses depend on the source of the enzyme. The potato tuber, spinach leaf and maize endosperm enzymes have small subunit masses of 50, 51 and 54 kDa, respectively, and large subunit masses of 51, 54 and 60 kDa, respectively. The small and large subunits have about 50–60% identity with each other and have about 30–40% identity with the procaryotic ADPGlcPPases. Because of the identity between the large and small subunits and the procaryotic subunit it has been hypothesized that the large and small subunits have evolved from the bacterial enzyme subunit via gene duplication and divergence. As will be shown later, there is evidence to suggest that the plant small subunit is the catalytic subunit and the large subunit is a regulatory subunit.

4.3.2 Regulatory properties

The major allosteric activator of the plant ADPGlcPPase is 3-phosphoglycerate (PGA) and the major inhibitor, Pi. By 1982, ADPGlcPPases from several species, 13 from leaf and 9 from non-photosynthetic tissues, were shown to be activated by PGA (Preiss, 1982). PGA also causes the enzyme to have higher affinity for its substrates, ATP and glucose 1-P, and reverses the inhibition caused by the inhibitor, Pi. In the non-photosynthetic tissues studied, e.g. maize endosperm, potato tuber, cassava root and recently, rice endosperm (Zhang et al., 1996), the ADPGlcPPase activity is highly dependent on the presence of PGA and is inhibited by Pi.

However, it has been reported that pea embryo (Hylton and Smith, 1992) and barley endosperm ADPGlcPPases are not highly activated by PGA (Kleczkowski et al., 1993). Activation of 1.5- to 3-fold were observed. Also activation of the bean cotyledon ADPGlcPPase by PGA was only 1.5-fold if the activity was measured in the pyrophosphorolysis direction (Weber et al., 1995). It has been shown that both plant and bacterial ADPGlcPPases are usually much less activated by allosteric activators in the pyrophosphorolysis direction than in the synthesis direction (Ghosh and Preiss, 1966; Preiss et al., 1966). Thus, higher activation for the bean cotyledon enzyme may be observed in the synthesis direction which is the physiological direction. It was also believed at first, that maize endosperm ADPGlcPPase was insensitive to PGA activation and Pi inhibition (Dickinson and Preiss, 1969a,b). However, it was found later, that if protease inhibitors were added to the maize endosperm extracts the ADPGlcPPase activity was very sensitive to activation by PGA and to inhibition by Pi (Plaxton and Preiss, 1987). It was also shown that in the absence of protease inhibitor, the 54 kDa subunit was reduced to 53 kDa. Thus partial proteolysis during enzyme isolation can strongly affect ADPGlcPPase regulatory properties, and proteolysis, which can be prevented by including appropriate inhibitors in the extraction buffer, may be the reason for allosteric insensitivity found for the pea embryo, barley endosperm and other ADPGlcPPases. The properties of the proteolyzed and non-proteolyzed enzymes can be very different (Dickinson and Preiss, 1969a,b; Plaxton and Preiss, 1987) where the allosteric effects of the proteolyzed enzyme become minimal.

Many experiments have been cited in reviews (Preiss, 1982, 1988, 1991; Preiss and Levi, 1980) which show a direct correlation between the concentration of PGA and starch accumulation, and an inverse one between Pi concentration and starch content. This is true for photosynthetic tissues, where Pi and PGA concentrations within the chloroplast are good indicators of energy and carbon status, and in this way the regulation of ADPGlcPPase provides a good mechanism for modulating flux of photosynthate into starch. Although it has been found that the regulatory properties of the enzyme of non-photosynthetic tissue such as potato tuber and maize endosperm are such that the ADPGlc synthesis is almost completely dependent on the presence of activator, it is still uncertain how PGA and Pi are signals of the availability of carbon and energy for starch synthesis since transport of carbon in the amyloplast is via glucose metabolites rather than by triose-phosphates as seen in chloroplasts (Hill and Smith, 1991; Neuhaus et al., 1993; Tyson and ap Rees, 1988). It is still possible that PGA in the amyloplast is an indicator of carbon excess and Pi concentration is an indicator of the energy status.

Large effects can be seen in Figure 4.1 on the rate of ADP-glucose synthesis with relatively small changes in the PGA and Pi concentrations, particularly at low concentrations of PGA where the activation is minimal, and in the presence of Pi. At 1.2 mM Pi and 0.2 mM PGA, ADPGlc synthesis is inhibited by more than 95%. However if the Pi concentration decreases by 33% to 0.8 mM and the PGA concentration increases by 50% to 0.3 mM, there is an 8.5-fold

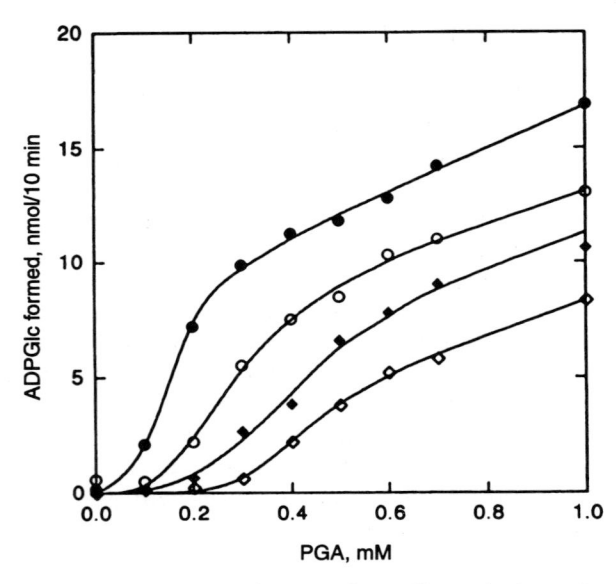

Figure 4.1 The effect of Pi and PGA on the rate of ADPGlc synthesis catalyzed by potato tuber ADPGlcPPase. -●- represents the PGA curve, generated in the presence of 0.4 mM Pi, -○- represents 0.8 mM Pi, -◆- represents 1.2 mM Pi and -◇- represents 1.6 mM Pi

increase in the rate of ADPGlc synthesis. Conversely, at 0.4 mM PGA and 0.8 mM Pi, the rate of ADPGlc synthesis is 7.5 nmol per 10 min. This is reduced to 2.2 nm (70% decrease) if only the PGA concentration decreases 50% to 0.2 mM. If the Pi concentration also increases to 1.2 mM, the synthetic rate is reduced to 0.65 nmol which is a reduction of ADPGlc synthesis of 91%. The reason for the small changes in the effector concentrations giving such large effects in the synthetic rate is due to the sigmoidal nature of the curves, particularly at the low concentrations of PGA.

4.3.3 *Evidence for* in vivo *regulation of ADPGlcPPase*

Some evidence has been obtained suggesting that the ADPGlcPPase is an important regulatory enzyme for starch synthesis *in vivo*. In the Kacser–Burns control and analysis methods (Kacser, 1987; Kacser and Burns, 1973) enzyme activity is varied, either by using mutants deficient in that enzyme or by varying the physiological conditions, and correlating the effect of these changes on the rate of a metabolic process, for example, starch synthesis. If the enzyme activity is rate limiting or regulatory of the metabolic process, then a large effect on that process should be seen. Conversely, if there is no effect, the enzyme level of activity is not considered to be rate-limiting for the metabolic process being measured. The effect is measured as a flux coefficient ratio. If the ratio correlates well with variation of enzyme activity with change in rate of the process measured, a ratio close to one should be observed.

This analysis on photosynthate partitioning was done by Stitt's group (Neuhaus and Stitt, 1990; Neuhaus et al., 1989). It was shown that in *Arabidopsis thaliana*, the leaf ADPGlcPPase is a major site of regulation for starch synthesis (Neuhaus and Stitt, 1990) and that regulation of the enzyme by PGA is an important determinant of the rate of starch synthesis *in vivo* (Neuhaus et al., 1989). *A. thaliana* mutant strains containing only 7% of the normal activity of ADPGlcPPase and a hybrid strain between the mutant and normal strain having 50%, had 90% and 39% reduction in the starch synthetic rate respectively, at a high level of light, as compared to the wild type (Neuhaus and Stitt, 1990). Thus, there is fairly good correlation between the activity of the ADPGlcPPase and the rate of starch synthesis. The flux control coefficient was determined to be 0.64.

Despite the fairly high value seen for the ADPGlcPPase flux control coefficient it still may be under-estimated due to the allosteric properties of the enzyme. In flux control analysis the maximal enzyme activity is measured. In the case of an allosteric enzyme the potential maximal enzyme activity may not be as critical as the allosteric effector concentrations that establish the enzyme activity. Therefore, one may not observe with an allosteric enzyme a valid flux control coefficient just based on potential maximum activity. With ADPGlcPPases, activation by PGA can be anywhere from 10- to 100-fold. Moreover, inhibition by the allosteric inhibitor, Pi and variations in the [PGA]/[Pi] ratio could cause greater fluctuations in the potential maximal activity. Thus, a flux coefficient control value based on only the potential maximal activities of the *Arabidopsis thaliana* mutants and normal ADPGlcPPases can underestimate the regulatory potential of the ADPGlcPPase step.

In experiments using a mutant of *Clarkia xantiana* deficient in leaf cytosolic phosphoglucoseisomerase (with only 18% of the activity seen in the wild type) sucrose synthetic rates were lower and the rate of starch synthesis was increased (Neuhaus et al., 1989). The chloroplast concentration of PGA increased about twofold, suggesting that the increase of starch synthetic rate measured in the mutant deficient in cytosolic phosphoglucoseisomerase is due to activation of the ADPGlcPPase by the increased PGA concentration and the PGA/Pi ratio.

The best type of evidence to determine whether the *in vitro* activation of the ADPGlcPPase is truly functional *in vivo* would come from isolation of a class of mutants where the mutation directly affects the allosteric properties of the ADPGlcPPase. Such mutants were found easily for the bacteria, *E. coli* and *Salmonella typhimurium* (Preiss, 1984, 1996; Preiss and Romeo, 1994). Recently, such types of mutants have been found for *Chlamydomonas reinhardtii* and for maize endosperm. A significant finding was made by Ball et al. (1991) who isolated a starch deficient mutant of *C. reinhardtii* in which the defect was shown to be in the ADPGlcPPase, which could not be effectively activated by PGA. The inhibition by Pi was similar to the wild type (unpublished results). The starch deficiency was observed in the mutant whether the organism was grown photoautotrophically with CO_2 or in the dark with acetate as the carbon source. Thus, the allosteric mechanism seems to be operative for photosynthetic or non-photosynthetic starch biosynthesis.

Recently, another putative ADPGlcPPase allosteric mutant isolated from a mutant maize endosperm, which had 15% more dry weight than the normal endosperm, has been described (Giroux et al., 1996). The mutant allosteric ADPGlcPPase was less sensitive to Pi inhibition than the normal enzyme. Thus, the *Chlamydomonas* starch deficient mutant and higher dry weight maize endosperm mutant studies strongly suggest that the *in vitro* regulatory effects observed with the photosynthetic and non-photosynthetic plant ADPGlcPPases are highly functional *in vivo* and that ADPGlc synthesis is rate-limiting for starch synthesis.

Obviously data continue to accumulate showing the importance of the plant ADPGlcPPase in the regulation of starch synthesis, and that 3PGA and Pi are important allosteric effectors *in vivo*, in photosynthetic as well as in non-photosynthetic plant tissue.

4.3.4 *Function of the higher plant ADPGlcPPase subunits*

Two cDNAs encoding the mature large subunit and small subunits of the potato tuber (*Solanum tuberosum* L.) ADPGlcPPase have been expressed in *Escherichia coli* (Ballicora et al., 1995; Iglesias et al., 1993). The large subunit and small subunits could be expressed separately as well as together. As seen in Table 4.1, considerable activity of ADPGlcPPase is obtained when the cDNA of the large subunit is expressed along with the cDNA of the small subunit enzyme in an *E. coli* mutant devoid of ADPGlcPPase activity.

If the large subunit is expressed alone very little activity is observed. Thus, for optimal activity, the small and large subunits must be expressed together. However, expression of the small subunit alone gave rise to significant ADPGlcPPase activity (Ballicora et al., 1995). The activity of this enzyme has been purified almost to homogeneity with a specific activity of 50 µmol min^{-1} mg^{-1} in the presence of 4 mM PGA. The purified recombinant enzyme, having both the large and small subunits, has a specific activity of 64 µmol min^{-1} mg^{-1} in the presence of 3 mM PGA. As shown in Table 4.2, the small subunit activity has a lower apparent affinity ($A_{0.5}$ = 2.4 mM) for the activator, PGA, than the enzyme having both the large and small subunits. The enzyme with only the small subunit is also more sensitive to Pi inhibition ($I_{0.5}$ of 0.08 mM in the presence of 3 mM PGA) as compared to the heteromeric enzyme ($I_{0.5}$ value of 0.63 mM). The K_m values for the substrates and Mg^{2+} are essentially the same whether the enzyme is composed of only one subunit, the small subunit, or two

Table 4.1 Expression of the potato tuber ADPGlcPPase subunits in *E. coli* ADPGlcPPase deficient mutant, AC70R1 extracts. The plasmids containing the cDNAs encoding the large subunit and the procedures used for expression have been described (Ballicora et al., 1995) and are pMLaugh10, encoding the small subunit and pMON17336, encoding the large subunit

Subunits expressed	Enzyme activity (µmol min^{-1} mg^{-1})
None	< 0.0003
Small + large	1.8
Small	0.17
Large	0.002

Table 4.2 The properties of the transgenic ADPGlcPPases with the native potato tuber enzyme. The ADPGlc synthesis assay has been described by Iglesias et al. (1993). The data for the native potato tuber enzyme are from Sowokinos and Preiss (1982). $A_{0.5}$ and $I_{0.5}$ are concentrations of activator needed for 50% of maximal activation and concentrations of inhibitor giving 50% inhibition, respectively

Source of enzyme	$A_{0.5}$	$I_{0.5}$		K_m at 3 mM PGA		
		0.25 mM PGA	3.0 mM PGA	ATP	Glc1P	Mg^{2+}
	(mM)	(mM)	(mM)	(mM)	(mM)	(mM)
pMLaugh10 + pMON17336 (large and small subunits)	0.16	0.07	0.63	0.12	0.04	2.0
Potato tuber	0.40	0.12	0.33	0.19	0.14	*
pMLaugh10 (small subunit alone)	2.4	–	0.08	0.20	0.03	2.2

*not determined

subunits, small and large (Table 4.2). In both cases the native enzyme is a tetramer: a homotetramer in the case of the small subunit alone and a heterotetramer in the case of the large and small subunits (Ballicora et al., 1995).

These data suggest that the small subunit is primarily involved in catalysis. It has substantial activity in the absence of the large subunit if the concentration of PGA, the activator, is high. The large subunit, when expressed alone, has very little activity, but if expressed with the small subunit, the resulting enzyme has similar regulatory kinetic constants as seen for the native potato enzyme. This suggests that the prime function of the large subunit is to regulate the activity of the small subunit. It increases the apparent affinity for the activator and decreases the affinity for inhibition by Pi. These results are in agreement with the results showing that the *Arabidopsis* mutant ADPGlcPPase lacking the large subunit still had activity, except that it had lower affinity for the activator, PGA, and higher affinity for Pi than the wild-type heterotetrameric enzyme (Li and Preiss, 1992).

As previously shown (Smith-White and Preiss, 1992), the small subunit of the higher plant ADPGlcPPases is highly conserved (85–95% identity), while the large subunit is less conserved (50–60% identity). Possibly the differences seen in the large subunit sequence reflect differences in modulating small subunit sensitivity to allosteric activation and inhibition. Expression of various large subunits could differ during development or in their occurrences in different plants and tissues (e.g. leaf, stem, guard cells, tuber, endosperm, root, embryo) and provide ADPGlcPPases with differing sensitivities to regulation.

4.3.5 *Identification of the glucose-1-phosphate and ATP binding sites*

The enzyme contains ligand-binding sites for the activator, PGA, and inhibitor, Pi, as well as for the binding sites for the substrates, ATP and glucose-1-P. It is possible that the effector and substrate binding sites may be located on different subunits. Chemical modification has been used to obtain information on the catalytic and allosteric effector sites (Ball and Preiss, 1994; Charng et al., 1994; Morell et al., 1988; Smith-White and Preiss, 1992). In addition, site-directed mutagenesis of residues to be involved in substrate binding have been confirmed in a number of studies (Charng et al., 1994, 1995; Hill et al., 1991; Kumar et al., 1989; Sheng et al., 1996).

These studies have provided information on the catalytic and regulatory sites of the spinach ADPGlcPPase and on the role of the large and small subunits. They have also shown that many of the studies initiated with the bacterial ADPGlcPPases are highly relevant for studies on the higher plant enzyme (Charng et al., 1994; Hill et al., 1991; Kumar et al., 1989; Sheng et al., 1996).

Lys residue 195 of the *E. coli* ADPGlcPPase has been shown to be the binding site for the phosphate of the substrate, glucose-1-P (Hill et al., 1991) and tyrosine residue 114 has been identified as part of the binding site for the adenosine portion of the substrate, ATP (Lee and Preiss, 1986). The amino acid sequence identity of the *E. coli* enzyme when aligned with the plant and cyanobacterial ADPGlcPPases ranges from 30 to 33% (Smith-White and Preiss, 1992). There is greater sequence identity though when the *E. coli* ATP and glucose-1-P binding sites (Table 4.3) are compared with the aligned sequences of the plant and cyanobacterial enzymes, suggesting that those sequences may be important in the plant enzyme, with the same function. A recent result with the potato tuber

Table 4.3 The glucose-1-phosphate (Hill et al., 1991) and the ATP (Preiss and Romeo, 1994) binding sites of *E. coli* ADPGlcPPase. References to sequences of the plant ADPGlcPPases are in Smith-White and Preiss (1992). Sequences for the *Anabaena* enzyme are in Charng et al. (1992), for the *Synechocystis* enzyme in Kakefuda et al. (1992) and for the wheat endosperm small subunit, in Ainsworth et al. (1993). Lys195 and Tyr114 of the *E. coli* enzyme are the Glc-1-P and ATP binding sites, respectively. * signifies the same amino acid as in the *E. coli* enzyme

Source	Glucose-1-P site	ATP site
Prokaryotes	195	114
E. coli	I I E F V E K P – A N	W Y R G T A D A V
S. typhimurium	* * D * * * * _ * *	* * * * * * * * *
Anabaena	V * D * S * * * K G E	* F Q * * * * * *
Synechocystis	* T D * S * * * Q G E	* F Q * * * * * *
Plant small subunit		
Spinach leaf 51 kDa	* * * * A * * * K G E	* F Q * * * * * *
Potato tuber 50 kDa	* * * * A * * * Q G E	* F Q * * * * * *
Rice seed (small subunit)	* V * * A * * * K G E	* F Q * * * * * *
Maize endosperm (54 kDa)	* * * * A * * * K G E	* F Q * * * * * *
A. thaliana (small subunit)	* * * * A * * * K G E	* F Q * * * * * *
Wheat endosperm (small subunit)	* * * * A * * * K G E	* F Q * * * * * *
Plant large subunit		
Spinach leaf 54 kDa	V L S * S * * * K G D	* F Q * * * * * *
Potato tuber 51 kDa	V V Q * A * * * K G F	* F Q * * * * * *
Maize endosperm 60 kDa	V L Q * F * * * K G A	* F Q * * * * S I
Wheat endosperm (large subunit)	V V Q * S * Q * K G D	* F R * * * * * W

ADPGlcPPase expressed in *E. coli* (Iglesias *et al.*, 1993), site-directed mutagenesis on the lysine residue K198 of the 50 kDa subunit (equivalent to the *E. coli* ADPGlcPPase K195) to a glutamate residue, increased the K_m for glucose-1-P from 40 µM to over 7.7 mM without any large change in the K_m or K_a for the other substrates, Mg^{2+}, ATP or for the activator, PGA or K_i for the inhibitor, Pi (Fu and Preiss, unpublished results). These results indicate involvement of the equivalent Lys residue of the plant ADPGlcPPase in the binding of glucose-1-P. In the case of the ATP binding site, instead of Tyr there is a Phe residue in the corresponding sequences of the plant and cyanobacterial enzymes. Future site-directed mutagenesis and chemical modification studies are in progress to see if the WFQGTADAV region of the plant enzyme is indeed a portion of the ATP binding site or whether conservative changes of two amino acids in the sequence have affected the function of that portion of the protein.

4.3.6 Identification of the activator sites

Chemical modification with pyridoxal phosphate (Table 4.4) indicated a Lys residue close to the C-terminus of the spinach leaf ADPGlcPPase being important for PGA activation (Morell *et al.*, 1988). When pyridoxal-P is covalently bound, the plant ADPGlcPPase does not require PGA for activation. The PLP binding is also prevented by the allosteric effectors, PGA and Pi. These observations, that the modified enzyme no longer needs activator for maximal activity and covalent modification is prevented by the presence of allosteric effectors, are strong evidence that PLP, the activator analogue, is binding at the activator site. Preiss *et al.* (1992) and Ball and Preiss (1994) showed also that at least two Lys residues of the spinach leaf large subunit are also involved or close to the binding site of pyridoxal-P and, presumably, of the activator, PGA (Table 4.4). The chemical modification of these two Lys residues (sites 1 and 2) by pyridoxal-P was prevented by the presence of PGA and Pi during the reductive pyridoxylation process. Similar results were obtained with the *Anabaena* ADPGlcPPase (Charng *et al.*, 1994). The modified Lys residue was identified as Lys419 and the sequence adjacent to that residue is very similar to that observed for site 1 sequences of the higher plants. Site-directed mutagenesis of Lys419 to either Arg, Ala, Gln, or Glu, produced mutant enzymes with lowered affinities, 25- to 150-fold lower than that of wild-type enzyme. No other kinetic constants for substrates and the inhibitor, Pi, were appreciably affected, nor was the heat stability or the catalytic efficiency of the enzyme affected. These mutant enzymes were still activated to a great extent at higher concentrations of PGA suggesting that an additional site was involved in the binding of the activator. The Lys419 Arg mutant was chemically modified with PLP. Modification of Lys382 in the Arg mutant was observed and also caused a dramatic alteration in the allosteric properties of the enzyme which could be prevented by the modification process. Thus, Lys382 was identified as the additional site involved in the binding of the activator and as seen in Table 4.4, the surrounding sequence about Lys382 in the *Anabaena* enzyme is very similar to that seen for site 2 of the higher plants.

Site 1 corresponds to the lysyl residue near the C-terminus, Lys440, that is phosphopyridoxylated in the spinach leaf small subunit (Morell *et al.*, 1988), and

Table 4.4 Plant and cyanobacterial ADP-glucose pyrophosphorylase activator binding sites. The sequences listed in one-letter code are from Smith-White and Preiss (1992) and from references indicated in Table 4.3. The sequences of the barley endosperm enzyme are from Villand et al. (1992). The Lys residues underlined indicate that they are covalently modified by pyridoxal-P and the chemical modification of the Lys residue is protected by PGA and Pi or that site-directed mutagenesis has identified them to be involved in binding of the activator. The number 440 corresponds to the Lys residue in the spinach leaf ADPGlcPPase small subunit. Site 1 is present both in the large and in the small subunits of the plant ADPGlcPPase while site 2 is only in the large subunit even though very similar sites are observed in the small subunit

	Activator site 1	Activator site 2
	440	
Spinach 51 kDa (small)	S G I V T V I K DAL I P S G T V I	I K R A I I D K N A R
Potato tuber 50 kDa	S G I V T V I K DAL I P S G I I I	I K R A I I D K N A R
Maize 54 kDa	G G I V T V I K DALL P S G T V I	I R R A I I D K N A R
Wheat seed (small)	S G I V T V I K DALL P S G T V I	I K R A I I D K N A R
Anabaena	S G I V V V L K N A V I T D G T I I	Q R R A I I D K N A R
Synechocystis	N G I V V V I K N V T I A D G T V I	I R R A I I D K N A R
Spinach 54 kDa (large)	S G I T V I F K Q A T I K D G V V	I K D A I I D K N A R
Potato 51 kDa (large)	S G I I I I L E K A T I R D G T V I	I R K C I I D K N A K
Maize 60 kDa (large)	S G I V V I L K N A T I N E C L V I	I R N C I I D M N A R
Wheat seed (large)	S G I V V I Q K N A T I K D G T V V	I Q N C I I D K N A R
Barley endosperm (large)	S G I V V I Q K N A T I K D G T V V	I S N C I I D M N A R

corresponds to Lys468 in the rice seed small subunit and Lys441 in the potato tuber ADPGlcPPase small subunit. Site 2 is also situated close to the C-terminus, equivalent to Lys382 in the *Anabaena* ADPGlcPPase and Lys417 of the potato tuber large subunit. Table 4.4 also shows that the amino acid sequence of the spinach leaf small subunit peptide containing the modified lysyl residue of site 1 is highly conserved in the rice seed, potato tuber, maize (Bae et al., 1990) and wheat endosperm small subunits (Ainsworth et al., 1993) and the *Anabaena* (Charng et al., 1992) and *Synechocystis* (Kakefuda et al., 1992) ADPGlcPPase subunits. Similarly, the amino acid sequence of site 2 of the spinach leaf large subunit is highly conserved in the large subunits of the potato tuber, maize (Bhave et al., 1990) and barley endosperm (Villand et al., 1992), wheat seed (Olive et al., 1989) and wheat leaf ADPGlcPPases (Olive et al., 1989). These comparisons suggest that from the lysyl residues modified via reductive phoshopyridoxylation, those in sites 1 and 2 are highly conserved and, therefore, more likely to be essential for the binding of the activator, PGA.

Recently, cDNA clones encoding the putative mature forms of the large and small subunits of the potato tuber ADPGlcPPase have been expressed together, using compatible vectors, in an *E. coli* mutant devoid of ADPGlcPPase activity (Ballicora et al., 1995; Iglesias et al., 1993). The ADPGlcPPase activity expressed displayed catalytic and allosteric kinetic properties very similar to the ADPGlcPPase purified from potato tuber (Ballicora et al., 1995). This expression system has been used to characterize further the allosteric function of the lysyl residues identified via chemical modification with pyridoxal-P of the spinach enzyme. Site directed mutagenesis of Lys441 of the potato ADPGlcPPase small subunit to Glu and Ala results in mutant enzymes affected in their activation by PGA (Preiss *et al.*, 1995). Thus, as shown for Lys440 of the spinach ADPGlcPPase small subunit via chemical modification, Lys441 of the small subunit of the potato enzyme appears to be important for the binding of the activator.

4.4 Are starch synthases or branching enzymes rate limiting in starch synthesis?

Can starch synthase and branching enzyme be rate limiting under certain physiological conditions? The wrinkled pea has a reduced starch content: about 66–75% of that seen in the round seed. The amylose content is about 33% in the round normal form and is 60–70% in the wrinkled pea seed. The activities of several enzymes involved in starch metabolism were measured in wrinkled pea at four different developmental stages (Edwards *et al.*, 1988). It was found that branching enzyme activity was, at its highest, only 14% of that seen for the round seed in the wrinkled seed. Other starch biosynthetic enzymes and phosphorylase had similar activities in the wrinkled and round seeds. Smith (1988) showed that the *r* (*rugosus*) lesion (as found in the wrinkled pea of genotype *rr*) was associated with the absence of one isoform of branching enzyme. Edwards *et al.* (1988) suggested that the reduction in starch content observed in the mutant seeds was caused indirectly by the reduction in SBE activity through an effect on the starch synthase. The authors proposed that, in the absence of branching enzyme activity, the starch synthase forms an α-1->4-glucosyl elongated chain which becomes a poor glucosyl-acceptor (primer) for the starch synthase substrate, ADPGlc, thus decreasing the rate of α-1->4-glucan synthesis. A previous study of rabbit muscle glycogen synthase (Carter and Smith, 1978) showed that continuous elongation of the outer chains of glycogen causes it to become an ineffective primer, and decreases the apparent activity of the glycogen synthase. ADPGlc in the wrinkled pea accumulated to higher concentrations than in round or normal pea. This was considered evidence that activity of the starch synthase was less *in vivo* than observed in *in vitro* conditions, where a suitable primer like amylopectin or glycogen was added and starch synthase activity in the wrinkled pea was equivalent to that found in the wild type. Smith *et al.* (1989) showed that in mutant *rr* leaves, in high light intensity there was a 40% decrease in the rate of starch synthesis. Control coefficient analysis (Smith *et al.*, 1990) showed essentially no effect on the rate of starch synthesis in low light intensity while in high light intensity a small value of the flux control coefficient (meaning very little control) of 0.13 was obtained. An 86% reduction of branching enzyme activity had a small effect on regulation of starch synthesis. It should be noted that the SBE control coefficient was only 20% of the value found for the ADPglucose pyrophosphorylase which was 0.64 (Neuhaus and Stitt, 1990).

At temperatures above 30°C both maize (Singletary *et al.*, 1994) and wheat endosperm (Hawker and Jenner, 1993; Jenner, 1994; Keeling *et al.*, 1993, 1994) had a reduction of starch deposition compared to lower temperatures. In wheat the starch biosynthetic enzyme affected was soluble starch synthase (Hawker and Jenner, 1993; Jenner, 1994; Keeling *et al.*, 1993, 1994). Using flux control coefficient analysis, Keeling *et al.* (1993) showed a control coefficient of 1.15 between rate of starch synthesis and the level of starch synthase activity in the wheat endosperm extracts. *In vitro*, the endosperm starch synthase activity was sensitive to heat treatment in the range of 30°C to 40°C if the treatment was for longer than 15 min. A similar study with maize endosperm showed a reduction of starch synthetic rate and a decrease in starch synthase activity in the heat-stressed maize endosperm (Singletary *et al.*, 1994). However, in the heat-stressed maize, the endosperm ADPGlcPPase activity was also reduced and to an even greater extent than the soluble starch synthase (Singletary *et al.*, 1994).

Thus, in wheat and maize there may be a relationship, under some environmental conditions, between reduction of starch synthase activity and decreased starch synthesis. However, as the maize data suggest (Singletary *et al.*, 1994), other factors besides starch synthase activity, not yet studied, may be the primary reason for the reduction of starch synthesis in the heat stressed plants. In the case of maize endosperm, another enzyme involved in starch synthesis, ADPGlcPPase, is also affected in the heat-stressed plant. It is also quite possible that other critical steps leading to starch biosynthesis are also affected in both plants, such as carbon flow (sucrose transport?) from source to sink tissues. These processes were not studied in the heat-stressed plants. Thus, it is premature on the basis of the above experiments to designate starch synthase as a major control point and other data should be obtained. Flux control coefficients for an enzyme can only be determined for any process if only that enzyme's activity is affected. In the case of the heat stressed plants it has not yet been shown that only the starch synthase activity is affected. Could starch synthetic rate be increased by over-expressing soluble starch synthase activity in the amyloplast? As shown below starch accumulation can be increased by expressing a bacterial ADPGlcPPase allosteric mutant in plants (Stark *et al.*, 1992).

4.5 Increasing starch content by transforming plants with an allosteric mutant ADPGlcPPase gene

A preponderance of evidence strongly suggests that a rate-limiting and regulatory enzyme of starch biosynthesis in algae or higher plants is ADPGlcPPase. Also, control analysis experiments have shown that ADPGlcPPase is rate limiting and important for regulation of leaf starch synthesis (Neuhaus and Stitt, 1990). Reduced ADPGlcPPase activity in mutants also leads to a reduction in the rate of starch synthesis in *Arabidopsis* leaves (Lin *et al.*, 1988a,b) and in potato tubers (Müller-Röber *et al.*, 1992).

It was of interest to see if starch content in a plant could be augmented by increased expression of activity of one of the enzymes involved in starch biosynthesis. Over-expression of a plant ADPGlcPPase activity however, would require expression of two distinct genes to reconstitute its ADPGlcPPase activity. It is also possible that the plant would compensate for the over-expression by altering the ratio of the effector metabolites, PGA and Pi, to a value at which starch synthesis would not be elevated. A different strategy was therefore chosen. An *E. coli* ADPGlcPPase, *glg* C gene of allosteric mutant 618, labelled as *glg* C16 (Leung *et al.*, 1986), which encodes an enzyme independent of the presence of activator for activity, was used for the transformation. Expression of this bacterial mutant gene would have two advantages. First, only one gene needs be expressed for ADPGlcPPase activity and, second, the mutant enzyme would be less sensitive to inhibition by its allosteric inhibitor, 5'AMP, insensitive to the plant enzyme's inhibitor, Pi, and independent of activator for good activity (Leung et al., 1986). Since starch synthesis occurs in the plastid, a nucleotide sequence encoding the transit peptide of the Arabidopsis ribulose 1,5-bisphosphate carboxylase chloroplast transit peptide was fused to the translation initiation site of the glg C16 gene (Figure 4.2). The chimeric gene was then cloned behind either a tuber-specific patatin promoter or a cauliflower mosaic virus (CaMV) enhanced 35S promoter, or, in the case of tomato plants, the Arabidopsis plant promoter from the rbcS gene (Stark et al., 1992; Figure 4.2). The chimeric gene containing promoter was placed in a cloning vector with a 35S-neomycin phosphotransferase gene as a selectable marker (Stark et al., 1992) and used for transformation of various plants (Stark et al., 1992).

Starch levels were increased over the controls lacking the glg C gene product by about 1.7–8.7-fold in tobacco calli where the glgC gene product activity was detected (Stark et al., 1992). When the CaMV-chimeric gene was electroporated into tobacco protoplasts (Table 4.5), extracts of the transformed protoplasts gave rise to ADPGlc synthesis resistant to Pi inhibition and activated by fructose 1,6 bis-P. Almost all plant ADPGlcPPases are most sensitive to inhibition by Pi and ADPGlc synthesis in the control protoplast extract was totally inhibited by Pi as expected. Comparison of transgenic tobacco with control calli shows a very large increase in the number of starch granules (Stark et al., 1992). Similarly, tomato shoots excised from the transformed calli containing the transit peptide-glg C16 gene, stained black with I_2 while the controls were essentially negative (Stark et al., 1992).

Similar results have been obtained for transformed Russet–Burbank potato tubers where the chimeric gene with transit peptide under control of a tuber-specific patatin promoter, increased starch in the tuber 25–60% over controls not containing the bacterial enzyme (Stark et al., 1992; Table 4.6). Bacterial ADPGlcPPase glg C16 gene expressed in the tuber lacking the transit peptide

		Cleavage site	Additional cleavage site	
Promotor	Arabidopsis Rubisco small subunit transit peptide	23 amino acids of N-terminus of Rubisco	ADPGlcPPase glg C16 gene	Nos Terminator

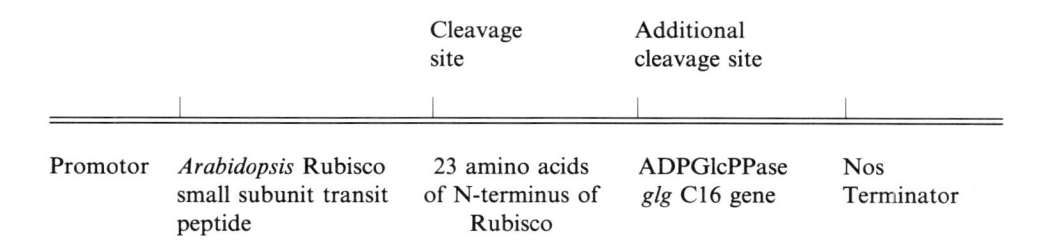

Figure 4.2 Synthetic promoter-plastid transit peptide-glg C16 ADPGlcPPase gene. The chimeric gene is composed of the Arabidopsis thaliana chloroplast transit peptide portion of the ribulose P_2 carboxylase gene modified to have an extra cleavage site to eliminate the 23 amino acids of the N-terminal of the small subunit (Stark et al., 1992) to prevent its possible interference with the catalytic or regulatory activity of the glg C16 gene product. The Nos terminator is the nopaline synthase 3' poly A signal. The promoter can be either a constitutive promoter or a tissue specific one as indicated in the text

Table 4.5 Activation and inhibition of ADPGlcPPase synthesis in transgenic tobacco protoplasts containing glg C16[a]

Protoplast extracts from	ADPGlc synthesized (nmol)
Non-transformed cells + 10 mM inorganic phosphate	0.0
Transformed cells + 2.5 mM Fru1,6P$_2$	20.2
Transformed cells + 2.5 mM Fru1,6P$_2$ + 10 mM inorganic phosphate	18.0
Transformed cells + 20 mM PGA	18.4
Transformed cells + 10 mM inorganic phosphate	6.4

[a]Data from Stark et al. (1992)

Table 4.6 Starch content of transgenic potato tubers with the glg C16 and glg C genes[a]

		Average starch content (% wet wt)
A.	Control; untransformed	12.3 ± 1.15
	Transit peptide-glg C16	16.0 ± 2.00
	glg C16, no transit peptide	12.4 ± 0.24
B.	Control; untransformed	13.2 ± 0.12
	Transit peptide-glg C	13.1 ± 0.07

[a]Data from Stark et al. (1992)

gene portion, gave no increase in starch content (Table 4.6). ADPGlcPPase activity was expressed but presumably was not present in the amyloplast and was not able to supply ADPGlc to the amyloplast localized starch synthases. When wild-type *E. coli* glg C gene was used for transformation, no increase in starch was observed (Stark et al., 1992). Thus, the allosteric properties of the ADPGlcPPase in the normal situation are important for regulation and alteration of the allosteric properties was needed to increase starch levels.

A relationship between the expression levels of the ADPGlcPPase of glg C16 as measured by Western blotting of the potato extracts and the increase in starch content was observed, particularly in tubers at lower range of starch content (Stark et al., 1992). With lower levels of expressed ADPGlcPPase, increases of 21–63% in starch were seen. Intermediate levels of the expressed ADPGlcPPase gave 33–118% increases in starch and the high expressed levels of the transit peptide-glg C16 gave increases of 33–167%. Thus, increasing starch synthesis is possible by transforming the tuber with an ADPGlcPPase having allosteric properties minimized to permit higher rates of ADPGlc synthesis under physiological conditions.

These results also strongly suggest that the ADPGlcPPase is a rate limiting enzyme for starch synthesis even in non-photosynthetic plant tissues. Further studies are currently being carried out to determine the relationship between increased enzyme activity (due to the transformation of the bacterial gene) and rate of starch synthesis by feeding labelled glucose or sucrose to potato slices. The possibility of other genes involved in starch metabolism being indirectly affected by the *glg* C16 transformation is also being investigated.

It is conceivable therefore, that similar methods can be used to change, in addition to quantity, starch quality via expression/transformation of the isoforms of starch synthase and branching enzymes in plants. These 'new starches' may have greater usefulness in food and industrial processes. The production of modified 'specialty' starches via molecular biology techniques is promising.

4.6 A hypothesis assigning specific roles for starch synthase, branching isozymes and debranching enzyme in *in vivo* synthesis of amylopectin

The cluster model of amylopectin structure as postulated by Hizukuri (1986) is the currently accepted structure. A feature of the model is the clustering of the α-1,6 linkage branch points in certain regions of the amylopectin. What specific roles would the starch synthase (SS) and branching enzyme (SBE) isozymes have in forming the crystalline starch granule and amylopectin structures? Why starch granules from different species differ in their size, number per cell and composition may be related to the specificities of the starch synthases in chain elongation and in the branching enzymes in chain size transfer in the site of α-1,6 formation. Recently maize endosperm SBEI and SBEII isozymes have been studied with respect to their mode of chain transfer, their substrate specificity and possible roles in the synthesis of amylopectin (Guan and Preiss, 1993; Guan *et al.*, 1995; Takeda *et al.*, 1993). SBEI can transfer long chains (DP 40 to >100) while SBEII transfers shorter size chains (DP 6 to 14). The preferred substrate for SBEI is amylose while the preferred substrate for SBEII is amylopectin. It was postulated that SBEI was more involved in synthesis of the interior B chains while SBEII were involved in the synthesis of exterior A and shorter B chains. When expressed in *E. coli*, maize SBEI transferred longer chains than SBEII while SBEII preferably transferred shorter chains.

In order to understand the various functions of the different starch synthases, Ball and associates isolated various mutants of *Chlamydomonas* deficient in starch synthase activities: a soluble starch synthase II (SSSII) deficient mutant (Fontaine *et al.*, 1993) and a double mutant deficient both in granule starch-bound starch synthase (GBSS) and in SSSII (Maddelin *et al.*, 1994). The SSSII mutant contained only 20–40% of the starch seen in the wild-type alga and the amylose portion of the total starch increased from 25% to 55%. This mutant also contained a modified amylopectin which had an increased amount of very short chains (2 to 7 DP) and a concomitant decrease of intermediate size chains (8 to 60 DP). This suggested that the SSSII was involved in the synthesis or maintenance of the intermediate size chains (mainly B chains) in amylopectin. The higher amylose content could be explained because of the failure of the SSSII mutant to make extended chains.

The double mutants defective in SSSII and a GBSS (Maddelin *et al.*, 1994) had an even lower starch content, 2% to 16% of the wild-type, and there was an inverse correlation with the severity of the GBSS defect of the double mutant and the amount of starch present in the double mutant. It is suggested that the GBSS is required to form the basic internal structure of the amylopectin and that these effects of the GBSS absence are exacerbated due to the diminished SSSII activity.

These *Chlamydomonas* mutant studies have been quite informative in that they provide good evidence for involvement of the GBSS in amylopectin synthesis as well as in amylose synthesis and suggest that an important function for SSSII would be in its involvement in synthesis of the intermediate size (β) branches in amylopectin.

A tentative scheme of how amylose and amylopectin are synthesized is proposed in Figure 4.3, based on the information discussed above.

It is quite possible that initiation of α-1,4-glucan synthesis occurs via synthesis on an acceptor protein (Tandecarz and Cardini, 1978). It still remains to be determined whether only UDPGlc or both UDPGlc and ADPGlc are the glucosyl donors in this process. This step is the least known of all the reactions of starch synthesis. After formation of the unbranched maltodextrin of undetermined size, there would be a path leading towards amylopectin synthesis. A higher rate of polysaccharide formation would occur at the surface of the developing granule, where SSSII and starch branching enzyme I and GBSS are now interacting with the maltodextrin/amylose product synthesized to form a branched long chain polysaccharide also containing intermediate size chains.

This hypothesis is based on the studies of the polysaccharide structures

1. Initiation of synthesis of unbranched maltodextrins (bound to protein?)

    GBSS + SSSII + SBEI
2. Unbranched maltodextrin + ADPGlc -------------> Long, intermediate size chains of glucan

    SBEII + SSSI
3. Long, intermediate size chain glucan + ADPGlc -------------> Synthesis of A and shorter B chains to finish cluster structure
4. Repeat of reactions 2 and 3 to form the complete pre-amylopectin polysaccharide

    Debranching enzyme
5. Pre-amylopectin structure ------------> amylopectin + pre-amylose chains

    GBSS
6. Pre-amylose chains ------------> amylose

Figure 4.3 A proposed scheme for synthesis of amylose and amylopectin. Initiation may involve synthesis of a maltodextrin attached covalently to a protein. This putative protein-α-glucan can then accept glucose from ADPGlc either via GBSS catalysis to form an amylose structure or in combination with SBEI, SSSII and (possibly GBSS) form a polysaccharide having the internal structure of the final amylopectin product. SBEII and SSSI carry out the reactions to form the exterior of the amylopectin structure. The enlargement of the amylopectin could proceed further by continuing participation of SBEI, SSSII and possibly GBSS by repeat of reactions 2 and 3. Production of amylopectin in reaction 5 is caused by debranching enzyme which also generates oligosaccharide chains which are elongated by GBSS to form the amylose fraction

synthesized by the *Chlamydomonas* SSSII and GBSS mutants as well as the *ae* mutants of rice (Mizuno *et al.*, 1993) and maize (Boyer and Preiss, 1981) which are defective in SBEII. SBEII deficient mutants have altered oligosaccharides with fewer branches and longer chains. SSSI and SBEII then become involved in the synthesis of the α-chains and exterior β-chains to produce an amylopectin precursor, pre-amylopectin or phytoglycogen, which is more highly branched and water soluble and non-crystalline. A debranching enzyme (an isoamylase?) then debranches the pre-amylopectin to form amylopectin which is now able to crystallize. The chains which are liberated by debranching from the pre-amylopectin are used as primers by GBSS in reaction 6 to form amylose. In the synthesis of amylose only GBSS is involved and at the site of amylose, inside the granule, branching enzyme activity is restricted. The amylose would only be slightly branched and possibly that branching had been done previously before the debranching reaction 5. In other words, the primer for GBSS was already branched.

The postulation of a water soluble pre-amylopectin polysaccharide is based on the effects of the *sugary* 1 mutation observed in maize endosperm which has reduced amylopectin content, starch granules, and accumulates up to 35% of its dry weight, a highly branched water soluble polysaccharide, referred to as phytoglycogen (Pan and Nelson, 1984). It has been shown that the *sugary* 1 mutation has a debranching enzyme activity deficiency (Pan and Nelson, 1984). That the *sugary* 1 mutation is the structural gene for a debranching enzyme is supported by the isolation of a cDNA of the *su* 1 gene and by the deduced amino acid sequence similarity with an *E. coli* isoamylase (James *et al.*, 1995). It remains to be shown whether the *su* 1 gene product debranching enzyme activity is actually an isoamylase, or pullulanase or an R enzyme.

These reactions do not have to occur in perfect sequence. In other words, there could be an overlap of reactions 2, 3 and 4 and possibly even 5 and 6. However, the present evidence (intermediate products formed by the various starch synthase and SBE mutants of *Chlamydomonas* and of the higher plants) does suggest that this is the sequence leading towards amylopectin and amylose biosynthesis. Further experiments are certainly required to test this hypothesis scheme. At least the initial results obtained with mutants and biochemical experiments support this initial proposal.

References

AINSWORTH, C., TRAVIS, M. and CLARK, J. (1993) Isolation and analysis of a cDNA clone encoding the small subunit of ADP-glucose pyrophosphorylase from wheat. *Plant Mol. Biol.* **23**, 23–33.

BAE, J.M., GIROUX, J. and HANNAH, L.C. (1990) Cloning and characterization of the *brittle*-2 gene of maize. *Maydica* **35**, 317–322.

BALL, S. (1995) Recent views on the biosynthesis of the plant starch granule. *Trends Glycosci. Glycotechnol.* **7**, 405–415.

BALL, K.L. and PREISS, J. (1994) Allosteric sites of the large subunit of the spinach leaf adenosine diphosphate glucose pyrophosphorylase. *J. Biol. Chem.* **269**, 24706–24711.

BALL, S., MARIANNE, T., DIRICK, L., FRESNOY, M., DELRUE, B. and DECQ, A. (1991) A *Chlamydomonas reinhardtii* low-starch mutant is defective for 3-phosphoglycerate activation and orthophosphate inhibition of ADP-glucose pyrophosphorylase. *Planta* **185**, 17–26.

BALLICORA, M.A., LAUGHLIN, M.J. FU, Y., OKITA, T.W., BARRY, G.F. and PREISS, J. (1995) ADPglucose from potato tuber. Significance of the N-terminal of the small subunit for catalytic properties and heat stability. *Plant Physiol.* **109**, 245–251.

BHAVE, M.R., LAWRENCE, S., BARTON, C. and HANNAH, L.C. (1990) Identification and molecular characterization of *shrunken*-2 cDNA clones of maize. *Plant Cell* **2**, 581–588.

BOYER, C.D. and PREISS, J. (1981) Evidence for independent genetic control of the multiple forms of maize endosperm branching enzymes and starch synthases. *Plant Physiol.* **67**, 1141–1145.

CARTER, J. and SMITH, E.E. (1978) Actions of glycogen synthase and phosphorylase of rabbit-skeletal muscle on modified glycogens. *Carbohydrate Res.* **61**, 395–406.

CASPAR, C., HUBER, S.C. and SOMERVILLE, C. (1985) Alterations in growth, photosynthesis and respiration in a starch mutant of *Arabidopsis thaliana* (L.) deficient in chloroplast phosphoglucomutase activity. *Plant Physiol.* **79**, 1–7.

CHARNG, Y.-Y., KAKEFUDA, G., IGLESIAS, A.A., BUIKEMA, W.J. and PREISS, J. (1992) Molecular cloning and expression of the gene encoding ADP-glucose pyrophosphorylase from Cyanobacterium *Anabaena* sp. strain PCC 7120. *Plant Mol. Biol.* **20**, 37–47.

CHARNG, Y.-Y., IGLESIAS, A.A. and PREISS, J. (1994) Structure-function relationships of cyanobacterial ADP-glucose pyrophosphorylase: site-directed mutagenesis and chemical modification of the activator-binding sites of ADP-glucose pyrophosphorylase from *Anabaena* PCC 7120. *J. Biol. Chem.* **269**, 24107–24113.

CHARNG, Y.-Y., SHENG, J. and PREISS, J. (1995) Mutagenesis of an amino acid residue in the activator-binding site of ADP-glucose pyrophosphorylase causes alteration in activator specificity. *Arch. Biochem. Biophys.* **318**, 476–480.

DICKINSON, D.B. and PREISS, J. (1969a) ADPglucose pyrophosphorylase from maize endosperm. *Arch. Biochem. Biophys.* **130**, 119–128.

DICKINSON, D.B. and PREISS, J. (1969b) Presence of ADP-glucose pyrophosphorylase in *shrunken*-2 and *brittle*-2 mutants of maize endosperm. *Plant Physiol.* **44**, 1058–1062.

EDWARDS, J., GREEN, J.H. and AP REES, T. (1988) Activity of branching enzyme as a cardinal feature of the Ra locus in *Pisum sativum*. *Phytochemistry* **27**, 1615–1620.

FONTAINE, T., D'HULST, C., MADDELEIN, M.-L., ROUTIER, F., PEPIN, T.M., DECQ, A., WIERUSZESKI, J.-M., DELRUE, B., VAN DEN KOORNHUYSE, N., BOSSU, J.-P., FOURNET, B. and BALL, S. (1993) Toward an understanding of the biogenesis of the starch granule. Evidence that Chlamydomonas soluble starch synthase II controls the synthesis of intermediate size glucans of amylopectin. *J. Biol. Chem.* **268**, 16223–16230.

GIROUX, M.J., SHAW, J., BARRY, G., COBB, B.G., GREENE, T., OKITA, T. and HANNAH, L.C. (1996) A single mutation that increases maize seed weight. *Proc. Nat. Acad. Sci. USA* **93**, 5824–5829.

GHOSH, H.P. and PREISS, J. (1966) Adenosine diphosphate glucose pyrophosphorylase. A regulatory enzyme in the biosynthesis of starch in spinach chloroplasts. *J. Biol. Chem.* **241**, 4491–4505.

GUAN, H.P. and PREISS, J. (1993) Differentiation of the properties of the branching isozymes from maize (Zea mays). *Plant Physiol.* **102**, 1269–1273.

GUAN, H., KURIKI, T., SIVAK, M. and PREISS, J. (1995) Maize branching enzyme catalyzes synthesis of glycogen-like polysaccharide in glgB-deficient *Escherichia coli*. *Proc. Natl Acad. Sci. USA* **92**, 964–967.

HAWKER, J.S. and JENNER, C.F. (1993) High temperature affects the activity of enzymes in the committed pathway of starch synthesis in developing wheat endosperm. *Austr. J. Plant Physiol.* **20**, 197–209.

HILL, L.M. and SMITH, A.M. (1991) Evidence that glucose-6-phosphate is imported as the substrate for starch synthesis by the plastids of developing pea embryos. *Planta* **185**, 91–96.

HILL, M.A., KAUFMANN, K., OTERO, J. and PREISS, J. (1991) Biosynthesis of bacterial glycogen: mutagenesis of a catalytic site residue of ADPglucose pyrophosphorylase from *Escherichia coli*. *J. Biol. Chem.* **266**, 12455–12460.

HIZUKURI, S. (1986) Polymodal distribution of the chain lengths of amylopectins and its significance. *Carbohydrate Res.* **147**, 342–347.

HIZUKURI, S. (1995) Starch: analytical aspects. In: *Carbohydrates in Food* (ELIASON, A.-C., ed.), pp. 347–429. Marcel Dekker, New York, Basel, Hong Kong.

HYLTON, C. and SMITH, A.M. (1992) The *rb* mutation of peas causes structural and regulatory changes in ADPGlc pyrophosphorylase from developing embryos. *Plant Phsyiol.* **99**, 1626–1634.

IGLESIAS, A.A., BARRY, G.F., MEYER, C., BLOKSBERG, L., NAKATA, P.A., GREENE, T., LAUGHLIN, M.J., OKITA, T.W., KISHORE, G.M. and PREISS, J. (1993) Expression of the potato tuber ADP-glucose pyrophosphorylase in *Escherichia coli*. *J. Biol. Chem.* **268**, 1081–1086.

JAMES, M.G., ROBERTSON, D.S. and MEYERS, A.M. (1995) Characterization of the maize gene *sugary* 1, a determinant of starch composition in kernels. *Plant Cell* **7**, 417–429.

JENNER, C.F. (1994) Starch synthesis in the kernel of wheat under high temperature conditions. *Austr. J. Plant Physiol.* **21**, 791–806.

KACSER, H. (1987) Control of metabolism. In: *The Biochemistry of Plants*. Vol. 11 (DAVIES, D.D., ed.), pp. 39–67. Academic Press, New York.

KACSER, H. and BURNS, J.A. (1973) Control of flux. *Symp. Soc. Exp. Biol.* **27**, 65–107.

KAKEFUDA, G., CHARNG, Y.-Y., IGLESIAS, A.A., McINTOSH, L. and PREISS, J. (1992) Molecular cloning and sequencing of ADP-glucose pyrophosphorylase from *Synechocystis* PCC6803. *Plant Physiol.* **99**, 344–347.

KEELING, P.L., BACON, P.J. and HOLT, D.C. (1993) Elevated temperature reduces starch deposition in wheat endosperm by reducing the activity of soluble starch synthase. *Planta* **191**, 342–348.

KEELING, P.L., BANISADR, R., BARONE, L., WASSERMAN, B.P. and SINGLETARY, G.W. (1994) Effect of temperature on enzymes in the pathway of starch biosynthesis in developing wheat and maize grain. *Austr. J. Plant Physiol.* **21** 807–827.

KLECZKOWSKI, L.A., VILLAND, P., LÜTHI, E., OLSEN, O.-A. and PREISS, J. (1993) Insensitivity of barley endosperm ADPGlcPPase pyrophosphorylase to 3-phosphoglycerate and orthophosphate regulation. *Plant Physiol.* **101**, 179–186.

KUMAR, A., GHOSH, P., LEE, Y.M., HILL, M.A. and PREISS, J. (1989) Biosynthesis of bacterial glycogen: determination of the amino acid changes that alter the regulatory properties of a mutant *Escherichia coli* ADPglucose synthetase. *J. Biol. Chem.* **264**, 10464–10471.

LEE, Y.M. and PREISS, J. (1986) Covalent modification of substrate binding sites of *E. coli* ADPglucose synthetase: isolation and structural characterization of 8-azido ADPglucose incorporated peptides. *J. Biol. Chem.* **261**, 1058–1064.

LEUNG, P., LEE, Y.M., GREENBERG, E., ESCH, K., BOYLAN, S. and PREISS, J. (1986) Cloning and expression of the *Escherichia coli glgC* gene from a mutant containing an ADPglucose pyrophosphorylase with altered allosteric properties. *J. Bacteriol.* **167**, 82–88.

LI, L. and PREISS, J. (1992) Characterization of ADPglucose pyrophosphorylase from a starch-deficient mutant of *Arabidopsis thaliana* (L.). *Carbohydrate Res.* **227**, 227–239.

LIN, T.P., CASPAR, T., SOMERVILLE, C. and PREISS, J. (1988a) Isolation and characterization of a starchless mutant of *Arabidopsis thaliana* L. Henyh lacking ADPglucose pyrophosphorylase activity. *Plant Physiol.* **86**, 1131–1135.

LIN, T.P., CASPAR, T., SOMERVILLE, C. and PREISS, J. (1988b) A starch deficient mutant of *Arabidopsis thaliana* with low ADPglucose pyrophosphorylase activity lacks one of the two subunits of the enzyme. *Plant Physiol.* **88**, 1175–1181.

MADDELIN, M.-L., BELLANGER, F., DELRUE, B., LIBESSART, N., D'HULST, C., VAN DEN KOORNHUYSE, N., FONTAINE, T., WIERUSZESKI, J.-M., DECQ, A. and BALL, S. (1994) Genetic dissection of starch metabolism in the monocellular alga *Chlamydomonas reinhardtii*: determination of granule-bound and soluble starch synthases functions in amylopectin biosynthesis. *J. Biol. Chem.* **269**, 25150–25157.

MANNERS, D.J. and MATHESON, N.K. (1981) The fine structure of amylopectin. *Carbohydrate Res.* **90**, 99–110.

MARTIN, C. and SMITH, A. (1995) Starch biosynthesis. *Plant Cell* **7**, 971–985.

MIZUNO, K., KAWASAKI, T., SHIMADA, H., SATOH, H., KOBAYASHI, E., OKAMURA, S., ARAI, Y. and BABA, T. (1993) Alteration of the structural properties of starch components by the lack of an isoform of starch branching enzyme in rice seeds. *J. Biol. Chem.* **268**, 19084–19091.

MORELL, M., BLOOM, M. and PREISS, J. (1988) Affinity labeling of the allosteric activator site(s) of spinach leaf ADPglucose pyrophosphorylase. *J. Biol. Chem.* **263**, 633–637.

MORRISON, W.R. and KARKALAS, J. (1990) Starch. In: *Methods in Plant Biochemistry*, Vol. 2 (DEY, P.M., ed.), pp. 323–352. Academic Press, San Diego.

MÜLLER-RÖBER, B. and KOSSMAN, J. (1994) Approaches to influence starch quantity and starch quality in transgenic plants. *Plant Cell Environ.* **17**, 601–613.

MÜLLER-RÖBER, B.T., SONNEWALD, U. and WILLMITZER, L. (1992) Inhibition of ADPglucose pyrophosphorylase in transgenic potatoes leads to sugar-storing tubers and influences tuber formation and expression of tuber storage protein genes. *EMBO J.* **11**, 1229–1238.

NEUHAUS, H.E. and STITT, M. (1990) Control analysis of photosynthate partitioning: impact of reduced activity of ADPglucose pyrophosphorylase or plastid phosphoglucomutase on the fluxes to starch and sucrose in *Arabidopsis*. *Planta* **182**, 445–454.

NEUHAUS, H.E., KRUCKEBERG, A.L., FEIL, R. and STITT, M. (1989) Reduced activity mutants of phosphoglucose isomerase in the cytosol and chloroplast of *Clarkia xantiana* II. Studies of the mechanisms which regulate photosynthate partitioning. *Planta* **178**, 110–122.

NEUHAUS, H.E., HENRICHS, G. and SCHEIBE, R. (1993) Characterization of glucose-6-P incorporation into starch by isolated intact cauliflower-bud plastids. *Plant Physiol.* **101**, 573–578.

OKITA, T. (1992) Is there an alternative pathway for starch synthesis? *Plant Physiol.* **100**, 560–564.

OLIVE, M.R., ELLIS, R.J. and SCHUCH, W.W. (1989) Isolation and nucleotide sequence of cDNA clones encoding ADPglucose pyrophosphorylase polypeptides from wheat leaf and endosperm. *Plant Mol. Biol.* **12**, 525–538.

OOSTERGETEL, G.T. and VAN BRUGGEN, E.F.J. (1993) The crystalline domains in potato starch granules are arranged in a helical fashion. *Carbohydrate Polymers* **21**, 7–12.

PAN, D. and NELSON, O.E. (1984) A debranching enzyme deficiency in endosperms of the *sugary* 1 mutants of maize. *Plant Physiol* **74**, 324–328.

PLAXTON, W.C. and PREISS, J. (1987) Purification and properties of nonproteolytically degraded ADPglucose pyrophosphorylase from maize endosperm. *Plant Physiol.* **83** 105–112.

PREISS, J. (1982) Regulation of the biosynthesis and degradation of starch. *Annu. Rev. Plant Physiol.* **54**, 431–454.

PREISS, J. (1984) Bacterial glycogen and its regulation. *Annu. Rev. Microbiol.* **38** 419–458.

PREISS, J. (1988) Biosynthesis of starch and its degradation. In: *The Biochemistry of Plants*, Vol. 14 (PREISS, J., ed.), pp. 181–254. Academic Press, San Diego.

PREISS, J. (1991) Starch biosynthesis and its regulation. In: *Oxford Surveys of Plant Molecular and Cell Biology*, Vol. 7 (MIFLIN, B.J. ed.), pp. 59–114. Oxford University Press, Oxford.

PREISS, J. (1996) ADPglucose pyrophosphorylase: basic science and applications in biotechnology. In: *Biotechnology Annual Review* Vol. II (EL-GEWELY, M.R., ed.), pp. 259–279. Elsevier Science, Amsterdam.

PREISS, J. and LEVI, C. (1980) Starch biosynthesis and degradation. In: *The Biochemistry of Plants*, Vol. 3 (PREISS, J., ed.), pp. 371–423. Academic Press, New York.

PREISS, J. and ROMEO, T. (1994) Molecular biology and regulatory aspects of glycogen biosynthesis in bacteria. In: *Prog. Nucl. Acid Res. Molec. Biol.*, Vol. 47 (COHN, W.E. and MOLDAVE, K., eds), pp. 299–329. Academic Press, San Diego.

PREISS, J. and SIVAK, M.N. (1996) Starch synthesis in sinks and sources. In: *Photoassimilate Distribution in Plants and Crops, Source-Sink Relationships* (ZAMSKI, E. and SCHAFFER, A.A., eds), pp. 63–96. Marcel Dekker, New York, Basel, Hong Kong.

PREISS, J., SHEN, L., GREENBERG, E. and GENTNER, N. (1966) Biosynthesis of bacterial glycogen. IV. Activation and inhibition of the adenosine diphosphate glucose pyrophosphorylase of *Escherichia coli* B. *Biochemistry* **5**, 1833–1845.

PREISS, J., BALL, K., CHARNG, Y.Y. and IGLESIAS, A. (1992) Structure-function relationships of ADPglucose pyrophosphorylase regulatory sites and *in vivo* evidence that ADPglucose is synthesized only in the chloroplast via ADPglucose pyrophosphorylase. In: *Research in Photosynthesis*, Vol. III (MURATA, N., ed.), pp. 697–700. Kluwer Academic Publishers, Dordrecht, The Netherlands.

PREISS, J., SHENG, J., FU, Y. and BALLICORA, M.A. (1995) Studies of the catalytic and regulatory sites of the cyanobacterial and higher plant ADP-glucose pyrophosphorylase; a regulatory enzyme of starch synthesis. In: *Proceedings of the Xth International Photosynthesis Congress*, Vol. 5 (MATHIS, P., ed.), pp. 47–52. Kluwer Academic Publishers, Dordrecht, The Netherlands.

ROBIN, J.P., MERCIER, C., CHARBONNIERE, R. and GUIBOT, A. (1974) Lintnerized starches. Gel filtration and enzymatic studies of insoluble residues from prolonged acid treatment of potato starch. *Cereal Chem.* **51**, 389–406.

SHENG, J., CHARNG, Y.-Y. and PREISS, J. (1996) Site-directed mutagenesis of Lysine^{382}, the activator binding site, of ADP-glucose pyrophosphorylase from *Anabaena* PCC 7120. *Biochemistry* **35**, 3115–3121.

SINGLETARY, G.W., BANISADR, R. and KEELING, P.L. (1994) Heat stress during grain filling in maize: effects on carbohydrate storage and metabolism. *Austr. J. Plant Physiol.* **21**, 829–841.

SIVAK, M.N. and PREISS, J. (1995) Starch synthesis in seeds. In: *Seed Development and Germination* (KIGEL, J. and GALILI, G. eds), pp. 139–168. Marcel Dekker, New York, Basel and Hong Kong.

SMITH, A.M. (1988) Major differences in isoforms of starch-branching enzyme between developing embryos of round and wrinkled-seeded peas (*Pisum sativum* L.). *Planta* **175**, 270–279.

SMITH, A.M., BETTEY, M. and BEDFORD, I.D. (1989) Evidence that the *rb* locus alters the starch content of developing pea embryos through an effect on ADPglucose pyrophosphorylase. *Plant Physiol.* **89**, 1279–1284.

SMITH, A.M., NEUHAUS, H.E. and STITT, M. (1990) The impact of decreased activity of starch-branching enzyme on photosynthetic starch synthesis in leaves of wrinkled-seeded peas. *Planta* **181**, 310–315.

SMITH, A., DENYER, K. and MARTIN, C.R. (1995) What controls the amount and structure of starch in storage organs? *Plant Physiol.* **107**, 673–677.

SMITH-WHITE, B.J. and PREISS, J. (1992) Comparison of proteins of ADP-glucose pyrophosphorylase from diverse sources. *J. Mol. Evol.* **34**, 449–464.

SOWOKINOS, J.R. and PREISS, J. (1982) Pyrophosphorylases in *Solanum tuberosum*. III. Purification, structural and catalytic properties of potato tuber ADP-glucose pyrophosphorylase. *Plant Physiol.* **69**, 1459–1466.

STARK, D.M., TIMMERMAN, K.P., BARRY, G.F., PREISS, J. and KISHORE, G.M. (1992) Role of ADPglucose pyrophosphorylase in regulating starch levels in plant tissues. *Science* **258**, 287–292.

TAKEDA, Y., GUAN, H.P. and PREISS, J. (1993) Branching of amylose by the branching isoenzymes of maize endosperm. *Carbohydrate Res.* **240**, 253–263.

TANDECARZ, J.S. and CARDINI, C.E. (1978) A two-step enzymatic formation of a glucoprotein in potato tuber. *Biochim. Biophys. Acta* **543**, 423–429.

TSAI, C.Y. and NELSON, O.E. (1966) Starch-deficient maize mutant lacking adenosine diphosphate glucose pyrophosphorylase activity. *Science* **151**, 341–343.

TYSON, R.H. and AP REES, T. (1988) Starch synthesis by isolated amyloplasts from wheat endosperm. *Planta* **175**, 33–38.

VILLAND, P., OLSEN, O.-A., KILIAN, A. and KLECZKOWSI, L.A. (1992) ADP-glucose pyrophosphorylase large subunit cDNA from barley endosperm. *Plant Physiol.* **100**, 1617–1618.

WEBER, H., HEIM, U., BORISJUK, L. and WOBUS, U. (1995) Cell-type specific, coordinate expression of two ADPglucose pyrophosphorylase genes in relation to starch biosynthesis during seed development in *Vicia faba* L. *Planta* **195**, 352–361.

YANG, H., LIU, M.Y. and ROMEO, T. (1996) Coordinate genetic regulation of glycogen catabolism and biosynthesis in Escherichia coli via the CSRA gene product. *J. Bacteriol.* **178**, 1012–1017.

ZHANG, Y., CHANTLER, S.E., GUPTA, S., ZHAO, Y., LEISY, D., HANNAH, L.C., MEYER, C., WESTON, J., WU, M.-X., PREISS, J. and OKITA, T.W. (1997) Molecular approaches to enhance rice productivity through manipulations of starch metabolism during seed development. In: *International Rice Research Institute Rice Genetics III*. PO Box 933, Manila (in press).

Molecular crosstalk and the regulation of C- and N-responsive genes

KAREN E. KOCH

Changes in gene expression can contribute to adjustment in the balance of C/N assimilation and storage balance in plants on a time scale that lies between that of rapid, fine tuning of metabolism, and long-term changes in morphology and development. Gene expression also integrates signals from these other levels of control (Koch, 1996). For short-term adjustments, it can be effectively argued that even rapid changes in gene expression, such as for NR (Crawford, 1995; Foyer et al., 1993; Vincentz et al., 1993), are likely to have less regulatory influence than direct effectors of fine control via allosteric modulation or enzyme phosphorylation (Foyer et al., 1993; Oaks, 1994; Quick and Schaffer, 1996). However, the ultimate amplitude of change over an extended period is greater at the level of gene expression, and thus provides a means of coarse control for extending the outside limits of more fine-tuned mechanisms for metabolic balance. In general, responses at the level of gene expression also reflect similar changes in enzyme regulation, and further amplify effects of these over time. Resulting influence on still longer-term changes in plant form can, in turn, favour acquisition versus utilization of C and N resources.

At least three special features are inherent in regulatory contributions at the level of gene expression relative to other means for adjusting C/N balance in plants.

1. Capacity for feed-forward, up-regulation of key genes is facilitated by their positive responses to reaction products of enzymes they encode. For C metabolism, this includes up-regulation of certain mRNAs for sucrose cleaving enzymes (*Ivr2* invertases and *Sus1* sucrose synthases in maize by hexose products of their own reactions; Koch et al., 1996; Xu et al., 1996), providing a means of increasing cellular capacity for sucrose import (Koch et al., 1996). A comparable mechanism for expansion of N-assimilation systems may lie in the positive responsiveness of specific glutamine synthetase genes (GS2 isozymes from several species) to amino acids produced during the GS reaction itself (Lam et al., 1996). Both instances involve potential modulation of pivotal steps in C- and N-assimilation, by a

self-amplifying (feed-forward) interface between gene expression and metabolic signals.

2. Progressive responses over an extended period can also contribute to C/N adjustment at the level of gene expression. Although some gene responses can occur within minutes (Crawford, 1995; Foyer et al., 1993; Sheen, 1990, 1994), metabolite induced changes are typically not evident for 8 to 12 hours or longer (Crawford, 1995; Geiger et al., 1996; Koch, 1996; Lam et al., 1996). These can continue to increase or decrease over an extended timespan, leading to ultimate physiological changes of significant magnitude.

This relatively slow overall response also provides a greater opportunity for integration of collective signal input over time. Among such signals would presumably be information on whether or not fine control systems had been operating at or near the limits of their dynamic range for a given period. The temporal progression of such responses is consistent with the positive features of a coarse control system as described by Farrar and Williams (1991). To this is added input from still longer term signals of plant morphological and developmental status (Geiger et al., 1996). Together these processes contribute to similar ends at the physiological level (Foyer et al., 1993), and have received increasing attention as joint effectors of assimilation, partitioning, and use of C- and N-resources in plants (Geiger et al., 1996; Koch, 1996; Pollock and Farrar, 1996).

3. Integration of signals from diverse sources is also facilitated at the level of gene expression. For C- and N-responsive genes, metabolites provide an important but not exclusive source of input. NO_3, NH_4, and P status are also sources of signals, and light is a strong environmental effector of numerous genes for N and C assimilation (see subsequent sections). Several of the light effects have been distinguished as responses not mediated exclusively through photosynthetic products alone (Lam et al., 1996). Other environmental signals having marked effects on expression of genes for C and N metabolism include drought stress, wounding, and pathogen responses (Koch, 1996; Lam et al., 1996). Plant growth regulators further influence the expression of many of these genes; these include cytokinins (especially for NR), ABA and methyl jasmonate (for storage proteins), auxin, ethylene, and others. Limitations imposed by overriding effects of intrinsic developmental signals may define windows of opportunity for signalling action (Bihn and Koch, 1997).

5.1 Metabolite responsive genes in higher plants

Specific changes in gene expression that can occur in response to C or N availability are examples of an ancient mechanism for essential adjustments to changing nutrient environments. Classic examples include the *lac* operon in *E. coli* (Saier, 1989), and the glucose-responsive genes in yeast (Carlson, 1987; Gancedo, 1992). N-modulated gene systems have also been investigated in the same organisms (Hinnebusch, 1984), as well as in algae and fungi (Huppe and Turpin, 1994). Evidence is now emerging for comparable and widespread responses among higher plant genes (Koch, 1996; Lam et al., 1996).

Special attributes of metabolite regulated gene expression in multicellular plants extend beyond their significance to molecular crosstalk (noted above). First, as sessile organisms, plants face extreme demands for adjustment to their environments. There are few other options for survival, so sensitivity and responses are often far greater than those observed in more mobile systems. A second and closely related feature is that plant systems are often less homeostatic, thus cells experience a greater variation of internal as well as external signals. Levels of C- and N-assimilates can vary markedly in plants compared to those in the mammalian bloodstream. These changes in metabolite levels could be a vital constituent of the ultimate mechanism for adjusting gene expression. Third, the structural complexity of multicelled plants requires long distance signals for co-ordination of responses by distant cells, tissues, and even organs to changing environments. In this context, transported sugars and amino acids become part of the environment for importing cells remote from sites of assimilation, and can provide valuable information on whole plant C and N status. Co-ordination of responses includes those of cells, tissues and organs specialized for key aspects of assimilation or use, and can progress to the level of long term developmental changes in plant form. C- and N-responsive genes can affect changes in plant development as well as resource partitioning among parts.

Resource balance can have marked effects on expression of genes for assimilation and use, especially for two such essential cellular constituents as C and N. Many aspects of their uptake and metabolism are closely interrelated, so it is not surprising that there should be extensive interface at the level of gene expression as well. Mechanisms for transducing signals of C and N availability may include changes in uptake, metabolites, calcium levels, kinase/phosphorylase activities, and a range of other effectors. Genetic and biochemical evidence indicates that signalling pathways may overlap to such a degree that the 'molecular crosstalk' between them creates an interactive system or network (Parks and Hangarter, 1994). There are typically multiple sources of input, shared steps, and branch points.

'Sensitivity' of a given signalling system can thus be regulated by downstream input from others. This is consistent with many aspects of plant signal perception *in vivo*, such as the importance of sugar levels and other 'critical molecules' to a range of light responses (Chorey, 1993). A point of downstream commonality is proposed in this instance and could readily affect similar changes in sensitivity of other sensing systems. An understanding of such interfaces also allows perturbations of one system to be manipulated at key points of interaction with a second. Alternatives may be useful either for applied ends, or for *in vivo* experimental purposes.

Actual signals regulating gene responses to C and N metabolism may be quite different from those most important to co-ordination of balance at the enzyme level. For example, initial data indicate that membrane signals may originate from sugars, but that actual transfer through the membrane may not be necessary to affect the sensing system (Stitt *et al.*, 1995). A similar situation may exist for N forms or N-assimilates. In addition, studies of strong metabolic regulators such as Fru2,6P_2 and ATP/ADP indicate that they may not have direct roles in altered gene expression (Stitt *et al.*, 1995). Furthermore, the reciprocal influence of glutamate and glutamine on protein kinases may well not extend to the level of gene expression (Vincentz *et al.*, 1993), despite the widespread influence of

Gene Responses to Shifts in C/N Balance	
↓C:↑N	↑C:↓N
▸ **Shoot development favoured**	▸ **Root development favoured**
▸ **Priority given to photosynthesis and remobilization of stored C**	▸ **Priority given to N-assimilation and remobilization of stored N**
• *Photosynthesis genes* · Rubisco (rbcS, rbcL) sug repr. N stim. · PEPcase sug repr. aa stim. · (lt. reactions) sug repr. N stim. · (chlorophyll) sug repr. N stim.	• *N-assimilation genes* · NR lt, sug stim. NO_3 stim., aa repr. · NiR lt stim. NO_3 stim., aa repr. · GS2 lt, sug stim. NO_3, aa stim., NH_4 repr. · GOGAT (fd) lt, sug stim. NO_3 stim., NH_4 repr. · GDH2 undef.

• *C-remobilization genes*
· α amylases sug repr. undef.
· (lipid breakdown) sug repr. undef.
· endoglycosidases sug repr. undef.
· proteases (non psyn) sug repr. undef.
· N-detox, cycling (non psyn) undef.
  - GDH1 sug repr. NH_4 stim.
  - ASN1 lt, sug repr. aa stim.
	• *N-remobilization genes*
	· proteases (psyn. cells) sug stim.

• *C-export genes* · SPS (leaf mesophyll) sug repr. undef. · SS (SH1, phloem) sug repr. undef. · GS1 (phloem) sug repr. NO_3, NH_4 stim.	• *N-export genes* · GS2 lt, sug stim. NO_3, aa stim., NH_4 repr. (senescing leaves)
▸ **N-sinks favoured**	▸ **C-sinks favoured**

• *N-import/metabolism*
· aa transporters? undef.
 	
• *N use*
· leaf storage proteins low sug N stim.
· Rubisco (storage) sug repr. N stim.
· leaf protease inhibitors
· ASN1 (ASN storage) lt, sug repr. aa, NH_4 stim.
	• *Sucrose import/cleavage*
	· sugar transporters sug stim. undef.
	· SUS1 and IVR2 sug stim. undef.
	
	• *C use*
	· starch formation
	  - ADPGlcPPase sug stim. NH_4, aa repr.
	  - SBE sug stim. undef.
	  - starch phosphorylase sug stim. undef.
	· sink storage proteins
	  - VSPs sug stim. N subordinate
	  - patatin sug stim. N subordinate
	  - sporamin sug stim. N subordinate
	  - seed storage sug stim. C/N coordinate
	· (lipid synthesis) sug stim. undef.
	· respiratory genes sug stim. undef.

Figure 5.1 Gene responses to shifts in C/N balance. Expression patterns are used to divide gene responses to C/N balance between those enhanced by relative decreases in C/N (comparative C depletion and N excess), and those up-regulated by relative increases in C/N (comparative C excess and N depletion). Note that these are shifts in C/N balance and not C/N ratios *per se*, which may be inherently high or low in given tissues. Also, the framework presented here is a

glutamine there (Lam *et al.*, 1996). The intricate mechanisms for fine control and balance in C and N metabolism may differ from those employed in the signalling systems for coarse regulation of relevant genes.

5.2 Gene responses to C/N balance and availability

Figure 5.1 divides plant gene responses to shifts in C/N balance on the basis of their possible biological significance, in an effort to provide a general framework for thought. In many organisms, a relative deficiency in either C or N resources tends to up-regulate genes facilitating its acquisition, whereas abundance favours expression of those for storage and use (Koch, 1996; Lam *et al.*, 1996). This view can be potentially useful in appraising gene responses ranging from those affecting uptake, assimilation and storage, to long-term alterations in plant form (e.g. roots vs. shoots). In some instances, a relative excess of either C or N can initiate varying degrees of deficiency responses in genes for acquisition of the other. Alternatively, sensitivity of genes to either C or N availability can sometimes be suppressed if signals indicate one of the two is present in relative excess. Developmental phase changes can also be involved and include alterations in flower induction, fruit set, juvenility, etc. (Bernier *et al.*, 1993; Koch, 1996). Programmes for reproductive development, however, once initiated, can have potentially overriding effects on signals of C and N availability (Koch, 1996; Xu *et al.*, 1996). So too can certain stresses; thus the overall outline presented here will hopefully be utilized as a non-exclusive one.

5.3 Genes up-regulated by low C/N, C-starvation, and/or N excess

Starvation induction often results from derepression of given genes as levels of preferred substrates for cellular metabolism drop. Preferred substrates tend to be those that can be acquired or metabolized at least cost, such as sugars vs. CO_2, and amino acids vs. NO_3 (see also the section on metabolite effectors). Photosynthesis is generally repressed in favour of heterotrophic growth by microorganisms, cell cultures, and even cells of intact plants. Functional significance is broader in multicelled organisms, however, and regulation is more complex. In higher plants these responses can aid optimization of protein investment in photosynthesis (typically high N cost) and facilitate a balanced distribution of N among plant parts and processes. In addition, localized changes in expression of starvation-responsive genes could prioritize vital cells or organs for import of C- (or possibly N-) assimilates during depletion stress (Koch and Nolte, 1995; Koch *et al.*, 1992). Export of essential resources from other cells or organs could also be stimulated by localized responses of C- or N-modulated genes (Thomas and Rodriguez, 1994).

---

general one, and not without exceptions (see text). Information is compiled from Koch (1996) and references therein, Lam *et al.* (1996) and references therein, unpublished data of U. Sonnewald, and citations appearing in Figure 5.2; protein products of genes are listed. Abbreviations are as follows: stim., stimulated; repr., repressed; undef., undefined; psyn, photosynthetic; lt, light; sug, sugar; aa, amino acid

Genes for photosynthesis are typically up-regulated by decreases in either carbohydrate levels or ratios of C-/N-assimilates, and provide a means of adjusting both. The mRNAs affected include those encoding the primary carboxylation enzymes in C_3 and C_4 plants, and extend to proteins for light-harvesting reactions (see Koch, 1996 for references). Nuclear as well as plastid genes can be affected, although changes in the latter are often slower (Criqui *et al.*, 1992; Shih and Goodman 1988; Van Oosten and Bessford, 1994).

Effects of assimilate C/N balance on photosynthetic genes are consistent with the high N cost of their expression (especially for Rubisco) and potential impact on N-partitioning (Koch, 1996; Sugiharto *et al.*, 1990). Structural components of photosynthesis (e.g. chlorophyll and Rubisco) are remobilized to a lesser degree when N-resources are abundant (Lam *et al.*, 1996), and is evident in the delayed yellowing of older leaves on N-rich plants. Conversely, repression of photosynthetic genes by sugars is enhanced when N resources are limited (Stitt *et al.*, 1995). In addition, although sugars repress PEPcarboxylase expression in maize leaves (Jang and Sheen, 1994; Sheen, 1990, 1994), these mRNAs can also be strongly induced by Gln (Sugiharto and Sugiyama, 1992).

The form of both N and C can also be important in the response of photosynthetic genes to C/N balance, particularly since sugars and amino acids can have markedly different effects than unassimilated CO_2, NO_3, or even metabolically generated NH_4 and NO_3 (see section on signalling molecules). The latter may exert especially strong influence as potentially toxic indicators of N cycling, which for photorespiration can exceed by tenfold the magnitude of primary N assimilation (Keys *et al.*, 1978; Lam *et al.*, 1996). Evidence also suggests that the extent of relatively direct effects of NO_3 on gene expression (such as for nitrate reductase (Galangau *et al.*, 1988), and for primary N assimilation (Crawford, 1995) may have been underestimated (see Chapter 6). Finally, glutamine influence on gene expression may occur via a mechanism different from that of its balance with glutamate at the level of metabolic regulation (Foyer *et al.*, 1993; Vincentz *et al.*, 1993).

Genes for C retrieval from stored reserves are typically induced as levels of soluble sugars and C/N ratios fall. These include genes for C remobilization from starch, lipid, and many non-photosynthetic proteins. In multicellular systems, mRNAs for synthesis of transport compounds, also allow export of vital resources during shortage. Overall changes in expression can progress to such a degree that previous C-importing cells (sinks) can become C-exporters (sources). This 'cellular altruism' is an essential characteristic of multicellular vs. unicellular organisms. Under extreme conditions, survival of some plant cells may even come at the expense of others, such as those of the root cortex (He *et al.*, 1994; Koch and Nolte, 1995).

For breakdown of C-rich reserves such as starch, related mRNAs are induced by sugar depletion in most structures (Koch, 1996; Thomas and Rodriguez, 1994) including leaves, tubers, endosperm and cotyledons. Utilization of vacuolar fructan reserves appears to involve a similar scenario (Pollock *et al.*, 1996), and genes for remobilization of lipid reserves are up-regulated by carbohydrate depletion in cotyledons of germinating cucurbits (Graham *et al.*, 1994a,b). The latter may involve breakdown of membrane constituents under severe C starvation (Brouquisse *et al.*, 1992).

Retrieval of C from storage proteins under C-limited conditions is also facilitated by derepression of genes for proteases, endoglycosidases (for glycosyl moieties), associated N cycling, and detoxification of N-intermediates (Brouquisse et al., 1992; Koch, 1996; Lam et al., 1996). The sugar-repressed mRNAs for N handling are likely to be especially important in these instances and glutamate dehydrogenase (*GDH1*) and asparagine synthetase (*ASN1*) are predominant among them. The light repression of the latter (Melo-Oliveira et al., 1996) might also help restrict this aspect of C-resource scavenging to cells and proteins not directly involved in acquisition of new C via photosynthesis.

Genes for export are typically co-ordinated with those for remobilization in multicellular structures. Both are up-regulated by decreased carbohydrate availability (Koch, 1996), and may also be affected by imbalances on the low side of the C/N ratio. SPS genes are typically induced when sugar levels drop, providing a means of synthesizing sucrose for export (Hesse et al., 1995; Klein et al., 1993). Genes for synthesis of transport amino acids are also up-regulated under these conditions. Those derepressed include *GS1* (glutamine synthases – localized in phloem), *GLU2* (glutamate synthases), and *ASN1* (asparagine synthases – C/N responsive) (Chevallier et al., 1996; Koch, 1996; Lam et al., 1996; see also Figure 5.1).

Genes for N storage are often sensitive to changes in C/N balance; however the degree and direction of response can vary depending on the protein and tissue involved (see Koch for references). Although protein remobilization is common under C-starvation stress (Brouquisse et al., 1992), photosynthetic proteins are minimally affected (Koch, 1996; Lam et al., 1996) allowing them to function simultaneously as both N sinks and C-fixation sources. In addition, glutamine and especially asparagine can provide stable, readily available forms of high N:C storage in a range of tissues. Respective genes for GS1, GLU2, GDH1, and ASN1 are responsive to decreasing sugar levels and low C/N ratios (Koch, 1996; Lam et al., 1996). Some of the vegetative storage proteins (in addition to the predominant Rubisco) can also be induced under conditions of elevated N availability even when C supplies are low (J. Davis, unpublished). In other instances, however, N-stimulated genes such as those for VSPs in sink cells and organs, can be up-regulated by rising sugar levels even when N levels are low (Grimes et al., 1993). A strong positive response to carbohydrate availability is also evident for many other storage proteins and may predominate over N signals for sporamins and some patatins (Frommer et al., 1994; Hattori et al., 1990; Peña-Cortez et al., 1992). A more balanced supply of both C and N for synthesis can be required in other instances, especially where overriding signals from injury or developmental programmes may be involved (Giroux et al., 1994; Koch, 1996; Stepien et al., 1994).

5.4 Genes up-regulated by high C/N, C-abundance, and/or N-depletion

Genes for primary N-assimilation are often up-regulated by sugar abundance in leaves, or high ratios of assimilated C/N (Lam et al., 1996; Vincentz et al., 1993). This reflects a pattern similar to that of the carefully integrated control of C and N metabolism at the enzyme level, aiding allocation of cellular resources to C- vs. N-assimilation (Foyer et al., 1993; Lam et al., 1995; Oaks, 1994; see also related chapters in this book). Sugars can induce genes for NR (nitrate reductase), NiR (nitrite reductase), chloroplastic GS (glutamine synthetase), and Fd-GOGAT (glutamate synthase) (Cheng et al., 1992; Faure et al., 1994; Lam et al., 1996; Vincentz et al., 1993). However, similar responses can also be obtained in instances of altered C/N balance, such as depression of NR in the absence of glutamine, and in the reciprocal effects of sucrose and glutamine (Vincentz et al., 1993). Effects on NiR are comparable, although less pronounced, and transcriptional influence of sugars is not evident (Vincentz et al., 1993).

Genes for photorespiratory N recycling are difficult to categorize in terms of interactive C/N metabolite regulation, because the identities of the most important genes have not been fully resolved. In addition, each of the probable candidates is represented by one or more genes encoding isozymes with contrasting patterns of N and C regulation (Lam et al., 1996; Melo-Oliveira et al., 1996). Some lines of evidence favour a role for glutamate dehydrogenase in photorespiratory N recycling (Lam et al., 1996). This would be compatible with effects of C depletion and elevated NH_4 availability on expression of a GDH gene in Chlorella (P. Miller and R. Schmidt, unpublished), Arabidopsis (Melo-Oliveira et al., 1996), and possibly in maize (Pryor, 1990). Although photorespiration would be minimal in the C_4 maize and CO_2-concentrating algae, observed characteristics of gene expression are similar to those of photosynthetic genes and consistent with the possible GDH role. However, data from GS mutants in barley (Wallsgrove et al., 1987) and Fd-GOGAT mutants in Arabidopsis (Sommerville and Ogren, 1980) barley (Bright et al., 1984), and pea (Blackwell et al., 1987) indicate that the GS-GOGAT pathway may have a previously underestimated role in photorespiratory N cycling. Expression of respective genes is also co-ordinated with those of photosynthesis in leaves, despite the contrasting inductive effects of sugars on these and other genes for N assimilation (Lam et al., 1996). Furthermore, GS2 expression increases under conditions of elevated photorespiration (Cock et al., 1991; Edwards and Coruzzi, 1989). Either or both pathways and sets of genes could be involved in this interface between C and N metabolism, particularly if their reciprocal regulation by metabolites overlaps enough to facilitate a shifting balance between their contributions.

Genes for N-remobilization and redistribution could predictably be up-regulated by N limitation and/or concomitant C/N imbalance (relative C excess). Breakdown of photosynthetic proteins can be especially pronounced if soluble sugars rise to high levels (Koch, 1996) and can provide a significant N source under these conditions. Sugar-enhanced senescence also involves increased expression of genes for GS2 and Fd-GOGAT, presumably aiding N scavenging, detoxification of N-intermediates, and synthesis of transport amino acids under these conditions (Lam et al., 1996). A putative gene for s-adenosyl methionine (SAM) synthase can also be up-regulated under these conditions, contributing further to N cycling and its interface with other processes under relative C excess (Winters et al., 1995). The balance between synthesis and remobilization of various proteins in response to changes in C/N ratios appears to differ with the protein and tissue involved, however, so that one type may be synthesized in the same tissue at the expense of the other (sporamin vs. photosynthetic proteins in sugar-fed leaves, for example).

Genes for C use/storage can respond to C/N balance, but are only recently being recognized as such. Storage proteins were among the first mRNAs found to be sugar-inducible (Hattori *et al.*, 1990; Johnson and Ryan, 1990; Liu *et al.*, 1991; Wenzler *et al.*, 1989), and this feature now appears to be shared by a number of storage proteins (Koch, 1996). Most of these proteins are also N responsive, however, so that maximum expression may occur under conditions of relative balance when metabolites alone are considered. The nature of this balance may differ in leaves, stems, tubers, and seeds; again, however, depending on the protein and situation involved.

Polysaccharide and lipid synthesis is enhanced as carbohydrate availability increases, and sugars are clear effectors of mRNAs for starch synthase (Kossmann *et al.*, 1991), starch phosphorylase (St Pierre and Brisson, 1995), ADPGlcPPase (Krapp and Stitt, 1994; Müller-Röber *et al.*, 1990), and branching enzyme (Kossmann *et al.*, 1991). Little direct analysis has addressed the role of C/N interactions in expression of these genes, however starch deposition and storage protein formation parallel one another closely enough that co-ordinated regulation of the respective genes has been proposed in association with seed storage programmes (Giroux *et al.*, 1994). In other instances, emerging data from leaf disk experiments indicate that NH_4 and glutamine can both counter sugar-induced increases in ADPGlcPPase transcript abundance (U. Sonnewald, unpublished data cited with permission). In this and other instances noted above, nitrogen availability has the potential to alter sugar-responsiveness of given genes and thus overall balance between acquisition and storage programmes for N and C resources.

Respiratory use of C is often favoured under conditions of carbohydrate abundance or elevated C/N balance. Related changes at the level of gene expression can occur in response to increases in C availability; however less attention has been directed towards N effects on transcription of respiratory genes not linked to amino acid biosynthesis.

Numerous genes for transporters of C- and N-assimilates as well as inorganic forms have been cloned recently, and analyses of their responses to metabolic signals are in progress (Bush, 1993; Crawford, 1995; Grimes *et al.*, 1993; Lauter *et al.*, 1996). Results are likely to add an important contribution to the overall understanding of how C/N balance is achieved, particularly at the whole plant level.

5.5 Contrasting C/N responses of isozyme genes

Figure 5.2 examines the similar dichotomies in gene expression potentially affecting key steps in assimilation and partitioning of C and N resources. Contrasting, even reciprocal, responses to assimilated C and/or N are involved. Conditions of cellular 'feast or famine' for C-assimilates each up-regulate genes for a different set of isozymes, and further, these are typically accompanied by inverse responses to supplies of N-assimilates.

The possible significance of these contrasts appears to be twofold. First, and most important, is the capacity for optimizing overall balance in response to different signals. This becomes particularly valuable where the ultimate adjustments to extremes of either signal alone could be detrimental under conditions represented by the other. Prioritization of cellular resources for assimilation of C vs. N is a prime example, and the balance between programmes for aquisition vs. storage is another. The extent of overlap between opposing signals

Key Gene Families that Respond to "Feast-or-Famine" Extremes in C/N Balance	
(↓C/↑N) Relative N Excess Priority given to photosynthesis, C remobilization, N sinks, shoot development	Relative C Excess (↑C/↓N) Priority given to N assimilation N remobilization, C sinks, root development
Sucrose Synthases: SH1 Koch et al., 1992 (*Asus1*)(*Sh1*) Martin et al., 1993 Sugar repressed Root tips, phloem, reprod. tissues	SUS1 (*Sus1*) Zea, Arabidopsis Sugar enhanced Broadly distributed in C-importing tissues
Invertases: IVR1 Xu et al., 1996 (*Ivr1*) Sugar repressed Root tips, reprod. tissues	IVR 2 (*Ivr2*) Zea Sugar enhanced Broadly distributed in C-importing tissues
Glutamine Synthetases: GS1 Lam et al., 1996 (*Gs1-1,2*)(*GLN1*) Crawford, 1995 Sugar repressed NO_3, NH_4 stim. Cytoplasm in roots, phloem, senescing leaves, shoots, germ seeds, legume nodules	GS 2 (*GLN2, Gln2*) Arabidopsis, Zea Sugar and light enhanced Amino acids and NO_3 induced, NH_4 repressed Plastids of roots, leaf mesophyll, esp. senescing leaves Photorespiratory role implicated in other species
Glutamate Synthases: NADH-GOGAT Lam et al., 1996 (*GLT1*) Crawford, 1995 (Light insensitive) NO_3 and NH_4 stim.? Root plastids, legume nodules, early seed germination	GOGAT-fd (*GLU1*) Arabidopsis Sucrose and light enhanced NO_3 induced, NH_4 repressed Leaf chloroplasts, roots Photorespiratory role implicated in other species
Glutamate Dehydrogenases: GDH1 Melo-Oliveira et al., 1996 (*GDH1, Gdh1*) Prunkard et al., 1986 Light and sugar repressed Magalhães et al., 1990 NH_4 can counter lt. repression Leaves, roots, bundle sheath Implicated in N-assim. at high NH_4 (*Arabidopsis*) Enhanced C partitioning to shoots (*Zea*)	GDH2 (*GDH2, Gdh2*) Arabidopsis, Lupine, Chlorella, Zea (sugar reg. undefined) Induced at low NH_4?
Asparagine Synthetases: ASN1 Chevallier et al. 1996 (*ASN1, Asn1*) Oaks & Ross, 1984 Light and sugar repressed GLU, GLN, ASN, NH_4 counter Roots and leaves, asparagus spears	ASN2 (*ASN2, Asn2*) Arabidopsis, Zea Sugar and light induced ASN repressed Leaves, roots Photorespiratory role suggested in other species
Storage Proteins: CLASS II-PATATINS Peña-Cortez et al., 1992 Sugar insensitive Frommer et al., 1994 N response? Grimes et al. 1993 Root tips, vascular bundles, reprod. tissues	PATATINS-CLASS I Sugar induced GLN induced if sugars present Stems (tubers), sugar-induced leaves

Figure 5.2 Key gene families that respond to 'feast-or-famine' extremes in C/N balance. Isozymes encoded by genes with opposite responses to C/N balance are grouped together in adjacent panels, and provide a basis for consideration of their collective physiological significance under contrasting conditions. One category is distinguished by up-regulation under conditions of relative C limitation and/or N excess, the other by induction under relative C excess and/or N limitation. For each gene family, enzyme acronyms are given first, respective genes second, followed by currently available information on responses of each to C- and N-effectors. Tissue or organ localization is also noted where known. Primary sources of information are designated beneath each gene family, with additional citations and discussion in the text.

thus facilitates a shift in balance across a region typically centred on the physiological midrange.

Second, this differential responsiveness of gene family members to C and/or N availability can extend to the level of cell and tissue specificity, further distinguishing their functional significance. This is consistent with instances of cellular specialization for concurrent, but contrasting functions (Koch et al., 1992; Lam et al., 1996; Martin et al., 1993). In addition, the potential exists for prioritization of some cells and tissues at the expense of others in instances of extreme C- or N-limitation. Differential maintenance of capacity for import or assimilation of either resource could readily exert such an effect at the cell, tissue, or organ level (Koch and Nolte, 1995; Koch et al., 1992; Xu et al., 1996).

5.6 Metabolite effectors of crosstalk in expression of C- and N-responsive genes

Metabolite signals can often override others originating from environmental and developmental cues. Sugars repress photosynthetic genes despite inductive light signals (Jang and Sheen, 1994; Sheen, 1990, 1994). A dominant, downstream role has also been proposed for metabolites in signal transduction models for several light sensing systems (Chory, 1993). In addition, sugar signals supercede input from development and other sources of signals, to induce genes for patatin storage protein at atypical sites in sweet potato plantlets (Hattori et al., 1990). N-metabolites such as glutamine can also override environmental signals such as light induction of nitrate reductase, and interestingly, also the up-regulation of the same gene by glucose (Vincentz et al., 1993). However, despite the obvious importance of metabolite signals in these and many instances, other signals can predominate for specific situations and genes, most notably related to programmes of reproductive priority (Koch, 1996).

Interactions at the level of metabolism and assimilate uptake can be difficult to separate from those occurring through molecular crosstalk between systems transducing signals for gene expression. Experimental introduction of given metabolites into a system often leads to rapid production of others and/or a quick series of adjustments in metabolic balance. Sugar feeding experiments can alter endogenous levels of certain amino acids (Foyer et al., 1993; Oaks, 1994), opening questions regarding actual effectors of signalling systems. Metabolism can also occur prior to entry into the system as for sucrose breakdown in the extracellular space by apoplastic, cell-wall forms of invertase (Koch, 1996). In addition, transporter characteristics can differ markedly and may affect entry of some effectors into the cellular symplast (Bush, 1993; Koch et al., 1992).

The physical path of arrival for a given metabolic signal has the potential to affect strongly its signal generation (Herbers et al., 1996; Koch, 1996). Many metabolites are generated internally, but many are also taken up from the external environment. In higher plants, this external environment is largely affected by translocation of sugars and amino acids from other sites. As signals from long-distance sources, there could be advantages in distinguishing them from similar but endogenously generated counterparts. Additional membrane input into signalling could occur in association with import of the effector, or via other interactions not involving actual transfer (Koch, 1996). Multicellular plants also have the capacity to deliver phloem-born C- and N-assimilates directly into the interior of some importing cells via plasmodesmatal connections (Koch, 1996). Analysis of transgenic plants with invertase over-expression targeted to cell wall, vacuolar, and cytoplasmic locales showed that resultant hexose production appeared to affect sugar signalling systems only when transfer across either tonoplast or plasma membrane was involved (Herbers et al., 1996). Both vacuolar and cell wall compartments represent sites of probable accumulation for C- or N-assimilates when either is in relative excess.

Effectors of metabolic signals to responsive genes are generally not the same as the regulatory metabolites and control mechanisms most prominent in rapid fine tuning. Important co-ordinators at the enzyme level, such as $Fru2,6P_2$, and ATP/ADP ratios have been tested for effects on gene expression and thus far appear to have indirect influence if any at all (Sheen, 1990, 1994; Stitt et al., 1995). So, too, do many effectors of N metabolism have variable apparent effects on the signalling systems for C- and N-responsive genes (Crawford, 1995; Vincentz et al., 1993). Links between metabolism and gene expression clearly occur at key points, but as yet these seem not to overlap directly with prime effectors of fine control systems. Instead, signals for C- and N-responsive genes appear to function as reporters of metabolic status rather than as mechanisms that overlap significantly with the metabolic regulatory interaction at the enzyme level.

Gene responses to different forms of C and N tyically differ. Assimilated C and N (sugars, acetate, amino acids) are often preferred substrates for cellular metabolism relative to raw resources (CO_2, NO_3, and exogenous NH_4), which require significant cost for acquisition. For carbohydrate sensitive genes, hexoses and acetate tend to generate the strongest signals (Graham et al., 1994a,b; Jang and Sheen, 1994; Koch, 1996; Sheen, 1994). For N-responsive genes, glutamine has a strong influence, yet perhaps not as an actual initiator of signals (Crawford, 1995; Crawford and Arst, 1993; Lam et al., 1996; Vincentz et al., 1993). Other N forms can also have pronounced effects, such as the NH_4 generated endogenously during photorespiration, NO_3 assimilation, and protein remobilization (Crawford, 1995; Lam et al., 1996). In this instance its effect may possibly be related to inhibition of the pyruvate dehydrogenase function (Oaks, 1994). NO_3 itself is also a source of signals, yet these are probably transduced via different mechanisms (Crawford, 1995).

Figure 5.3 shows the essential features of sugar signal transduction in yeast, which can be a source of testable hypotheses in plant systems, and clues for sites of crosstalk. The primary attributes pictured are: a role for hexokinase as both a catalytic enzyme of carbohydrate metabolism and an initiating agent in the signalling system; a protein kinase cascade; and a dual function for SNF1 through its effects on this kinase cascade and also through its direct interaction with a DNA-binding transcription factor. Thus far the first two characteristics appear to have important similarities in plant systems (Alderson et al., 1991; Banno et al., 1993; Halford et al., 1992; Koch, 1996; Muranaka et al., 1994; Sheen, 1994; Stitt et al., 1995). In addition Nakamura and co-workers have identified the involvement of Ca^{2+} and a calcium dependent CDPK in sugar signalling for higher plants (Ohto and Nakamura, 1995; Ohto et al., 1995). Comparable information for N signalling systems is still more sparse (Crawford,

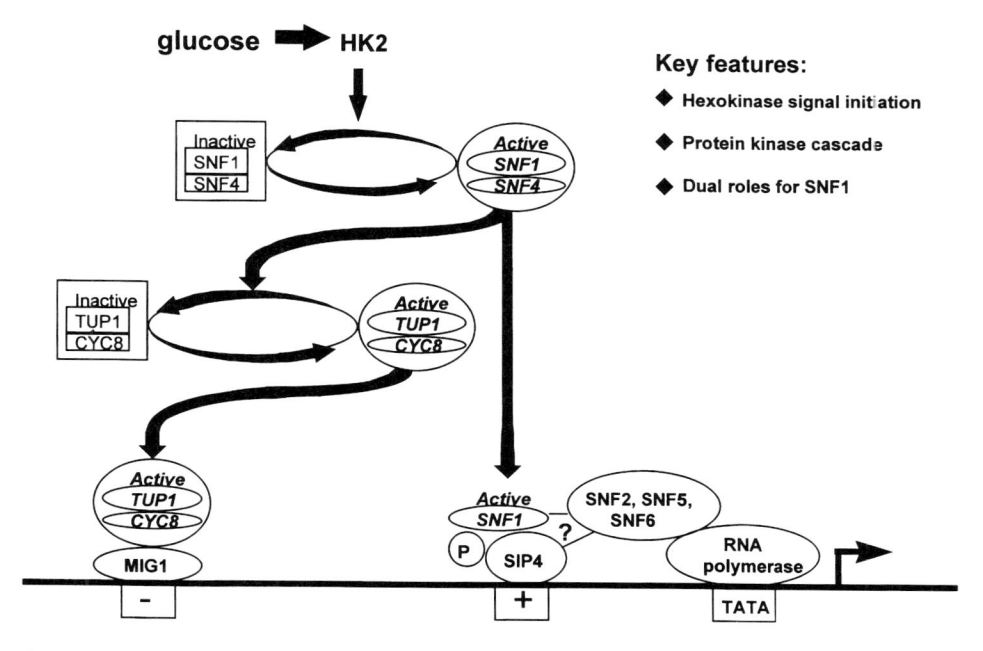

Figure 5.3 Key features of sugar signal transduction in yeast: a source of testable hypotheses for plants. Mechanisms of sugar signal transduction and possible points of interface with N-signalling systems in plants remain unclear, however clues may be available in scenarios from other organisms. The model shown here for yeast is redrawn from Bihn and Koch (1997), and is based on a combination of information from Gancedo (1992), Trumbly (1992), Lesage et al. (1996), Yang et al. (1992). Essential tennants of this scenario are signal generation by a specific hexokinase (encoded here by HK2), a protein kinase cascade (comprised of a two-level series for phosphorylation and dephosphorylation by SNF1:SNF4 and by TUP1:CYC8), and dual roles for SNF1 affecting both the kinase cascade and a DNA-binding transcription factor

1995; Hinnebusch, 1984; Huppe and Turpin, 1994; Vincentz et al., 1993) and currently provides little basis for speculation on points for crosstalk other than possible interface between protein kinases and/or calcium influence. Interaction through metabolism prior to signal generation is likely to be a significant source of interactive input.

5.7 Closing comments

Abundant evidence supports interactive effects of C and N availability on expression of specific genes, however it is unclear where most important points of interface or 'crosstalk' occur. N availability can affect C-responsive genes and in many instances, C availability can affect N-responsive genes. Currently available information on actual metabolite signals and transduction systems indicates separate mechanisms with little if any points of overlap. However, much of the crosstalk may very well occur at the level of metabolic adjustment, with gene-level responses driven largely by these.

References and further reading

ALDERSON, A., SABELLI, P.A., DICKINSON, J.R., COLE, D., RICHARDSON, M., KREIS, M., SHEWRY, P.R. and HALFORD, N.G. (1991) Complementation of snf1, a mutation affecting global regulation of carbon metabolism in yeast, by a plant protein kinase cDNA. *Proc. Natl. Acad. Sci. USA* **88**, 8602–8605.

BANNO, H., MURANAKA, T., ITO, Y., MORIBE, T., USAMI, S., HINATA, K. and MACHIDA, Y. (1993) Isolation and characterization of cDNA clones that encode protein kinases of *Nicotinia tabacum*. *J. Plant Res.* **3**, 181–192.

BERNIER, G., HAVELANGE, A., HOUSSA, C., PETITJEAN, A. and LEJEUNE, P. (1993) Physiological signals that induce flowering. *Plant Cell* **5**, 1147–1155.

BIHN, E.A. and KOCH, K.E. (1997) A role for invertase in sugar signalling and plant development. *Photosynth. Res.* (in press).

BLACKWELL, R.D., MURRAY, A.J.S. and LEA, P.J. (1987) The isolation and characterization of photorespiratory mutants of barley and pea. In: *Progress in Photosynthesis Research* (BIGGINS, J., ed.), pp. 625–628. Nijhoff, Dordrecht.

BOWLER, C. and CHUA, N.H. (1994) Emerging themes of plant signal transduction. *Plant Cell* **6**, 1529–1541.

BRIGHT, S.W.J., LEA, P.J., ARRUDA, P., HALL, N.P., KENDALL, A.C. *et al.* (1984) Manipulation of key pathways in photorespiration and amino acid metabolism by mutation and selection. In: *The Genetic Manipulation of Plants and Its Application to Agriculture* (LEA, P.J. and STEWART, G.R., eds), pp. 141–169. Oxford University Press, Oxford.

BROUQUISSE, R., PRADET, JAMES F., PRADET, A. and RAYMOND, P. (1992) Asparagine metabolism and nitrogen distribution during protein degradation in sugar-starved maize root tips. *Planta* **188**, 384–395.

BUSH, D.R. (1993) Proton-coupled sugar and amino acid transporters in plants. *Annu. Rev. Plant Physiol. Plant Mol. Biol.* **44**, 513–542.

CARLSON, M. (1987) Regulation of sugar utilization in *Saccharomyces* species. *J. Bacteriol.* **169**, 4873–4877.

CARVALHO, H., PEREIRA, S., SUNKEL, C. and SALEMA, R. (1992) Detection of cytosolic glutamine synthetase in leaves of *Nicotiana tabacum* L. by immunocytochemical methods. *Plant Physiol.* **100**, 1591–1594.

CHENG, C.L., ACEDO, G.N., CRISTINSIN, M. and CONKLING, M.A. (1992) Sucrose mimics the light induction of *Arabidopsis* nitrate reductase gene transcription. *Proc. Natl. Acad. Sci. USA* **89**, 1861–1864.

CHEVALLIER, C., BOURGEOIS, E., JUST, D. and RAYMOND, P. (1996) Metabolic regulation of asparagine synthetase gene expression in maize (*Zea mays* L.) root tips. *Plant J.* **9**, 1–11.

CHOREY, J. (1993) Out of darkness: mutants reveal pathways controlling light-regulated development in plants. *Trends Genet.* **9**, 167–172.

COCK, J.M., BROCK, I.W., WATSON, A.T., SWARUP, R., MORBY, A.P. and CULLIMORE, J.V. (1991) Regulation of glutamine synthetase genes in leaves of *Phaseolus vulgaris*. *Plant Mol. Biol.* **17**, 761–771.

CRAWFORD, N.M. (1995) Nitrate: nutrient and signal for plant growth. *Plant Cell* **7**, 859–868.

CRAWFORD, N.M. and ARST, H.N.J. (1993) The molecular genetics of nitrate assimilation in fungi and plants. *Annu. Rev. Genet.* **27**, 115–146.

CRIQUI, M.C., DURR, A., PARMENTIER,, Y., MARBACH, J., FLECK, J. and JAMET, E. (1992) How are photosynthetic genes repressed in freshly-isolated mesophyll protoplasts of *Nicotiana sylvestris*? *Plant Physiol. Biochem.* **30**, 597–601.

DENG, X.W. (1994) Fresh view of light signal transduction. *Cell* **76**, 423–426.

EDWARDS, J.W. and CORUZZI, G.M. (1989) Photorespiration and light act in concert to regulate the expression of the nuclear gene for chloroplast glutamine synthetase. *Plant Cell* **1**, 241–248.

FARRAR, J.F. and WILLIAMS, J.H.H. (1991) Control of the rate of respiration in roots: compartmentation, demand, and the supply of substrate. In: *Compartmentation of Metabolism* (EMMS, M.J., ed.), pp. 167–188. Cambridge University Press, New York.

FAURE, J.D., JULLIEN, M. and CABOCHE, M. (1994) *Zea3*: a pleiotropic mutation affecting cotyledon development, cytokinin resistance and carbon-nitrogen metabolism. *Plant J.* **5**, 481–491.

FOYER, C.H., LEFEBVRE, C., PROVOT, M., VINCENTZ, M. and VAUCHERET, H. (1993) Modulation of nitrogen and carbon metabolism in transformed *Nicotiana plumbaginifolia* mutant E23 lines expressing either increased or decreased nitrate reductase activity. In: *Aspects of Applied Biology*, Volume 34, Physiology of Varieties (WHITE, E., KETTLEWELL, P.S., PARRY, M.A. and ELLIS, R.P., eds), pp. 137–145. Association of Applied Biologists, Wellesbourne, Warwick.

FROMMER, W.B., MIELCHEN, C. and MARTIN, T. (1994) Metabolic control of patatin promoters from potato in transgenic tobacco and tomato plants. *Life Sci. Adv.* **13**, 329–334.

GALANGAU, F., DANIEL-VEDELE, F., MOUREAUX, T., DORBE, M.F., LEYDECKER, M.T. and CABOCHE, M. (1988) Expression of leaf nitrate reductase genes from tomato and tobacco in relation to light-dark regimes and nitrate supply. *Plant Physiol.* **88**, 383–388.

GANCEDO, J.M. (1992) Carbon catabolite repression in yeast. *Eur. J. Biochem.* **206**, 297–313.

GEIGER, D.R., KOCH, K.E. and SHIEH, W.J. (1996) Effect of environmental factors on whole plant assimilate partitioning and associated gene expression. *J. Exp. Bot.* **47**, 1229–1238.

GIROUX, M.J., BOYER, C., FEIX, G. and HANNAH, L.C. (1994) Coordinated transcriptional regulation of storage product genes in the maize endosperm. *Plant Physiol.* **106**, 713–722.

GRAHAM, I.A., BAKER, C.J. and LEAVER, C.J. (1994a) Analysis of the cucumber malate synthase gene promotor by transient expression and gel retardation assays. *Plant J.* **6**, 893–902.

GRAHAM, I.A., DENBY, K.J. and LEAVER, C.J. (1994b) Carbon catabolite repression regulates glyoxylate cycle gene expression in cucumber. *Plant Cell* **6**, 761–772.

GRIMES, H.D., TRANBARGER, T.J. and FRANCESCHI, V.R. (1993) Expression and accumulation patterns of nitrogen-responsive lipoxygenase in soybeans. *Plant Physiol.* **103**, 457–466.

HALFORD, N.G., VINCENTE-CARBAJOSA, J., SABELLI, P.A., SHREWRY, P.R., HANNAPPEL, U. and KREIS, M. (1992) Molecular analysis of a barley multigene family homologous to the yeast protein kinase gene SNFI. *Plant J.* **2**, 791–797.

HATTORI, T., NAKAGAWA, S. and NAKAMURA, K. (1990) High-level expression of tuberous root storage protein genes of sweet potato in stems of plantlets grown in vitro on sucrose medium. *Plant Mol. Biol.* **14**, 595–604.

HE, C.J., DREW, M.E. and MORGAN, P. (1994) Induction of enzymes associated with lysigenous aerenchyma formation in roots of *Zea mays* during hypoxia or nitrogen starvation. *Plant Physiol.* **105**, 861–865.

HERBERS, K., MEUWLY, P., FROMMER, W.B., METRAUX, J.P. and SONNEWALD, U. (1996) Systemic aquired-resistance mediated by the ectopic expression of invertase: possible hexose sensing in the secretory pathway. *Plant Cell* **8**, 793–803.

HESSE, H., SONNEWALD, U. and WILLMITZER, L. (1995) Cloning and expression analysis of sucrose-phosphate-synthase from sugar beet (*Beta vulgaris* L.). *Mol. Gen. Genet.* **247**, 515–520.

HINNEBUSCH, A.G. (1984) Evidence for translational regulation of the activator of general amino acid control in yeast. *Proc. Natl. Acad. Sci. USA* **81**, 6442–6446.

HINNEBUSCH, A.G. (1985) A hierarchy of trans-acting factors modulate translation of an activator of amino acid biosynthetic genes in yeast. *Mol. Cell. Biol.* **5**, 2349–2360.

HINNEBUSCH, A.G. (1986) The general control of amino acid biosynthetic genes in the yeast *Saccharomyces cerevisiae*. *Crit. Rev. Biochem.* **21**, 277–317.

HINNEBUSCH, A.G. and FINK, G.R. (1983a) Repeated DNA sequences upstream from HISI also occur at several other co-regulated genes in Saccharomyces cerevisiae. *J. Biol. Chem.* **258**, 5238–5247.

HINNEBUSCH, A.G. and FINK, G.R. (1983b) Positive regulation in the general amino acid control of Saccharomyces cerevisiae. *Proc. Natl. Acad. Sci. USA* **80**, 5374–5378.

HINNEBUSCH, A.G., LUCCHINI, G. and FINK, G.R. (1985) A synthetic HIS4 regulatory element confers general amino acid control on the cytochrome c gene (CYCl) of yeast. *Proc. Natl. Acad. Sci. USA* **82**, 498–502.

HIREL, B., BOUET, C., KING, B., LAYZELL, B., JACOBS, F. and VERMA, D.P.S. (1987) Glutamine synthetase genes are regulated by ammonia provided externally or by symbiotic nitrogen fixation. *EMBO J.* **6**, 1167–1171.

HUPPE, H.C. and TURPIN, D.H. (1994) Integration of carbon and nitrogen metabolism in plant and algal cells. *Annu. Rev. Plant Physiol. Plant Mol. Biol.* **45**, 577–607.

JANG, J.-C. and SHEEN, J. (1994) Sugar sensing in higher plants. *Plant Cell* **6**, 1665–1679.

JOHNSON, R. and RYAN, C.A. (1990) Wound-inducible potato inhibitor II genes: enhancement of expression by sucrose. *Plant Mol. Biol.* **14**, 527–536.

KAMACHI, K., YAMAYA, T., MAE, T. and OJIMA, K. (1991) A role for glutamine synthetase in the remobilization of leaf nitrogen during natural senescence in rice leaves. *Plant Physiol.* **96**, 411–417.

KAMACHI, K., YAMAYA, T., HAYAKAWA, T., MAE, T. and OJIMA, K. (1992a) Changes in cytosolic glutamine synthetase polypeptide and its mRNA in a leaf blade of rice plants during natural senescence. *Plant Physiol.* **98**, 1323–1329.

KAMACHI, K., YAMAYA, T., HAYAKAWA, T., MAE, T. and OJIMA, K. (1992b) Vascular bundle-specific localization of cytosolic glutamine synthetase in rice leaves. *Plant Physiol.* **99**, 1481–1486.

KAWAKAMI, N. and WATANABLE, A. (1988) Senescence-specific increase in cytosolic glutamine synthetase and its mRNA in radish cotyledons. *Plant Physiol.* **98**, 1323–1329.

KEYS, A.J., BIRD, I.F., CORNELIUS, M.J., LEA, P.J., WALLSGROVE, R.M. and MIFLIN, B.J. (1978) Photorespiratory nitrogen cycle. *Nature* **275**, 741–743.

KLEIN, R.R., CRAFTS-BRANDNER, S.J. and SALVUCCI, M.E. (1993) Cloning and developmental expression of the sucrose-phosphate-synthase gene from spinach. *Planta* **190**, 498–510.

KOCH, K.E. (1996) Carbohydrate-modulated gene expression in plants. *Annu. Rev. Plant Physiol. Plant Mol. Biol.* **47**, 509–540.

KOCH, K.E. and NOLTE, K.D. (1995) Sugar modulated expression of genes for sucrose metabolism and their relationship to transport pathways. In: *Carbon Partitioning and Source Sink Interactions in Plants* (MADORE, M.M. and LUCAS, W.L., eds), pp. 68–77. American Society of Plant Physiology, Rockville.

KOCH, K.E., NOLTE, K.D., DUKE, E.R., McCARTY, D.R. and AVIGNE, W.T. (1992) Sugar levels modulate differential expression of maize sucrose synthase genes. *Plant Cell* **4**, 59–69.

KOCH, K.E., XU, J., DUKE, E.R., McCARTY, D.R., YUAN, C-X., TAN, B.C. and AVIGNE, W.T. (1995) Sucrose provides a long distance signal for coarse control of genes affecting its metabolism. In: *Sucrose Metabolism, Biochemistry, Physiology and Molecular Biology*, Vol. 14 (PONTIS, H.G., SALERNO, G. and ECHEVERRIA, E.J., eds), pp. 266–277. American Society of Plant Physiology, Rockville.

KOCH, K.E., WU, Y. and XU, J. (1996) Sugar and metabolic regulation of genes for sucrose metabolism: potential influence of maize sucrose synthase and soluble invertase responses on carbon partitioning and sugar sensing. *J. Exp. Bot.* **47**, 1179–1186.

KOSSMANN, J., VISSER, R.G.F., MÜLLER-RÖBER, B.T., WILLMITZER, L. and SONNEWALD, U. (1991) Cloning and expression analysis of a potato cDNA that encodes branching enzyme: evidence for coexpression of starch biosynthetic genes. *Mol. Gen. Genet.* **230**, 39–44.

KRAPP, A., HOFFMANN, B., SCHAFFER, C. and STITT, M. (1993) Regulation of the expression of rbcS and other photosynthetic genes by carbohydrates: a mechanism for the 'sink regulation' of photosynthesis? *Plant J.* **3**, 817–828.

KRAPP, A. and STITT, M. (1994) Influence of high carbohydrate content on the activity of plastidic and cytosolic isoenzyme pairs in photosynthetic tissues. *Plant Cell Environ.* **17**, 861–866.

LAM, H.M., COSCHIGANO, K.T., SCHULTZ, C., MELO-OLIVEIRA, R. and TJADEN, G. *et al.* (1995) Use of Arabidopsis mutants and genes to study amide amino acid biosynthesis. *Plant Cell* **7**, 887–898.

LAM, H.M., COSCHIGANO, K.T., OLIVEIRA, I.C., MELO-OLIVEIRA, R. and CORUZZI, G.M. (1995) The molecular-genetics of nitrogen assimilation into amino acids in higher plants. *Annu. Rev. Plant Physiol. Plant Mol. Biol.* **47**, 569–593.

LAUTER, F.R., NENNEMANN, O., RIESMEIER, J.W. and FROMMER, W.B. (1996) Preferential expression of an ammonium transporter and of two putative nitrate transporters in root hairs of tomato. *Proc. Natl. Acad. Sci. USA* **93**, 8139–8144.

LIU, X.Y., ROCHA-SOSA, M., HUMMEL, S., WILLMITZER, L. and FROMMER, W.B. (1991) A detailed study of the regulation and evolution of the two classes of patatin genes in *Solanum tuberosum* L. *Plant Mol. Biol.* **17**, 1139–1154.

LESAGE, P., YANG, X. and CARLSON, M. (1996) Yeast SNF1 protein kinase interacts with SIP4, a C6 zinc cluster transcriptional activator: a new role for SNF1 in the glucose response. *Mol. Cell. Biol.* **16**, 1921–1928.

MAGALHÃES, J.R., JU, G.C., RICH, P.J. and RHODES, D. (1990) Kinetics of $^{15}NH_4^+$ assimilation in *Zea mays*. *Plant Physiol.* **94**, 647–656.

MARTIN, T., FROMMER, W.B., SALANOUBAT, M. and WILLMITZER, L. (1993) Expression of an Arabidopsis sucrose synthase gene indicates a role in metabolization of sucrose both during phloem loading and in sink organs. *Plant J.* **4**, 367–377.

Melo-Oliveira, R., Oliveira, I.C. and Coruzzi, G.M. (1996) *Arabidopsis* mutant analysis and gene regulation define a nonredundant role for glutamate dehydrogenase in nitrogen assimilation. *Proc. Natl Acad. Sci. USA* **93**, 4718–4723.

Müller-Röber, B.T., Kossmann, J., Hannah, L.C., Willmitzer, L. and Sonnewald, U. (1990) One of two different ADP-glucose pyrophosphorylase genes from potato responds strongly to elevated levels of sucrose. *Mol. Gen. Genet.* **224**, 136–146.

Muranaka, T., Banno, H. and Machida, Y. (1994) Characterization of tobacco protein-kinase npk5, a homolog of *Saccharomyces cerevisiae* snf1 that constitutively activates expression of the glucose-repressible suc2 gene for secreted invertase of *Saccharomyces cerevisiae*. *Mol. Cell. Biol.* **14**, 2958–2965.

Oaks, A. (1994) Efficiency of nitrogen utilization in C_3 and C_4 cereals. *Plant Physiol.* **106**, 407–414.

Oaks, A. and Ross, D.W. (1984) Asparagine synthetase in *Zea mays*. *Can. J. Bot.* **62**, 68–73.

Oaks, A., Stulen, I., Jones, K., Winspear, M.J., Misra, S. and Boesel, I.L. (1980) Enzymes of nitrogen assimilation in maize roots. *Planta* **148**, 477–484.

Ohto, M.-A., Hayashi, K., Isobe, M. and Nakamura, K. (1995) Involvement of Ca^{2+} signalling in the sugar-inducible expression of genes coding for sporamin and -amylase of sweet potato. *Plant J.* **7**, 297–307.

Ohto, M.-A. and Nakamura, K. (1995) Sugar-induced increase of calcium-dependent protein kinases associated with the plasmamembrane in leaf tissues of tobacco. *Plant Physiol.* **109**, 973–981.

Overvoorde, P.J., Frommer, W.B. and Grimes, H.D. (1996) A soybean sucrose binding-protein independently mediates nonsaturable sucrose uptake in yeast. *Plant Cell* **8**, 271–280.

Parks, B.M. and Hangarter, R.P. (1994) Blue light sensory systems in plants. *Cell Biol.* **5**, 347–353.

Peña-Cortez, H., Liu, X., Sanchez-Serrano, J., Schmid, R. and Willmitzer, L. (1992) Factors affecting gene expression of patatin and proteinase-inhibitor-II gene families in detached potato leaves: implications for their co-expression in developing tubers. *Planta* **186**, 495–502.

Pollock, C.J. and Farrar, J. (1996) Source-sink relations: the role of sucrose. In: *Environmental Stress and Photosynthesis* (Baker, N.R., ed.), pp. 261–279. Kluwer, The Netherlands.

Pollock, C.J., Winters, A.L., Gallagher, J. and Cairns, A.J. (1996) Sucrose and the regulation of fructan metabolism in leaves of temperate gramineae. In: *Sucrose Metabolism, Biochemistry, Physiology, and Molecular Biology* (Pontis, H.G., ed.), **14**, 167–178.

Prunkard, D.E., Bascomb, N.F., Molin, W.T. and Schmidt, R.R. (1986) Effect of different carbon sources on the ammonium induction of different forms of NADP-specific glutamate dehydrogenase in Chlorella sorokiniana cells cultured in the light and dark. *Plant Physiol.* **81**, 413–422.

Pryor, A.J. (1990) A maize glutamic dehydrogenase null mutant is cold temperature sensitive. *Maydica* **35**, 367–372.

Quick, W.P. and Schaffer, A.A. (1996) Sucrose metabolism in sources and sinks. In: *Photoassimilate Distribution in Plants and Crops: Source-Sink Relationships* (Zamski, E. and Schaffer, A.A, eds). Marcel Dekker, New York.

Saier, M.H. (1989) Protein phosphorylation and allosteric control of inducer exclusion and catabolite repression by the bacterial phosphoenol pyruvate: sugar phosphotransferase system. *Microbiol. Rev.* **53**, 109–120.

SHEEN J. (1990) Metabolic repression of transcription in higher plants. *Plant Cell* **2**, 1027–1038.
SHEEN J. (1994) Feedback-control of gene-expression. *Photosyn. Res.* **39**, 427–438.
SHIH, M.C. and GOODMAN, H.M. (1988) Differential light regulated expression of nuclear genes encoding chloroplast and cytosolic glyceraldehyde-3-phosphate dehydrogenase in *Nicotiana tabacum*. *EMBO J.* **7**, 893–898.
SOMERVILLE, C.R. and OGREN, W.L. (1980) Inhibition of photosynthesis in Arabidopsis mutants lacking leaf glutamate synthase activity. *Nature* **286**, 257–259.
ST PIERRE, B. and BRISSON, N. (1995) Induction of the plastidic starch-phosphorylase gene in potato storage sink tissue – effect of sucrose and evidence for coordinated regulation of phosphorylase and starch biosynthetic genes. *Planta* **195**, 339–344.
STEPIEN, V., SAUTER, J.J. and MARTIN, F. (1994) Vegetative storage proteins in woody plants. *Plant Physiol. Biochem.* **32**, 185–192.
STITT, M., KRAPP, A., KLEIN, D., RÖPER-SCHWARZ, U. and PAUL, M. (1995) Do carbohydrates regulate photosynthesis and allocation by altering gene expression? In: *Carbon Partitioning and Source Sink Interactions in Plants* (MADORE, M.M. and LUCAS, W.L., eds), pp. 68–77. American Society of Plant Physiology, Rockville.
SUGIHARTO, B., MIYATA, K., NAKAMOTO, H., SASAKAWA, H. and SUGIYAMA, T. (1990) Regulation of expression of carbon-assimilating enzymes by nitrogen in maize leaf. *Plant Physiol.* **92**, 963–969.
SUGIHARTO, B. and SUGIYAMA, T. (1992) Effects of nitrate and ammonium on gene expression of phosphoenolpyruvate carboxylase and nitrogen metabolism in maize leaf tissue during recovery from nitrogen stress. *Plant Physiol.* **98**, 1403–1408.
SUKANYA, R., LI, M.-G. and SNUSTAD, D.P. (1994) Root- and shoot-specific responses of individual glutamine synthetase genes of maize to nitrate and ammonium. *Plant Mol. Biol.* **5**, 1935–1946.
TANIGUCHI, M., KOBE, A., KATO, M. and SUGIYAMA, T. (1995) Aspartate aminotransferase isozymes in Panicum miliaceum L., an NAD-malic enzyme-type C4 plant: comparison of enzymatic properties, primary structures, and expression patterns. *Arch. Biochem. Biophys.* **318**, 295–306.
THOMAS, B.R. and RODRIGUEZ, R.L. (1994) Metabolite signals regulate gene expression and source/sink relations in cereal seedlings. *Plant Physiol.* **106**, 1235–1239.
TRUMBLY, R.J. (1992) Glucose repression in the yeast Saccharomyces cerevisiae. *Mol. Microbiol.* **6**, 15–21.
VAN OOSTEN, J.J. and BESSFORD, R.T. (1994) Sugar feeding mimics effect of acclimation to high CO_2: rapid downregulation of RuBisCo small subunit transcripts but not of the large subunit transcripts. *J. Plant Physiol.* **143**, 306–312.
VINCENTZ, M., MOUREAUX, T., LEYDECKER, M.T., VAUCHERET, H. and CABOCHE, M. (1993) Regulation of nitrate and nitrite reductase expression in *Nicotiana plumbaginifolia* leaves by nitrogen and carbon metabolites. *Plant J.* **3**, 315–324.
WALLSGROVE, R.M., TURNER, J.C., HALL, N.P., KENDALL, A.C. and BRIGHT, S.W.J. (1987) Barley mutants lacking chloroplast glutamine synthetase – biochemical and genetic analysis. *Plant Physiol.* **83**, 155–158.
WENZLER, H.C., MIGNERY, G.A., FISHER, L.M. and PARK, W.D. (1989) Analysis of a chimeric class-I patatin-GUS gene in transgenic potato plants: high-level expression in tubers and sucrose-inducible expression in cultured leaf and stem explants. *Plant Mol. Biol.* **12**, 41–50.
WINTERS, A.L., GALLAGHER, J.A., POLLOCK, C.J. and FARRAR, J.F. (1995) Isolation of a gene expressed during sucrose accumulation in leaves of *Lolium temulentum* L. *J. Exp. Bot.* **46**, 1345–1350.

XU, J., AVIGNE, W.T., MCCARTY, D.R. and KOCH, K.E. (1996) A similar dichotomy of sugar modulation and developmental expression affects both paths of sucrose metabolism: evidence from a maize invertase gene family. *Plant Cell* **8**, 1209–1220.

YANG, X., HUBBARD, J.A. and CARLSON, M. (1992) A protein kinase substrate identified by the two-hybrid system. *Science* **257**, 680–682.

6

Manipulation of the pathways of sucrose biosynthesis and nitrogen assimilation in transformed plants to improve photosynthesis and productivity

SYLVIE FERRARIO-MÉRY, ERIK MURCHIE, BERTRAND HIREL, NATHALIE GALTIER, W. PAUL QUICK AND CHRISTINE H. FOYER

6.1 Introduction

Photosynthesis drives carbon (C) and nitrogen (N) assimilation by the provision of assimilatory power (reduced ferredoxin, NADPH, ATP). The assimilation of N requires not only energy but also C skeletons (Figure 6.1). These are provided by C metabolism; triose phosphate produced as a result of photosynthetic C assimilation in the leaves can either be used to make carbohydrates (mainly sucrose and starch) or it can be directed through the anapleurotic pathway to provide skeletons for N assimilation. Sucrose exported from the leaves provides the energy and C skeletons to drive N assimilation in the roots and associated tissues. C assimilation, C partitioning and N assimilation are highly co-ordinated and there are many points of reciprocal control (Foyer et al., 1994a, 1996).

Nitrogen is mainly absorbed as nitrate which is reduced in leaf cells to ammonia by the consecutive action of two enzymes, nitrate reductase (NR), a cytosolic enzyme and nitrite reductase (NiR) localized in the chloroplast (Figure 6.2). NR is considered to be the rate limiting enzyme for N-assimilation (Lillo, 1984). It catalyzes the two-electron reduction of nitrate to nitrite using NADH or NADPH as electron donor (Hoff et al., 1994). The NR gene (nia) has been isolated and characterized in barley (Miyazaki et al., 1991), squash (Crawford et al., 1986), tomato (Daniel-Vedele et al., 1989), rice (Choi-Hong et al., 1989), tobacco (Vaucheret et al., 1989), bean (Hoff et al., 1991), and spinach (Prosser and Lazarus, 1990). NR activity is induced by NO_3^- in all plant species examined. Many other factors, such as light, growth regulators, environmental stresses, the availability of CO_2, oxygen, temperature, and the physiological and developmental stage of the plant, modulate NR at the transcriptional and/or post-translational level (Deng et al., 1993; Huppe and Turpin, 1994; Kaiser and Huber, 1994). In higher plants, the tissue distribution of NR is also regulated by factors such as developmental state, the external nitrate concentration, the presence of competing sinks and irradiance (Andrews, 1986). Although most higher plants express NR in the photosynthetic tissues, there is much variation

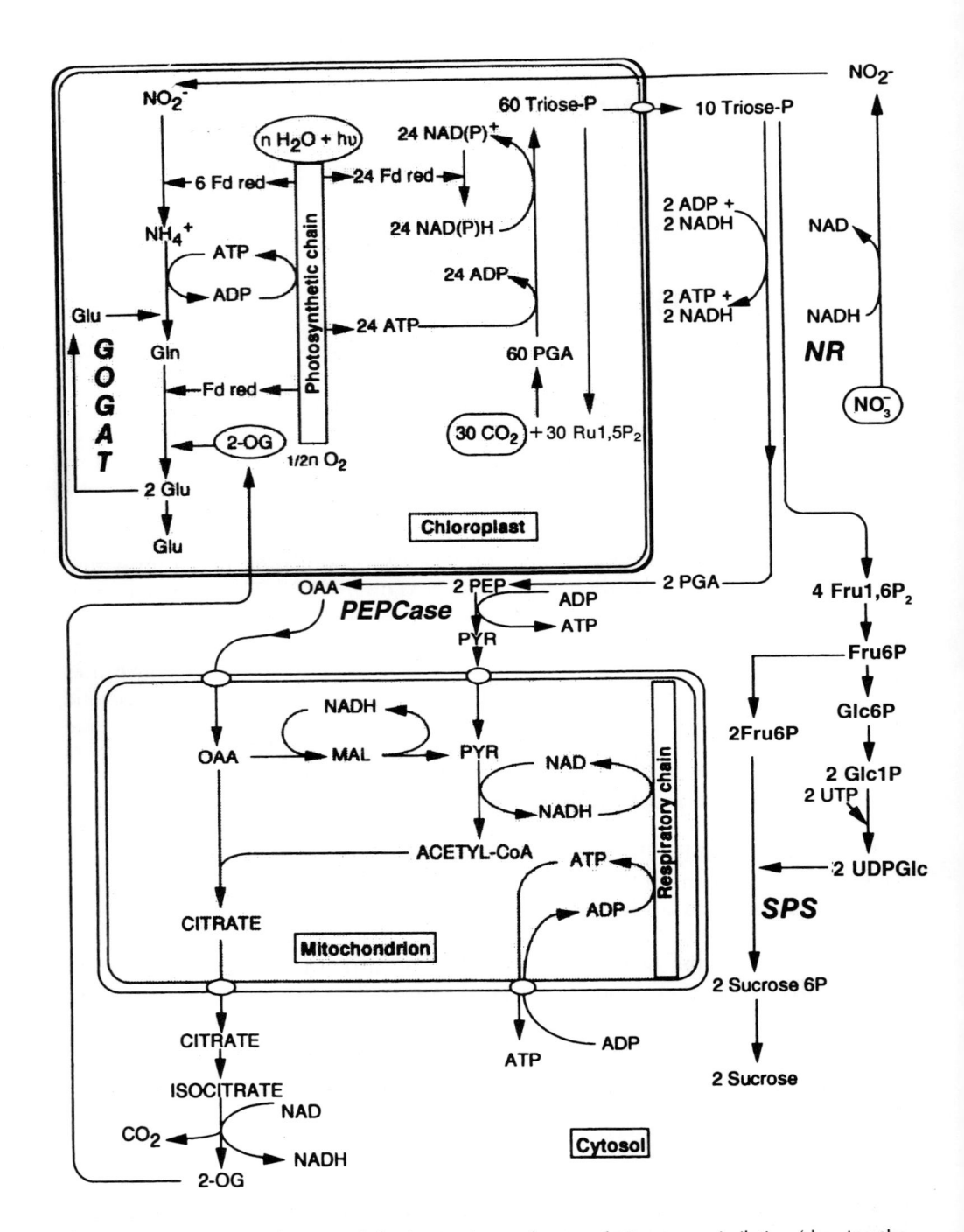

Figure 6.1 Schematic diagram of the interacting pathways of nitrogen assimilation (showing the key regulatory enzyme nitrate reductase, NR) sucrose biosynthesis (showing the key regulatory enzyme sucrose phosphate synthase, SPS) and the anapleurotic pathway (showing the key regulatory enzyme phosphoenolpyruvate carboxylase, PEPCase) 2-PGA, 2 phosphoglycerate

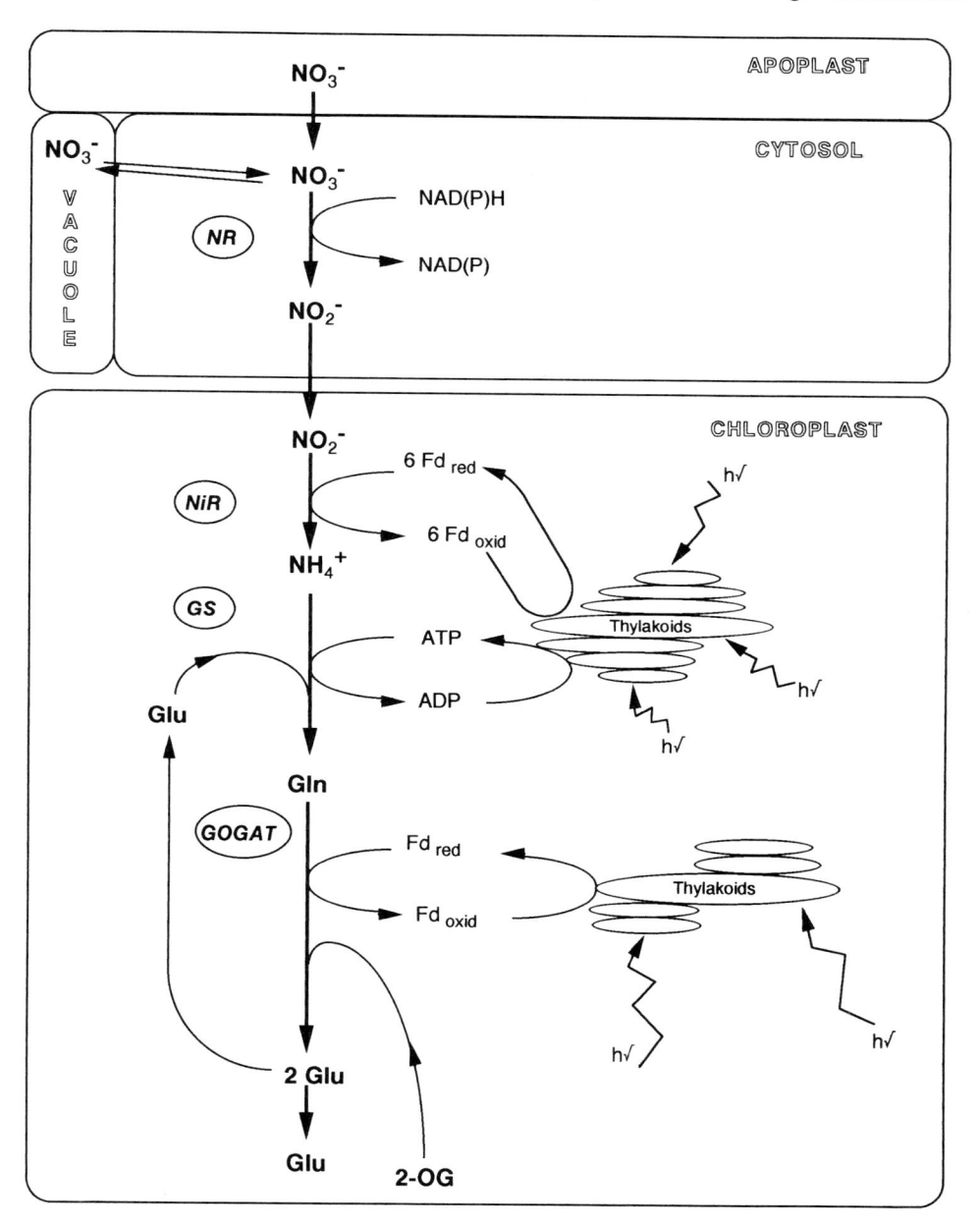

Figure 6.2 Schematic representation of the pathway of nitrogen assimilation and its energy requirements

in the partitioning of nitrate assimilation between root and shoot (Andrews, 1986). NR is a phosphorylated protein; it is phosphorylated in the dark resulting in decreased NR activity. Dephosphorylation in the light results in increased activity.

Phosphoenolpyruvate carboxylase (PEPCase) is also regulated by reversible protein phosphorylation (MacKintosh and MacKintosh, 1993; Nimmo, 1993). Activation of PEPCase is enhanced by light in C_4 and C_3 plants (Budde and Chollet, 1986; Duff and Chollet, 1995; Li and Chollet, 1993; Pierre et al., 1992). Changes in phosphorylation state are generally measured by assaying the enzymes under non-optimal conditions, i.e. in the presence of either Mg^{2+} or EDTA for NR or in the absence or presence of malate for PEPCase (Jiao and Chollet, 1991; MacKintosh et al., 1995). PEPCase catalyzes the carboxylation of phosphoenolpyruvate (PEP) to form oxaloacetate (OAA) utilizing atmospheric CO_2 (as HCO_3^-). This cytosolic enzyme has multiple roles. It fixes CO_2 during photosynthesis in C_4 and CAM plants (O'Leary, 1982; Muller and Kluge, 1983). It is a key enzyme of the anapleurotic pathway that replenishes levels of OAA in the TCA cycle during N assimilation (Melzer and O'Leary, 1987). In C_3, CAM and C_4 plants, PEPCase is sensitive to allosteric regulation. Malate is a particularly powerful inhibitor of PEPCase activity (Schuller and Werner, 1993). In C_4 plants PEPCase shows strong light/dark modulation. This is achieved by phosphorylation of the enzyme protein. Phosphorylation is regulated directly through photosynthesis. It results in an increase in maximal catalytic activity and a decrease in the sensitivity to inhibition by malate (Bakrim et al., 1993; Jiao and Chollet, 1991, 1992; Job et al., 1978). In C_3 plants light/dark modulation has also been demonstrated (Duff and Chollet, 1995; Rajagopalan et al., 1993; Van Quy et al., 1991) but it has not been found in all cases (Chastain and Chollet, 1989; Leport et al., 1996). Feeding NO_3^- to both C_3 and C_4 leaves in the light causes an increase in PEPCase activity, an increased C flow into amino acid synthesis and a concomitant reduction of sucrose synthesis via regulation of sucrose phosphate synthase (SPS) activity (Foyer et al., 1994a,c; Van Quy et al., 1991). The precise mechanisms of N-mediated control are uncertain but changes in the regulation of protein kinase activity have been implicated (Champigny and Foyer, 1992; Duff and Chollet, 1995). In wheat, Duff and Chollet (1995) reported that NO_3^- feeding to illuminated leaves caused a several-fold greater induction in PEPCase kinase activity than in leaves given light alone, suggesting N may amplify any light-induced effect. The PEPCase kinase activity, which controls the phosphorylation state of PEPCase, is regulated by a circadian oscillator in CAM plants (Carter et al., 1991) and by illumination in C_4 plants (Jiao and Chollet, 1991; McNaughton et al., 1991). In both cases the appearance of kinase activity requires protein synthesis.

SPS, NR and PEPCase are located in the cytosol. NR and PEPCase are linked to amino acid synthesis while SPS is involved in sucrose synthesis. All these enzymes respond to changes in irradiance and N-supply. Co-ordinated activation of NR and PEPCase and inactivation of SPS, essentially prevents unregulated competition for C skeletons and energy between the pathways of sucrose and amino acid biosynthesis. The hypothesis that competition for available energy and C resources dominates the pathways of C- and N-assimilation is widely accepted (for reviews, see Huppe and Turpin, 1994; Oaks, 1994). While there is strong evidence to support this view in algae (Huppe and Turpin, 1994), competition does not appear to be a dominant feature of the C/N interaction in higher plants. Phosphorylation of NR, PEPCase and SPS co-ordinates nitrogen assimilation, CO_2 fixation and carbon partitioning, and essentially prevents competition for resources (Foyer et al., 1996).

6.2 Manipulation of nitrogen assimilation in transformed plants

6.2.1 Manipulation of NR

NR activity is controlled at both transcriptional and post-transcriptional levels. NR is induced at the transcriptional level by NO_3^-, light and carbohydrates (Cheng *et al.*, 1992; Vincentz *et al.*, 1993; Warner and Kleinhofs, 1992) and inhibited by glutamine (Deng *et al.*, 1991). The quantity of NR message and protein vary diurnally in the cell. Maximum NR message levels are observed at the end of the night while the quantity of NR protein is maximal during the photoperiod (Galangau *et al.*, 1988). In addition to this transcriptional regulation NR activity is regulated by covalent modulation involving phosphorylation/dephosphorylation of the enzyme protein in response to light/dark transitions or variations of CO_2 availability (Kaiser and Huber, 1994). When NR is phosphorylated it remains active but phosphorylation permits the binding of a protein inhibitor (NIP; see Chapter 8) in the presence of Mg^{2+} ions. Consequently, in *in vitro* measurements, it is more inhibited in the presence of Mg^{2+} than the dephosphorylated protein. The *in vitro* NR activation state is defined as the ratio of the activity measured in the presence of Mg^{2+} to the activity measured in the presence of EDTA. NR inactivation is therefore a multiple step process involving first the phosphorylation of the protein followed by binding of an inhibitor protein (Glaab and Kaiser, 1995; MacKintosh *et al.*, 1995). NR inactivation occurs during the night, and in situations where CO_2 is depleted or the cytosolic pH increased (Kaiser and Brendle-Behnisch, 1991; Kaiser and Forster, 1989; Pace *et al.*, 1990). It correlates with phosphorylation of a serine residue which has been identified in spinach and *Arabidopsis* (Bachmann *et al.*, 1996; Su *et al.*, 1996). Two NR protein kinase(s) which catalyze the phosphorylation of the NR protein have been partially identified. The activity of these kinases is inhibited by photosynthetic metabolites (DHAP, Glc6P, Fru1,6P$_2$). Both kinases are calcium-dependent and have similar specificity determinants but are distinct in their immunological properties (Bachmann *et al.*, 1995, 1996).

The introduction of a 35S-NR gene construct into an NR-deficient mutant of *Nicotiana plumbaginifolia* has allowed the production of tobacco plants where expression of NR is constitutive (Vincentz and Caboche, 1991). In these plants NR activity is no longer regulated at the transcriptional level by NO_3^-, light or carbohydrates (Vincentz *et al.*, 1993), but is still regulated at the post-transcriptional level by protein turnover (Vincentz and Caboche, 1991) and post-translational phosphorylation/dephosphorylation regulation (Nussaume *et al.*, 1995). Constitutive NR expression has allowed confirmation of the regulation of the NR gene at the post-transcriptional level by NO_3^-, light, carbohydrates and glutamine. Experiments with these plants have demonstrated that light is involved in the regulation of translation of NR mRNA and in the stability of the NR protein. Post-transcriptional regulation of NR by phosphorylation/dephosphorylation modulation of the NR protein induced by light still occurs in these transformed plants (Ferrario *et al.*, 1996). These plants have provided an ideal system in which to study the relative roles of transcriptional and post-transcriptional regulation of NR and its relationship to the N and C interaction in plants exposed to different environmental conditions.

6.2.2 35S-NR expression

When the NR-deficient mutant of *N. plumbaginifolia* was transformed with a full-length tobacco NR cDNA fused to the CaMV 35S promoter, the transformed plants were phenotypically comparable to the untransformed controls (Foyer *et al.*, 1993, 1994b; Vincentz and Caboche, 1991). Constitutive NR expression caused a twofold increase in maximal extractable NR activity, a decrease in the foliar NO_3^- content and an increase in the total amino acid (largely due to an increase in the glutamine pool) contents without any changes in total N, soluble sugars or starch (Foyer *et al.*, 1994b; Quilleré *et al.*, 1994). This would suggest that an increase in NO_3^- assimilation in these plants *in situ* led to greater NH_4^+ incorporation into glutamine but did not result in increased total soluble protein or total N contents; further experimental verification of an increased nitrate assimilation rate using ^{15}N-labelling is necessary to verify this.

Tobacco plants transformed with a 35S-NR construct contain both NR protein expressed by the constitutive promoter as well as that produced by the native endogenous promoter which is subject to metabolic regulation (Dorlhac de Borne *et al.*, 1994). These plants showed an increase in the $^{15}NO_3^-$ assimilation rate when supplied with 1 mM $^{15}NO_3^-$ (Gojon and Touraine, 1995). In this case, the increased efficiency in NO_3^- reduction was partially negated, however, by a decrease in NO_3^- uptake by the roots. Consequently, the rate of amino acid synthesis was unchanged. The observed increase in the glutamine to glutamate ratio and the glutamine to sucrose ratio in the 35S-NR plants had no apparent effect on other enzymes such as SPS or PEPCase, or on photosynthesis (Foyer *et al.*, 1994b). Glutamine is a signal molecule in plants. It is cycled between leaves and roots. An increase in root glutamine content will cause a decrease in NO_3^- uptake (Gojon and Touraine, 1995) if the hypothesis of Cooper and Clarkson (1989) is correct.

Nitrogen deficiency

The effect of nitrogen supply on NR activity in the 35S-NR *N. plumbaginifolia* was studied because N-deficiency should have no direct effect on NR gene expression in these plants. However, NR activity was found to decrease not only in the untransformed controls but also in the 35S-NR plants as a result of N limitation (Ferrario *et al.*, 1995). This led to the conclusion that NR turnover is specifically increased by N-deficiency (Ferrario *et al.*, 1995). Nitrate is required to stabilize the NR protein since NR protein levels and NR activity decrease within 24 h of N-deprivation (Galangau *et al.*, 1988). The NR message levels decreased much more slowly (after 14 days) as a result of N-deprivation (Galangau *et al.*, 1988). Photosynthesis and growth were both inhibited by N limitation in a similar manner in both 35S-NR and untransformed controls. Our results indicate that constitutive NR expression in *N. plumbaginifolia* did not confer any great advantage to tobacco plants faced with N-deficiency.

CO_2 enrichment

N assimilation is tightly linked to C metabolism in higher plants because C skeletons produced by photosynthesis are used for the assimilation of inorganic

N into amino acids (Figure 6.1). An increase in atmospheric CO_2 generally results in an increase in CO_2 assimilation in C_3 plants because of the suppression of photorespiration. Carbohydrate accumulation (sucrose or starch) is frequently observed in leaves following CO_2 enrichment and there is generally an increase in the C/N ratio. The role of NR in determining the C/N ratio in plants grown at elevated CO_2 was studied using the 35S-NR transformants. The systems used to cultivate plants subjected to CO_2 enrichment (pots, hydroponics or field conditions) are very important in determining the outcome of such experiments since plants grown at high CO_2 are prone to N-deficiency. 35S-NR and untransformed controls were grown for two weeks in conditions of CO_2 enrichment (1000 ppm) in pots or in hydroponic culture. In pot-grown plants, the leaf NO_3^- content decreased considerably in both 35S-NR plants and untransformed controls following CO_2 enrichment. Foliar NR activity and amino acid levels (particularly glutamine) also decreased suggesting that NO_3^- was depleted in these experiments following CO_2 enrichment (Ferrario-Méry et al., 1997). Leaf NR message levels also decreased in both types of plants. Since NR expression in the 35S-NR transformants is no longer responsive to nitrate supply, this suggests that NR protein and NR message turnover are influenced by NO_3^- and/or amino acid levels (Ferrario-Méry et al., 1997). In untransformed plants this type of regulation occurs in parallel to the induction of the NR gene transcription by NO_3^- and its repression by glutamine. This phenomenon appears to be specific for NR in these conditions since the activities of other enzymes (SPS, PEPCase, GS) were not modified (Ferrario-Méry et al., 1997). In hydroponically-grown plants restrictions to growth and nutrient uptake are less severe than in those occurring in pots. In this situation NR activity, NO_3^- and amino acid levels were not decreased in leaves and in root tissues following CO_2 enrichment. The NR mRNA was decreased, however, and the glutamine (Gln) level was increased in all plants exposed to high CO_2 even though NR activity and NO_3^- levels were unchanged. 35S-GUS expression was used in these experiments to determine the sensitivity of the 35S promoter to high CO_2 conditions (Ferrario-Méry et al., 1997). GUS activity and GUS mRNA were not modified by CO_2 enrichment. This suggests a specific effect of CO_2 enrichment on the turnover of NR mRNA rather than a change in expression.

The NR activation state did not vary in response to CO_2 enrichment suggesting that the phosphorylation/dephosphorylation state of the protein was not responsive to CO_2 enrichment (Ferrario-Méry et al., 1997). High CO_2-induced increases in biomass were similar in all plants and the C/N ratio was increased to the same extent suggesting that NR activity is not limiting N assimilation in tobacco in high CO_2 conditions.

Water stress

Since N assimilation in leaves requires energy, reducing power and C skeletons derived from photosynthesis, any stress which decreases photosynthetic electron transport and C assimilation rates would be expected to lead to a decrease in N assimilation (Kaiser and Forster, 1989). When tobacco plants were deprived of water for four days, NR activity declined within 48 h in the untransformed controls but not in the 35S-NR plants. NR message levels declined more slowly than NR activity in the untransformed plants and appeared to be more stable than NR protein in drought conditions. NR mRNA levels increased transiently in the 35S-NR plants following water deprivation before decreasing within four days. Rehydration after three days of water stress restored the leaf NR activity of the untransformed controls to a level similar to that of the water-replete plants within 24 h. NR message levels increased more slowly than NR activity. The NR activation state was not modified by water stress, suggesting that increases in NR activity following water stress in tobacco are due to decreased turnover of the NR protein. This may involve increased translation of existing NR mRNA since there was a delay in NR mRNA loss compared to that in NR activity as a result of water deprivation. Water stress was accompanied by a decrease in NO_3^- content in the leaves, a less pronounced decrease in total amino acids, and an increase in foliar hexoses. These compounds are considered to be involved in the regulation of NR protein turnover. Such regulation would serve to decrease NO_3^- assimilation when photosynthesis is decreased as a result of water deficit. Different strategies may be employed to limit N assimilation in response to water deprivation in different plant species, for example in maize leaves subjected to water deprivation NR message levels decreased as quickly as the NR activity, whereas in tobacco there was a lag between the decrease of NR protein and NR message levels.

6.2.3 Constitutive expression of an NR protein involving a 5′ end deletion

Transformation of the NR-deficient mutant with a construct composed of a full-length tobacco NR cDNA with an internal deletion of 168 pb in the 5′ end under the control of 35S CaMV promoter has produced transformants which lack transcriptional regulation and also post-transcriptional regulation since the phosphorylation of the NR protein is impaired (Nussaume et al., 1995). The NR protein contains three highly conserved functional domains (FAD, haem, and molybdenum cofactor). These are involved in the transfer of electrons from NADH to NO_3^-. The N-terminal region is not particularly well conserved and its function remains equivocal. The presence of acidic residues in the regions could provide sites of electrostatic interaction with other proteins. An internal deletion of 56 amino acids in the N terminal MoCo domain of the NR protein (ΔNR) was performed to determine whether this region is involved in the post-translational regulation of the NR protein, for example, in proteolysis, inactivation by protein phosphorylation, or association with the NR protein inhibitor. The resultant transformants contained substantial NR activity and displayed normal growth characteristics comparable to the untransformed controls but, similar to the 35S-NR transformants, the ΔNR gene was not regulated at the transcriptional level. In addition foliar NR activity was no longer regulated by light at the post-transcriptional level, the activation state of the enzyme remaining constant during light/dark transitions. The NR protein was always in the activated form (Nussaume et al., 1995). Interestingly, the ΔNR deletion does not involve the serine residue involved in protein phosphorylation. This site occurs in the hinge region of the NR protein which separates the molybdenum cofactor and the haem domains in spinach (Ser 543) and in *Arabidopsis* (Ser 534) (Bachmann et al., 1996; Su et al., 1996). Whether phosphorylation of the protein still occurs in the ΔNR protein or whether linkage to the NR protein inhibitor is impaired by the N terminal deletion remains unresolved. Diurnal fluctuations in total foliar amino acid pools, particularly Gln, which are observed in the untransformed plants, are damped in the ΔNR plants. Experiments using ^{15}N incorporation to establish nitrogen assimilation rates have not as yet been performed but nitrogen assimilation at night in the ΔNR plants may be limited by the availability of reducing power in the absence of photosynthetic electron transport.

Such studies demonstrate the complexity of NR regulation at the transcriptional, post-transcriptional and post-translational levels. These regulatory mechanisms act in parallel to determine the overall rate of NR activities. The metabolic factors acting at each of these steps have not been clearly identified. NO_3^-, N metabolites such as glutamine, and carbohydrates such as sucrose, are involved in induction of NR gene transcription and may also be involved in turnover of the NR mRNA and NR protein. The relative importance of each of these regulatory devices may vary from species to species; NR is inactivated in the absence of NO_3^- in transformed *Chlamydomonas* constitutively expressing NR (Navarro *et al.*, 1996) and in tobacco (Ferrario *et al.*, 1995). In our experiments we have demonstrated the importance of NO_3^- on the turnover of NR mRNA and NR protein (Ferrario *et al.*, 1995, 1996; Ferrario-Méry *et al.*, 1997). These studies also demonstrate that over-expression of a single component of a metabolic pathway (even a key enzyme in control) may have a relatively small effect on overall flux if the enzyme activity is highly regulated. Post-transcriptional regulation involving mRNA stability, translation efficiency, and protein stability in enzymes such as NR, have a profound effect on the outcome of experiments involving constitutive or other types of over-expression.

6.2.4 Transformed plants with altered activities of glutamine synthetase and glutamate synthase

The reduction of NO_3^- by NR and NiR results in the formation of ammonia (Figure 6.2). This is not the only mechanism of ammonia production in plants. The photorespiratory pathway liberates ammonia at high rates from the conversion of glycine to serine (Hausler *et al.*, 1994a,b; Leegood *et al.*, 1995). In addition, ammonia is released during some specific interconversions of some amino acids and also in reactions associated with both the formation and breakdown of transport compounds. Ammonia is thus formed continuously and ubiquitously throughout the plant (Figure 6.3). Free ammonia in plant tissues is a valuable resource but it is also toxic (Mehrer and Mohr, 1989). Its recovery, therefore, is of great importance. This is particularly important in the leaves of C_3 plants where rates of ammonia re-assimilation following photorespiration can be several-fold that of primary assimilation following the reduction of nitrate.

There is now substantial evidence that glutamine synthetase (GS) activity is the sole means by which higher plants incorporate ammonia into amino acids. GS catalyzes the ATP-dependent conversion of glutamate into glutamine (Figure 6.2) and is found in all plant tissues. GS exists as a small multi-gene family in higher plants (Hirel *et al.*, 1993). In leaves there are two isoenzymes localized in the cytoplasm (GS1) and the chloroplast (GS2) (McNally *et al.*, 1983), the relative

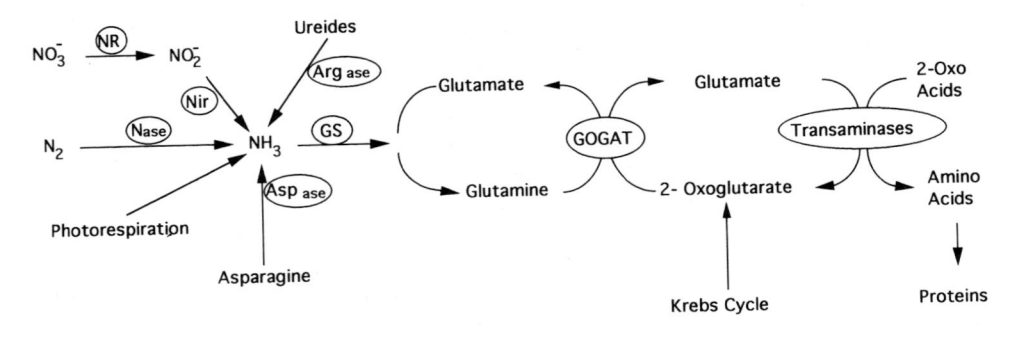

Figure 6.3 The pathways of ammonia production in plants and the reassimilation of ammonia by glutamine synthetase and glutamate synthetase

proportion of which being dependent on the species of plant studied and the developmental status, for example the level of expression of cytosolic GS increases during leaf senescence in rice (Kamachi et al., 1991). Additionally GS1, where present, has recently been found to be exclusively located in the vascular system (Brears et al., 1991; Carvalho et al., 1992). In non-photosynthetic tissues GS1 is encoded by different genes expressed differentially in roots, stems and flowers. N-fixing root nodules of leguminous species have been shown to contain very high activities of GS, the result of the expression of a nodule-enhanced plant encoded gene (Cullimore and Bennett, 1989; Marsolier et al., 1995).

Most of the glutamine formed in the GS reaction is channelled into the formation of other amino acids via the synthesis of glutamate, a reaction catalyzed by glutamate synthase (GOGAT). In this reaction, the amide group of glutamine is transferred to 2-oxoglutarate, forming two molecules (Figure 6.2). There are two forms of GOGAT present in higher plants. They differ in their requirement for a source of reductant. Ferredoxin-dependent GOGAT is located in the chloroplast and is the more predominant form in leaves. NADH-dependent GOGAT has also been found in leaves of higher plants, although in much smaller quantities than the ferredoxin-dependent form, and its importance is considered to be more related to its presence in non-photosynthetic tissues, where it exists as the sole form.

Much of the earlier work exploring the function of the GS/GOGAT system was prompted by the involvement of these two enzymes in the recovery of ammonia synthesized by photorespiratory reactions. Mutants deficient in enzymes of the photorespiratory pathway were selected on the basis of exhibiting stress symptoms when transferred to ambient CO_2 conditions following growth under high CO_2 (Somerville and Ogren, 1981). The physiological properties of these mutants are directly relevant to later experiments attempting to transform plants genetically to contain altered levels and patterns of activities of GS and GOGAT and the results will be summarized before discussing the application of transformation technology in this area.

Barley mutants deficient in GS2 (though containing normal levels of GS1) had enhanced ammonia accumulation in leaves in air (Blackwell et al., 1987). This was subsequently found to be mostly due to photorespiratory release of ammonia in the mitochondria (Oliver et al., 1990). In addition, levels of principal amino acids of the photorespiratory cycle declined in air and the rate of CO_2 assimilation also declined. Mutants deficient in Fd-dependent GOGAT have been isolated for *Arabidopsis thaliana* (Somerville and Ogren, 1980), barley (Blackwell *et al.*, 1988) and pea (Lea *et al.*, 1992). The NADH-dependent form was present at normal levels in all these lines whilst the Fd-dependent form was almost absent. As in the GS mutants, rates of CO_2 assimilation fell when plants were transferred to air from a CO_2-enriched environment. On exposure to air, glutamine levels increased promptly in the mutant barley leaves, although levels of ammonia showed little change. In the same experiment, amounts of other amino acids decreased. A comprehensive examination of barley mutants containing intermediate levels of chloroplastic GS and Fd-GOGAT was made by Hausler *et al.* (1994a,b). For the GS mutants, rates of CO_2 fixation were lower than for the untransformed plants regardless of conditions.

The effects of a decreased GS activity were more pronounced under photorespiratory conditions (1000 µE m^{-2} s^{-1}, 145 ppm CO_2) when leaf ammonia concentrations increased and total amino acid pools decreased. Additionally, the quantum yield of PSII declined and there was a lower rate of linear electron transport when GS activities were less than 50% (Hausler *et al.*, 1994a,b). The activation state of Rubisco was also slightly higher in GS mutants under photorespiratory conditions. The authors concluded that GS can exert a high degree of control on the electron requirement for CO_2 assimilation when flux through the photorespiratory pathway is high. Ratios of serine to glycine indicated some possible negative feedback on the activity of glycine decarboxylase. For the Fd-GOGAT mutants, however, little change was noted although there were signs that the extent of cyclic electron transport changed to accommodate the differing requirements for ATP versus NADPH synthesis. In GS mutants with a 50% decrease in GS activity, a 20% reduction in total protein per unit leaf area was observed but no changes in chlorophyll per unit leaf area, chlorophyll *a/b* ratio or specific leaf fresh weight were noted. In GOGAT-deficient plants with only 35% of the GOGAT activity of the untransformed controls the effects were more pronounced. A 30% reduction in total protein per unit leaf area, a 23% reduction in chlorophyll per unit leaf area, a 17% reduction in chlorophyll *a/b* and a slightly lower specific leaf fresh weight and fresh weight/dry weight ratio were observed in these GOGAT-deficient plants. Therefore, a reduction in the activities of these enzymes can have a significant impact on the protein content of leaves and this presumably influences morphological characteristics. The changes in protein content correlated well with changes in activities of Rubisco through the leaf. Hausler *et al.* (1994a) suggest that in the case of the GS plants this is a result of a limitation on ammonia re-assimilation which may result in ammonia loss from the leaf. However, they also noted that the amino acid pools were partially restored during the dark period. In the case of the GOGAT mutants, observed changes in quantities of amino acids probably caused the observed decreases in protein and Rubisco.

GS is a target for a number of commercial herbicides, and this has allowed the selection of an alfalfa cell line which was resistant to L-phosphinothricin (Donn *et al.*, 1984), the resistance mechanism for which was found to be several-fold amplification of the cytosolic form of GS. However, no physiological or growth analysis resulted. The first attempt to produce elevated levels of GS activity in plants using genetic transformation methods was made by Eckes *et al.* (1989). These workers cloned the amplified GS gene of alfalfa and introduced this into tobacco, using a construct which replaced the native promoter with the 35S promoter of the cauliflower mosaic virus (CaMV) in order to obtain high levels of expression. Indeed, levels of the alfalfa GS1 mRNA in the transformed tobacco reached ten times those normally found in alfalfa. Leaf GS activities were five times those seen in untransformed tobacco, and the amounts of GS protein accounted for up to 5% of total leaf soluble protein. Significantly, levels of free ammonia were reduced sevenfold for plants grown on synthetic media. However, although no data were shown, these authors reported no significant alteration in amino acid composition and no pleiotropic effects were reported (Eckes *et al.*, 1989). Temple *et al.* (1993; see Chapter 7) obtained transformed tobacco lines using the cDNA corresponding to the alfalfa GS1 gene, also driven by the 35S promoter. In this case only a relatively modest increase in total leaf activity per unit fresh weight was seen (up to 25%) which nevertheless resulted in plants containing 40% more protein when expressed on a leaf area basis; the leaves of the transformants were 'visibly greener' than the controls, implying a greater concentration of chlorophyll per unit leaf area. Temple *et al.* (1993) also report the effects of introducing the same construct into tobacco plants but with the alfalfa GS1 gene in antisense orientation. These plants had up to a 40% reduction in total leaf GS activity per unit leaf area and a decrease in total soluble leaf protein to a similar magnitude. The antisense RNA appeared to down-regulate both the GS1 and GS2 transcripts, which makes an interesting comparison with the over-expressors where levels of GS2 protein actually exceeded those of the controls (Temple *et al.*, 1993). These authors also performed Western analysis on the transformed plants for two enzymes which have key roles in plant N metabolism: phosphoenolpyruvate carboxylase, which catalyzes the anapleurotic fixation of CO_2 to provide C skeletons for amino acid synthesis, and hydroxypyruvate reductase which catalyzes the conversion of hydroxypyruvate to glycerate in the photorespiratory pathway. Whilst no apparent alteration was seen for the GS over-expressors, reduced amounts of both enzymes were found in the GS antisense plants, which may be caused by a general down-regulation of N metabolism in the leaf due to the lower capacity for ammonia assimilation. However, no further biochemical or physiological analysis was performed. An interesting approach to exploring the various roles of GS in different cellular compartments was taken by Hemon *et al.* (1990) who attempted to direct GS activity to the mitochondria by the use of an appropriate targeting sequence. Unfortunately the plants when fully grown did not express the introduced gene and physiological analysis was not possible.

Also using the 35S promoter, Hirel *et al.* (1992) transformed tobacco plants with a soybean root genomic GS1 gene and obtained plants with an overproduction of cytosolic GS in both roots and leaves of transformed plants. In this case, the cellular localization of GS in leaves was investigated by immunocytochemistry and GS was found to be present in both the cytosol and the chloroplast of leaf mesophyll cells in transformants, whilst in the untransformed controls GS was seen only in the chloroplast. The soybean GS was functional, and represented between 15% and 20% of the total GS activity in both leaves and roots. The novel compartmental distribution had no obvious effects on growth or morphology in greenhouse-grown plants. Interestingly, the expression of GS in the cytosol appeared to induce the expression of an endogenous GS1 gene. The authors speculated that an alteration in cytosolic metabolism may be responsible.

The transformed tobacco plants expressing the root genomic GS1 were grown in the greenhouse at an irradiance of 150–300 µE m^{-2} s^{-1} with a full nutrient solution containing ammonia as the sole N source (1 mM fed every 20 minutes). The total GS activity was increased by between 10% and 30% in the transformed plants compared to the controls (Figure 6.4). No changes in total soluble leaf protein or chlorophyll content were observed (Figure 6.4B,C). However, the leaves of the transformants contained a slightly lower ammonia concentration. The leaves of transformed plants also contained slightly lower

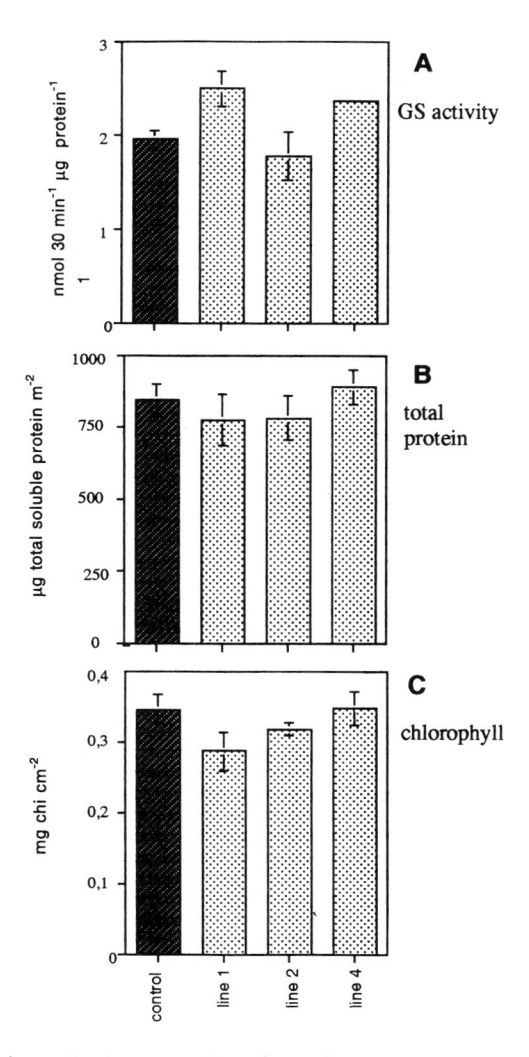

Figure 6.4 The effect of constitutive expression of a soybean root genomic GS1 gene in tobacco on total glutamine synthetase activity (A); total leaf protein (B); and total leaf chlorophyll (C). Three lines of transformants (1, 2 and 4) are compared with the untransformed control

levels of total free amino acids than those of the untransformed controls (Figure 6.5A). No significant differences in the amounts of individual amino acids in the leaves between the transformed plants and the controls were found. The ratio of glutamine to glutamate and serine to glycine was unchanged. The plants were also grown on a higher concentration of ammonia (8 mM) but again little difference was observed between the transformed and the controls. The total amino acid content was somewhat higher in the leaves of transformed plants (Figure 6.5B) suggesting that when leaf ammonia concentrations are high, the supplementary GS activity may result in a higher pool of free amino acids but there was no difference in the actual amounts of ammonia in the leaves of transformed and control plants. In conclusion, the effects of a cytosolic GS activity in leaf mesophyll cells appears to confer little change to growth or leaf protein content. However, there still remains the intriguing possibility that these plants may display unique characteristics under conditions that favour a higher degree of oxygenation of Rubisco and hence greater photorespiratory flux. The reaction that results in the release of ammonia by the photorespiratory pathway occurs in the mitochondria whilst ammonia is re-fixed into amino acids by GS in the chloroplast. The presence of GS activity in the cytosol may result in a

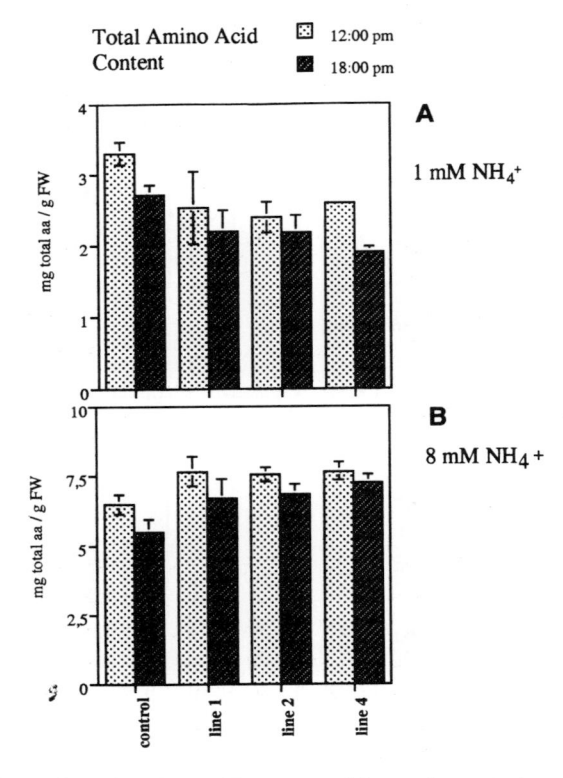

Figure 6.5 A comparison of total amino acid content of leaves harvested at midday or at the end of the photoperiod from untransformed tobacco (control) and tobacco constitutively overexpressing a soybean root glutamine synthetase gene grown on either 1 mM NH_4^+ (A) or 8 mM NH_4^+

higher recovery of ammonia, since it is known that some ammonia is lost from the leaf during photorespiration. Additionally the supplementary GS activity may influence the export of nitrogenous compounds from the leaf during senescence.

6.3 Sucrose biosynthesis

The hypothesis that some degree of metabolic co-ordination exists between source and sink tissues involving synthesis and demand for sucrose, is widely accepted (Foyer and Galtier, 1996). The essential elements of the component pathways have largely been identified but the molecular mechanisms that allow reciprocal control are not fully understood. One important putative control point is the enzyme sucrose-phosphate synthase (SPS). SPS has long been considered to be a key enzyme in the regulation of C assimilation and export from the leaf. The arguments supporting this hypothesis are as follows:

1. The observed positive correlation between the capacity of a leaf to export sucrose and its SPS activity (Huber and Israel, 1982; Rocher, 1988).
2. The induction of SPS activity in leaves following modulation of the source/sink ratio (Rufty and Huber, 1983).
3. The complex regulation of this enzyme involving both direct metabolic regulation, via the levels of Glc6P and Pi, and covalent modulation of the enzyme involving phosphorylation/dephosphorylation modifications responsive to metabolic signals (Doehlert and Huber, 1983; Huber and Huber, 1991; Huber et al., 1987, 1989, 1991; Kalt-Torres and Huber, 1987; Sicher and Kremer, 1984; Stitt et al., 1987).
4. The precise co-ordination of photosynthetic C assimilation and sucrose synthesis involving regulation of SPS activity (Battistelli et al., 1991; Stitt, 1989; Stitt et al., 1988).

SPS is, thus, suggested to be responsive to metabolic regulation by both source (the rate of photosynthesis) and sink (sucrose export and utilization). Regulation of SPS also serves to maintain the level of metabolites in the stroma allowing optimal rates of Calvin cycle activity (Stitt and Quick, 1989).

6.3.1 Over-expression of SPS

Various enzymes of C metabolism (SPS, fructose-1,6-bisphosphate, invertase, ADP-glucose pyrophosphorylase) have been considered to be major control points in the source–sink relationship plants (Foyer, 1987; Geiger, 1987; Herold, 1980; Huber, 1983). Consequently, analysis of the role of these enzymes in transformed plants with modified levels of the activities of these enzymes has recently been undertaken (Galtier et al., 1993; Heineke et al., 1992; Sonnewald, 1992; Sonnewald et al., 1991; von Schaewen et al., 1990). Transformed tomato plants expressing both native SPS and an SPS gene from maize (Worrell et al., 1991) have been used to analyze the role of this enzyme in C assimilation and C partitioning (Galtier et al., 1993, 1995). Regardless of the promoter used, the transformed tomatoes expressing the maize leaf SPS cDNA construct exhibited a higher maximal photosynthetic rate than the untransformed controls. This corroborated previous evidence suggesting a role for SPS in the feedback regulation of C assimilation (Foyer, 1987; Stitt, 1986). Furthermore, there was a strong positive correlation between the increase in the ratio of sucrose to starch in the leaves and SPS activity (Galtier *et al.*, 1993, 1995). The marked increase in the maximum extractable SPS activity in these plants was suggested to be caused by the absence of complete regulation of the introduced SPS protein; for example, the introduced maize enzyme did not appear to be inactivated to a great extent in darkness (Galtier *et al.*, 1993, 1995; Worrell *et al.*, 1991). However, the published sequences of the maize and the spinach SPS cDNA clones indicate that the SPS proteins from C_3 and C_4 plants are remarkably similar with 74% homology at the amino acid level (Klein *et al.*, 1993; Salvucci *et al.*, 1990; Sonnewald *et al.*, 1993). The nature of the difference between the native and the introduced SPS proteins that allows their differential regulation is, as yet, unknown.

Recently, there has been much interest in the effect of high concentrations of sucrose and other carbohydrates on the regulation of leaf metabolism. The rate of CO_2 fixation has been shown to decrease in response to accumulating leaf carbohydrate but this is not always the case and the precise mechanisms and sequence of events remain a matter of debate. High concentrations of carbohydrate may act both directly at a metabolic level (Foyer, 1987), or indirectly on gene transcription (Krapp *et al.*, 1991, 1993; Sheen, 1994). Photosynthetic metabolism was modified in tobacco plants which had been transformed to express yeast invertase in their cell walls (Stitt *et al.*, 1990; von Schaewen *et al.*, 1990). Following accumulation of carbohydrate, photosynthesis was inhibited and respiration enhanced as a result of a decrease in Calvin cycle enzymes and an increase of glycolytic enzymes. Of particular interest is the recent finding that sucrose is involved in the modulation of gene expression in leaves (Krapp *et al.*, 1993; Sheen, 1990). Important assimilatory enzymes may be affected by sucrose at the level of gene expression, e.g. the light-regulated expression of NR has been shown to be mediated by sucrose (Cheng *et al.*, 1992; Vincentz *et al.*, 1993). Furthermore, metabolic repression of the transcription of seven photosynthetic gene promoters by sucrose has been documented (Sheen, 1990, 1994).

Galtier *et al.* (1993) studied the SPS activity of different tissues within tomato plants when the maize SPS gene was expressed under the control of rbcS promoter. In this case, SPS activities were highest in the leaves, the total extractable SPS activity in roots and petioles being ten times less than that in the leaves. The level of SPS activity in the roots of the rbcS transformants was three times higher than in the untransformed controls. In the second type of transformed plant the maize leaf SPS cDNA construct was expressed under the control of the CaMV 35S promoter. The CaMV 35S promoter is considered to produce constitutive gene expression in all cell types. Williamson *et al.* (1989) have shown that promoter activity depends primarily on the age of the tissues. The rates of photosynthesis and sucrose synthesis were increased in the high SPS expressors compared to the untransformed controls and the sucrose/starch ratios were highest during the photoperiod in the leaves of all plants expressing high SPS activity regardless of the type of promoter (Galtier *et al.*, 1995).

6.3.2 Relative growth rates (RGR) and biomass production in tomato plants expressing a maize SPS cDNA

DW accumulation and average RGR of the untransformed control plants and the rbcS-SPS and 35S-SPS lines were measured during the first six weeks of vegetative development (Figure 6.6). The increase in RGR observed in 35S-SPS line 13 was slight but was sustained throughout the growth period (Figure 6.6A). Due to the exponential increase in growth this resulted in a considerable increase in biomass accumulation at the time of harvest when the plants were close to attaining full size (Figure 6.6B). The relative accumulation of biomass was compared in two independent rbcS lines (9 and 11) to that in untransformed controls (Figure 6.7). The plants were grown in the greenhouse during the summer of 1994 and harvested after four months at the end of the growing period. The biomass distribution within the populations is shown in Figure 6.7. When biomass accumulation is plotted against SPS activity for these plants there is no correlation between foliar SPS activity and biomass accumulation (Figure 6.8).

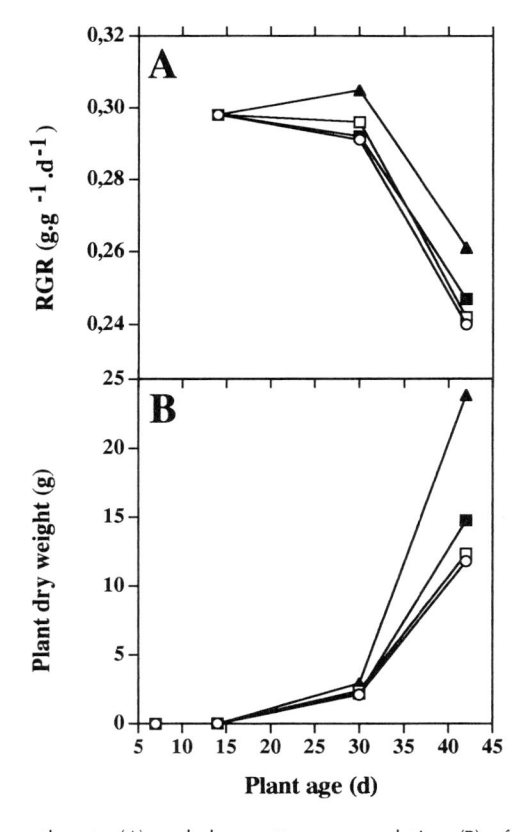

Figure 6.6 Relative growth rate (A) and dry matter accumulation (B) of untransformed tomato plants (○) and rbcS lines 18 (□) and 9 (■) and 35S-SPS line 13 (▲). Measurements were made during the first six weeks of vegetative growth. Ten plants were used in each line

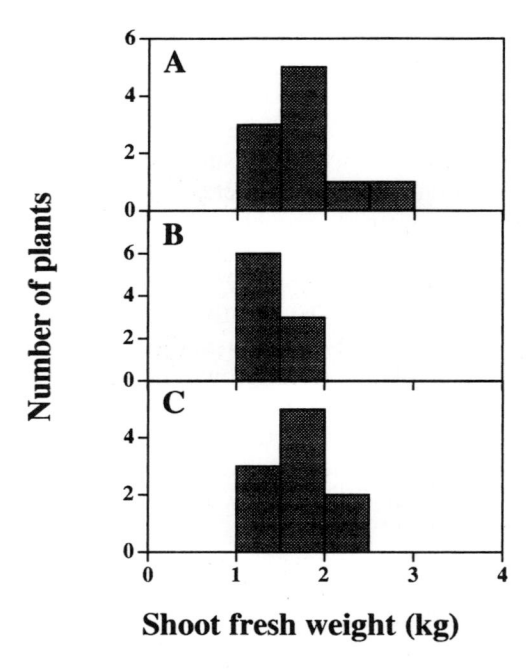

Figure 6.7 Distribution of biomass in four-month-old plants from rbcS lines 11 (A) and 9 (B) and untransformed controls (C). Plants (ten per line) were grown in the greenhouse in Versailles during the summer of 1994

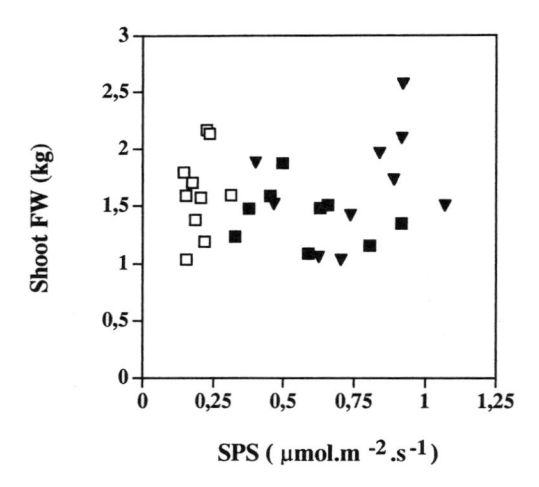

Figure 6.8 The relationship between SPS accumulated biomass in two rbcS-SPS lines 9 (■) and 11 (▼) and untransformed controls (□) measured in the same study as Figure 6.7

In contrast to the rbcS lines, the 35S-SPS lines 12 and 13 could be visually distinguished from the untransformed controls at an early stage. The biomass distribution within the populations of 35S-SPS lines 12 and 13 and the untransformed controls after four weeks growth in controlled conditions is given in

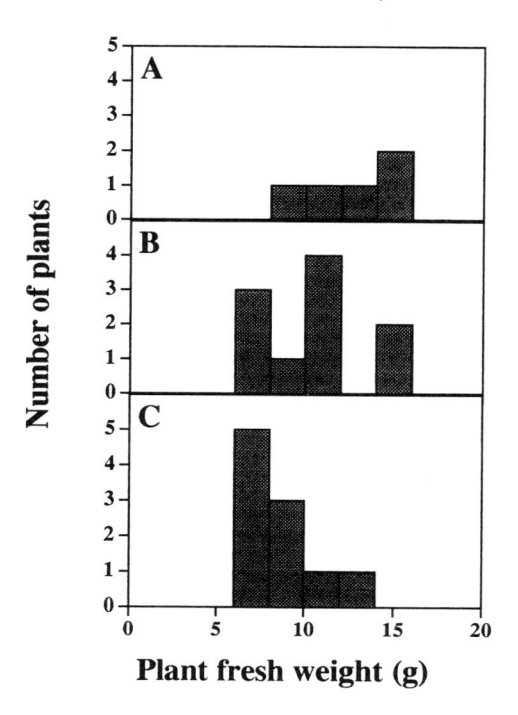

Sucrose biosynthesis and nitrogen assimilation

Figure 6.9 Distribution of biomass in populations of four-week-old plants (ten per line) from 35S-SPS 13 (A), 12 (B) and untransformed controls (C). In this experiment plants were grown at 450 µmol m^{-2} s^{-1} with a 15-hour photoperiod

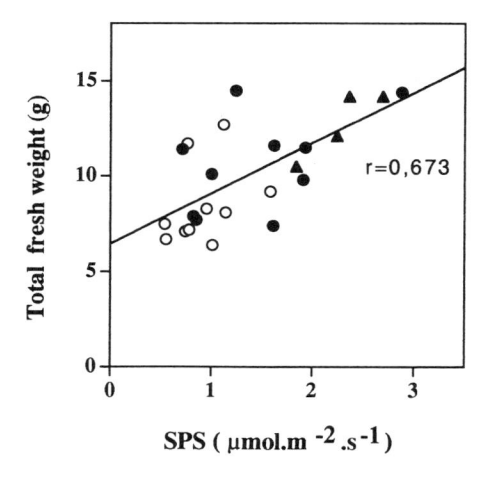

Figure 6.10 The relationship between SPS activity and accumulated biomass for two 35S-SPS lines: 12 (●), and 13 (▲), and the untransformed controls (○) measured in the same study as Figure 6.9. The correlation coefficient is given on the figure

Figure 6.9. In this case when biomass accumulation is plotted against SPS activity there is a strong positive correlation between the two parameters (Figure 6.10).

The biomass accumulated by untransformed controls and rbcS-SPS line 9 and 35S-SPS line 13 was compared after six weeks growth under a range of conditions (Figure 6.11). In these lines, the leaf SPS activity of the population is similar (Galtier et al., 1995) but the promoter was different. The fresh weight (FW)/DW ratio was constant in all plant types (Figure 6.10). Biomass accumulation was comparable in transformed controls and the untransformed controls regardless of the growth conditions (Figure 6.9) but there was always a marked increase in biomass in the lines. Line 18 is an rbcS line where the SPS gene is not expressed and the leaf SPS activity is comparable to the untransformed controls. In contrast to line 13, lines 18 and 9 showed no tendency to increasing biomass accumulation with increasing foliar SPS activity (Figure 6.12).

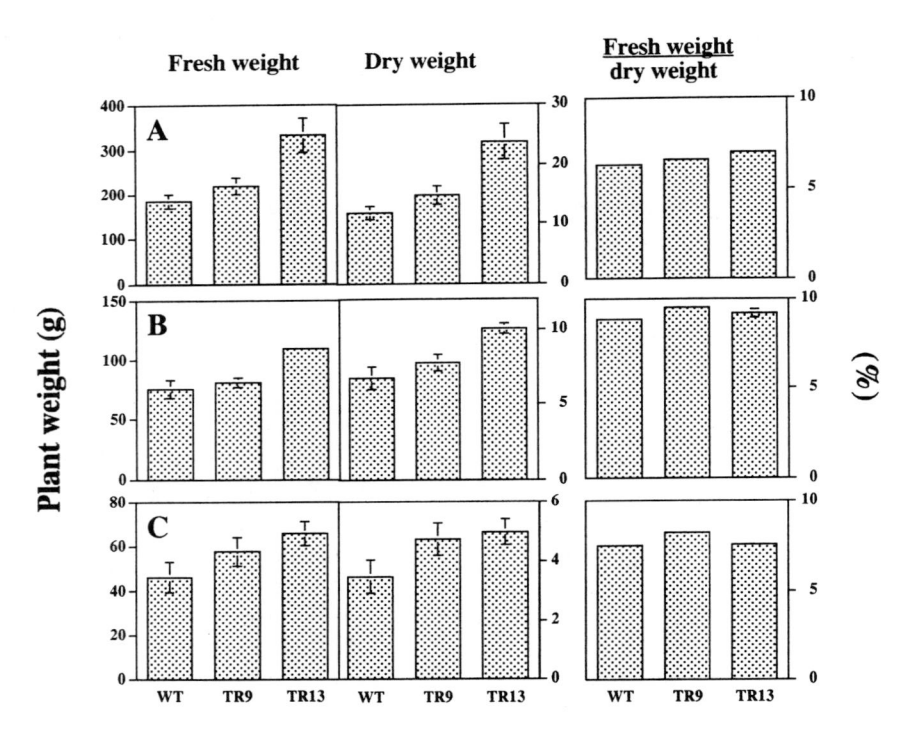

Figure 6.11 Biomass production in control (WT) and high rbcS-SPS line 9 and 35S-SPS line 13 grown under differing conditions of irradiance and photoperiod (300 µmol m^{-2} s^{-1}, or 450 µmol m^{-2} s^{-1} for a 15 h photoperiod, A and B respectively, and 450 µmol m^{-2} s^{-1} with an 8 h photoperiod, C). Plants were harvested six weeks after transfer to the growth chamber. Ten plants were used for each line. Mean plus standard error values are given in all cases but were frequently too small to be visible in the figure. Biomass accumulation cannot be compared between A, B and C because these experiments were not performed simultaneously. Seedlings were moved into the chambers at different stages of development

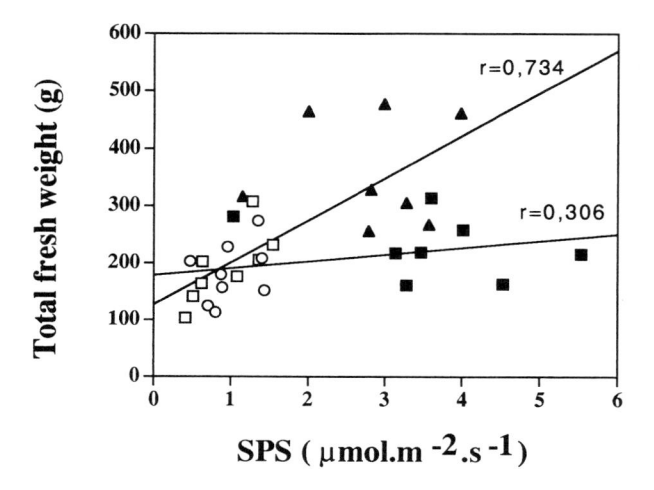

Figure 6.12 The relationship between the SPS activity of the leaves and accumulated biomass of the whole plants for plants grown with a 15 h photoperiod at an irradiance of 300 µmol m^{-2} s^{-1} for 42 days. Untransformed controls (□) and transformed lines 18 (○) and 9 (■) where the maize SPS gene was expressed under the control of the rbcS promoter are shown as well as line 13 (▲) where the SPS gene was expressed under the control of the CaMV promoter. Correlation coefficients are given on the figure

6.4 Conclusions and perspectives

The assimilation and re-assimilation of ammonia in the plant is a process fundamental to plant growth and, due to the role played by nitrogenous compounds in the photorespiratory cycle, is involved in determining the capacity to withstand light and temperature associated stress. Indeed photorespiratory conditions are known to induce the expression of GS genes (Edwards and Coruzzi, 1989). The work using mutants deficient in GS activity (Hausler et al., 1994a,b) has shown that this enzyme is involved in the integration of a number of cellular processes. However, to date the use of transformation technology in the investigation of ammonia assimilation and physiology in general has been limited (see Lea and Forde (1994) for a more detailed discussion) and there is particularly a paucity of data concerning the use of transformation technology to explore GOGAT, and the roles played by the various forms of GS in the plant.

In greenhouse and in controlled environments the constitutive over-expression of SPS dramatically increased plant biomass production while over-expression of SPS in photosynthetic tissues alone had only a marginal effect. The accumulation of biomass was increased in the 35S-SPS plants in all experiments in all growth conditions. This was not the case in the rbcS-SPS plants where biomass accumulation was comparable to the untransformed plants. We have, thus far, fully characterized two lines of transformants of the 35S-SPS type and two lines of the rbcS type.

Improved rates of photosynthesis in air and at elevated CO_2, and modification in foliar C partitioning in favour of sucrose are common to all the transformed tomato plants expressing the maize SPS (Galtier et al., 1995). These new characteristics are, therefore, not the result of somaclonal variation consecutive to tissue culture but the result of increased SPS activity in the leaves.

The quantity of sucrose in a tissue other than the photosynthetic cells can limit the accumulation of C by other tissues. Whether this effect is on export or on C accumulation in the sink tissues we have not yet determined. We can conclude that increases in biomass are possible via genetic manipulation of SPS activity when SPS is expressed constitutively and the activity present is not deactivated *in situ* to a level similar to that of the untransformed controls. A possible candidate for the site of a beneficial effect of increased SPS activity would be the perivascular parenchyma. SPS, over-expressed in these cells, may promote the higher cytoplasmic sucrose contents needed to drive increased phloem loading and export. If these cells have generally a low level of SPS they might not be able to sustain the high sucrose gradients necessary for increased phloem loading.

References and further reading

ANDREWS, M. (1986) The partitioning of nitrate assimilation between root and shoot of higher plants. *Plant Cell Environ.* **9**, 511–519.

BACHMANN, M., McMICHAEL, R.W., HUBER, J.L. JR, KAISER, W.M. and HUBER, S.C. (1995) Partial purification and characterization of a calcium-dependent protein kinase and an inhibitor protein required for inactivation of spinach leaf nitrate reductase. *Plant Physiol.* **108**, 1083–1091.

BACHMANN, M., SHIRAISHI, N., CAMPBELL, W.H., YOO, B.C., HARMON, A.C. and HUBER, S.C. (1996) Identification of Ser-543 as the major regulatory phosphorylation site in spinach leaf nitrate reductase. *Plant Cell* **8**, 505–517.

BAKRIM, N., PRIOUL, J.-L., DELEENS, E., ROCHER, J.-P., ARRIO-DUPONT, M., VIDAL, J., GADAL, P. and CHOLLET, R. (1993) Regulatory phosphorylation of C_4 phospho*enol*pyruvate carboxylase: a cardinal event influencing the photosynthesis rate in *Sorghum* and maize. *Plant Physiol.* **101**, 891–897.

BATTISTELLI, A., ADCOCK, M.D. and LEEGOOD, R.C. (1991) The relationship between the activation state of sucrose-phosphate synthase and the rate of CO_2 assimilation in spinach leaves. *Planta* **183**, 620–622.

BLACKWELL, R.D., MURRAY, A.J.S. and LEA, P.J. (1987) Inhibition of photosynthesis in barley with decreased levels of chloroplastic glutamine synthetase activity. *J. Exp. Bot.* **38**, 1799–1809.

BLACKWELL, R.D., MURRAY, A.J.S., LEA, P.J. and JOY, K.W. (1988) Photorespiratory amino donors, sucrose synthesis and the induction of CO_2 fixation in barley deficient in glutamine synthetase and glutamate synthase. *J. Exp. Bot.* **39**, 845–858.

BREARS, T., WALKER, E.L. and CORUZZI, G.M. (1991) A promoter sequence involved in cell-specific expression of the pea glutamine synthetase gene GS3A in organs of transgenic tobacco and alfalfa. *Plant J.* **1**, 235–240.

BUDDE, R.J.A. and CHOLLET, R. (1986) *In vitro* phosphorylation of maize leaf phosphoenolpyruvate carboxylase. *Plant Physiol.* **82**, 1107–1114.

CARTER, P.J., NIMMO, H.G., FEWSON, C.A. and WILKINS, M.B. (1991) Circadian rhythms in the activity of a plant protein kinase. *EMBO J.* **10**, 2063–2068.

CARVALHO, E., PEREIRA, S., SUNKEL, C. and SALEMA, R. (1992) Detection of cytosolic glutamine synthetase in leaves of *Nicotiana tabacum* L. by immunocytochemical methods. *Plant Physiol.* **100**, 1591–1594.

CHAMPIGNY, M.-L. and FOYER, C.H. (1992) Nitrate activation of cytosolic protein kinases diverts photosynthetic carbon from sucrose to amino acid biosynthesis. *Plant Physiol.* **100**, 7–12.

CHASTAIN, C.J. and CHOLLET, R. (1989) Interspecific variation in assimilation of $^{14}CO_2$ into C_4 compensation concentration. *Planta* **179**, 81–88.

CHENG, C.L., ACEDO, G.N., CRISTINSIN, M. and CONKLING, M.A. (1992) Sucrose mimics the light induction of *Arabidopsis* nitrate reductase gene transcription. *Proc. Natl Acad. Sci. USA* **89**, 1861–1864.

CHOI-HONG, K., KLEINHOFS, A. and AN, G. (1989) Nucleotide sequence of rice nitrate reductase genes. *Plant Mol. Biol.* **13**, 731–733.

COOPER, H.D. and CLARKSON, D.T. (1989) Cycling of amino-nitrogen and other nutrients between shoots and roots in cereals. A possible mechanism integrating shoot and root in the regulation of nutrient uptake. *J. Exp. Bot.* **40**, 753–762.

CRAWFORD, N.M., CAMPBELL, W.H. and DAVIS, R.W. (1986) Nitrate reductase from squash: cDNA cloning and nitrate regulation. *Proc. Natl Acad. Sci. USA* **83**, 8073–8076.

CULLIMORE. J.V. and BENNETT, M.J. (1989) The molecular biology and biochemistry of plant glutamine synthetase from root nodules of *Phaseolus vulgaris* L. and other legumes. *J. Plant Physiol.* **132**, 387–393.

DANIEL-VEDELE, F., DORBE, M.F., CABOCHE, M. and ROUZE, P. (1989) Cloning and analysis of the nitrate reductase gene from tomato: a comparison of nitrate reductase protein sequences in higher plants. *Gene* **85**, 371–380.

DENG, M.D., MOUREAUX, T., CHÉREL, I., BOUTIN, J.P. and CABOCHE, M. (1991) Effects of nitrogen metabolites on the regulation and circadian expression of tobacco nitrate reductase. *Plant Physiol. Biochem.* **29**, 239–247.

DENG, M.D., FAURE, J.D. and CABOCHE, M. (1993) The molecular aspects of nitrate reductase expression in higher plants. In: *Control of Gene Expression* (VERMA, D.P.S., ed.), pp. 425–441. Springer, Wien, New York.

DOEHLERT, D.C. and HUBER, S.C. (1983) Regulation of spinach leaf sucrose-phosphate synthase by glucose 6-phosphate, inorganic phosphate, and pH. *Plant Physiol.* **73**, 989–994.

DONN, G., TISCHER, E., SMITH, J.A. and GOODMAN, H.M. (1984) Herbicide-resistant alfalfa cells: an example of gene amplification in plants. *J. Mol. Appl. Genet.* **2**, 621–635.

DORLHAC DE BORNE, F., VINCENTZ, M., CHUPEAU, Y. and VAUCHERET, H. (1994) Co-suppression of nitrate reductase host genes and transgenes in transgenic tobacco plants. *Mol. Gen. Genet.* **243**, 613–621.

DUFF, S.M.G. and CHOLLET, R. (1995) *In vivo* regulation of wheat-leaf phospho*enol*pyruvate carboxylase by reversible phosphorylation. *Plant Physiol.* **107**, 775–782.

ECKES, P., SCHMITT, P., DAUB, W. and WENGENMAYER, F. (1989) Overproduction of alfalfa glutamine synthetase in transgenic tobacco plants. *Mol. Gen. Genet.* **217**, 263–268.

EDWARDS, J.W. and CORUZZI, G.M. (1989) Photorespiration and light act in concert to regulate the expression of the nuclear gene for chloroplast glutamine synthetase. *Plant Cell* **1**, 241–248.

FERRARIO, S., VALADIER, M.H., MOROT-GAUDRY, J.F. and FOYER, C.H. (1995) Effects of constitutive expression of nitrate reductase in transgenic *Nicotiana plumbaginifolia* in response to varying nitrogen supply. *Planta* **196**, 288–294.

FERRARIO, S., VALADIER, M.-H. and FOYER, C.H. (1996) Short-term modulation of nitrate reductase activity by exogenous nitrate in *Nicotiana plumbaginifolia* and *Zea mays* leaves. *Planta* **199**, 366–371.

FERRARIO-MÉRY, S., THIBAUD, M.C., BETSCHE, T., VALADIER, M.-H., FOYER, C.H. and FERRARIO, S. (1997) Modulation of carbon and nitrogen metabolism and nitrate reductase in untransformed and transformed *Nicotiana plumbaginifolia* during CO_2 enrichment of plants grown in pots and in hydroponic culture. *Planta* (in press).

FORDE, B.J., DAY, H.M., TURTON, J.F., WEN-JUN, S., CULLIMORE, J.V. and OLIVIER, J.E. (1989) Two glutamine synthetase genes from *Phaseolus vulgaris* display contrasting developmental and spacial patterns of expression in transgenic *Lotus corniculatus* plants. *Plant Cell* **1**, 391–401.

FOYER, C.H. (1987) The basis for source-sink regulation in leaves. *Plant Physiol. Biochem.* **25**, 649–657.

FOYER, C.H. and GALTIER, N. (1996) Source sink interaction and communication in leaves. In: *Photoassimilate Distribution in Plants and Crops: Source Sink Relationships* (ZAMSKI, E. and SCHAFER, A.A., eds), pp. 311–340. Marcel Dekker, New York.

FOYER, C.H., LEFEBVRE, C., PROVOT, M., VINCENTZ, M. and VAUCHERET, H. (1993) Modulation of nitrogen and carbon metabolism in transformed *Nicotiana plumbaginifolia* mutant E23 lines expressing either increased or decreased nitrate reductase activity. In: *Aspects of Applied Biology* (WHITE, E., KETTLEWELL, P.S., PARRY, M.A. and ELLIS, R.P., eds) No. 34 Physiology of Varieties, pp. 137–145. Association of Applied Biologists. Wellesbourn, Warwick.

FOYER, C.H., VALADIER, M.H. and FERRARIO, S. (1994a) Co-regulation of nitrogen and carbon assimilation in leaves. In: *Environment and Plant Metabolism, Flexibility and Acclimation* (SMIRNOFF, N., ed.), pp. 17–33. Bios Scientific Publishers, Guildford, UK.

FOYER, C.H., LESCURE, J.C., LEFEBVRE, C., VINCENTZ, M. and VAUCHERET, H. (1994b) Adaptations of photosynthetic electron transport, carbon assimilation and carbon partitioning in transgenic *Nicotiana plumbaginifolia* plants to changes in nitrate reductase activity. *Plant Physiol.* **104**, 171–178.

FOYER, C.H., NOCTOR, G., LELANDAIS, M., LESCURE, J.C., VALADIER, M.H., BOUTIN, J.P. and HORTON, P. (1994c) Short-term effects of nitrate, nitrate and ammonia assimilation on photosynthesis, carbon partitioning and protein phosphorylation in maize. *Planta* **192**, 211–220.

FOYER, C.H., CHAMPIGNY, M.L., VALADIER, M.H. and FERRARIO, S. (1996) Partitioning of photosynthetic carbon: the role of nitrate activation of protein kinases. In: *Protein Phosphorylation in Plants. Proceedings of the Phytochemical Society of Europe* (SHEWRY, P., HALFORD, N. and HOOLEY, R., eds), pp. 35–51. Clarendon Press, Oxford.

GALANGAU, F., DANIEL-VEDELE, F., MOUREAUX, T., DORBE, M.F., LEYDECKER, M.T. and CABOCHE, M. (1988) Expression of leaf nitrate reductase genes from tomato and tobacco in relation to light–dark regimes and nitrate supply. *Plant Physiol.* **88**, 383–388.

GALTIER, N., FOYER, C.H., HUBER, J., VOELKER, T.A. and HUBER, S.C. (1993) Effects of elevated sucrose-phosphate synthase activity on photosynthesis, assimilate partitioning and growth in tomato (*Lycopersicon esculentum* var. UC82B). *Plant Physiol.* **101**, 535–543.

GALTIER, N., FOYER, C.H., MURCHIE, E., ALRED, R., QUICK, P., VOELKER, T.A., THÉPENIER, C., LASCÈVE, G. and BETSCHE, T. (1995) Effects of light and atmospheric carbon dioxide enrichment on photosynthesis and carbon partitioning in leaves of tomato (*Lycopersicon esculentum* L.) plants over-expressing sucrose phosphate synthase. *J. Exp. Bot.* **46**, 1335–1344.

GEIGER, D.R. (1987) Understanding interactions of source and sink regions of plants. *Plant Physiol. Biochem.* **25**, 659–666.

GLAAB, J. and KAISER, W.M. (1995) Inactivation of nitrate reductase involves NR protein phosphorylation and subsequent 'binding' of an inhibitory protein. *Planta* **195**, 514–518.

GOJON, A. and TOURAINE, B. (1995) Effects of NR gene over-expression on $^{15}NO_3^-$ uptake and reduction in transgenic tobacco. *4th International Symposium on Inorganic Nitrogen Assimilation*, p. 49. Dormstadt, Germany.

HAUSLER, R.E., BLACKWELL, R.D., LEA, P.J. and LEEGOOD, R.C. (1994a) Control of photosynthesis in barley leaves with reduced activities of glutamine synthetase or glutamate synthase. I. Plant characteristics and changes in nitrate, ammonium and amino acids. *Planta* **194**, 406–417.

HAUSLER, R.E., LEA, P.J. and LEEGOOD, R.C. (1994b) Control of photosynthesis in barley leaves with reduced activities of glutamine synthetase or glutamate synthase. II. Control of electron transport and CO_2 assimilation. *Planta* **194**, 418–435.

HEINEKE, D., SONNEWALD, U., BÜSSIS, D., GÜNTER, G., LEIDREITER, K., WILKE, I., RASCHKE, K., WILLMITZER, L. and HELDT, H.W. (1992) Apoplastic expression of yeast-derived invertase in potato. Effects on photosynthesis, leaf solute composition, water relations and tuber composition. *Plant Physiol.* **100**, 301–308.

HEMON, P., ROBBINS, M.P. and CULLIMORE, J.V. (1990) Targeting of glutamine synthetase to the mitochondria of transgenic tobacco. *Plant Mol. Biol.* **15**, 895–904.

HEROLD, A. (1980) Regulation of photosynthesis by sink activity – the missing link. *New Phytol.* **86**, 131–144.

HIREL, B., MARSOLIER, M.C., HOARAU, A., HOARAU, J., BRANGEON, J., SCHAFER, R. and VERMA, D.P.S. (1992) Forcing expression of a soybean root glutamine synthetase gene in tobacco leaves induces a native gene encoding cytosolic enzyme. *Plant Mol. Biol.* **20**, 207–218.

HIREL, B., MIAO, G.-H. and VERMA, D.P.S. (1993) Metabolic and developmental control of glutamine synthetase genes in legume and non-legume plants. In: *Control of Plant Gene Expression* (VERMA, D.P.S., ed.), pp. 443–458. CRC Press, Boca Raton, Florida.

HOFF, T., STUMMANN, B.M. and HENNINGSEN, K.W. (1991) Cloning and expression of a gene encoding a root specific nitrate reductase in bean (*Phaseolus vulgari*). *Physiol. Plant.* **82**, 197–204.

HOFF, T., TRUONG, H.-N. and CABOCHE, M. (1994) The use of mutants and transgenic plants to study nitrate assimilation. *Plant Cell Environ.* **17**, 489–506.

HUBER, S.C. (1983) Role of sucrose-phosphate synthase in partitioning of carbon in leaves. *Plant Physiol.* **71**, 818–821.

HUBER, S.C. and HUBER, J.L. (1991) Regulation of maize leaf sucrose-phosphate synthase by protein phosphorylation. *Plant Cell Physiol.* **32**, 319–326.

HUBER, S.C. and ISRAEL, D.W. (1982) Biochemical basis for partitioning of photosynthetically fixed carbon between starch and sucrose in soybean (*Glycine max* Merr.) leaves. *Plant Physiol.* **69**, 691–696.

HUBER, S.C., OHSUGI, R., USUDA, H. and KALT-TORRES, W. (1987) Light modulation of maize leaf sucrose-phosphate synthase. *Plant Physiol. Biochem.* **25**, 515–523.

HUBER, J.L., HUBER, S.C. and NIELSEN, T.H. (1989) Protein phosphorylation as a mechanism for regulation of spinach leaf sucrose-phosphate synthase activity. *Arch. Biochem. Biophys.* **270**, 681–690.

HUBER, J.L., HITE, D.R.C., OUTLAW, W.H. JR and HUBER, S.C. (1991) Inactivation of highly activated spinach leaf sucrose-phosphate synthase by dephosphorylation. *Plant Physiol.* **95**, 291–297.

HUPPE, H.C. and TURPIN, D.H. (1994) Integration of carbon and nitrogen metabolism in plant and algae cells. *Annu. Rev. Plant Physiol. Plant Mol. Biol.* **45**, 577–607.

JIAO, J.-A. and CHOLLET, R. (1991) Post-translational regulation of phospho*enol*pyruvate carboxylase in C_4 and crassulacean acid metabolism plants. *Plant Physiol.* **95**, 981–985.

JIAO, J.-A. and CHOLLET, R. (1992) Light activation of maize phospho*enol*pyruvate carboxylase protein-serine kinase activity is inhibited by mesophyll and bundle sheath-directed photosynthesis inhibitors. *Plant Physiol.* **98**, 152–156.

JOB, D., COCHET, C., DHIEN, A. and CHAMBAZ, E. (1978) A rapid method for screening inhibitor effects: determination of I_{50} and its standard deviation. *Ann. Bio. Chem.* **84**, 68–77.

KAISER, W.M. and BRENDLE-BEHNISCH, E. (1991) Rapid modulation of spinach leaf nitrate reductase by photosynthesis. I. Modulation *in vivo* by CO_2 availability. *Plant Physiol.* **96**, 363–367.

KAISER, W.M. and FÖRSTER, J. (1989) Low CO_2 prevents nitrate reduction in leaves. *Plant Physiol.* **91**, 970–974.

KAISER, W.M. and HUBER, S.C. (1994) Post-translational regulation of nitrate reductase in higher plants. *Plant Physiol.* **106**, 817–821.

KALT-TORRES, W. and HUBER, S.C. (1987) Diurnal changes in maize leaf photosynthesis. *Plant Physiol.* **83**, 294–298.

KAMACHI, K., YAMAYA, T., MAE, T. and OJIMA, K. (1991) A role for glutamine synthetase in the remobilisation of leaf nitrogen during natural senescence in rice leaves. *Plant Physiol.* **96**, 411–417.

KLEIN, R.R., CRAFTS-BRANDNER, S.J. and SALVUCCI, M.E. (1993) Cloning and developmental expression of the SPS gene from spinach. *Planta* **190**, 498–510.

KRAPP, A., QUICK, W.P. and STITT, M. (1991) Ribulose-1,5-bisphosphate carboxylase-oxygenase, other enzymes and chlorophyll decrease when glucose is supplied to mature spinach leaves via the transpiration stream. *Planta* **186**, 58–69.

KRAPP, A., HOFFMANN, B., SCHAFFER, C. and STITT, M. (1993) Regulation of the expression of rbcS and other photosynthetic genes by carbohydrates: a mechanism for the 'sink regulation' of photosynthesis? *Plant J.* **3**, 817–828.

LEA, P.J. and FORDE, B.G. (1994) The use of mutants and transgenic plants to study amino acid metabolism. *Plant Cell Environ.* **17**, 541–556.

LEA, P.J., BLACKWELL, R.D. and JOY, K.W. (1992) Ammonia assimilation. In: *Nitrogen Metabolism in Plants* (MENGEL, K. and PILLBEAM, D.J., eds), pp. 153–186. Clarenden Press, Oxford.

LEEGOOD, R.C., LEA, P.J., ADCOCK, M.D. and HAUSLER, R.E. (1995) The regulation and control of photorespiration. *J. Exp. Bot.* **46**, 1397–1414.

LEPORT, L., KANDLBINDER, A., BAUR, B. and KAISER, W.M. (1996) Diurnal modulation of phospho*enol*pyruvate carboxylation in pea leaves and roots as related to tissue malate concentrations and to the nitrogen source. *Planta* **198**, 495–501.

LI, B. and CHOLLET, R. (1993) Resolution and identification of C_4 phospho*enol*pyruvate-carboxylase protein-kinase polypeptides and their reversible light activation in maize leaves. *Arch. Biochem. Biophys.* **307**, 416–419.

LILLO, C. (1984) Diurnal variations of nitrite reductase, glutamine synthetase, glutamate synthase, alanine aminotransferase and aspartate aminotransferase in barley leaves. *Physiol. Plant.* **61**, 214–218.

MACKINTOSH, R.W. and MACKINTOSH, C. (1993) Regulation of plant metabolism by reversible protein (serine/threonine) phosphorylation. In: *Post-translational Modification in Plants* (BATTEY, N.H., DICKINSON, H.G. and HETHERINGTON, A.M., eds), pp. 197–212. Cambridge University Press, Cambridge.

MACKINTOSH, C., DOUGLAS, P. and LILLO, C. (1995) Identification of a protein that inhibits the phosphorylated form of nitrate reductase from spinach leaves. *Plant Physiol.* **107**, 451–457.

MARSOLIER, M.C., DEBROSSES, G. and HIREL, B. (1995) Identification of several soybean cytosolic glutamine synthetase transcripts highly or specifically expressed in nodules: expression studies using one of the corresponding genes in transgenic *Lotus corniculatus*. *Plant Mol. Biol.* **27**, 1–15.

MCNALLY, S.F., HIREL, B., GADAL, P., MANN, F. and STEWART, G.R. (1983) Glutamine synthetases of higher plants: evidence for a specific isoform content related to their possible physiological function within the leaf. *Plant Physiol.* **72**, 22–25.

MCNAUGHTON, G.A.L., MACKINTOSH, C., FEWSON, C.A., WILKINS, M.B. and NIMMO, H.G. (1991) Illumination increases the phosphorylation state of maize leaf phospho*enol*pyruvate carboxylase by causing an increase in the activity of a protein kinase. *Biochim. Biophys. Acta* **1093**, 189–195.

MEHRER, I. and MOHR, H. (1989) Ammonium toxicity: description of the syndrome in *Sinapis alba* and the search for its causation. *Physiol. Plant* **77**, 545–554.

MELZER, E. and O'LEARY, M. (1987) Anapleurotic fixation by phospho*enol*pyruvate carboxylase in C_3 plants. *Plant Physiol.* **84**, 58–60.

MIYAZAKI, J., JURICEK, M., ANGELIS, K., SCHNORR, K.M., KLEINHOFS, A. and WARNER, R.L. (1991) Characterization and sequence of a novel nitrate reductase from barley. *Mol. Gen. Genet.* **228**, 329–334.

MULLER, D. and KLUGE, M. (1983) Immunological evidence for a crassulacean acid metabolism specific phospho*enol*pyruvate carboxylase in *Sedum* and *Kalanchoe* species. *Physiol. Veg.* **21**, 919–926.

NAVARRO, M.T., PREITO, R., FERNANDEZ, E. and GALVAN, A. (1996) Constitutive expression of nitrate reductase changes the regulation of nitrate and nitrite transporters in *Chlamydomonas reinhardtii*. *Plant J.* **9**(6), 819–827.

NIMMO, H.G. (1993) The regulation of phospho*enol*pyruvate carboxylase by reversible phosphorylation. In: *Post-translational Modification in Plants*. (BATTEY, N.H., DICKINSON, H.G. and HETHERINGTON, A.M., eds), pp. 161–170. Cambridge University Press, Cambridge.

NUSSAUME, L., VINCENTZ, M., MEYER, C., BOUTIN, J.-P. and CABOCHE, M. (1995) Post-transcriptional regulation of nitrate reductase by light is abolished by an N-terminal deletion. *Plant Cell* **7**, 611–621.

OAKS, A. (1994) Efficiency of nitrogen utilization in C_3 and C_4 cereals. *Plant Physiol.* **106**, 407–414.

O'LEARY, M.H. (1982) Phospho*enol*pyruvate carboxylase: an enzymologist's view. *Annu. Rev. Plant Physiol.* **33**, 297–315.

OLIVER, D.J., NEUBURGER, M., BOURGUIGNON, J. and DOUCE, R. (1990) Glycine metabolism by plant mitochondria. *Physiol. Plant.* **80**, 487–491.

PACE, G.H., VOLK, R.J. and JACKSON, W.A. (1990) Nitrate reduction in response to CO_2-limited photosynthesis. Relationship to carbohydrate supply and nitrate reductase activity in maize seedlings. *Plant Physiol.* **92**, 286–292.

PIERRE, J.N., PACQUIT, V., VIDAL, J. and GADAL, P. (1992) Regulatory phosphorylation of phospho*enol*pyruvate carboxylase in protoplasts from *Sorghum* mesophyll cells and the role of pH and Ca^{2+} as possible components of the light-transduction pathway. *Eur. J. Biochem.* **210**, 531–537.

PROSSER, I.M. and LAZARUS, C.M. (1990) Nucleotide sequence of a spinach nitrate reductase cDNA. *Plant Mol. Biol.* **15**, 187–190.

QUILLERÉ, I., DUFOSSÉ, C., ROUX, Y., FOYER, C.H., CABOCHE, M. and MOROT-GAUDRY, J.F. (1994) The effects of the deregulation of NR gene expression on growth and nitrogen metabolism of winter-grown *Nicotiana plumbaginifolia*. *J. Exp. Bot.* **45**, 1205–1212.

RAJAGOPALAN, A.V., TIRUMALA, D.M. and RAGHAVENDRA, A.S. (1993) Patterns of phospho*enol*pyruvate carboxylase activity and cytosolic pH during light activation and dark deactivation in C_3 and C_4 plants. *Photosyn. Res.* **38**, 51–60.

ROCHER, J.P. (1988) Comparison of carbohydrate compartmentation in relation to photosynthesis assimilate export and growth in a range of maize genotypes. *Austr. J. Plant Physiol.* **15**, 677–685.

RUFTY, T.W. JR and HUBER, S.C. (1983) Changes in starch formation and activities of sucrose-phosphate synthase and cytoplasmic fructose-1,6-bisphosphatase in response to source-sink alterations. *Plant Physiol.* **72**, 474–480.

SALVUCCI, M.E., DRAKE, R.R. and HALEY, B.E. (1990) Purification and photoaffinity labelling of sucrose-phosphate synthase from spinach leaves. *Arch. Biochem. Biophys.* **281**, 212–218.

SCHULLER, K.A. and WERNER, D. (1993) Phosphorylation of soybean (*Glycine max* L.) nodule phospho*enol*pyruvate carboxylase *in vitro* decreases sensitivity to inhibition by malate. *Plant Physiol.* **101**, 1267–1273.

SHEEN, J. (1990) Metabolic repression of transcription in higher plants. *Plant Cell* **2**, 1027–1038.

SHEEN, J. (1994) Feedback control of gene expression. *Photosyn. Res.* **39**, 427–438.

SICHER, R.C. and KREMER, D.F. (1984) Changes in sucrose-phosphate synthase activity in barley primary leaves during light-dark transitions. *Plant Physiol.* **76**, 910–912.

SOMERVILLE, C.R. and OGREN, W.L. (1980) The inhibition of photosynthesis in *Arabidopsis* mutants lacking glutamate synthase activity. *Nature* **286**, 257–259.

SOMERVILLE, C.R. and OGREN, W.L. (1981) Photorespiration-deficient mutants of *Arabidopsis thaliana* lacking mitochondrial serine transhydroxymethylase activity. *Plant Physiol.* **67**, 666–671.

SONNEWALD, U. (1992) Expression of *E. coli* inorganic pyrophosphatase in transgenic plants alters photoassimilate partitioning. *Plant J.* **2**, 571–581.

SONNEWALD, U., BRAUER, M., VON SCHAEWEN, A., STITT, M. and WILLMITZER, L. (1991) Transgenic tobacco plants expressing yeast-derived invertase in either the cytosol, vacuole or apoplast: a powerful tool for studying sucrose metabolism and sink-source interactions. *Plant J.* **1**, 95–106.

SONNEWALD, U., QUICK, W.P., MACRAE, E., KRAUSE, K.-P. and STITT, M. (1993) Purification, cloning and expression of spinach leaf sucrose-phosphate synthase in *Escherichia coli*. *Planta* **189**, 174–181.

STITT, M. (1986) Limitation of photosynthesis by carbon metabolism. I. Evidence for excess electron transport capacity in leaves carrying out photosynthesis in saturating light and CO_2. *Plant Physiol.* **81**, 1115–1122.

STITT, M. (1989) Control analysis of photosynthetic sucrose synthesis: assignment of elasticity coefficients and flux-control coefficients to the cytosolic fructose-1,6-bisphosphatase and sucrose-phosphate synthase. *Proc. R. Soc. UK B* **323**, 327–338.

STITT, M. and QUICK, W.P. (1989) Photosynthetic carbon partitioning: its regulation and possibilities for manipulation. *Physiol. Plant* **77**, 633–641.

STITT, M., HUBER, S. and KERR, P. (1987) Control of photosynthetic sucrose formation. In: *The Biochemistry of Plants* (HATCH, M.D. and BOARDMAN, N.K., eds), Vol. 10. Photosynthesis, pp. 327–409. Academic Press, New York.

STITT, M., WILKE, I., FEIL, R. and HELDT, H.W. (1988) Coarse control of SPS in leaves: alterations of the kinetic properties in response to the rate of photosynthesis and accumulation of sucrose. *Planta* **174**, 217–230.

STITT, M., VON SCHAEWEN, A. and WILLMITZER, L. (1990) 'Sink' regulation of photosynthetic metabolism in transgenic tobacco plants expressing yeast invertase in their cell wall involves a down-regulation of the Calvin cycle and an up-regulation of glycolysis. *Planta* **174**, 217–230.

SU, W., HUBER, S.C. and CRAWFORD, N.M. (1996) Identification *in vitro* of a post-transcriptional regulatory site in the hinge 1 region of *Arabidopsis* nitrate reductase. *Plant Cell* **8**, 519–527.

TEMPLE, S.J., KNIGHT, T.J., UNKEFER, P.J. and SENGUPTA-GOPALAN, C. (1993) Modulation of glutamine synthetase gene expression in tobacco by the introduction of an alfalfa glutamine synthetase gene in sense and antisense orientation: molecular and biochemical analysis. *Mol. Gen. Genet.* **236**, 315–325.

VAN QUY, L., FOYER, C. and CHAMPIGNY, M.-L. (1991) Effect of light and NO_3^- on wheat leaf phospho*enol*pyruvate carboxylase activity. *Plant Physiol.* **97**, 1476–1482.

VAUCHERET, H., VINCENTZ, M., KRONENBERGER, J., CABOCHE, M. and ROUZE, P. (1989) Molecular cloning and characterization of the two homologous genes coding for nitrate reductase in tobacco. *Mol. Gen. Genet.* **216**, 10–15.

VINCENTZ, M. and CABOCHE, M. (1991) Constitutive expression of nitrate reductase allows normal growth. *EMBO J.* **10**, 1027–1035.

VINCENTZ, M., MOUREAUX, T., LEYDECKER, M.T., VAUCHERET, H. and CABOCHE, M. (1993) Regulation of nitrate and nitrite reductase expression in *Nicotiana plumbaginifolia* leaves by nitrogen and carbon metabolites. *Plant J.* **3**, 315–324.

VON SCHAEWEN, A., STITT, M., SCHMITT, R., SONNEWALD, V. and WILLMITZER, L. (1990) Expression of yeast derived invertase in the cell wall of tobacco and *Arabidopsis* plants leads to accumulation of carbohydrate and inhibition of photosynthesis and strongly influences growth and phenotype of transgenic tobacco plants. *EMBO J.* **9**, 3033–3044.

WARNER, R.L. and KLEINHOFS, A. (1992) Genetics and molecular biology of nitrate metabolism in higher plants. *Physiol. Plant.* **85**, 245–252.

WILLIAMSON, J.D., HIRSCH-WYNCOTT, M.E., LARKINS, B.A. and GELVIN, S.B. (1989) Differential accumulation of a transcript driven by the CaMV 35S promoter in transgenic tobacco. *Plant Physiol.* **90**, 1570–1576.

WORRELL, A.C., BRUNEAU, J.-M., SUMMERFELT, K., BOERSIG, M. and VOELKER, T.A. (1991) Expression of a maize sucrose-phosphate synthase in tomato alters leaf carbohydrate partitioning. *Plant Cell* **3**, 1121–1130.

7

Manipulating amino acid biosynthesis

STEPHEN J. TEMPLE AND CHAMPA SENGUPTA-GOPALAN

The twenty L-amino acids are of central biological importance as they provide the building blocks of proteins. Considering this importance, it is perhaps surprising that mammals can synthesize only ten amino acids by *de novo* pathways. The remainder must be supplemented in their diet. This leads to significant nutritional problems to the human population of many regions of the world. Higher plants and many bacteria possess the ability for *de novo* synthesis of all protein amino acids. Details of the amino acid biosynthetic pathways can be found in any good biochemistry text.

Higher plants with mutations in enzymes of amino acid biosynthetic pathways have been available for many years, following the pioneering work of Somerville and Ogren (1980). There are several excellent recent reviews on these and many other plant mutants used to study amino acid biosynthesis and metabolism (Lam *et al.*, 1995; Lea and Forde, 1994; Radwanski and Last, 1995). With the development of molecular genetics and efficient plant transformation techniques it is now possible to produce transgenic plants containing altered levels of specific amino acid biosynthetic enzymes. In this chapter, we will review recent research in areas where modulation of amino acid metabolism has been attempted, including the modulation of several key enzymes from biosynthetic pathways whose end products are amino acids.

7.1 Ammonia assimilation

Higher plants acquire their nitrogen from two principal sources: the soil in the form of nitrate (or nitrite), which is converted to ammonia by the sequential reductive action of nitrate and nitrite reductases; and in legumes directly from the atmosphere through symbiotic nitrogen fixation (Miflin and Lea, 1980). For a discussion of recent work on the manipulation of nitrate reductase see Chapter 6. The ammonia generated from these sources and a variety of metabolic pathways such as photorespiration, catabolism of amino acids, and the metabolism of phenylpropanoids, is assimilated via the joint action of glutamine

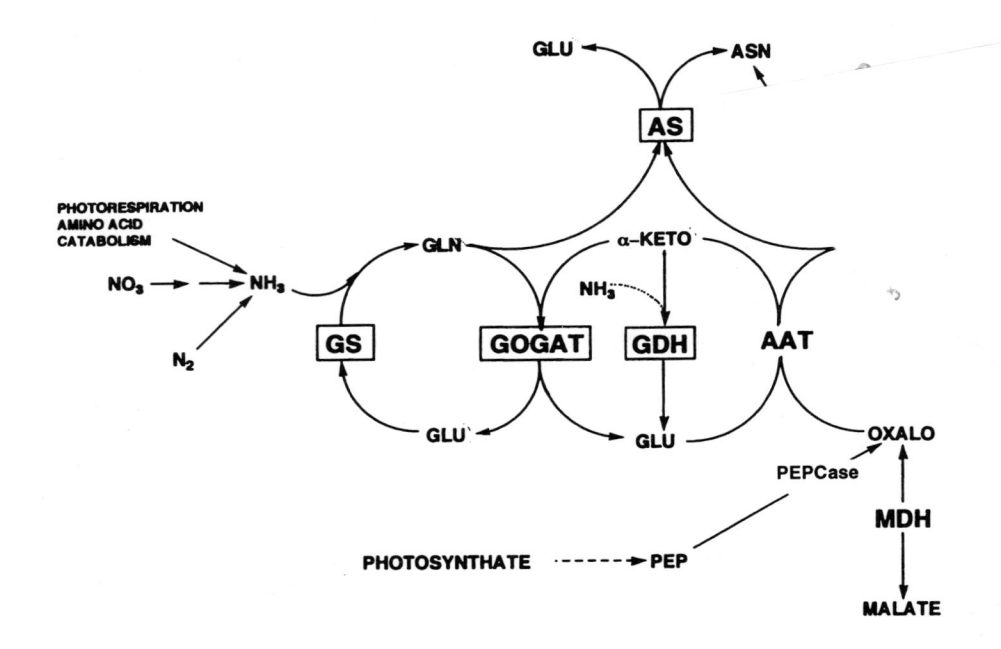

Figure 7.1 Biosynthetic pathways of ammonia assimilation. Where enzyme modulation has been attempted enzymes are in boxes. Dashed lines indicate a pathway consisting of multiple enzyme steps. Organic acid abbreviations: α-keto, α-ketoglutarate; OXALO, oxaloacetate

synthetase (GS) and glutamate synthase (GOGAT) (Lea et al., 1990). GS catalyzes the ATP-dependent condensation of ammonia with glutamate to yield glutamine. Subsequently, GOGAT transfers the amide group from glutamine to α-ketoglutarate producing glutamate, thus establishing GS as a key enzyme in the flow of nitrogen into organic acids. Other pathways that have been implicated in ammonia assimilation include aspartate and asparagine synthesis, the reactions being mediated by aspartate aminotransferase (AAT) and asparagine synthetase (AS), respectively. Phosphoenolpyruvate carboxylase (PEPCase) provides the carbon skeletons for loading up with assimilated nitrogen. Glutamate dehydrogenase (GDH) seems to play a minor role in glutamate biosynthesis in plants (Figure 7.1).

7.2 Manipulation of glutamine synthetase genes

In higher plants GS is an octameric enzyme of 320–380 kDa (Stewart et al., 1980) and is encoded by a small multigene family, whose members exhibit organ-specific patterns of expression (Bennett et al., 1989; Peterman and Goodman, 1991; Stanford et al., 1993; Temple et al., 1995; Tingey et al., 1987; Walker and Coruzzi, 1989). In all published reports, a single gene encodes the chloroplastic isoform (GS_2), which is present in leaf mesophyll cells (Edwards et al., 1990; Peterman and Goodman, 1991; Stanford et al., 1993). All plants have multiple genes encoding the cytosolic (GS_1) isoform that is expressed in the vascular system (Carvalho et al., 1992; Kamachi et al., 1992). The GS_1 genes are differentially regulated and assimilate ammonia derived from different physiological processes (Hirel et al., 1993). Little is known about the regulation of GS_1 gene expression in plants or the significance of the multiple GS_1 gene family members. In yeast, GS synthesis is regulated by three independent control pathways that respond to different metabolic signals. One pathway responds to intracellular concentrations of glutamine and glutamate, the second pathway involves general amino acid control, and the third responds to purine limitation (Benjamin et al., 1989). Considerable evidence is accumulating suggesting the involvement of metabolites in the expression of GS genes in plants (Kozaki et al., 1991; Miao et al., 1991; Sukanya et al., 1994). With the aim of improving our understanding of GS_1 gene regulation and function and with the long-term goal of improving the efficiency of plant nitrogen assimilation, several research groups have attempted to manipulate GS_1 levels by either over-expressing GS_1 genes or down-regulating them using antisense RNA technology. We will review the results of this work, the problems encountered, and discuss how alterations in GS_1 levels in different tissues/organs affect the nitrogen status and overall performance of the plant.

7.2.1 Over-expression of cytosolic glutamine synthetase

The first report of plants cells over-expressing GS was made by Donn et al. (1984). An alfalfa cell line able to grow in the presence of 0.6 mM concentrations of the competitive GS inhibitor, L-phosphinothricin (PPT), was selected. The cell line was found to contain a 3–7-fold elevation in the level of GS_1 activity resulting from a 4–11-fold amplification of a specific GS_1 gene and an eightfold increase in GS_1 mRNA levels. Unfortunately, no studies at the whole plant level were possible as the PPT resistant cell line could not be regenerated. Subsequently, the amplified GS_1 genomic sequence was cloned (Tischer et al., 1986) and the genomic fragment without the promoter engineered behind the CaMV 35S. Transgenic tobacco plants containing this construct showed a tenfold increase in GS_1 mRNA levels and a fivefold increase in specific GS activity. The GS_1 polypeptides accounted for up to 5% total soluble leaf protein in these transgenic tobacco plants and under in vitro tissue culture conditions they showed a 20-fold increase in resistance to PPT (Eckes et al., 1989). The amino acid composition of the over-expressing plants was not altered significantly, although there was a sevenfold reduction in free ammonia (Eckes et al., 1989). The authors concluded that high-level expression of a key metabolic enzyme such as GS does not interfere with the plant's growth and fertility.

The results of our initial attempts at constitutive over-expression of GS_1 in transgenic tobacco were less dramatic (Temple et al., 1993). The construct contained a full length GS_1 cDNA derived from the GS_1 genomic clone used in the previous study and was placed under the control of the CaMV 35S promoter. Only a moderate increase in GS_1 polypeptide level was observed, in spite of the significant accumulation of transcript corresponding to the transgene (Temple et al., 1993). These transformants showed only a 10–25% increase in total GS activity. A similar increase in GS_1 polypeptide level was observed in tobacco plants transformed with a nodule-enhanced alfalfa GS_1 cDNA driven by the CaMV 35S promoter (Temple and Sengupta-Gopalan, unpublished data).

In order to determine if GS targeted to the mitochondria would participate in the reassimilation of ammonia derived from the photorespiratory nitrogen cycle, Hemon et al. (1990) used a gene construct specifically aimed at targeting the GS polypeptide into the mitochondria. To achieve this, the mitochondrial targeting sequence of the N. plumbaginifolia b-f_1 ATPase was fused to the cDNA encoding the GS_1 γ polypeptide of Phaseolus vulgaris. As a control, just the γ GS_1 polypeptide coding sequence was used and both constructs were driven by the CaMV 35S promoter. Under tissue culture conditions, the γ isozyme in the transformants made up to 25% of the total GS activity and when the mitochondrial targeting sequence was included, this activity was associated with the mitochondria (Hemon et al., 1990). However, when re-examined as mature, phototropically grown, greenhouse plants, expression of the introduced GS gene at the protein and enzymatic level was greatly reduced, although high transgene mRNA levels (12–30 pg mg^{-1} total RNA) were maintained.

Hirel et al. (1992) introduced a soybean GS_1 genomic fragment containing the coding region and 3′ untranslated region fused to the CaMV 35S promoter into tobacco. The resulting soil-grown, mature, greenhouse adapted transgenic tobacco plants expressed the transgene in leaves and roots. Anion exchange chromatography indicated that GS_1 now represented 25–30% of total GS activity, the remainder being GS_2. Immunochemical analysis localized this expression to the cytosol of the leaf mesophyll cells of the transgenic plants (Hirel et al., 1992). A rather puzzling finding from this study was that mRNA for an endogenous tobacco GS_1 gene was detected specifically in the leaves of the transgenic plants. The authors speculated that an alteration in a leaf metabolite concentration resulting from the forced expression of the GS_1 gene in the mesophyll may be responsible for the induction of the native tobacco GS_1 gene (Hirel et al., 1992). However, overall the soybean GS protein and enzyme activities in transgenic tobacco tissues was found to be highly labile (Hirel et al., 1992).

Both the pea chloroplastic and cytosolic GS genes were expressed under the control of the CaMV 35S promoter in transgenic tobacco (Lam et al., 1995). While the data on detailed biochemical and molecular analysis are not yet available, the authors indicated that the ectiopic expression of the cytosolic GS had a considerable growth advantage when compared to the wild-type plants.

All the research discussed thus far has involved the constitutive expression of legume GS_1 genes in the model system, tobacco. Research in our laboratory, however, has focused mainly on manipulating GS_1 gene expression in the forage crop, alfalfa. Two full length GS_1 cDNA clones (pGS100 and pGS13) representing the two major classes of GS_1 genes in alfalfa have been characterized (Temple et al., 1993, 1995). Genomic Southern analysis suggests that multiple genes and/or alleles exist for each class of alfalfa GS_1 gene. They share 82% nucleotide homology in their coding regions, but have divergent 3′ untranslated regions. Both classes exhibit constitutive expression patterns, although at significantly varying levels, with the pGS13 class representing the nodule-enhanced class (Temple et al., 1995).

Gene constructs with either the pGS100 or pGS13 cDNA and driven by either the CaMV 35S promoter, a vasculature specific promoter (the acidic chitinase promoter from *Arabidopsis*, (Samac et al., 1990) or the soybean leghaemoglobin gene promoter (Stougaard et al., 1987), have been introduced into alfalfa. Alfalfa transformants containing the CaMV 35S promoter driving the pGS100 cDNA, showed no change in GS_1 polypeptide levels. Two-dimensional SDS-PAGE Western analysis indicated that the specific distribution of all GS polypeptides remained essentially unchanged (Temple and Sengupta-Gopalan, unpublished data). Transcript analysis of these plants revealed a significant reduction in total GS_1 transcript in the nodules. By using gene specific probes in Northern analysis, this reduction was shown to be specific to the pGS100 subclass related transcripts. The level of pGS13 subclass related transcripts remained unchanged (Temple and Sengupta-Gopalan, unpublished data). At present, it is not known what the relative contributions of the endogenous and transgene pGS100 subclass transcripts make to the total. The mechanism of this down-regulation does not appear to be classical co-suppression (Mol et al., 1994). It is interesting to note that this same gene construct (pGS100 cDNA driven by the CaMV 35S promoter), when introduced into tobacco, showed a significant accumulation of the corresponding transcript and peptide (Temple et al., 1993). We have not yet analyzed the alfalfa transformants containing gene constructs with the pGS13 cDNA. More recently, we have introduced a soybean GS_1 cDNA driven by the CaMV 35S promoter into alfalfa (Ortega and Sengupta-Gopalan, unpublished data). Analysis of these transformants should determine if the regulation in the pGS100 transcripts is due to metabolic repression of the native pGS100 gene promoter. Increased GS activity in tissues resulting from expressing GS_1 genes can have an effect on the relative levels of GS substrate and product. It is thus crucial to have a good understanding of the regulatory mechanism underlying expression of the native GS genes before designing additional gene constructs to up-regulate GS activity.

Alfalfa transformants containing a chimeric gene consisting of the acidic chitinase promoter from *Arabidopsis* (Samac et al., 1990) fused to the pGS100 GS_1 cDNA in sense orientation have been regenerated. This promoter functions in a vascular specific and root enhanced manner in tomato and *Arabidopsis* (Samac and Shah, 1991) and is expressed in the vasculature of alfalfa (Samac, unpublished data). A biochemical and molecular analysis of the physiologically mature, greenhouse grown, nodulated alfalfa transformants showed a significant decrease in the level of pGS100 related transcripts in the nodules, and a significant increase in the stem (Temple et al., 1994). We do not have any explanation for the reduction in pGS100 transcripts in the nodules of these transformants. Transcript levels of the pGS13 subclass remained unchanged in all tissues tested. Analysis of the GS_1 polypeptide levels showed a significant increase in the nodules and a slight increase in leaf, stem, and root tissue (Temple et al., 1994). *In situ* hybridization and immunolocalization experiments will be performed to try and determine the basis for the discrepancy between transcripts and polypeptide levels.

Only the experiments of Eckes et al. (1989) have resulted in the significant over-expression of GS_1 that one might expect from using the CaMV 35S promoter, although it remains unclear if these high GS activities were maintained in mature, soil-grown plants. It is clear that differences in plant background, perhaps relative to that of the transgene (homologous vs. heterologous), plant age, developmental stage, and growth conditions all appear to strongly affect the level of over-expression. The lack of correlation between transgene transcript level, accompanying polypeptide and activity attributable to the transgene (Hemon et al., 1990; Temple et al., 1993; Temple and Sengupta-Gopalan, unpublished data) support the idea of translational, post-translational or assembly mediated control. Evidence exists for an assembly related factor that may become limiting, resulting in the turnover of unassembled subunits (Temple et al., 1993). Chaperones related to Gro EL have been implicated to have a role in the folding/assembly of higher plant GS_2 (Lubben et al., 1989; Tsuprun et al., 1992) and bacterial GS (Fischer, 1992). Recently we have demonstrated that the assembly, stability or activation of the nodule-specific GS isozymes in soybean appears to be regulated by active nitrogen fixation or a product of the reaction possibly via the release of the holoenzyme from a GS-chaperone complex (Temple et al., 1996). Another report suggests that GS activity in the roots of soybean is regulated by an uncharacterized mechanism at the post-translational level (Hoelzle et al., 1992). The failure of two common bean genotypes to assemble specific GS_1 subunits has been ascribed to the lack of an assembly factor (Gao and Wong, 1994). If complex post-translational control of GS_1 exists in all plant systems it may explain the rather limited success in over-expressing GS_1. It is difficult to reconcile the dramatic differences seen between different plant backgrounds containing the same gene construct. If the assembly and/or activation of GS is mediated via a chaperone like protein that is rate limiting it may be advantageous to co-express a chaperone gene with a suitable GS gene in order to obtain high levels of over-expression. The success of such an approach would require extensive biochemical analysis of GS assembly and turnover.

7.2.2 Antisense modulation of GS activity

In order to determine the functional role of a particular GS_1 isozyme on the significance of tissue specific expression of the different GS_1 gene members, we have used antisense RNA technology to down-regulate GS_1 in a gene class or tissue specific manner. The rationale is to correlate specific down-regulation with the physiological and biochemical outcome. The feasibility of using antisense RNA technology to down-regulate GS_1 activity has been demonstrated (Temple et al., 1993). The pGS100 alfalfa GS_1 cDNA was placed in antisense orientation under the control of the CaMV 35S promoter and the construct introduced into tobacco. Leaf tissue from several transformants showed a significant decrease in the level of both GS_1 and surprisingly GS_2 polypeptides, and up to a 25–40% reduction in GS activity (on a per gram fresh weight basis). The same transformants also contained about 30% less soluble protein per gram fresh weight. These changes were not accompanied by any change in endogenous GS_1 mRNA level suggesting that the antisense down-regulation was at the level of translation (Temple et al., 1993). These transformants showed a reduction in GS_2 level. The limited nucleotide homology between alfalfa GS_1 and tobacco GS_2 genes (Temple et al., 1995) and the lack of in vitro hybridization between the two genes suggests that the observed reduction in GS_2 polypeptide along with that of PEPCase and hydroxypyruvate reductase in these antisense GS_1 tobacco transformants may be the result of an overall change in plant performance (Temple et al., 1993).

The same antisense GS construct (CaMV 35S promoter driving the pGS100 cDNA in antisense orientation) in alfalfa, showed a specific reduction in the pGS100 transcript, but no effect on the level of pGS13 class transcript. This is despite the 82% nucleotide homology in the coding region between the two genes. Similar results were obtained with antisense GS_1 constructs aimed at gene/subclass down-regulation. The gene class specific 3′ untranslated region of pGS100 was fused in antisense orientation behind a functional hygromycin phosphotransferase (HPT) gene, the construct being driven by the CaMV 35S promoter. Delauney et al. (1988) had previously demonstrated an increased effectiveness of a short antisense transcript following its fusion to a sense orientation transcript. In tissue culture, alfalfa transformants containing both these constructs appeared slightly stressed with a pale-yellow appearance, but they slowly recovered to produce normal looking plants when transferred to the soil. A large number of viable transformants were obtained for both constructs. In spite of the dramatic reduction in the pGS100 transcripts, no significant change in GS_1 polypeptide level was detected in these transformants. Antisense constructs consisting of the pGS13 cDNA and the 3′ untranslated region of the pGS13 gene driven by the CaMV 35S promoter have also been introduced into alfalfa. These transformants await molecular and biochemical analysis.

GS_1 activity has been localized in the vasculature and it has been suggested that it functions in the translocation of glutamine (Brears et al., 1991; Carvalho et al., 1992; Kamachi et al., 1991). To check the importance of GS activity in the vasculature, a gene construct aimed at specifically down-regulating GS_1 in the vascular tissue was introduced into alfalfa. This construct consisted of the acidic chitinase promoter from *Arabidopsis* driving the pGS100 cDNA in antisense orientation. Transgenic alfalfa containing this construct had an extremely high level of mortality during the later stages of tissue culture typified by the bleaching of the leaves. The few surviving transformants were pale yellow in tissue culture and showed a characteristic yellowing. Specifically, there was a gradual decrease in pigmentation starting from the mid-vein to the outside of the leaf. This yellowing has been maintained primarily in the growing tips of the mature, greenhouse maintained, nodulated plants. Analysis of the GS_1 transcripts of these plants showed a reduction in pGS100 related transcripts in all organs tested (Temple et al., 1994). As with the other constructs, the antisense transcript was not detectable, and there was no change in the level of pGS13 class transcripts. These transformants showed a dramatic decrease in GS_1 polypeptide subunit level in the nodules, a slight increase in the stems and leaves, and no change in the roots. The GS activity data in most cases correlated well with the changes in GS_1 subunit levels (Temple et al., 1994; Temple and Sengupta-Gopalan, unpublished data). The fact that vascular specific down-regulation of pGS100 transcripts produced plants that were inviable would suggest that vascular localized GS isoenzymes play a very crucial role in nitrogen assimilation.

Antisense RNA technology has been shown to be a powerful tool for studying the effects of down-regulating gene expression (Blokland et al., 1994; Bourque, 1995; Krol et al., 1988). All the antisense constructs described here effectively down-regulate their target transcript. However, in each case there is no direct correlation between the transcript reduction and the resulting GS_1 subunit level. Among the numerous explanations for this discrepancy are: regulation at the translational level; regulation at the level of holoenzyme assembly and protein turnover; down-regulation of one GS_1 gene member is compensated for by up-regulation of another; or a change in nitrogen status in one organ, tissue or cell type affects GS_1 gene expression in another. Experimental evidence appears to exist for all four explanations and further analysis of the transformants should elucidate which mechanism(s) function. The striking difference in plant phenotype observed between the constitutive (CaMV 35S) and vasculature specific (acidic chitinase) promoter driven constructs, demonstrates the value and need for using tissue or cell type specific promoters for metabolic engineering experiments. Due to the unexpected gene class specificity of the constitutive antisense construct, two additional constructs using the full length nodule-enhanced pGS13 alfalfa GS_1 cDNA and a 3' untranslated specific region both in antisense orientation and controlled by the CaMV 35S promoter have been introduced into alfalfa. A comparison of the physiological and biochemical effects and differences resulting from the down-regulation of both gene classes will be interesting. The analysis of plants obtained following crossing lines containing the various constructs will also add to our understanding of the regulatory mechanisms underlying the expression of the different GS_1 isoforms. These double mutants and the parental lines, including those containing the sense constructs described earlier, will be useful for studying the effects of metabolites in the nitrogen assimilatory pathway on the expression of the different alfalfa GS_1 genes. The transformants will also provide suitable material to study the implications of modulated ammonia assimilation on potential alternative routes of ammonia assimilation such as AS, GDH and alanine and aspartate dehydrogenases. The effects on the metabolic fluxes through enzymes of other associated pathways such as nitrate reductase, GOGAT, PEPC, and AAT will also be measured.

7.2.3 Alternative approaches to down-regulate glutamine synthetase

In an attempt to determine the specific role of individual GS genes in *P. vulgaris*, Bennett and Cullimore (1992) have used ribozymes. Besides having endo-ribonuclease activity, ribozymes also demonstrate a high degree of sequence-specificity. Three different ribozymes were designed which could discriminate between the four highly homologous GS gene members (α, β, γ, δ) *in vitro*, although only at elevated, non-physiological temperatures. It still remains to be seen whether these ribozymes have the potential to inactivate the function of specific gene members *in vivo*.

Another potential approach to inactivate plant GS would be to introduce transgenes that express truncated or mutated GS subunits. The unusual location of the GS active site at the interface of two neighbouring subunits (Almassy *et al.*, 1986) and the observation that cytosolic subunits of plant GS form holoproteins of mixed subunits (Bennett and Cullimore, 1989) would make this theoretically possible. The strong conservation in five regions of the GS sequences identified by phylogenetic analysis (Pesole *et al.*, 1991), findings from X-ray diffraction studies (Yamashita *et al.*, 1989) and results of active site mutation studies with prokaryotic GS (Alibhai and Villafranca, 1994) could be used to identify potential target sequences for mutation. The ability of an alfalfa GS_1 gene to complement successfully an *E. coli glnA* mutant (DasSarma *et al.*, 1986) should allow the rapid testing for potential negative complementing mutants in *E. coli* before their introduction into plants.

7.3 Modulation of other nitrogen assimilatory enzymes

Glutamate synthase (GOGAT) which functions in conjunction with GS to assimilate ammonia (Figure 7.1), occurs in two distinct forms in plants: a ferredoxin-dependent form (Fd-GOGAT) and an NADH-dependent form (NADH-GOGAT). Fd-GOGAT is predominantly localized in chloroplasts and is involved in the assimilation of ammonia derived from the reduction of nitrate and from photorespiration (Lea *et al.*, 1990). NADH-GOGAT is, however, found primarily in non-green tissues. The abundance of NADH-GOGAT transcripts increases substantially in developing nodules (Gregerson *et al.*, 1993). The enzyme exists as a monomer with a native and subunit mass of about 200 kDa. Despite the key role that GOGAT plays in ammonia assimilation, little work has been done on the manipulation of GOGAT levels. An alfalfa cDNA clone for NADH-GOGAT (Gregerson *et al.*, 1993) engineered in antisense orientation behind the constitutive CaMV 35S promoter and the nodule specific AAT-2 gene promoter has been introduced into alfalfa (Schoenbeck and Vance, personal communication). These transformants showed reduced GOGAT activity. Biochemical and molecular analysis of these transformants should determine the impact of NADH-GOGAT down-regulation on ammonia assimilation in different plant organs and the expression of other key enzymes in ammonia assimilation. In a different study, a construct containing a Fd-GOGAT sequence in antisense orientation behind the CaMV 35S promoter was introduced into tobacco. The primary transformants were regenerated under high CO_2 levels to repress photorespiration (Foyer and Hirel, personal communication). These transformants showed a reduction in GOGAT activity (20–80%) and transformants with less than 40% GOGAT activity levels (compared to WT) failed to grow in a normal atmosphere.

Asparagine synthetase (AS) catalyzes the formation of asparagine (Figure 7.1). The active enzyme is a dimer and in plants has been shown to utilize glutamine *in vitro*. However, in maize roots AS can utilize ammonia directly as a substrate at high ammonia concentrations (Oaks and Ross, 1984). AS genes from pea and tobacco are expressed exclusively in the dark. The cell specific expression of AS in the phloem (Tsai and Corruzzi, 1990), suggests that AS functions to generate asparagine for long-distance nitrogen transport. With the goal of increasing asparagine production, the pea AS1 cDNA driven by the CaMV 35S promoter was introduced into tobacco (Brears *et al.*, 1993). The transformants showed light-independent accumulation of AS mRNA. Neither the AS polypeptide level or AS activity was determined. However, the transformants showed a 10–100-fold increase in the level of free asparagine and a decrease in the AS substrates, glutamine and aspartate. The results demonstrate that the availability of glutamine is not limiting in the AS reaction. No statistically significant changes in vegetative growth were observed in spite of the dramatic increase in free asparagine levels (Brears *et al.*, 1993).

In the same study, Brears *et al.* (1993) constitutively expressed a gene encoding the AS enzyme in which the glutamine-binding domain had been deleted (glnΔ

AS1). These transformants showed a 3–19-fold increase in asparagine levels with little effect on aspartate, glutamate or glutamine levels, suggesting that glnΔ AS1 is able to use ammonia directly (Figure 7.1). Despite the high level accumulation of asparagine in the leaves, these transformants grew poorly when compared to the control plants. The authors speculated that the dimerization between the native WT AS subunits and the glnΔAS1 subunits in the phloem would produce inactive enzymes and as such the phloem cells would fail to function in asparagine transport (Brears et al., 1993).

Glutamate dehydrogenase (GDH) was originally thought to be the primary route of ammonia assimilation in plants. However, this idea was challenged with the discovery of the GS/GOGAT cycle (Figure 7.1). The high K_m of GDH for ammonia indicates a catabolic function for this enzyme, although the actual role of this enzyme in higher plants remains elusive (Stewart et al., 1980). In bacteria, both the GS/GOGAT and GDH/GOGAT pathways operate and an assimilatory GDH function has been demonstrated under conditions of high ammonia availability. A bacterial *gdhA* gene has been modified for expression in plants and introduced into tobacco, where it directed expression at specific activities comparable to GS (Long and Lightfoot, 1995). Ammonia assimilation rates were reported to be sufficient to increase ammonia tolerance and alter the protein content of seeds and leaves.

Our failed attempts to increase GS activity significantly in transgenic alfalfa, along with the fact that GS and GOGAT function in conjunction, makes us speculate that probably both GS and GOGAT have to be up-regulated simultaneously. In the same context, it is interesting to note that in the AS overexpressing plants, there was no change in plant growth in spite of the 10–100-fold increase in free asparagine levels (Brears et al., 1993). It is possible that in these transformants, glutamine levels are limiting relative to increases in growth. It would thus follow that both AS and GS levels have to be increased simultaneously to make an impact on growth. In collaboration with Carroll Vance's laboratory, we are now in the process of producing transformants that overexpress GS, GOGAT and AS simultaneously.

7.4 Modulation of the aspartate pathway

Among the essential amino acids, lysine and threonine are often considered to be among the most important as they are the most seriously limited in cereal grains (Bright and Shewry, 1983). As a result, the regulation of lysine and threonine metabolism has been extensively studied at both biochemical and molecular levels (Bryan, 1980; Galili, 1995). Higher plants and many bacteria synthesize lysine, threonine and methionine from aspartate using two branches of the aspartate family as shown schematically in Figure 7.2. From biochemical studies, it is evident that the aspartate family pathway in higher plants is regulated by end product feedback inhibition. The two major enzymes that have been used for genetic manipulation are: aspartate kinase (AK), which catalyzes the phosphorylation of aspartate to form 3-aspartyl phosphate; and dihydrodipicolinate synthase (DHPS), which catalyzes the first reaction unique to lysine biosynthesis, the condensation of 3-aspartic semialdehyde with pyruvate to form 2,3-dihydrodipicolinate (Figure 7.2). Higher plants possess at least two or three AK isozymes

Manipulating amino acid biosynthesis

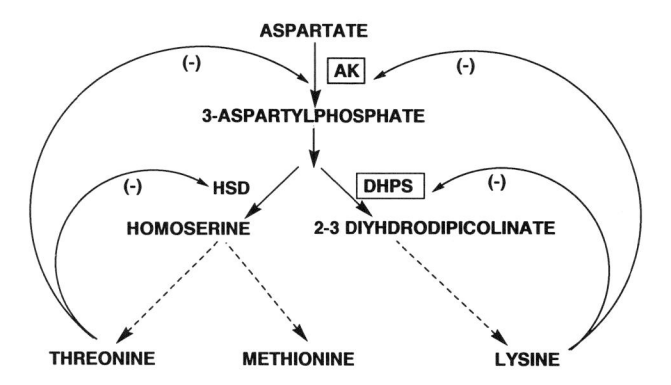

Figure 7.2 The aspartate biosynthetic pathway. Biosynthetic pathway for the amino acids threonine, methionine, lysine and isoleucine. Steps where enzyme modulation has been attempted are in boxes. Curved arrows and minus sign (–) represent feedback inhibition by end product amino acids. Dashed lines indicate pathways consisting of multiple enzyme steps. For the pathway from threonine to isoleucine see Figure 7.4

that are under feedback inhibition by lysine and threonine (Galili, 1995; Lea and Forde, 1994). The other key enzyme in the aspartate family pathway in plants, DHPS, is most sensitive to feedback inhibition by lysine (Bryan, 1980). Most of the enzymes in this pathway have been localized to the plastids (Galili, 1995). Despite the fact that cDNA clones encoding plant AK and DHPS are available (Ghislain et al., 1994; Kaneko et al., 1990), the feedback inhibition that they are subjected to prohibits their use for over-expression studies in transgenic plants. These limitations have been partly overcome by generating plant mutants (for review, see Galili, 1995) and more successfully by expressing bacterial genes encoding feedback-insensitive AK and DHPS enzymes in transgenic plants (Glassman, 1992; Perl et al., 1992; Shaul and Galili, 1992a,b).

Constructs containing the coding sequences of the bacterial AK and DHPS genes were fused to the CaMV 35S promoter and used to obtain transgenic plants. Additional constructs containing the pea *rbcS-3A* chloroplast transit sequence fused to the bacterial genes have been used to direct the bacterial enzymes into the organelles. Transgenic tobacco plants expressing the chloroplast localized *E. coli* DHPS contained free lysine at levels 40-fold higher than that found in the control plants. This resulted in a 56% elevation in protein lysine content. No changes in soluble methionine or threonine were detected (Shaul and Galili, 1992a). No increase in lysine was detected when expression was localized in the cytoplasm, thus confirming the chloroplast location of the pathway. The same plastid targeted DHPS gene construct has also been introduced into potato (Perl et al., 1992). Despite a 50-fold increase in the level of DHPS, the level of lysine overproduction in potato leaves was significantly lower than that seen for tobacco. The authors suggested that in potato, AK may play a more significant role in the regulation of lysine biosynthesis.

Constitutive expression of an *E. coli* AK gene in transgenic tobacco resulted in a significant overproduction of free threonine (Shaul and Galili, 1992b). Furthermore, it was shown that unlike DHPS, plastid localization was not essential for the overproduction of threonine. This would suggest that some metabolic intermediates in the aspartate family pathway can shuttle between plastids and the cytoplasm (Galili, 1995). The lysine and threonine levels of the transgenic plants varied in different tissues and at different stages of development suggesting that complex developmental regulatory signals lead to differential expression of the aspartate family genes (Galili, 1995). Transgenic plants constitutively co-expressing both a feedback-insensitive AK and feedback-insensitive DHPS (Frankard et al., 1992; Shaul and Galili, 1993) were found to contain free lysine levels that exceeded those in plants expressing only the insensitive AK or DHPS. The increase in lysine pools in these double transformants was accompanied by a significant decrease in threonine as compared to the AK overexpressors. These findings suggest that in these transgenic tobacco plants lysine and threonine synthesis is under the most stringent regulation. With deregulation of DHPS, lysine synthesis competes with the other branch of the pathway resulting in the preferential conversion of 3-aspartic semialdehyde into lysine at the expense of threonine (Galili, 1995). In addition, the synthesis of lysine and threonine appears to be determined by the competition between DHPS and homoserine dehydrogenase for their common substrate 3-aspartic semialdehyde (Figure 7.2), the level of which appears to be determined by AK activity (Galili, 1995).

To study the regulation of lysine synthesis and accumulation in plant seeds the bacterial feedback-insensitive DHPS gene was expressed by itself or together with the insensitive AK gene in a seed-specific manner in tobacco (Karchi et al., 1993). Both gene constructs were designed to target the enzyme into the plastids and were driven by the seed specific β-phaseolin gene promoter. Seed specific expression of the bacterial DHPS in transgenic tobacco plants showed increased levels of free lysine in the seeds, while seed specific coexpression of both DHPS and AK genes resulted in seed specific accumulation of both free lysine and threonine. However, free lysine failed to accumulate to high levels in mature seeds of these transformants. Several lines of evidence suggest that lysine catabolism is responsible for preventing the accumulation of lysine (Karchi et al., 1994). Co-ordinate expression of high levels of lysine-ketoglutarate reductase with the two bacterial transgenes suggests that lysine synthesis and catabolism are co-ordinately regulated during seed development. Co-expression of both bacterial enzymes in the same plant resulted in a significant increase in the proportion of lysine and threonine in seed albumins (Karchi et al., 1994). These findings demonstrated that the aspartate pathway is active in plant seeds and may be limiting for the synthesis of seed proteins that are rich in these amino acids. The possibility now clearly exists for producing crops with increased levels of essential amino acids in the seed grains.

7.5 Manipulation of aromatic amino acid biosynthesis

The aromatic amino acids tryptophan, phenylalanine and tyrosine are synthesized via the shikimate pathway (Herrmann, 1995). In addition, a range of secondary compounds such as lignin, flavonoids and indole acetic acid are also derived from this pathway. The biosynthesis of these aromatic amino acids has two

```
                PHOSPHOENOL PYRUVATE
                         +
                ERYTHROSE 4-PHOSPHATE
                         │
                         │    ┌──────────────┐
                         │    │ DAHP SYNTHASE │
                         ▼    └──────────────┘
                       DHAP
                         │
                         ▼
                     SHIKIMATE
                         │
                         │    ┌──────────────┐
                         │    │ EPSP SYNTHASE │
                         ▼    └──────────────┘
                       EPSP
                         │
                         ▼
                    CHORISMATE
              ╱         │         ╲  ASA
            ╱           │           ╲
PHENYLALANINE    TYROSINE       ANTHRANILATE
                                     │ PAT
                                     ▼
                              5'-PHOSPHORIBOSYL-
                                ANTHRANILATE
     ╲            ╲                  │ ┌─────┐
      ╲            ╲                 │ │ PAI │
       ▼            ▼                ▼ └─────┘
  LIGNIN + FLAVONOIDS         1-(o-CARBOXYPHENYLAMINO)-1-
                                DEOXYRIBULOSE-5-PHOSPHATE
                                     │
                                     ▼
                                 TRYPTOPHAN
```

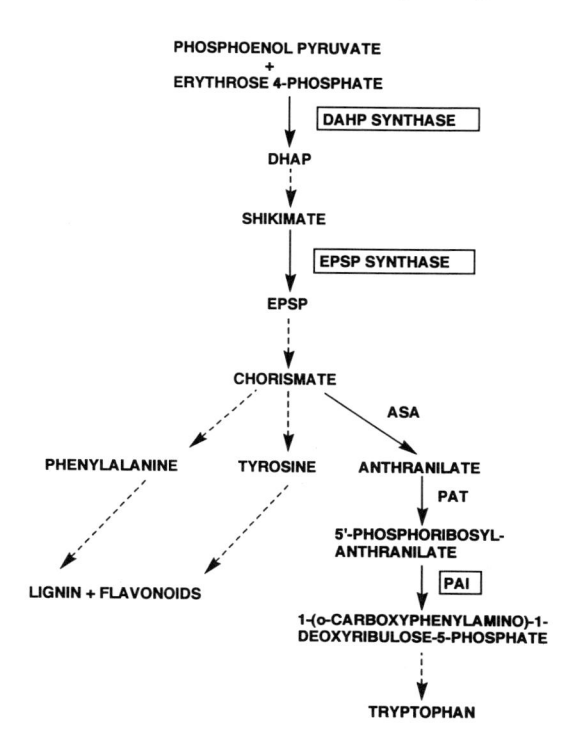

Figure 7.3 The shikimate and aromatic amino acid biosynthetic pathways. Biosynthetic pathway for the amino acids phenylalanine, tyrosine and tryptophan. Steps where enzyme modulation has been attempted are in boxes. Dashed lines indicate pathways consisting of multiple enzyme steps. Abbreviation: ASA, anthranilate synthase

parts: the shikimate pathway from phosphoenolpyruvate and erythrose 4-phosphate to chorismate; and the three specific pathways with chorismate as a common substrate for each leading to a specific aromatic amino acid (Figure 7.3). The first enzyme of the shikimate pathway is 3-deoxy-D-arabino-heptulosonate-7-phosphate (DAHP) synthase which catalyzes the condensation of phosphoenolpyruvate and erythrose 4-phosphate to form DAHP. In *E. coli*, the three DAHP isozymes are feedback regulated by the aromatic amino acids (Ogino *et al.*, 1982). Plant DAHP synthases are not subject to feedback inhibition by the aromatic amino acids, however, both tyrosine and tryptophan act as activating ligands of the enzyme. Environmental factors including wounding or microbial challenge are known to induce DAHPs. An antisense construct consisting of the 5' end of the wound-inducible DAHP synthase gene (coding sequence) from potato under the control of the CaMV 35S promoter has been used to transform potato (Jones *et al.*, 1995). A number of transgenic plants with impaired wound-induced DAHP synthase activity, polypeptide and mRNA were identified. Several of the plants had altered morphology including reduced plant height, stem diameter and reduced stem lignin levels (Jones *et al.*, 1995). No amino acid pool measurements were reported, and since the aromatic amino acids are precursors of other secondary metabolites it would be interesting to see the effects the DAHP antisense construct had on other plant processes involving these metabolites. It has been suggested that distinct DAHP synthase genes support separate pathways for the aromatic amino acids destined for proteins and those for secondary metabolites.

The penultimate step in the shikimate pathway is catalyzed by the enzyme 5-enolpyruvyl-shikimate-3-phosphate synthase (EPSPS). The enzyme catalyzes the addition of the enolpyruvyl moiety from phosphoenolpyruvate to shikimate-3-phosphate to produce chorismate (Figure 7.3), the substrate for the synthesis of all three aromatic amino acids (Herrmann, 1995). Being the target for the herbicide glyphosate, EPSPS has been the subject of extensive biochemical study (Steinrucken and Amrhein, 1980). For a recent review of the analysis of the numerous glyphosate resistant plants and cell lines, see Lea and Forde (1994). Glyphosate inhibits the activity of the plastid-localized enzyme EPSPS and thereby prevents synthesis of the chorismate derived aromatic amino acids and secondary metabolites. Much of the research effort to impart glyphosate tolerance to plants has concentrated on the introduction of genes encoding glyphosate tolerant EPSPS in transgenic plants (Barry et al., 1992; Comai et al., 1985; Fillatti et al., 1987). A number of EPSPS genes from different bacteria that are less sensitive to inhibition by glyphosate have been identified. However, the increased K_i for glyphosate seen with the mutant bacterial EPSPS was accompanied by an increased K_m for one of the substrates, phosphoenolpyruvate (Barry et al., 1992). A gene encoding a glyphosate resistant EPSPS (*aroA*) was isolated from an *Agrobacterium* sp. (strain CP4). The encoded enzyme maintained a low K_m for phosphoenolpyruvate despite the increased K_i for glyphosate. This CP4 EPSPS gene fused to the chloroplast targeting sequence and driven by the CaMV 35S promoter, has been introduced into petunia, *Arabidopsis*, canola and soybean (Barry et al., 1992; Klee et al., 1987; Shah et al., 1986). In context to this review, it would be interesting to analyze the EPSPS over-expressing plants with respect to the pools of aromatic amino acids and processes involving secondary products.

The tryptophan biosynthetic pathway has recently received considerable attention, the results of which now provide a model for the biochemical and genetic dissection of a complex, multi-step biochemical pathway (for a review, see Radwanski and Last, 1995). In addition to providing the essential amino acid tryptophan, the pathway also provides the precursors for the synthesis of auxin, phytoalexins, glucosinolates and both indole and anthranilate-derived alkaloids. The biosynthetic pathway for tryptophan is schematically shown in Figure 7.3. It is extremely well conserved in its biochemistry across kingdoms. This pathway has recently been dissected very thoroughly with the use of *Arabidopsis* mutants (Radwanski and Last, 1995). The genes for the different enzymes in the pathway have been isolated by complementation of *E. coli* mutants (Rose and Last, 1994). This wealth of plant mutants has limited the development of or need for genetically modulating enzymes in this pathway. However, to date no mutants have been isolated for phosphoribosylanthranilate isomerase (PAI), probably due to the presence of three PAI genes. This enzyme catalyzes the third step of the tryptophan biosynthetic pathway, the conversion of 5-phosphoribosylanthranilate to 1-(O-carboxyphenylamino)-1-deoxyribulose-5-phosphate (Radwanski and Last, 1995; Figure 7.3). Plants constitutively expressing an antisense PAI cDNA had as

Manipulating amino acid biosynthesis

little as 15% PAI activity, 10% PAI protein and phenotypes consistent with a block early in the tryptophan biosynthetic pathway: blue fluorescence under UV light due to the accumulation of anthranilate compounds and resistance to the anthranilate analogue 6-methylanthranilate, although no amino acid pool measurements were reported (Li et al., 1995).

7.6 Manipulation of branch chain amino acid biosynthesis

Research focused on the manipulation of the biosynthetic pathways for the branch chain amino acids leucine, isoleucine and valine has also been driven by the search for herbicide tolerant plants. Again the primary enzyme under manipulation is the site of herbicide action. The pathways leading to the three branch chain amino acids are biochemically parallel and are catalyzed by enzymes that possess dual substrate specificities. Acetohydroxyacid synthase (AHAS, also known as acetolactate synthase) appears to be the primary control point of the pathway and its regulation is achieved by feedback inhibition of AHAS by leucine and valine (Figure 7.4). Many plant mutants resistant to the various AHAS inhibitors have been isolated and their characterization has been discussed in detail (Lea and Forde, 1994; Mazur and Falco, 1989). A mutant AHAS gene (*csr*1) conferring resistance to chlorsulphuron was isolated from *Arabidopsis* and when expressed in transgenic tobacco using its own promoter was shown to confer herbicide resistance (Haughn et al., 1988). When the mutant *csr*1 gene was expressed in tobacco under the control of the CaMV 35S promoter, it conferred a 300-fold increase in resistance to chlorsulphuron. Interestingly, the 25-fold increase in AHAS transcript levels correlated to only a twofold increase in AHAS specific activity (Odell et al., 1990). The discrepancy between RNA levels

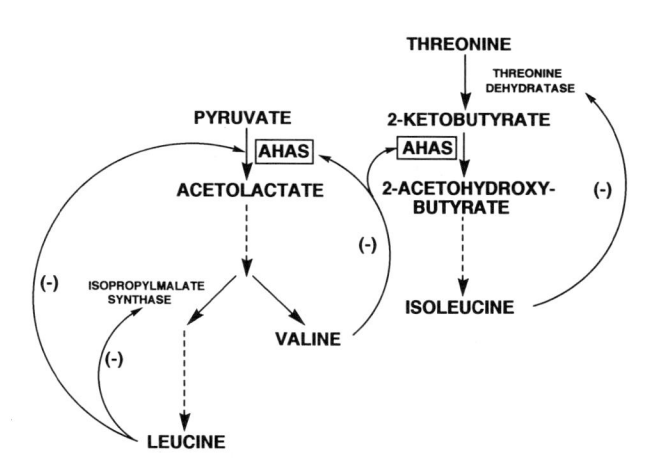

Figure 7.4 The branch chain amino acid biosynthetic pathway. Biosynthetic pathway for the amino acids valine, leucine and isoleucine. The enzyme AHAS is boxed to indicate that enzyme modulation has been attempted. Curved arrows and minus sign (−) represent feedback inhibition by the end product amino acids. Dashed lines indicate pathways consisting of multiple enzyme steps

and enzyme activity has been attributed to inefficient transport into the chloroplast, a translational block or protein instability. It has also been proposed that AHAS activity may involve two subunits – large and small, and that only the large subunit gene was being over-expressed in the transformants. The use of the CaMV 35S promoter with a duplicate enhancer (p70) to drive the *csr*1 gene resulted in a 1500-fold increase in chlorsulphuron resistance and 12-fold increase in AHAS activity (Tourneur *et al.*, 1993). Resistance to inhibition by exogenously supplied valine was now achieved, however, there was no significant increase in the content of free branched chain amino acids in leaves, suggesting that subsequent steps in the branch chain amino acid pathway may also regulate the synthesis of valine, leucine and isoleucine (Tourneur *et al.*, 1993).

Recently potato plants expressing a constitutive antisense AHAS have been generated and have been subjected to a comprehensive biochemical analysis (Hofgen *et al.*, 1995). A decrease in AHAS activity of up to 85%, with a corresponding reduction in transcript level was observed. This resulted in plants that were almost non-viable without amino acid supplements. Severe growth retardation and a strong phenotypic resemblance to plants treated with the AHAS-inhibiting herbicides were observed (Hofgen *et al.*, 1995). The overall level of individual amino acid pools increased in both source and sink tissue of the antisense plants, although not to the levels seen in plants treated with the AHAS herbicide Scepter. However, the increases differed among the various amino acids with no clear trends. The antisense plants, instead of showing a reduction in valine, leucine and isoleucine, actually had elevated levels of the three branch chain amino acids (Hofgen *et al.*, 1995). The authors speculated that the increase in free amino acids produces an imbalance in their relative proportions and this in turn produces a general deregulation of amino acid biosynthesis.

7.7 Modulation of proline synthesis

Besides being an essential component of proteins, proline is also thought to play an important role in the adaptation of plant cells to drought and salinity stress (Delauney and Verma, 1990). In plants, proline is synthesized from either glutamate or ornithine (Delauney and Verma, 1993; Figure 7.5). Both pathways lead to the common intermediate Δ-pyrroline-5-carboxylate (P5C) which is reduced to proline by Δ-pyrroline-5-carboxylate reductase (P5CR). The first step

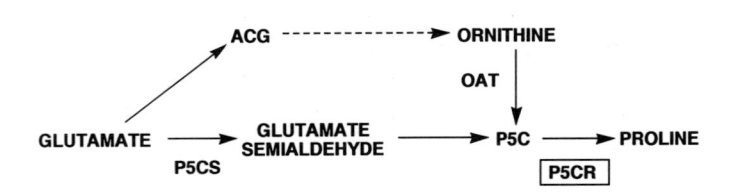

Figure 7.5 Proline biosynthetic pathway. The enzyme P5CR is boxed to indicate that enzyme modulation has been attempted. The dashed line indicates a pathway consisting of multiple enzyme steps

in proline synthesis involving the conversion of glutamate to glutamate semialdehyde is catalyzed by the bifunctional enzyme Δ-pyrroline-5-carboxylate synthetase (P5CS). The levels of both P5CR and P5CS are elevated in salt stressed plants (Delauney and Verma, 1990; Hu et al., 1992). Expression of a soybean nodule cytosolic P5CR in transgenic tobacco under the control of the CaMV 35S led to highly elevated levels of P5CR activity (50-fold) in the cytosol of the transformants. The transgenic plants, however, did not show any significant change in the levels of proline or the intermediate P5C over the control plants, even under conditions of salt stress, suggesting that P5CR activity may not be the limiting factor in proline production. An understanding of the regulation of this pathway is required for the successful engineering for drought and salinity tolerance in crop plants.

7.8 Conclusions

Even though we have a good understanding of the amino acid biosynthetic pathways in bacteria, elucidating the amino acid biosynthetic regulatory mechanisms of plants is of great interest. While the enzymes involved in the pathway are catalytically similar between plants and bacteria, the mechanisms regulating the expression and activity of these enzymes are quite distinct. This understanding will also have an impact on genetic engineering strategies in applied research such as increasing the potential yield, or improving protein quality. With the acceptance of *Arabidopsis* as a model plant, many mutants in the amino acid biosynthetic pathways have been isolated and characterized and they have added to our understanding of the amino acid biosynthetic regulatory mechanisms in plants. However, generating mutants is not as straightforward for polyploid plants and for specific members of multigene families. A genetic engineering approach to manipulate levels of key enzymes could thus have a broader impact. Moreover, this approach allows for modulating the expression to desired levels in a gene member specific and cell specific manner and for protein engineering. However, antisense RNA technology with its capability to down-regulate genes specifically to desired levels has some limitations. An alternative approach to antisense RNA technology is metabolic redirection. This approach has been very elegantly demonstrated by Chavadej et al. (1994) in their attempts to reduce indole glucosinolates in Canola. Canola plants transformed with a gene for tryptophan decarboxylase redirected tryptophan to tryptamine rather than to glucosinolate. Thus, removing a particular intermediate in an amino acid biosynthetic pathway would have the same impact as down-regulating the enzyme that uses the intermediate as a substrate.

Research in this field is still in its infancy and while we have not made great strides in crop improvement, we feel the analysis of the different transformants, with both sense and antisense constructs, has added to our understanding of the regulation of amino acid biosynthesis in plants.

Acknowledgments

The authors wish to acknowledge the communication of unpublished results from Carroll Vance, Chrisitine Foyer and Bertrand Hirel. This work was supported by a grant from the National Science Foundation (IBN-9220142). CSG also acknowledges support from the Agriculture Experiment Station at New Mexico State University.

References

ALIBHAI, M.F. and VILLAFRANCA, J.J. (1994) Kinetic and mutagenic studies of the role of the active site residues Asp-50 and Glu-327 of *Escherichia coli* glutamine synthetase. *Biochemistry* **33**, 675–681.

ALMASSY, R.J., JANSON C.A., HAMLIN, R., XUONG, N.-H. and EISENBERG, D. (1986) Novel subunit–subunit interactions in the structure of glutamine synthetase. *Nature* **323**, 304–309.

BARRY, G., KISHORE, G., PADGETTE, S., TAYLOR, M., KOLACZ, K., WELDON, M.R.D., EICHHOLTZ, D., FINCHER, K. and HALLAS, L. (1992) Inhibitors of amino acid biosynthesis: strategies for imparting glyphosate tolerance to crop plants. In: *Biosynthesis and Molecular Regulation of Amino Acids in Plants* (SINGH, B.K., FLORES, H.E. and SHANNON, J.C., eds), pp. 139–143. American Society of Plant Physiologists, Rockville, MD.

BENJAMIN, P.M., WU, J.-L., MITCHELL, A.P. and MAGASANIK, B. (1989) Three regulatory systems control expression of glutamine synthetase in *Saccharomyces cerevisae* at the level of transcription. *Mol. Gen. Genet.* **217**, 370–377.

BENNETT, M.J. and CULLIMORE, J.V. (1989) Glutamine synthetase isoenzymes of *Phaseolus vulgaris* L.: subunit composition in developing root nodules and plumules. *Planta* **179**, 433–440.

BENNETT, M.J. and CULLIMORE, J.V. (1992) Selective cleavage of closely-related mRNAs by synthetic ribozymes. *Nucl. Acid. Res.* **20**, 831–837.

BENNETT, M.J., LIGHTFOOT, D.A. and CULLIMORE, J.V. (1989) cDNA sequence and differential expression of the gene encoding the glutamine synthetase γ polypeptide of *Phaseolus vulgaris* L. *Plant Mol. Biol.* **12**, 553–565.

BLOKLAND, R.V., DE GEEST, N.V., MOL, J.N.M. and KOOTER, J.M. (1994) Transgene-mediated suppression of chalcone synthase expression in *Petunia hybrida* results from an increase in RNA turnover. *Plant J.* **6**, 861–877.

BOURQUE, J.E. (1995) Antisense strategies for genetic manipulations in plants. *Plant Sci.* **105**, 125–149.

BREARS, T., WALKER, E.L. and CORUZZI, G.M. (1991) A promoter sequence involved in cell-specific expression of the pea glutamine synthetase GS3A gene in organs of transgenic tobacco and alfalfa. *Plant J.* **1**, 235–240.

BREARS, T., LIU, C., KNIGHT, T.J. and CORUZZI, G.M. (1993) Ectopic overexpression of asparagine synthetase in transgenic tobacco. *Plant Physiol.* **103**, 1285–1290.

BRIGHT, S.W.J. and SHEWRY, P.R. (1983) Improvement of protein quality in cereals. *Crit. Rev. Plant Sci.* **1**, 49–93.

BRYAN, J.K. (1980) Synthesis of the aspartate family and branched amino acids. In: *The Biochemistry of Plants*, Vol. 5 (MIFLIN, B.J., ed.), pp. 403–452. Academic Press, New York.

CARVALHO, H., PEREIRA, S., SUNKEL, C. and SALEMA, R. (1992) Detection of a cytosolic glutamine synthetase in leaves of *Nicotiana tabacum* L. by immunocytochemical methods. *Plant Physiol.* **100**, 1591–1594.

CHAVADEJ, S., BRISSON, N., MCNEIL, J.N. and DE LUCA, V. (1994) Redirection of tryptophan leads to production of low indole glucosinolate canole. *Proc. Natl Acad. Sci. USA* **91**, 2166–2170.

COMAI, L., FACCIOTTI, D., HIATT, W.R., THOMPSON, G., ROSE, R. and STALKER, D. (1985) Expression in plants of a mutant *aro*A gene from *Salmonella typhimurium* confers tolerance to glyphosate. *Nature* **317**, 741–744.

DASSARMA, S., TISCHER, E. and GOODMAN, H.M. (1986) Plant glutamine synthetase complements a *glnA* mutation in *Escherichia coli*. *Science* **232**, 1242–1244.

DELAUNEY, A.J. and VERMA, D.P.S. (1990) A soybean gene encoding Δ'-pyrroline-5-carboxylase reductase was isolated by functional complementation in *Escherichia coli* and is found to be osmoregulated. *Mol. Gen. Genet.* **221**, 299–305.

DELAUNEY, A.J. and VERMA, D.P.S. (1993) Proline biosynthesis and osmoregulation in plants. *Plant J.* **4**, 215–223.

DELAUNEY, A.J., TABAEIZADEH, Z. and VERMA, D.P.S. (1988) A stable bifunctional antisense transcript inhibiting gene expression in transgenic plants. *Proc. Natl Acad. Sci. USA* **85**, 4300–4304.

DONN, G., TISCHER, E., SMITH, J.A. and GOODMAN, H.M. (1984) Herbicide-resistant alfalfa cells: an example of gene amplification in plants. *J. Mol. Appl. Genet.* **2**, 621–635.

ECKES, P., SCHMITT, P., DAUB, W. and WENGENMAYER, F. (1989) Overproduction of alfalfa glutamine synthetase in transgenic tobacco plants. *Mol. Gen. Genet.* **217**, 263–268.

EDWARDS, J.W., WALKER, E.L. and CORUZZI, G.M. (1990) Cell-specific expression in transgenic plants reveals nonoverlapping roles for chloroplast and cytosolic glutamine synthetase. *Proc. Natl Acad. Sci. USA* **87**, 3459–3463.

FILLATTI, J.A.J., KISER, J., ROSE, R. and COMAI, L. (1987) Efficient transfer of a glyphosate tolerance gene into tomato using a binary *Agrobacterium tumefaciens* vector. *Biotechnology* **5**, 726–730.

FISCHER, M.T. (1992) Promotion of the *in vitro* renaturation of dodecameric glutamine synthetase from *Escherichia coli* in the presence of GroEL (chaperonin-60) and ATP. *Biochemistry* **31**, 3955–3963.

FRANKARD, V., GHISLAIN, M. and JACOBS, M. (1992) Two feedback-insensitive enzymes of the aspartate pathway in *Nicotiana sylvestris*. *Plant Physiol.* **99**, 1285–1293.

GALILI, G. (1995) Regulation of lysine and threonine synthesis. *Plant Cell* **7**, 899–906.

GAO, F. and WONG, P.P. (1994) Genotypes of the common bean (*Phaseolus vulgaris* L.) lacking the nodule-enhanced isoform of glutamine synthetase. *Plant Physiol.* **106**, 1389–1394.

GHISLAIN, M., FRANKARD, V., VANDENBOSSCHE, D., MATTHEWS, B.F. and JACOBS, B. (1994) Molecular analysis of the aspartate kinase–homoserine dehydrogenase gene from *Arabidopsis thaliana*. *Plant Mol. Biol.* **24**, 835–851.

GLASSMAN, K.F. (1992) A molecular approach to elevating free lysine in plants. In: *Biosynthesis and Molecular Regulation of Amino Acids in Plants* (SINGH, B.K., FLORES, H.E. and SHANNON, J.C., eds), pp. 217–228. American Society of Plant Physiologists, Rockville, MD.

GREGERSON, R.G., MILLER, S.S., TWARY, S.N., GANTT, J.S. and VANCE, C.P. (1993) Molecular characterization of NADH-dependent glutamate synthase from alfalfa nodules. *Plant Cell* **5**, 215–226.

HAUGHN, G.W., SMITH, J., MAZUR, B. and SOMERVILLE, C. (1988) Transformation with a mutant *Arabidopsis* acetolactate synthase gene renders tobacco resistant to sulfonylurea herbicides. *Mol. Gen. Genet.* **211**, 266–271.

HEMON, P., ROBBINS, M.P. and CULLIMORE, J.V. (1990) Targeting of glutamine synthetase to the mitochondria of transgenic tobacco. *Plant Mol. Biol.* **15**, 895–904.

HERRMANN, K.M. (1995) The shikimate pathway: early steps in the biosynthesis of aromatic compounds. *Plant Cell* **7**, 907–919.

HIREL, B., MARSOLIER, M.C., HOARAU, A., HOARAU, J., BRANGEON, J., SCHAFER, R. and VERMA, D.P.S. (1992) Forcing expression of a soybean root glutamine synthetase gene in tobacco leaves induces a native gene encoding cytosolic enzyme. *Plant Mol. Biol.* **20**, 207–218.

HIREL, B., MIAO, G.-H. and VERMA, D.P.S. (1993) Metabolic and developmental control of glutamine synthetase genes in legume and non-legume plants. In: *Control of Plant Gene Expression* (VERMA, D.P.S., ed.), pp. 443–458. CRC Press, Boca Raton, FL.

HOELZLE, I., FINER J.J., MCMULLEN M.D. and STREETER, J.G. (1992) Induction of glutamine synthetase activity in nonnodulated roots of *Glycine max*, *Phaseolus vulgaris* and *Pisum sativum*. *Plant Physiol.* **100**, 525–528.

HOFGEN, R., LABER, B., SCHUTTKE, I., KLONUS, A.-K., STREBER, W. and POHLENZ, H.-D. (1995) Repression of acetolactate synthase activity through antisense inhibition. *Plant Physiol.* **107**, 469–477.

HU C.A.A., DELAUNEY, A.J. and VERMA D.P.S. (1992) A bifunctional enzyme (Δ'-pyroline-5-carboxylate synthase) catalyzes the first two steps in proline biosynthesis in plants. *Proc. Natl Acad. Sci. USA* **89**, 9354–9358.

JONES, J.D., HENSTRAND, J.M., HANDA, A.K., HERRMANN, K.M. and WELLER, S.C. (1995) Impaired wound induction of 3-deoxy-D-arabino-heptulosonate-7-phosphate (DAHP) synthase and altered stem development in transgenic potato plants expressing a DAHP synthase antisense construct. *Plant Physiol.* **108**, 1413–1421.

KAMACHI, K., YAMAYA, T., MAE, T. and OJIMA, K. (1991) A role for glutamine synthetase in the remobilization of leaf nitrogen during senescence in rice leaves. *Plant Physiol.* **96**, 411–417.

KAMACHI, K., YAMAYA, T., HAYAKAWA, T., MAE, T. and OJIMA, K. (1992) Vascular bundle-specific localization of cytosolic glutamine synthetase in rice leaves. *Plant Physiol.* **99**, 1481–1486.

KANEKO, T., HASHIMOTO, T., KUMPAISAL, R. and YAMADA, Y. (1990) Molecular cloning of wheat dihydrodipicolinate synthase. *J. Biol. Chem.* **265**, 17451–17455.

KARCHI, H., SHAUL, O. and GALILI, G. (1993) Seed specific expression of bacterial desensitized aspartate kinase increases the production of seed threonine and methionine in transgenic tobacco. *Plant J.* **3**, 721–727.

KARCHI, H., SHAUL, O. and GALILI, G. (1994) Lysine synthesis and catabolism are coordinately regulated during tobacco seed development. *Proc. Natl Acad. Sci. USA* **91**, 2577–2581.

KLEE, H.J., MUSKOPF, Y.M. and GASSER, C.S. (1987) Cloning of an *Arabidopsis thaliana* gene encoding 5-enolpyruvylshikimate-3-phosphate synthase: sequence analysis and manipulation to obtain glyphosate tolerant plants. *Mol. Gen. Genet.* **210**, 437–442.

KOZAKI, A., SAKAMOTO, A., TANAKA, K. and TAKEBA G. (1991) The promoter of the gene for glutamine synthetase from rice shows organ-specific and substrate-induced expression in transgenic tobacco plants. *Plant Cell Physiol.* **32**, 353–358.

KROL, A.R. VAN DE, MOL, J.N.M. and STUITJE, A.R. (1988) Modulation of eukaryotic gene expression by complementary RNA or DNA sequences. *BioTechniques* **6**, 958–976.

LAM, H.-M., COSCHIGANO, K., SCHULTZ, C., MELO-OLIVEIRA, R., TJADEN, G., OLIVEIRA, I., NGAI, N., HSIEH, M.-H. and CORUZZI, G. (1995) Use of *Arabidopsis* mutants and genes to study amide amino acid biosynthesis. *Plant Cell* **7**, 887–898.

LEA, P.J. and FORDE, B.G. (1994) The use of mutants and transgenic plants to study amino acid metabolism. *Plant Cell Environ.* **17**, 541–556.

LEA, P.J., ROBINSON, S.A. and STEWART, G.R. (1990) The enzymology and metabolism of glutamine, glutamate and asparagine. In: *Biochemistry of Plants* (MIFLIN, B.J. and LEA, P.J., eds), pp. 121–159. Academic Press, New York.

LI, J., ZHAO, J., ROSE, A.B., SCHMIDT, R. and LAST, R.L. (1995) *Arabidopsis* phosphoribosylanthranilate isomerase: molecular genetic analysis of triplicate tryptophan pathway genes. *Plant Cell* **7**, 447–461.

LONG, L.M. and LIGHTFOOT, D.A. (1995) Assimilatory glutamate dehydrogenase (gdhA) expression in tobacco. *Plant Physiol.* **108**, 126.

LUBBEN, T.H., DONALDSON, G.H., VIITANEN, P.V. and GATENBY, A.A. (1989) Several proteins imported into chloroplasts form stable complexes with the GroEL related chloroplast molecular chaperone. *Plant Cell* **1**, 1223–1230.

MAZUR, B.J. and FALCO, S.C. (1989) The development of herbicide resistant crops. *Annu. Rev. Plant Physiol. Mol. Biol.* **40**, 441–470.

MIAO, G.H., HIREL, B., MARSOLIER, M.C., RIDGE, R.W. and VERMA, D.P.S. (1991) Ammonia-regulated expression of a soybean gene encoding glutamine synthetase in transgenic *Lotus corniculatus*. *Plant Cell* **3**, 11–22.

MIFLIN, B.J. and LEA, P.J. (1980) Ammonia assimilation. In: *The Biochemistry of Plants* (MIFLIN, B.J., ed.), Vol. 5, pp. 169–202. Academic Press, New York.

MOL, J.N.M., VAN BLOKLAND, R., DE LANGE, P., STAM, M. and KOOTER J.M. (1994) Post-transcriptional inhibition of gene expression: sense and antisense genes. In: *Homologous Recombination and Gene Silencing in Plants* (PASZKOWSKI, J., ed.), pp. 309–334. Kluwer, Dordrecht, The Netherlands.

OAKS, A. and ROSS, D.W. (1984) Asparagine synthetase in *Zea mays*. *Can. J. Bot.* **62**, 68–73.

ODELL, J.T., CAIMI, P.G., YADAV, N.S. and MAUVAIS, C.J. (1990) Comparison of increased expression of wild-type and herbicide-resistant acetolactate synthase genes in transgenic plants, and indication of posttranscriptional limitation on enzyme activity. *Plant Physiol.* **94**, 1647–1654.

OGINO, T., GARNER, C., MARKLEY, J.L. and HERRMANN, K.M. (1982) Biosynthesis of aromatic compounds: ^{13}C NMR spectroscopy of whole *Escherichia coli* cells. *Proc. Natl Acad. Sci. USA* **79**, 5828–5832.

PERL, A., SHAUL, O. and GALILI, G. (1992) Regulation of lysine synthesis in transgenic potato plants expressing a bacterial dihydrodipicolinate synthase in their chloroplasts. *Plant Mol. Biol.* **19**, 815–823.

PESOLE, G., BOZETTI, M.P., LANAVE, C., PREPARATA, G. and SACCONE, C. (1991) Glutamine synthetase gene evolution: a good molecular clock. *Proc. Natl Acad. Sci. USA* **88**, 522–526.

PETERMAN, T.K. and GOODMAN, H.M. (1991) The glutamine synthetase gene family of *Arabidopsis thaliana* light-regulation and differential expression in leaves, roots and seeds. *Mol. Gen. Genet.* **230**, 145–154.

RADWANSKI, E.R. and LAST, R.L. (1995) Tryptophan biosynthesis and metabolism: biochemical and molecular genetics. *Plant Cell* **7**, 921–934.

ROSE, A.B. and LAST, R.L. (1994) Molecular genetics of amino acid, nucleotide and vitamin biosynthesis. In: *Arabidopsis*. (MEYEROWITZ, E.M. and SOMERVILLE, C.R., eds), pp. 835–879. Cold Spring Harbor Press, Cold Spring Harbor, NY.

SAMAC, D.A. and SHAH, D.M. (1991) Developmental and pathogen-induced activation of the *Arabidopsis acidic* chitinase promoter. *Plant Cell* **3**, 1063–1072.

SAMAC, D.A., HIRONAKA, C.M., YALLAY, P.E. and SHAH, D.M.

(1990) Isolation and characterization of the genes encoding basic and acidic chitinase in *Arabidopsis thaliana*. *Plant Physiol.* **93**, 907–919.

SHAH, D., HORSCH, R., KLEE, H., KISHORE, G., WINTER, J., TURNER, N., HIRONAKA, C., SANDERS, P., GASSER, C., AYKENT, S., SLEGAL, N., ROGERS, S. and FRALEY, R. (1986) Engineering herbicide tolerance in transgenic plants. *Science* **233**, 478–481.

SHAUL, O. and GALILI, G. (1992a) Increased lysine synthesis in transgenic tobacco plants expressing a bacterial dihydrodipicolinate synthase in their chloroplasts. *Plant J.* **2**, 203–209.

SHAUL, O. and GALILI, G. (1992b) Threonine overproduction in transgenic tobacco plants expressing a mutant desensitized aspartate kinase from *Escherichia coli*. *Plant Physiol.* **100**, 1157–1163.

SHAUL, O. and GALILI, G. (1993) Concerted regulation of lysine and threonine synthesis in tobacco plants expressing bacterial feedback insensitive aspartate kinase and dihydrodipicolinate synthase. *Plant Mol. Biol* **23**, 759–768.

SOMERVILLE, C.R. and OGREN, W.L. (1980) The inhibition of photosynthesis in *Arabidopsis* mutants lacking leaf glutamate synthase activity. *Nature* **286**, 257–259.

STANFORD, A.C., LARSEN, K., BARKER, D.G. and CULLIMORE, J.V. (1993) Differential expression within the glutamine synthetase gene family of the model legume *Medicago truncatula*. *Plant Physiol.* **103**, 73–81.

STEINRUCKEN, H.C. and AMRHEIN, N. (1980) The herbicide glyphosate is a potent inhibitor of 5-enolpyruvyl-shikimic acid-3-phosphate synthase. *Biochem. Biophys. Res. Commun.* **94**, 1207–1212.

STEWART, G.R., MANN, A.F. and FENTEM, P.A. (1980) Enzymes of glutamate formation: glutamate dehydrogenase, glutamine synthase and glutamate synthase. In: *The Biochemistry of Plants* (MIFLIN, B.J., ed.), Vol. 5, pp. 271–327. Academic Press, New York.

STOUGAARD, J., SANDAL, N.N., GRON, A., KUHLE, A. and MARCKER, K.A. (1987) 5' analysis of the soybean leghemoglobin lbc3 gene: regulatory elements required for promoter activity and organ specificity. *EMBO J.* **6**, 3565–3569.

SUKANYA, R., LI, M.-G. and SNUSTAD, D.P. (1994) Root- and shoot-specific responses of individual glutamine synthetase genes of maize to nitrate and ammonium. *Plant Mol. Biol.* **5**, 1935–1946.

TEMPLE, S.J., KNIGHT, T.J., UNKEFER, P.J. and SENGUPTA-GOPALAN, C. (1993) Modulation of glutamine synthetase gene expression in tobacco by the introduction of an alfalfa glutamine synthetase gene in sense and anti-sense orientation: molecular and biochemical analysis. *Mol. Gen. Genet.* **236**, 315–325.

TEMPLE, S.J., BAGGA, S. and SENGUPTA-GOPALAN, C. (1994) Can glutamine synthetase activity levels be modulated in transgenic plants by the use of recombinant DNA technology? *Biochem. Soc. Trans.* **22**, 915–920.

TEMPLE, S.J., HEARD, J., GANTER, G., DUNN, K. and SENGUPTA-GOPALAN, C. (1995) Characterization of a nodule-enhanced glutamine synthetase from alfalfa: nucleotide sequence, *in situ* localization and transcript analysis. *Mol. Plant–Microbe Interac.* **8**, 218–227.

TEMPLE, S.J., KUNJIBETTU, S., ROCHE, D. and SENGUPTA-GOPALAN, C. (1997) Total glutamine synthetase activity during soybean nodule development is controlled at the level of transcription, and holoprotein turnover. *Plant Physiol.* **112**, 1723–1733.

TINGEY, S.V., WALKER, E.L. and CORUZZI, G.M. (1987) Glutamine synthetase genes of pea encode distinct polypeptides which are differentially expressed in leaves, roots and nodules. *EMBO J.* **6**, 1–9.

TISCHER, E., DASSARMA, S. and GOODMAN, H.M. (1986) Nucleotide sequence of an alfalfa glutamine synthetase gene. *Mol. Gen. Genet.* **203**, 221–229.

TOURNEUR, C., JOUANIN, L. and VAUCHERET, H. (1993) Over-expression of acetolactate synthase confers resistance to valine in transgenic tobacco. *Plant Sci.* **88**, 159–168.

TSAI, F.-Y. and CORUZZI, G.M. (1990) Dark-induced and organ-specific expression of two asparagine synthase genes in *Pisum sativum*. *EMBO J.* **9**, 323–332.

TSUPRUN, V.L., BOEKEMA, E.J., PUSHKIN, A.V. and TAGUNOVA, I.V. (1992) Electron microscopy and image analysis of the GroEL-like protein and its complexes with glutamine synthetase from pea leaves. *Biochem. Biophys. Acta* **1099**, 67–73.

WALKER, E.L. and CORUZZI, G.M. (1989) Developmentally regulated expression of the gene family for cytosolic glutamine synthetase in *Pisum sativum*. *Plant Physiol.* **91**, 702–708.

YAMASHITA, M.M., ALMASSY, R.J., JANSON, C.A., CASCIO, D. and EISENBERG, D. (1989) Refined atomic model of glutamine synthetase at 3.5 Å resolution. *J. Biol. Chem.* **264**, 17681–17690.

8

Regulation of C/N metabolism by reversible protein phosphorylation

CAROL MACKINTOSH

8.1 Introduction

The activities of several enzymes of C and N metabolism in the cytosol, mitochondria and plastids of plant cells are regulated by rapid changes in protein phosphorylation in response to environmental and metabolic stimuli. Recent experiments using transgenic plants have highlighted the practical importance of understanding the impact of these regulatory mechanisms on patterns of metabolic flux in plants. For example, over- or under-expressing SPS and NR in plants was found to have minimal effects in some cases because compensatory changes in the activation (phosphorylation) states of these enzymes took place that restored their activities towards the levels in wild-type plants (e.g. Sonnewald et al., 1994). In contrast, plants that constitutively over-expressed 'unregulated' SPS showed large increases in biomass (Chapter 6). Testing the physiological effects of mutating the regulatory phosphorylation sites on metabolic enzymes, and altering the activities of the relevant protein kinases and phosphatases by genetic and/or 'pharmaceutical' means are likely, therefore, to be important contributions to future developments in engineering plant metabolism.

Over the last six years or so, the realization that plants contain regulatory protein kinases and phosphatases that are closely related to their counterparts in other eukaryotes has provided a tremendous boost to investigations of plant metabolism. For example, probes derived from mammalian protein kinases and phosphatases have been used extensively to identify plant homologues by cDNA and PCR cloning. Potent inhibitors of the eukaryotic protein kinases and phosphatases have been vital tools for probing the control mechanisms for the cytosolic enzymes, Fru6P2K/Fru2,6P$_2$ase, SPS, NR, and PEPCase (e.g. Huber et al., 1992a; Siegl et al., 1990). Furthermore, there are excellent prospects that a wider range of specific inhibitors of eukaryotic protein kinases and phosphatases will come from screening programmes aimed at finding drugs to treat metabolic disorders in humans (Cuenda et al., 1995). Who knows? Maybe derivatives of these new compounds will prove useful as 'herbidrugs' to manipulate plant metabolism.

It is important, though, to allay suggestions that studies of protein phosphorylation in plants are simply following the lead of research on animals. Not true! In some cases, plant protein kinases and phosphatases have been found to be more analogous to bacterial signalling components (Theologis, 1996); others are completely novel, or are hybrids comprising both familiar and novel features (Braun and Walker, 1996). Moreover, it is particularly exciting to report that despite the relatively short history of research on protein phosphorylation in plants, novel or unusual twists on the well-known paradigms from mammalian studies are being discovered; this chapter highlights some examples, including a protein kinase (PEPCase kinase) that is regulated by protein synthesis and degradation, an enzyme (NR) whose phosphorylation causes binding to an inhibitor protein (NIP), and a kinase which uses Mg-ADP as substrate (pyruvate Pi dikinase (PPdK) kinase).

Plants and animals have different lifestyles and face different environmental challenges, so comparing metabolic regulation in these two kingdoms is likely to be mutually beneficial in understanding which are the important general principles of control, and which are specific to particular responses. Before tackling the regulation of plant C and N metabolism this chapter will, therefore, discuss reversible protein phosphorylation as a universal switching mechanism, and briefly survey the eukaryotic protein kinases and phosphatases.

8.2 Reversible protein phosphorylation

8.2.1 *Reversible protein phosphorylation as a regulatory device in eukaryotes*

Reversible phosphorylation, catalyzed by protein kinases and protein phosphatases, is the most widespread method of acute regulation of protein function in eukaryotic cells (Krebs, 1994), controlling the activities of enzymes, ion channels, transcription factors, components of the cell division apparatus, and structural and contractile proteins. Eukaryotic protein kinases transfer the gamma phosphate from Mg-ATP or Mg-GTP to specific serine, threonine or tyrosine residues in a target protein, causing the protein to change its shape and function. Protein phosphatases remove the phosphate groups by hydrolysis and the target protein reverts to its original state. Because protein phosphorylation and dephosphorylation are distinct reactions catalyzed by distinct enzymes, these systems are not constrained by the equilibrium rules that apply to processes (such as ligand binding) that are reversed by simple back reactions. This means that by controlling the rate of the kinase and/or the phosphatase reaction, the state of the target pool can be varied to any point between the extremes of 'completely on' or 'completely off'. If the activities of a protein kinase and its opposing phosphatase are not vastly different, the system can act analogously to a 'dimmer' switch, fine-tuning the activities of the target subtly. If either the kinase or the phosphatase dominate, the target is flipped completely to the 'on' or 'off' state.

The activities of protein kinases and phosphatases can be controlled by a variety of mechanisms including phosphorylation by a second protein kinase; interactions with second messengers, metabolites, and specific protein inhibitors; subcellular targeting; and (most notably in plants) protein synthesis and

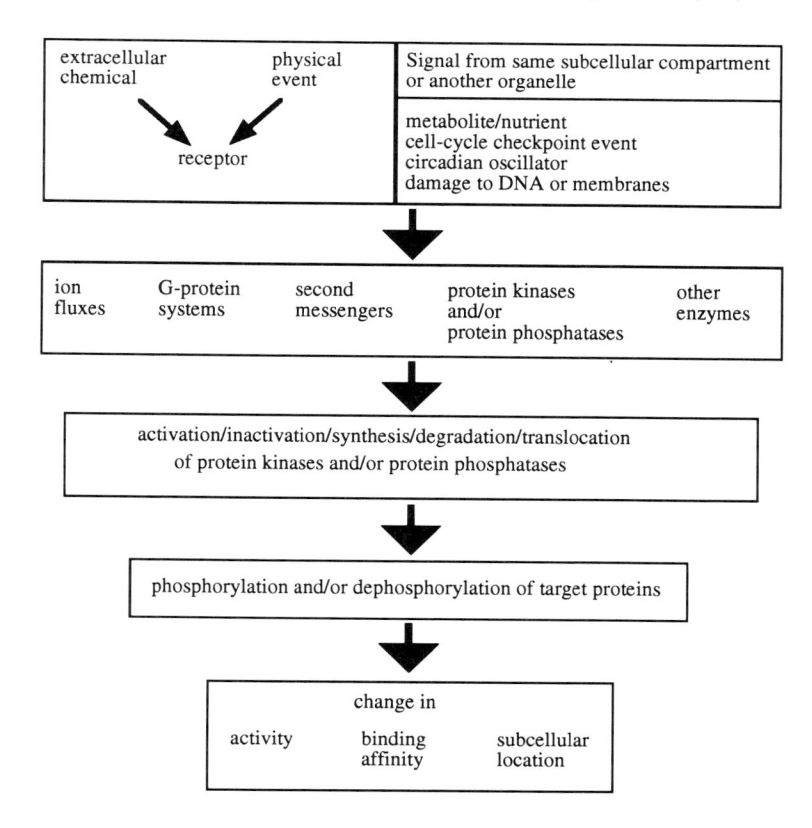

Figure 8.1 Schematic illustration of the central role of reversible protein phosphorylation in regulating cytoplasmic and nuclear processes in eukaryotic cells

degradation. These mechanisms are controlled by signals that can originate extracellularly; from a different subcellular compartment; from metabolic changes within the same compartment; from circadian oscillators; and/or from the checkpoint systems that control cell cycle events (Figure 8.1). Often, 'networks' of protein kinases and phosphatases that control more than one pathway are formed and the regulatory and amplifying potential of reversible protein phosphorylation can be further increased by multisite phosphorylation of a target enzyme by different kinases, or by linking two or more phosphorylation/dephosphorylation cycles in a cascade sequence. Examples of each of these 'network' motifs have been found in plants. For example, the recent discovery of a dual-specificity SPS/NR kinase (McMichael et al., 1995) provides a potential mechanism for co-ordinating sucrose synthesis and nitrate assimilation. There are indications that SPS, and possibly NR and PEPCase, may be phosphorylated at multiple sites. Protein kinases that are activated by phosphorylation by kinase kinases have been discovered in plants, namely the mitogen-activated protein (MAP) kinases (Hardie and Hanks, 1995), 3-hydroxy-3methyl glutacyl-CoA-reductase HMGR kinase (Dale et al., 1995), and two NR/SPS kinases (Douglas et al., 1997).

8.2.2 Eukaryotic protein kinases

That reversible protein phosphorylation is the major mechanism for switching cellular activity from one state to another is being witnessed by the discovery of more-and-more eukaryotic protein kinases forming an enormous 'superfamily' comprising 113 members in budding yeast and predicted to have up to 2000 members in vertebrate species (equivalent to 3% of the coding sequences in the genome) (Hunter, 1997). How many members of the protein kinase superfamily will be found in plants is anybody's guess. Over 50 different protein kinases have already been cloned from higher plants, including plant homologues of several vertebrate and yeast enzymes (Hardie and Hanks, 1995). Moreover, plants seem to have many isoforms of some protein kinases, for example, the MAP kinases (Hardie and Hanks, 1995). These early indications suggest that the plant contingent of the protein kinase superfamily may turn out to be very substantial.

All eukaryotic protein kinases have a common catalytic core of about 250–300 amino acids containing 11 highly conserved subdomains whose function is to transfer gamma phosphates from ATP or GTP to form phosphoesters with the hydroxyl groups of serine/threonine residues and tyrosine residues within target proteins. Protein (serine/threonine) kinases are termed PSKs, and the protein tyrosine kinases are PTKs. The three-dimensional structures of four PSK catalytic domains have been solved and they each reveal a small N-terminal lobe involved in ATP binding, and a larger C-terminal lobe providing the substrate binding site and the catalytic base (Hunter, 1994). Variable regions lying in loops on the outside of the globular core structure take part in a cleft that recognizes the sequence around the target residue; variable loops can also contain activation sites, autoinhibitory domains, intracellular localization motifs, and sites for binding to other signalling proteins (Hardie and Hanks, 1995).

The need to keep track of the protein kinase 'superfamily' has stimulated the compilation of computer databases, phylogenetic trees, and the first edition of *The Protein Kinase Facts Book* (Hardie and Hanks, 1995), a sequence catalogue with notes on the distinguishing properties of each enzyme. By comparing the variable regions, the protein kinase superfamily has been classified into four groups of PSKs (the AGC group, CaMK group, and CMGC group, and a larger group of 'miscellaneous' protein kinases), plus the PTKs (Hardie and Hanks, 1995) (Table 8.1). Representatives of all four PSK groups have been identified in plants, including one or two subfamilies that are, so far, unique to plants, and some enzymes which were first identified as plant developmental mutants (Table 8.1). Interest in cAMP, cGMP, and Ca^{2+}/phospholipid-derived signalling in plants has stimulated searches for plant versions of PKA, PKG and PKC (Table 8.1), but no direct plant homologues of these kinases have been found, to date. Likewise, there are no known PTKs in plants (nor in yeast). In animals, cell–cell adhesion, proliferation, differentiation, and motility are regulated by reversible tyrosine phosphorylation of critical substrates (Hardie and Hanks, 1995), but whether these processes are regulated by tyrosine (de)phosphorylation in plants is unknown.

The kinases that regulate the cytosolic enzymes $6PF2K/Fru2,6P_2ase$, SPS, NR, and PEPCase in plants have biochemical properties indicative of 'eukaryotic PSKs', but they have not yet been purified and sequenced (Section 8.3.5). So far, only two plant protein kinases have been well characterized by both activity and sequence, namely HMGR kinase (Dale et al., 1995) which has been identified as belonging to the RKIN1 branch of the CaMK subfamily (Table 8.1), and the calmodulin-domain protein kinase (CDPK) group of the CaMK subfamily whose candidate targets include the plasma membrane H^+-ATPase and the nodulin 26 protein of legumes (Table 8.1; Hardie and Hanks, 1995).

Table 8.1 Classification of the eukaryotic protein kinase superfamily (derived from Hardie and Hanks, 1995). Primary references and full details of the regulatory properties of, and phylogenetic relationships among these enzymes can be obtained from Hardie and Hanks (1995)

The AGC group are generally (but not all) 'basic amino acid-directed' meaning that they phosphorylate serine/threonine residues that have arginine and lysines in specific positions nearby. Different members of this family have different sequence preferences.
Well-characterized members include:
  Cyclic nucleotide-dependent enzymes (PKA and PKG)
  The protein kinase C family (PKC)
  The mammalian β-adrenergic receptor kinase (βARK) family
  ZmPK1 and related putative serine/threonine kinase receptors from higher plants

The CaMK group also tends to be basic amino acid-directed, and includes enzymes that are activated by binding Ca^{2+}/CaM and/or by phosphorylation.
Well-characterized subfamilies include:
  Kinases regulated by Ca^{2+}/CaM and in some cases, phosphorylation (CaMKI, CaMKII, phosphorylase kinase, myosin light chain kinases (MLCKs))
  The CDPKs, a subfamily of plant enzymes with a calmodulin-like domain on the same polypeptide as the kinase
  The Snf1/AMPK subfamily activated by phosphorylation (includes enzymes from mammals, yeast and the HMGR kinase from plants)

The CMGC group are generally 'proline-directed' enzymes that phosphorylate sites in proline-rich regions.
There are plant homologues of each of the following well-characterized subfamilies of this group:
  Cyclin-dependent protein kinases that control the cell-division cycle
  The Erk1 or MAP kinase family that are activated by protein kinase cascades
  Glycogen synthase kinase-3
  The casein kinase II subfamily

The 'miscellaneous' group is the largest group and comprises many small subfamilies.
Well-characterized members include:
  The *raf* subfamily which includes Ctr 1, a negative regulator of ethylene signalling in *Arabidopsis*

The PTK group of 'conventional protein-tyrosine' kinases.
(The term conventional is used to distinguish this family from other kinases, such as the MAP kinases, that can phosphorylate on serine/threonine and tyrosine.)
  Includes tyrosine kinase receptors and soluble enzymes from mammalian cells
  No known plant or yeast PTKs

8.2.3 Eukaryotic protein (serine/threonine) phosphatases

In contrast to the single protein kinase 'superfamily', there are four distinct gene families of protein phosphatases (PPs): two families of serine/threonine-specific enzymes; the tyrosine phosphatases and the dual-specificity phosphatases. One of the (serine/threonine) phosphatase gene families comprises enzymes that were first identified by their biochemical properties (PP1, PP2A, and PP2B), plus many related 'novel' enzymes that have been identified by cDNA and PCR cloning (Cohen, 1994).

The properties and structures of PP1 and PP2A have been more conserved during evolution than any other enzyme; for example, the amino acid sequences of the catalytic subunits of mammalian PP2As are over 70% identical to the plant PP2As (Casamayor et al., 1994; MacKintosh et al., 1991; Smith and Walker, 1993). PP1, PP2A, and related members of this family are the intracellular targets for a number of naturally occurring toxins, such as okadaic acid, the substance responsible for diarrhetic shellfish poisoning, and microcystins, the cyclic peptide liver toxins produced by blue-green algae (Cohen, 1994; MacKintosh and MacKintosh, 1994). Testing the effects of okadaic acid and microcystin on intact cells (MacKintosh and MacKintosh, 1994) has helped establish that members of this phosphatase family regulate many processes in the cytosol and nucleus of eukaryotic cells; a discovery that opens up some major challenges. Which protein phosphatase dephosphorylates which target substrate *in vivo*? And if, say, PP1 can act on many different substrates within a single cell, how does that allow for independent regulation of different cellular processes?

One emerging answer to these problems is that the catalytic subunits of PP1 and PP2A are never found free *in vivo*, but are bound to a variety of different regulatory and targeting subunits that direct the enzymes to particular subcellular locations and modify their substrate specificities. The activities of regulatory subunits of PP1 can be altered by phosphorylation or allosteric effectors, which means that PP1 in different locations within a single cell can be controlled selectively (Cohen, 1994). We know that the catalytic subunits of PP1 and PP2A in plant extracts are bound to proteins that are presumed to be regulatory subunits (Matthews and MacKintosh, 1995). However, no plant PP1 regulatory subunits have been purified and characterized. cDNA clones that are highly related to A and B regulatory subunits of the animal PP2A have been isolated from plant libraries (Slabas et al., 1994). In animal cells, the catalytic and A subunit of PP2A form a core heterodimer that binds to different B or 'variable' subunits that alter the substrate specificity of the enzyme. PP2A is of particular relevance to this chapter because PP2A, or one of the PP2A-like 'novel' enzymes, has been implicated in the control of the cytosolic enzymes that are discussed in Section 8.3.5.

The calcium-calmodulin dependent PP2B is inhibited specifically by the immunosuppressant cyclosporin when this drug is complexed to a protein called cyclophilin. Plants contain cyclophilins, and cyclosporin has been shown to affect the K^+ channels that regulate stomatal pore opening in plants (Luan et al., 1993), although whether this effect is mediated by inhibition of a plant PP2B is, as yet, uncertain.

Plant leaf extracts have relatively high PP2C activity (MacKintosh et al.,

1991). The magnesium-dependent PP2Cs are unrelated in sequence to the PP1/2A/2B gene family, and isoforms of PP2C exist as monomers in mammalian cells, but little is known about their physiological roles. It was, therefore, particularly exciting to learn that the ABI1 mutant of *Arabidopsis* is unable to respond to abscisic acid because of a mutation in a gene whose C-terminal end closely resembles PP2C and whose N-terminal domain contains a putative Ca^{2+}-binding motif (Leung *et al.*, 1994). Another plant PP2C, termed the kinase-associated protein phosphatase (KAPP), has been identified by its ability to bind to an orphan plant serine-threonine kinase receptor, RLK5 (Braun and Walker, 1996). It is hoped that identification of the substrates of these PP2Cs in plants might give clues to aid understanding of the function of this enzyme in mammalian cells.

8.2.4 Prokaryotic two-component systems

Many aspects of bacterial physiology (including chemotaxis, sporulation, osmolarity and nitrogen regulation) are regulated by 'two component signal response systems' in which a histidine kinase component becomes autophosphorylated in response to the external stimulus, and the phosphate is then transferred from the phosphohistidine to an aspartate in a cognate response protein (Swanson *et al.*, 1994). There are over 50 known examples of two-component systems in prokaryotes, and a few homologous proteins are now being discovered in eukaryotes. Examples include the mammalian mitochondrial pyruvate decarboxylase kinase (Popov *et al.*, 1992), and the ethylene sensing proteins of plants (Theologis, 1996).

8.3 Reversible protein phosphorylation and C/N metabolism in plants

8.3.1 Comparing metabolic and developmental control in plants and animals

Clearly plants contain regulatory protein kinases and phosphatases that are strikingly similar to those in mammalian cells. However, when considering the global strategies for controlling metabolism we soon run into difficulties in drawing too many analogies between plants and animals.

Dramatic changes in fluxes through the pathways of C and N metabolism in different organs of a plant occur diurnally and developmentally, and in response to nutrient supplies, environmental parameters and stress situations. The cumulative effects of metabolic flux patterns throughout a season are reflected in the overall growth of a plant, and in its changing morphology: by the changing distribution of number, size, composition and growth of various organs. Developmental plasticity linked to the metabolic status of plants contrasts markedly with mammals where development is highly constrained to a usual form and size, and metabolic patterns in different organs are co-ordinated by hormones and neurotransmitters produced in specialized locations.

In animals, strict limits are imposed on the levels of certain primary metabolites, and emergency signals override regular communications if these metabolites fall to critical concentrations, or rise to toxic levels. For example, death will be rapid if blood glucose drops much below 5 mM because the brain absolutely requires glucose for carbon and energy. Glucose sensors in the pancreas regulate the release of hormones such as insulin and glucagon that trigger changes in the phosphorylation state of regulatory enzymes and transport systems in liver, muscle and adipose tissue to gear metabolism in these organs towards maintenance of blood glucose levels. In plants too, minimum levels of certain metabolites are required to maintain respiratory and photosynthetic activities, and electrochemical gradients across membranes, and hormones play a role in their maintenance. The developmental plasticity of plants, however, often makes it difficult to assign limits to the allowable variations in plant metabolites. Thus, the identification of flux control points and regulatory interactions among different pathways, has relied upon measuring how enzyme activities, metabolites, and growth patterns change in response to different environmental, hormonal and metabolic stimuli. The regulatory capacities of particular enzymes are often tested experimentally by altering their amounts *in vivo* and by measuring resulting changes in flux. Several enzymes that have been proposed from such studies to be 'regulatory' in plant C and N metabolism are controlled by reversible protein phosphorylation. The following sections consider the evidence that reversible protein phosphorylation is a (more probably, *the*) major intracellular mechanism for controlling fluxes, switching metabolic branchpoints, co-ordinating different pathways, and preventing the build-up of toxic intermediates in plant cells.

8.3.2 Subcellular compartmentation and reversible protein phosphorylation

The distribution of the pathways of primary C and N metabolism among several specialized subcellular compartments – plastids, mitochondria, cytosol, and vacuoles (see Chapter 9) – has profound consequences for regulation. First, it appears that the protein kinases and phosphatases in different cellular compartments have distinct properties that probably reflect the prokaryotic and eukaryotic ancestries of organelles. Second, in addition to exchanging metabolites, these compartments must exchange signals because the metabolic state of one compartment can affect the activities of enzymes in another.

In photosynthetic cells, mitochondria and chloroplasts serve complementary roles in supplying both energy and substrates for synthesis, but chloroplasts use external energy to drive the process, whereas mitochondria use photosynthates (Huppe and Turpin, 1994). This means that mitochondria and chloroplasts are intimately linked in the demands for energy and substrates, and there are indications that these two compartments can 'sense' each other's metabolic state and respond to metabolic needs using mechanisms that involve protein phosphorylation.

8.3.3 Reversible protein phosphorylation in chloroplasts and other plastids

Chloroplasts operate a complex set of light-responsive regulatory mechanisms that seem to function to protect photosynthesis-related metabolism from transient 'flickers' in light intensity, and to prevent 'futile cycles' occurring between respiratory and catabolic processes, and the photosynthetic pathway, which all share the same biochemical intermediates (Furbank and Taylor, 1995). Rubisco activity is controlled by light by a specific Rubisco activase, and the effects of tight-binding inhibitors (Keys et al., 1995), which are responsive to changes in the chloroplast's energy status. The activities of a number of enzymes in the Calvin cycle respond to changes in light and could potentially limit photosynthetic flux. These responses are mediated by reduction/oxidation of enzymes by the thioredoxin system, a signal transduction system responsive to the redox state of photosystem I. Changes in the stromal pH and Mg^{2+} concentration, both of which increase on illumination, may also be involved in regulation of Calvin cycle enzymes (Furbank and Taylor, 1995). Reversible phosphorylation of Calvin cycle enzymes has been reported, but has usually been attributed to phospho-enzyme intermediates produced during catalysis from the phosphorylated sugar substrates of these enzymes.

Pyruvate Pi dikinase (PPdK) is the enzyme that regenerates PEP, the acceptor for CO_2 fixation (see section on PEPCase) in the mesophyll chloroplasts of C_4 plants, and the finding that its extractable activity is only just sufficient to account for observed rates of photosynthesis suggests that this enzyme is potentially rate limiting (Furbank and Taylor, 1995). PPdK is regulated by reversible phosphorylation in a light–dark manner. PPdK is inactivated in the dark by phosphorylation on a threonine that is distinct from the site that is phosphorylated during catalysis, which is a nearby histidine residue (Ashton et al., 1984). The enzyme that catalyzes regulatory phosphorylation of PPdK, the PPdK regulatory protein (PPRP), is unusual in that it uses ADP rather than ATP as the phosphate donor. In addition, the same enzyme catalyzes the removal of the phosphate group reactivating PPdK by phosphorolysis rather than by hydrolysis (Ashton et al., 1984). It is not clear how the activity of the PPRP itself is controlled, although high pyruvate levels appear to block its ability to phosphorylate and inactivate PPdK.

Up to 30 thylakoid membrane proteins are phosphorylated/dephosphorylated in response to changes in light and redox state (Bennett, 1991). Some are phosphorylated in the light, others phosphorylated in the dark. In particular, it has been established that reversible phosphorylation of the light harvesting chlorophyll a/b protein of photosystem II (LHCII) acts as a mechanism for optimizing the distribution of excitation energy between photosystem I and photosystem II (Allen, 1995). The protein kinase that phosphorylates LHCII and a number of other thylakoid proteins is membrane-bound and is activated under reducing conditions, in the light, by a mechanism involving the cytochrome b_6/f complex of the photosynthetic electron transport chain (Allen, 1995). Dephosphorylation of LHCII in cell-free extracts is redox-independent and stimulated by, but not dependent on Mg^{2+} ions (Cheng et al., 1994; Sun and Markwell, 1992); a property that distinguishes this activity from PP2C. The thylakoid protein phosphatase is inhibited by fluoride and molybdate (general non-specific protein phosphatase inhibitors) but not by okadaic acid, and is not, therefore, a type 1 or type 2A protein phosphatase (Cheng et al., 1994).

In summary, although no plastidic protein kinases or phosphatases have been sequenced, the limited information available suggests that the enzymes in plastids are quite distinct from the eukaryotic kinases and phosphatases, probably reflecting the prokaryotic origins of these organelles. Moreover, no plant 'eukaryotic' PSKs or PPs have been found to have plastid targeting sequences, and fractionations of leaf extracts have shown that HMGR, NR and SPS kinases (Carol MacKintosh, unpublished), and PP1, PP2A and PP2C (MacKintosh et al., 1991; Sun and Markwell, 1992), are absent from chloroplasts. Incubation of leaf discs with okadaic acid and microcystin did affect both phosphorylation of chloroplast proteins, and photosynthesis rate (Siegl et al., 1990). However, these toxins had no effect on photosynthesis in isolated chloroplasts (Carol MacKintosh, unpublished), which suggests that their effects on chloroplasts in leaves must have been indirect consequences of cytosolic changes in protein phosphorylation.

8.3.4 Reversible protein phosphorylation in mitochondria

In plants and animals, the pyruvate dehydrogenase multienzyme complex (PDC) in the mitochondrial matrix occupies a key position in energy and carbon metabolism by catalyzing the oxidative decarboxylation of pyruvate to acetyl CoA, thereby controlling entry of carbon into the TCA cycle. PDC is inactivated by PDC kinase (which phosphorylates serine residues on the E1α component of PDC) and reactivated by dephosphorylation. Both the PDC kinase and phosphatase are intrinsic parts of the PDC complex. The biological significance of PDC control concerns fuel selection, and in situ studies have shown that when pea leaf mitochondria are oxidizing glycine the PDC is phosphorylated and inactivated. PDC is also phosphorylated and inactivated in the light. The role of PDC phosphorylation in C_3 leaves, therefore, appears to be to inhibit carbon flow into the TCA cycle under conditions when the cell's demand for ATP can be met from other sources, such as photosynthesis and glycine oxidation. The signals that control the phosphorylation state of PDC may be metabolites; for example, in vitro, pyruvate inhibits and NH_4^+ stimulates PDC kinase activity (Randall et al., 1996).

Although it phosphorylates serine, the mammalian PDC kinase is more closely related to the histidine/aspartate PKs of the prokaryotic 'two-component systems' than to the eukaryotic PSKs (Popov et al., 1992). The mammalian PDC phosphatase is related to PP2C (Lawson et al., 1993). It seems reasonable to assume that the same will be true for the plant PDC kinase and phosphatase, respectively.

8.3.5 Reversible protein phosphorylation in the cytosol

In leaves, the pathways of starch, sucrose and amino acid synthesis, and nitrate assimilation are sometimes termed 'light-coupled' to indicate their intimate links with photosynthetic provision of carbon and energy. Fluxes through these 'light-coupled' pathways are sensitive to changes in light and dark that are mediated in part by fluctuations in photosynthate supply, and also by rapid changes in the phosphorylation states and activities of the cytosolic enzymes 6PF2K/Fru2,6P_2ase, SPS, NR and PEPCase. 'Rapid' generally means a time-scale of a few minutes. Where tested in C_3 leaves, the light effects on (de)phosphorylation of these enzymes have been found to be blocked by low CO_2 levels (Kaiser and

Förster, 1989) or photosynthesis inhibitors, suggesting that a signal reflecting the rate of photosynthesis must emanate from the chloroplast to alter the activities of the relevant protein kinase(s) and/or phosphatase(s) in the cytosol.

Superimposed on the light/dark and CO_2 effects, many other parameters (circadian rhythms, hormones, water status, nitrogen status, anoxia, build-up of photosynthetic end-products over the course of a day) modulate the activities and phosphorylation status of these enzymes in leaves (e.g. Huber et al., 1995), and in non-photosynthetic organs. Many of these parameters also alter the level of protein expression of these target enzymes, through signalling pathways that (probably) lead to changes in reversible phosphorylation of proteins involved in transcription, mRNA processing, protein synthesis, and protein degradation. Only the rapid post-translational changes in enzyme activities will be considered further in this chapter. However, when considering genetic manipulation of C and N metabolism, it is especially important to realize that there are probably mechanistic links between different levels of regulation for any individual enzyme. For example, in transgenic plants with altered post-translational control of NR, the activity state of the NR was found to influence the level of protein expression of the enzyme (Nussaume et al., 1995), and conversely, the expression level of NR affects the enzyme's activity state (Scheible et al., 1995).

The next section reports progress in understanding the mechanisms of control of 6PF2K/Fru2,6P_2ase, SPS, NR and PEPCase, with an emphasis on responses to changes in photosynthesis in leaves. Note the first indications of control networks (a dual-specificity NR/SPS kinase, and activation of protein kinases by phosphorylation) that are described in a separate section (8.3.6) on co-regulation of SPS, NR and PEPCase in C_3 leaves. In most cases, the effects of okadaic acid and microcystin were among the first pieces of evidence that these enzymes were regulated by reversible phosphorylation. Where investigated (for SPS, NR, PEPCase), a type 2A enzyme (either the archetypal PP2A or one of the novel PP2A-like enzymes) has been implicated as the regulatory protein phosphatase (Huber et al., 1992a; MacKintosh et al., 1995; Siegl et al., 1990). Currently, efforts are focused on purifying and characterizing the regulatory properties of the relevant protein kinases and (type 2A) protein phosphatases, and identification of sites phosphorylated by the kinases in vitro. No in vivo sites have yet been rigorously identified. Nevertheless, the effects on metabolic flux of deregulating the phosphorylation and activity states of NR, 6PF2K/Fru2,6P_2ase, SPS and PEPCase are starting to be examined quantitatively in transgenic plants.

Sucrose synthesis

In source leaves, triose-Ps derived directly from photosynthesis, or from remobilization of starch, are exported to the cytosol in exchange for Pi, and used as precursors mainly for sucrose synthesis, but also for synthesis of organic acids, amino acids, and other compounds. Pi is regenerated (mostly) during sucrose synthesis and must be returned to the chloroplast to maintain the Calvin cycle.

In the sucrose synthesis pathway, Fru1,6Pase and SPS have been pinpointed as important control enzymes whose activities have been implicated in co-ordination of sucrose synthesis with carbon dioxide fixation and partitioning between sucrose, starch, and organic acid synthesis. Briefly, it is proposed that sucrose synthesis must be regulated so that flux is fast enough to ensure that sufficient Pi is returned to the chloroplast, but without being so rapid that the chloroplast is depleted of triose-P. In addition, sucrose synthesis is controlled so that partitioning of carbon between export (sucrose) and storage (starch) forms can be altered without necessarily altering the rate of CO_2 fixation (Huber et al., 1992a; Stitt, 1990). Starch synthesis does not seem to be directly regulated by reversible protein phosphorylation which, perhaps, supports the idea that control of sucrose synthesis has a dominant role in determining sucrose/starch partitioning.

6PF2K/Fru2,6P_2ase

The activity of Fru1,6Pase, the first committed step in the cytosolic conversion of triose-P to sucrose is regulated allosterically by pathway intermediates, and by the potent inhibitor Fru2,6P_2 (Stitt, 1990). Fru2,6P_2 is a signal metabolite generated when triose-P concentrations are high. Fru2,6P_2 is synthesized and degraded by 6PF2K and Fru2,6P_2ase, respectively, and these enzymes are, in turn, regulated by triose-Ps and metabolites generated from triose-Ps (3PG, Fru6P and Pi) (Stitt, 1990), and by reversible protein phosphorylation. Many plants contain a bifunctional 6PF2K/Fru2,6P_2ase and an additional Fru2,6P_2ase.

That Fru2,6P_2 does indeed control carbon partitioning in plants was demonstrated directly by constructing transgenic tobacco plants expressing modified copies of the rat liver bifunctional 6PF2K/Fru2,6P_2ase (Kruger and Scott, 1995). Plants with higher amounts of Fru2,6P_2 exhibited low rates of flux to sucrose, organic acids and amino acids at the beginning of the photoperiod, with an accompanying increase in starch accumulation. In addition, the reduction in sucrose synthesis due to high Fru2,6P_2 was demonstrated to have a limiting effect on photosynthesis. Plant lines with reduced Fru2,6P_2 were found to accumulate sucrose more rapidly than untransformed plants, while the rate of starch biosynthesis was lower (Kruger and Scott, 1995).

It has recently been discovered that the dramatic decrease in 6PF2K/Fru2,6P_2ase ratio which occurs in wild-type plants at the onset of the night period is blocked by okadaic acid. Although there is some evidence that the phosphorylation state of the spinach 6PF2K/Fru2,6P_2ase responds, at least partly, to the carbohydrate status of the leaf, the molecular details and the identities of the relevant protein kinase(s) and phosphatase(s) are unknown.

Sucrose phosphate synthase

In many plants, SPS is phosphorylated and converted into a lower-activity-state form within minutes of the beginning of the dark period (Huber et al., 1992b, 1995). Depending on the plant species, inactivation is manifested either by a decrease in V_{max}, or an increase in K_m and modified sensitivity to the allosteric activator Glu6P, and the inhibitor Pi (Huber et al., 1989). Reactivation of SPS in the light is due to dephosphorylation by a PP2A-like enzyme (Siegl et al., 1990).

A controlling role for SPS has been inferred from the findings that changes in SPS activity often correlate with changes in the rate of sucrose synthesis and export, and with changes in partitioning during the photoperiod (Huber et al., 1995; Sonnewald et al., 1994). The rates of sucrose export and plant growth correlate with SPS activity in different genotypes of maize. Thus, increasing the activity of SPS as a possible means of increasing the growth of plants has been a major goal of several laboratories (Sonnewald et al., 1994).

Lowering the levels of SPS protein in transgenic potatoes by antisense technology caused an inhibition of sucrose synthesis, with a concomitant stimulation of starch and amino acid synthesis (Geigenberger et al., 1995). However, the full impact of reducing SPS activity was not achieved in these plants because the reduction in SPS protein was partly compensated for by an increase in the activation state (reduction in phosphorylation) of the SPS, compared with the SPS in wild-type plants (Geigenberger et al., 1995). Over-expressing spinach SPS in tobacco plants, or over-expressing tomato SPS in tomato, had no effect, and it was discovered that most of the excess SPS protein in these plants was inactivated, most probably by protein phosphorylation (Sonnewald et al., 1994). In contrast, over-expressing maize SPS in tomato leaves did alter leaf carbohydrate levels, and if the maize SPS was constitutively expressed throughout tomato plants under control of a CaMV promoter there were large increases in plant biomass (Chapter 6). It was discovered that the maize SPS was not being inactivated by phosphorylation in the tomato, which is puzzling because this type of regulation does occur in maize plants. The reasons why the tomato SPS kinase(s) could not act on the maize SPS are unclear.

SPS seems to be phosphorylated at multiple sites in vivo (Huber and Huber, 1990). Serine-158 on spinach SPS (McMichael et al., 1993) (serine-162 in maize; Huber et al., 1995) is phosphorylated in vitro by SPS kinase, and after mutation of serine-158 to alanine the spinach SPS could no longer be inactivated in vitro. Although it is not certain that phosphorylation of serine-158 underlies the acute inactivation of SPS in vivo in the dark, plants transformed with the $_{ser}158_{ala}$ SPS mutant are being tested to discover whether this mutation will allow more precise manipulation of sucrose synthesis.

The regulatory properties of two SPS kinase(s) from spinach leaf extracts are described in Section 8.3.6.

Nitrate reductase

NR is rapidly inactivated in leaves in the dark (Huber et al., 1995), and in low CO_2 (Kaiser and Brendle-Behnisch, 1991). The inactivation is an unusual two-step mechanism in which phosphorylation of NR has no direct affect on the enzyme's activity, but instead allows the NR to interact (in the presence of Mg^{2+} or Ca^{2+} ions) with an inhibitor protein, termed NIP (Huber et al., 1995; MacKintosh et al., 1995; Spill and Kaiser, 1994). NIP was discovered independently in two ways: the inactivation of NR in vitro was found to require Mg-ATP plus two proteins (Spill and Kaiser, 1994) that were subsequently identified as NR kinase and NIP (Glaab and Kaiser, 1995; MacKintosh et al., 1995); and the low-activity form of NR from leaves harvested in the dark became activated during purification due to separation from NIP. NIP activity was identified in other fractions from the purification by its ability to inhibit the phosphorylated form of NR (MacKintosh et al., 1995).

Protein phosphorylation often changes the affinity of an enzyme for its substrates and allosteric regulators. There are no known allosteric regulators of NR. Is NIP a substitute allosteric regulator, or are there other physiological reasons for this mechanism to have evolved? Are other processes regulated by NIP? Answers to these questions await purification and further characterization of this protein (see Note added in proof).

Phosphorylation *in vitro* of a single serine residue in NR (serine-543 in hinge 1 between the haem and molybdenum cofactor domains) is sufficient to allow inhibition by NIP (Douglas *et al.*, 1995). However, NIP cannot bind (even to phosphorylated NR) if the N-terminal tail of the NR has been lost by proteolysis (Douglas *et al.*, 1995) or mutagenesis (Nussaume *et al.*, 1995). These findings suggest that phosphorylation of serine-543 causes a conformational change in the N-terminal tail of NR to reveal the NIP binding site. When plants were transformed with NR carrying an N-terminal tail deletion (designated ΔNR), the ΔNR was not inactivated by NIP upon transfer from light to the dark (Nussaume *et al.*, 1995), just as expected from the *in vitro* data. The night-time concentrations of glutamine and asparagine were found to be higher in the ΔNR transformants than the controls, which might mean that NR activity normally limits flux to these amino acids in the dark. Nussaume *et al.* (1995) also made the interesting observation that when the ΔNR plants were left in the dark for 70 h, the level of ΔNR protein remained relatively high. In contrast, control plants showed the sharp decrease in NR protein that was expected because expression of NR protein is usually regulated by phytochrome (see Chapter 6). These observations suggest either that formation of the complex between phosphorylated NR and NIP is an essential step in the NR degradation pathway, and/or that the phosphoNR–NIP complex is a negative regulator of phytochrome-induced NR expression.

PEPCase in CAM, C_4 and C_3 metabolism

PEPCase catalyzes the fixation of CO_2 (as HCO_3^-) into oxaloacetate. In the CAM adaption to drought, CO_2 is taken in at night and fixed by PEPCase to yield oxaloacetate, which is converted to malate and stored in vacuoles until the morning when it provides carbon for the Calvin cycle. Thus CAM avoids the need to open stomata during the day. In leaves of the CAM plant, *Bryophyllum fedtschenkoi*, CO_2 metabolism and PEPCase activity are both regulated by circadian rhythm(s), such that, under normal diurnal conditions, PEPCase becomes phosphorylated and activated (meaning that its sensitivity to the inhibitor malate is reduced) just before nightfall (Nimmo *et al.*, 1995). The coincidence of timing suggests that the phosphorylation state of PEPCase is (at least partly) responsible for the rhythm in CO_2 metabolism. Unusually, a circadian rhythm in synthesis and degradation of PEPCase kinase (or a kinase activator) underlies the control of PEPCase activity (Nimmo *et al.*, 1995).

C_4 plants photosynthesize efficiently at high temperatures without major losses to photorespiration, because PEPCase, not Rubisco, is the primary fixer of CO_2 in the mesophyll cells. The resulting oxaloacetate is converted into malate for transport into the internal bundle sheath cells where it provides carbon for the Calvin cycle. In line with the diurnal pattern of CO_2 fixation, the C_4 PEPCase is activated by phosphorylation during the day (opposite to the CAM pattern) (Nimmo *et al.*, 1987). Serine-8 has been identified as a regulatory phosphorylation site on the *Sorghum* PEPCase, and after mutation and modification of this residue, recombinant PEPCase cannot be activated *in vitro* (Duff *et al.*, 1993). Like the CAM enzyme, the C_4 PEPCase kinase is regulated by synthesis and degradation (Jiao *et al.*, 1991), but not by a circadian rhythm. Instead, the C_4 PEPCase kinase levels increase an hour or so after illumination (McNaughton *et al.*, 1991). Light does not affect PEPCase kinase in isolated mesophyll cells, however, and current evidence suggests that the light effect is mediated by a signal derived from Calvin cycle metabolism in the bundle sheath cells (Jiao and Chollet, 1992).

The slowness of PEPCase activation/phosphorylation after illumination raised doubts about whether phosphorylation of PEPCase was controlling C_4 photosynthesis. However, the physiological effects of altering phosphorylation of PEPCase in transgenic plants have not yet been tested. Therefore, to assess the importance of PEPCase phosphorylation in C_4 metabolism, an alternative strategy was devised using the protein synthesis inhibitor, cyclohexamide, to block the light-induced appearance of PEPCase kinase (Bakrim *et al.*, 1993). In these experiments, cyclohexamide prevented PEPCase activation and inhibited C_4 photosynthesis, but did not affect the activities of any other C_4 enzymes (Bakrim *et al.*, 1993). These effects, therefore, strongly supported the hypothesis that synthesis of PEPCase kinase is essential for activation of C_4 photosynthesis during the day.

The CAM and C_4 PEPCases are special isozymes, but other PEPCase isoforms are common to CAM, C_4 and C_3 leaves and play an anaplerotic role in making organic acids for use in amino acid synthesis. The C_3 PEPCase is activated by phosphorylation when plants are supplied with nitrogen (Van Quy *et al.*, 1991), conditions where there is a concomitant activation (dephosphorylation) of NR and inactivation (phosphorylation) of SPS (Huber *et al.*, 1995; see next section), and it has been proposed that these modulations direct carbon flow towards amino acid synthesis. Changes in PEPCase activity have also been observed during cation uptake and pH regulation. In guard cells, light and fusicoccin (which both promote malate accumulation and stomatal opening) promoted phosphorylation of PEPCase, which reduced both the K_m and V_{max} of the enzyme (Nimmo *et al.*, 1995). A physiological link between these changes in PEPCase activity and malate accumulation in guard cells has been inferred, but has not been tested experimentally.

8.3.6 *Evidence for co-regulation of SPS, NR and PEPCase in C_3 leaves*

Three peaks of SPS and/or NR kinase activity have been identified after anion-exchange chromatography of spinach leaf extracts (McMichael *et al.*, 1995):

1. 'Activity peak 1' phosphorylates both NR and SPS (McMichael *et al.*, 1995) which suggests a role in co-ordinating nitrate assimilation and sucrose synthesis. This activity is regulated both by Ca^{2+}/Mn^{2+} (McMichael *et al.*, 1995), and by phosphorylation (Pauline Douglas and Carol MacKintosh, unpublished).

2. 'Activity 2' acts on NR, but not SPS, and is Ca^{2+}-dependent (McMichael *et al.*, 1995).

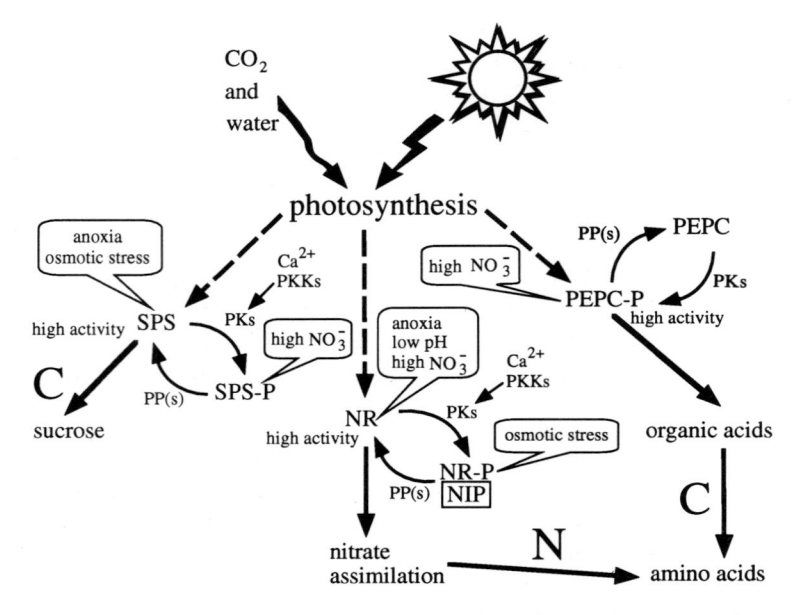

Figure 8.2 A simplified scheme of the post-translational regulation of SPS, NR and PEPCase in C_3 leaves. Activation of photosynthesis initiates signalling pathway(s) (hatched arrows) that lead to activation/dephosphorylation of SPS and NR, and activation/phosphorylation of PEPCase (PEPC). The factors enclosed in 'balloons' override the photosynthesis signals to promote the inactivation/activation of the enzymes as indicated. For example, anoxia and osmotic stress promote the activation/dephosphorylation of SPS, and high external NO_3^- promotes inactivation/phosphorylation of SPS. The changes in activities of SPS, NR and PEPCase will be mediated by changes in the activities of the relevant protein kinases or phosphatases or both. SPS and NR kinases that are regulated by Ca^{2+} and/or phosphorylation have been identified in leaf extracts, but which kinase is active under which cirsumstances has not yet been determined. PK, protein kinase; PKK, protein kinase kinase; PP, protein phosphatase; NIP, nitrate reductase inhibitor protein

3. 'Activity 3' has been reported to act on SPS but not NR (McMichael et al., 1995). This third enzyme is only active when it is phosphorylated, and can be inactivated *in vitro* by dephosphorylation with PP2A or PP2C (P. Douglas et al., 1997).

These findings raise some intriguing questions, particularly about the co-regulation of SPS and NR in leaves. Does the 'dual-specificity' NR/SPS kinase (activity 1) provide a mechanism for parallel regulation of NR and SPS, and the other protein kinases allow independent regulation of NR and SPS in different circumstances? NR and SPS are both activated roughly in parallel in the light, in response to CO_2, and under anoxic conditions. However, in some situations these enzymes are activated/inactivated in an opposite manner. For example, SPS is activated by osmotic stress, whereas NR is not. SPS is progressively inactivated (phosphorylated) when end-products of photosynthesis are allowed to accumulate in a leaf (even in the light), whereas NR is activated (dephosphorylated) under these conditions (Huber et al., 1995), and these differences in the activity states of SPS and NR become even more pronounced in plant fed with high (10 mM) nitrate (Huber et al., 1995). Concomitant with these changes in SPS and NR, the

C₃ PEPCase is activated by phosphorylation when plants are supplied with high nitrate (Champigny and Foyer, 1992), and the net result of these modulations of SPS, NR and PEPCase would tend to direct carbon flow away from sucrose synthesis and towards amino acid synthesis (Champigny and Foyer, 1992; Huppe and Turpin, 1994). Glutamine and glutamate have been proposed as possible mediators of the 'high nitrate' effects (Champigny and Foyer, 1992).

Taken together, the patterns of responses to different stimuli suggest the model shown in Figure 8.2, that under normal diurnal conditions when photosynthesis commences, nitrate assimilation, sucrose synthesis and amino acid synthesis are stimulated by dephosphorylation (SPS and NR) and phosphorylation (PEPCase), respectively. However, other stimuli, such as high nitrate, can override the 'photosynthesis signals' to affect the distribution of C and N within the cell.

The activities of SPS, NR and PEPCase are also regulated by reversible phosphorylation in non-photosynthetic tissues. For example, activation of SPS in potato tubers has been implicated in the conversion of starch to sugar that is stimulated by cold storage (Geigenberger et al., 1995; Reimholtz et al., 1994). NR in roots is activated by anoxia (Glaab and Kaiser, 1993). Other enzymes of C and N metabolism, including sucrose synthase in maize leaves and roots (Koch et al., 1995), and phosphoenolpyruvate carboxykinase in cucumber cotyledons (Walker and Leegood, 1994), also undergo reversible phosphorylation *in vivo* and *in vitro*, but the regulatory significance of these effects is not yet clear.

Understanding the networks of protein phosphorylation events that underlie these metabolic changes will require answers to many questions. Which sites are phosphorylated *in vivo* under each physiological circumstance? What are the physiological effects of mutating putative regulatory sites to deregulate individual enzymes? Which protein kinases act under different circumstances? Does a single PP2A-like protein phosphatase act on SPS, NR and PEPCase, and is it regulated? Are there SPS, NR and PEPCase kinases that are regulated by nitrogen metabolites or pH changes (Kaiser and Brendle-Behnisch, 1995), in addition to those that have been identified to be controlled by Ca^{2+}, phosphorylation, and protein synthesis? PEPCase from C₃ plants can be phosphorylated by Ca^{2+}-dependent and Ca^{2+}-independent kinases. Do any of the SPS/NR kinases act on PEPCase? What are the molecular identities of the SPS/NR protein kinases? Are they related to any of the enzymes in the CaMK protein kinase subfamily (Table 8.1) that are known to be regulated by Ca^{2+} and/or phosphorylation? For example, is 'SPS kinase activity 3' the same as the HMGR-inactivating kinase (Dale et al., 1995; see also Table 8.1), and if so, does this enzyme provide a regulatory link between C and N assimilation, and isoprenoid synthesis? Are osmotic stress effects on SPS and NR mediated by homologues of the protein kinase cascades that mediate stress responses of mammalian cells (Cuenda et al., 1995)?

8.4 Perspectives for future developments

It is becoming evident that primary C/N metabolism in plants is co-ordinated by networks of reversible protein phosphorylation that respond to environmental and metabolic factors. Whether the same protein kinases and phosphatases that regulate SPS, NR and PEPCase also regulate other cellular events in plants is not known at this stage, but it seems reasonable to assume that reversible protein phosphorylation will be central to the control of many aspects of plant metabolism, growth, and development. How does the concept of global control networks affect prospects for engineering plants with desirable patterns of metabolism and growth? In animals, uncontrolled growth is often caused by protein kinases or phosphatases becoming constitutively active. A plant oncogene (*rolB* from *Agrobacterium rhizogenes*) has recently been found to have phosphatase activity (Filippini *et al.*, 1996). Is it too wild a speculation to suggest that one day it may be possible to channel nitrogen and carbon selectively into tubers, fruits or seeds by modulating the protein kinases and phosphatases that control the growth of these organs? In some cases, pleiotrophic changes could be damaging to plants, but it should also be possible to devise ways to achieve specificity and subtlety by making small perturbations in protein phosphorylation patterns. Listed below are some possible strategies for filling in gaps in our understanding, and for manipulating protein phosphorylation patterns to different degrees:

1. Identifying the sites that are phosphorylated *in vivo* under different physiological circumstances:
   - labelling plant tissues with ^{32}P, followed by phosphoamino acid analysis and mass spectrometry of labelled proteins;
   - mutating putative regulatory sites;
   - using antibodies that recognize only the phosphorylated forms of target proteins as probes.
2. Establishing the molecular identities of protein kinases and phosphatases that act on specific target proteins:
   - screening peptide libraries to establish the consensus recognition sequences for novel kinases;
   - testing whether well-characterized mammalian protein kinases will phosphorylate plant proteins *in vitro* (and then searching for plant homologues);
   - purifying and characterizing the regulatory properties of protein kinases and protein phosphatases;
   - testing the effects of specific inhibitors of well-characterized protein kinases and phosphatases.
3. Identifying whether other aspects of plant C and N metabolism, such as metabolite transport systems, are regulated by reversible phosphorylation.
4. Mutating phosphorylation sites so that target enzymes become fixed in their low or high activity states.
5. Modifying the flanking sequences near to phosphorylated residues to make the target a better or worse substrate for its kinase(s) and/or phosphatase(s). This approach might change the target activity more subtly than all-or-none activation/inactivation.
6. Modifying the expression of different protein kinases and protein phosphatases in transgenic plants. For example, expressing 'dominant-negative' activity-dead protein kinases and phosphatases that would compete with the endogenous counterparts for binding to effector molecules.

7. Designing drugs that specifically block the interactions between protein kinases and phosphatases, and individual regulatory proteins.

Acknowledgments

The author extends apologies to all those whose work she has had to cite through reviews because of space limitations. The author thanks the Biotechnology and Biological Sciences Research Council UK for financial support.

Note added in proof

NIP has been purified to homogeneity, and comprises one or more 14-3-3 proteins that bind directly to the 'consensus' RTAS*TP serine-543 phosphorilation site on NR. (Bachman *et al.*, 1996; Moorhead *et al.*, 1996). 14-3-3s are ubiquitous in eukarides and interact with many signalling proteins, reinforcing speculations (page 192) that the NR-NIP-14-3-3 complex may itself be recognized as a signal that interacts with other cellular processes.

NR kinase 'activity3' has been identified positively as belonging to the SNF-1 family of protein kinases (Douglas *et al.*, 1997).

References and further reading

ALLEN, J.F. (1995) Thylakoid protein phosphorylation, state 1-state 2 transition, and photosystem stoichiometry adjustment: redox control at multiple levels of gene expression. *Plant Physiol.* **93**, 96–105.

ASHTON, A.R., BURNELL, J.N. and HATCH, M.D. (1984) Regulation of C4 photosynthesis: inactivation of pyruvate, Pi dikinase by ADP-dependent phosphorylation and activation by phosphorolysis. *Arch. Biochem. Biophys.* **230**, 492–503.

BAKRIM, N., PRIOUL, J.-L., DELEENS, E., ROCHER, J.-P., ARRIO-DUPONT, M., VIDAL, J., GADAL, P. and CHOLLET, R. (1993) Regulatory phosphorylation of C4 phosphoenolpyruvate carboxylase. *Plant Physiol.* **101**, 891–897.

BENNETT, J. (1991) Protein phosphorylation in green plant chloroplast. *Annu. Rev. Plant Physiol. Plant Mol. Biol.* **42**, 281–331.

BRAUN, D.M. and WALKER, J.C. (1996) Plant transmembrane receptors: new pieces in the signaling puzzle. *Trends Biochem. Sci.* **21**, 70–73.

CASAMAYOR, A., PEREZ-CALLEJON, E., PUJOL, G., ARINO, J. and FERRER, A. (1994) Molecular characterization of a fourth isoform of the catalytic subunit of protein phosphatase 2A from *Arabidopsis thaliana*. *Plant Mol. Biol.* **26**, 523–528.

CHAMPIGNY, M.L. and FOYER, C.H. (1992) Nitrate activation of cytosolic protein kinases diverts photosynthetic carbon from sucrose to amino acid biosynthesis. Basis for a new concept. *Plant Physiol.* **100**, 7–12.

CHENG, L., SPANGFORT, M.D. and ALLEN, J.F. (1994) Substrate specificity and kinetics of thylakoid phosphoprotein phosphatase reactions. *Biochem. Biophys. Acta* **1188**, 151–157.

COHEN, P. (1994) The discovery of protein phosphatases: from chaos and confusion to an understanding of their role in cell regulation and human disease. *BioEssays* **16**, 583–588.

CORUM, J.W., HARTUNG, A.J., STAMEY, R.T. and RUNDLE, S.J. (1996) Characterization of DNA sequences encoding a novel isoform of the 55kDaB regulatory subunit of the type 2A protein serine/threonine phosphatase of *Arabidopsis thaliana*. *Plant Mol. Biol.* **31**, 419–427.

CUENDA, A., ROUSE, J., DOZA, Y.N., MEIER, R., COHEN, P., GALLAGHER, T.F., YOUNG, P.R. and LEE, J.C. (1995) SB 203580 is a specific inhibitor of a MAP kinase homologue which is stimulated by cellular stresses and interleukin-1. *FEBS Lett.* **364**, 229–233.

DALE, S., ARRO, M., BECERRA, B., MORRICE, N.G., BORONAT, A., HARDIE, D.G. and FERRER, A. (1995) Bacterial expression of the catalytic domain of 3-hydroxy-3-methylglutaryl-CoA reductase (HMGR1) from *Arabidopsis thaliana*, and its inactivation by phosphorylation at Ser577 by *Brassica oleracea* 3-hydroxy-3-methylglutaryl-CoA reductase kinase. *Eur. J. Biochem.* **233**, 506–513.

DOUGLAS, P., MORRICE, N. and MACKINTOSH, C. (1995) Identification of a regulatory phosphorylation site in the hinge 1 region of nitrate reductase from spinach (*Spinacea oleracia*) leaves. *FEBS Lett.* **377**, 113–117.

DOUGLAS, P., PIGAGLIO, E., FERRER, A. and MACKINTOSH, C. (1997) Three spinach leaf nitrate reductase/3-Hydroxy-3-methylglutacyl-CoA reductase kinases that are regulated by reversible phosphorylation and/or calcium ions. *Biochem J.* **325**, 101–109.

DUFF, S.M.G., LEPINIEC, L., CRETIN, C., ANDREO, C., CONDON, S.A., SARATH, G., VIDAL, J., GADAL, P. and CHOLLET, R. (1993) An engineered change in the L-malate sensitivity of a site directed mutant of sorghum phosphoenolpyruvate carboxylase: the effect of sequential mutagenesis and S-carboxymethylation at position 8. *Arch. Biochem. Biophys.* **199**, 216–221.

FILIPPINI, F., ROSSI, V., MARIN, O., TROVATO, M., COSTANTINO, P., DOWNEY, P.M., SCHIAVO, F.L. and TERZI, M. (1996) A plant oncogene as a phosphatase. *Nature* **379**, 499–500.

FOYER, C.H., VALADIER, M.H. and FERRARIO, S. (1995) Co-regulation of nitrogen and carbon assimilation in leaves. In: *Environment and Plant Metabolism, Flexibility and Acclimation* (SMIRNOFF, N., ed.). Bios Scientific Publishers, Oxford.

FURBANK, R.T. and TAYLOR, W.C. (1995) Regulation of photosynthesis in C3 and C4 plants: a molecular approach. *Plant Cell* **7**, 797–807.

GEIGENBERGER, P., KRAUSE, K.-P., HILL, L.M., REIMHOLZ, R., MACRAE, E., QUICK, P., SONNEWALD, U. and STITT, M. (1995) The regulation of sucrose synthesis in leaves and tubers of potato plants. In: *Sucrose Metabolism, Biochemistry, Physiology and Molecular Biology* (PONTIS, H.G., SALERNO, G.L. and ECHEVERRIA, E.J., eds). American Society of Plant Physiologists, MD.

GERHARDT, R., STITT, M. and HELDT, H.W. (1987) Subcellular metabolite levels in spinach leaves. Regulation of sucrose synthesis during diurnal alterations in photosynthetic partitioning. *Plant Physiol.* **83**, 399–407.

GLAAB, J. and KAISER, W.M. (1993) Rapid modulation of nitrate reductase in pea roots. *Planta* **191**, 173–179.

GLAAB, J. and KAISER, W.M. (1995) Inactivation of nitrate reductase involves NR protein phosphorylation and subsequent 'binding' of an inhibitory protein. *Planta* **195**, 514–518.

HARDIE, G. and HANKS, S. (eds) (1995) *The Protein Kinase Facts Book*. Academic Press, London.

HUBER, S.C. and HUBER, J.L.A. (1990) Regulation of spinach leaf sucrose-phosphate synthase by multisite phosphorylation. *Curr. Top. Plant Biochem. Physiol.* **9**, 329–343.

HUBER, S.C., NIELSEN, T.H., HUBER, J.L.A. and PHARR, D.M. (1989) Variation among species in light activation of sucrose-phosphate synthase. *Plant Cell Physiol.* **30**, 277–285.

HUBER, J.L., HUBER, S.C., CAMPBELL, W.H. and REDINBAUGH, M.G. (1992a) Reversible light/dark modulation of spinach leaf nitrate reductase activity involves protein phosphorylation. *Arch. Biochem. Biophys.* **296**, 58–65.

HUBER, S.C., HUBER, J.L.A. and MCMICHAEL, R.W. (1992b) The regulation of sucrose synthesis in leaves. In: *Carbon Partitioning* (POLLOCK, C.J., FARRAR, J.F. and GORDON, A.J., eds). Bios, UK.

HUBER, S.C., BACHMANN, M., MCMICHAEL, R.W. and HUBER, J.L. (1995) Regulation of sucrose-phosphate synthase by reversible protein phosphorylation: manipulation of activation and inactivation in vivo. *Int. Symp. Sucrose Met.* (PONTIS, H.G., SALERNO, G.L. and ECHEVERRIA, E.J., eds). American Society of Plant Physiologists, Maryland, USA.

HUNTER, T. (1994) 1001 protein kinases redux – towards 2000. *Seminars Cell Biol.* **5**, 367–376.

HUNTER, T. and PLOWMAN, G.D. (1997) The protein kinases of budding yeast: six score and more. *Trends Biochem. Sci.* **22**, 18–22.

HUPPE, H.C. and TURPIN, D.H. (1994) Integration of carbon and nitrogen metabolism in plant and algal cells. In: *Annu. Rev. Plant Physiol. Plant Mol. Biol.* (JONES, R.L., SOMERVILLE, C.R. and WALBOT, V., eds). Annual Reviews, California.

JIAO, J.-A. and CHOLLET, R. (1992) Light activation of maize phosphoenolpyruvate carboxylase protein-serine kinase activity is inhibited by mesophyll and bundle sheath-directed photosynthesis inhibitors. *Plant Physiol.* **98**, 152–156.

JIAO, J.-A., ECHEVARRIA, C., VIDAL, J. and CHOLLET, R. (1991) Protein turnover as a component in the light/dark regulation of phosphoenolpyruvate carboxylase protein-serine kinase activity in C4 plants. *Proc. Natl Acad. Sci. USA* **88**, 2712–2715.

KAISER, W.M. and BRENDLE-BEHNISCH, E. (1991) Rapid modulation of spinach leaf nitrate reductase by photosynthesis. I. Modulation *in vivo* by CO_2 availability. *Plant Physiol.* **96**, 363–367.

KAISER, W.M. and BRENDLE-BEHNISCH, E. (1995) Acid-base-modulation of nitrate reductase in leaf tissues. *Planta* **196**, 1–6.

KAISER, W.M. and FÖRSTER, J. (1989) Low CO_2 prevents nitrate reduction in leaves. *Plant Physiol.* **91**, 970–974.

KAISER, W.M. and SPILL, D. (1991) Rapid modulation of spinach leaf nitrate reductase by photosynthesis. II. *In vivo* modulation by ATP and AMP. *Plant Physiol.* **96**, 368–375.

KAISER, W.M., SPILL, D. and GLAAB, J. (1993) Rapid modulation of nitrate reductase in leaves and roots: indirect evidence for the involvement of protein phosphorylation/dephosphorylation. *Physiol. Plant* **89**, 557–562.

KEYS, A.J., MAJOR, I. and PARRY, M.A.J. (1995) Is there another player in the game of Rubisco regulation?, *J. Exp. Bot.* **46**, 1245–1251.

KOCH, K.F., XU, J., DUKE, E.R., MCCARTY, D.R., YUAN, C.-X., TAN, B.-C. and AVIGNE, W.T. (1995) Sucrose provides a long distance signal for course control of genes affecting its metabolism. *Int. Symp. Sucrose Met.* (PONTIS, H.G., SALERNO, G.L. and ECHEVERRIA, E.J., eds). American Society of Plant Physiologists, Maryland, USA.

KREBS, E.G. (1994) The growth of research on protein phosphorylation. *Trends Biochem. Sci.* **19**, 439.

KRUGER, N.J. and SCOTT, P. (1995) Integration of cytosolic and plastidic carbon metabolism by fructose 2,6-bisphosphate. *J. Exp. Bot.* **46**, 1325–1333.

LAWSON, J.E., NIU, X.-D., BROWNING, K.S., TRONG, H.L., YAN, J. and REED, L.J. (1993) *Biochemistry* **32**, 8987–8993.

LEUNG, J., BOUVIER-DURAND, M., MORRIS, P.-C., GUERRIER, D., CHEFDOR, F. and GIRAUDAT, J. (1994) Arabidopsis ABA response gene ABI1: features of a calcium-modulated protein phosphatase. *Science* **264**, 1448–1452.

LUAN, S., LI, W., RUSNAK, F., ASSMAN, S.M. and SCHREIBER, S. (1993) Immunosuppressants implicate protein phosphatase regulation of K+ channels in guard cells. *Proc. Natl Acad. Sci USA* **90**, 2202–2206.

MACKINTOSH, C. and MACKINTOSH, R.W. (1994) Inhibitors of protein kinases and phosphatases. *Trends Biochem. Sci.* **19**, 444–447.

MACKINTOSH, C., COGGINS, J.R. and COHEN, P. (1991) Plant protein phosphatases; subcellular distribution, detection of protein phosphatase 2C and identification of protein phosphatase 2A as the major quinate dehydrogenase phosphatase. *Biochem J.* **273**, 733–738.

MACKINTOSH, C., DOUGLAS, P. and LILLO, C. (1995) Identification of a protein that inhibits the phosphorylated form of nitrate reductase from spinach leaves. *Plant Physiol.* **107**, 451–457.

MATTHEWS, H.R. and MACKINTOSH, C. (1995) Protein histidine phosphatase activity in rat liver and spinach leaves. *FEBS Lett.* **364**, 51–54.

MCMICHAEL, R.W., KLEIN, R.R., SALVUCCI, M.E. and HUBER, S.C. (1993) Identification of the major regulatory phosphorylation site in sucrose-phosphate synthase. *Arch. Biochem. Biophys.* **307**, 248–252.

MCMICHAEL, R.W., BACHMANN, M. and HUBER, S.C. (1995) Spinach leaf sucrose-phosphate synthase and nitrate reductase are phosphorylated/inactivated by multiple protein kinases in vitro. *Plant Physiol.* **108**, 1077–1082.

MCNAUGHTON, G.A.L., MACKINTOSH, C., FEWSON, C.A., WILKINS, M.B. and NIMMO, H.G. (1991) Illumination increases the phosphorylation state of maize leaf phosphoenolpyruvate carboxylase by causing an increase in the activity of a protein kinase. *Biochem. Biophys. Acta* **1093**, 189–195.

NIMMO, G.A., MCNAUGHTON, G.A.L., FEWSON, C.A., WILKINS, M.B. and NIMMO, H.G. (1987) Changes in the kinetic properties and phosphorylation state of phosphoenolpyruvate carboxylase in *Zea mays* leaves in response to light and dark. *FEBS Lett.* **213**, 18–22.

NIMMO, H.G., CARTER, P.J., FEWSON, C.A., NELSON, J.P.S., NIMMO, G.A. and WILKINS, M.B. (1995) Regulation of malate synthesis in CAM plants and guard cells; effects of light and temperature on the phosphorylation of phosphoenolpyruvate carboxylase. In: *Environment and Plant Metabolism, Flexibility and Acclimation* (SMIRNOFF, N., ed.). Bios Scientific Publishers, Oxford.

NUSSAUME, L., VINCENTZ, M., MEYER, C., BOUTIN, J.-P. and CABOCHE, M. (1995) Post-transcriptional regulation of nitrate reductase by light is abolished by an N-terminal deletion. *Plant Cell* **7**, 611–621.

POPOV, K.M., KEDISHVILI, N.Y., ZHAO, Y., SHIMOMURA, Y., CRABB, D.W. and HARRIS, R.A. (1992) Primary structure of pyruvate dehydrogenase kinase establishes a new family of eukaryotic protein kinases. *J. Biol. Chem.* **268**, 26602–26606.

QUICK, P., SIEGL, G., NEUHAUS, H.E., FEIL, R. and STITT, M. (1989) Short term water stress leads to a stimulation of sucrose synthesis by activating sucrose phosphate-synthase. *Planta* **177**, 536–546.

RANDALL, D.D., MIERNYK, J.A., DAVID, N.R., GEMEL, J. and LUETHY, M.H. (1996) Regulation of leaf mitochondrial pyruvate dehydrogenase complex activity by reversible phosphorylation. In: *Protein Phosphorylation in Plants* (SHEWRY, P., HALFORD, N. and HOOLEY, R., eds). London 87–104.

REIMHOLZ, R., GEIGENBERGER, P. and STITT, M. (1994) Sucrose phosphate synthase is regulated via metabolites and protein phosphorylation in potato tubers, in a manner analogous to the enzyme in leaves. *Planta* **9**, 480–488.

SCHEIBLE, W.-R., GONZALES-FONTES, A. and STITT, M. (1995) Effects of decreased nitrate reductase activity on the metabolism and growth of tobacco plants. Fourth International Symposium on Inorganic Nitrogen Assimilation, Abstract 111P.

SIEGL, G., MACKINTOSH, C. and STITT, M. (1990) Sucrose-phosphate synthase is dephosphorylated by protein phosphatase 2A in spinach leaves; evidence from the effects of okadaic acid and microcystin. *FEBS Lett.* **270**, 198–202.

SLABAS, A., FORDHAM-SKELTON, A., FLETCHER, D., MARTINEX-RIVAS, J., SWINHOE, R., CROY, R. and EVANS, M. (1994) Characterization of cDNA and genomic clones encoding homologues of the 65 kDa regulatory subunit of protein phosphatase 2A in Arabidopsis thaliana. *Plant Mol. Biol.* **26**, 1125–1129.

SMITH, R. and WALKER, J. (1993) Expression of multiple type 1 phosphoprotein phosphatases in *Arabidopsis thaliana*. *Plant Mol. Biol.* **21**, 307–316.

SONNEWALD, U., LERCHL, J., ZRENNER, R. and FROMMER, W. (1994) Manipulation of sink-source relations in transgenic plants. *Plant Cell Environ.* **17**, 649–658.

SPILL, D. and KAISER, W.M. (1994) Partial purification of two proteins (100 kDa and 67 kDa) cooperating in the ATP-dependent inactivation of spinach leaf nitrate reductase. *Planta* **192**, 183–188.

STITT, M. (1990) Fructose-2,6-bisphosphate as a regulatory molecule in plants. *Annu. Rev. Plant Physiol. Plant Mol. Biol.* **41**, 153–185.

SUN, G. and MARKWELL, J. (1992) Lack of types 1 and 2A protein serine (P)/threonine(P) phosphatase activities in chloroplasts. *Plant Physiol.* **100**, 620–624.

SWANSON, R.V., ALEX, L.A. and SIMON, M.I. (1994) Histidine and aspartate phosphorylation: two-component systems and the limits of homology. *Trends Biochem. Sci.* **19**, 485–490.

THEOLOGIS, A. (1996) Plant hormones: more than one way to detect ethylene. *Curr. Biol.* **6**, 144–145.

VAN QUY, L., FOYER, C.H. and CHAMPIGNY, M.L. (1991) Effect of light and NO_3^- on wheat leaf phosphoenolpyruvate carboxylase activity. Evidence for covalent modulation of the C3 enzyme. *Plant Physiol.* **97**, 1476–1482

WALKER, R.P. and LEEGOOD, R.C. (1994) Purification and phosphorylation *in vivo* and *in vitro*, of phosphoenolpyruvate carboxykinase from cucumber cotyledons. *FEBS Lett.* **362**, 70–74.

WARREN, D.M. and WILKINS, M.B. (1961) An endogenous rhythm in the rate of dark-fixation of carbon dioxide in leaves of *Bryophyllum fedtschenkoi*. *Nature* **191**, 686–688.

WORRELL, A.C., BRUNEAU, J.-M., SUMMERFELT, K., BOERSIG, M. and VOELKER, T.A. (1991) Expression of a maize sucrose-phosphate synthase in tomato alters leaf carbohydrate partitioning. *Plant Cell* **3**, 1121–1130.

Compartmentation, Transport and Whole Plant Interactions

9

Compartmentation of C/N metabolism

DIETER HEINEKE, GERTRUD LOHAUS AND HEIKE WINTER

9.1 Introduction

Carbon and nitrogen are important components of plants: carbon amounts to 50% and nitrogen to 2–5% of the dry matter. Whereas carbon is assimilated as CO_2 from the atmosphere, nitrogen can be taken up as ammonium and nitrate from the soil and as N_2 and ammonia from the gas phase. Carbon reduction is restricted to the green parts of plants, but nitrogen incorporation in organic components can take place in the roots and in the leaves, depending on the nitrogen source, plant species, and nutrient availability. N_2-utilization is only found in plants which are living in a symbiotic relationship with bacteria, such as *Rhizobia* or cyanobacteria. If leaves are able to fix more CO_2 during the light period than they need for their metabolism, the surplus carbon is exported to support other tissues. Such leaves are called 'source' leaves. 'Sink' tissues such as growing leaves, flowers, seeds, roots and storage organs are net importers of photoassimilates (Ho, 1988). The mesophyll cell is the dominant cell type of leaves and it contains most of the chloroplasts. Other abundant tissues are epidermis cells and the sieve tube complex. The primary CO_2 and nitrate fixation reactions and the syntheses of the transported assimilates occur in the mesophyll cells; these contain subcellular compartments which have different functions in carbon and nitrogen metabolism. Their export to the sink tissues proceeds by mass flow through the sieve tubes. The production and export of assimilates involves a co-operation of various cell types and requires several transfer steps across membranes. To understand the complex interactions between all these compartments, subcellular concentrations of intermediates and products must be known.

9.2 Determination of subcellular metabolite concentrations of source leaves

9.2.1 Subcellular volumes

To determine subcellular metabolite concentrations in a leaf, the following must be known: the volume of the specific compartment; and the subcellular distribution of metabolites in the leaf. For measuring these parameters specific techniques had to be developed. Determination of the volumes of cellular and subcellular compartments of source leaves from spinach, barley and potato was carried out using light- and electron-microscopy techniques (Leidreiter et al., 1995; Winter et al., 1993, 1994). For this leaves are fixed under conditions minimizing volume shrinkage, cut into sections, and the relative areas occupied by each compartment are evaluated by planimetric analysis. Light microscopy is used to determine the relative size of mesophyll and epidermis cells and the veins; the subcellular volumes of the cytosol, mitochondria, nuclei, and vacuole are derived from electron microscopy. The results are summarized in Table 9.1. As expected from the phenotype of the leaves there are considerable differences between the three species. In barley the epidermis occupies 27% of the whole leaf volume; in the dicotyledoneous species spinach (3%) and potato (12%) this value is considerably lower. The relative compartmentation of mesophyll cells, however, is rather similar. In all three species the chloroplasts occupy about

Table 9.1 Cellular and subcellular volumes of mature spinach, barley and potato leaf cells determined from light- and electron-micrographs and by the infiltration technique. Data obtained from Winter et al. (1993, 1994) and Leidreiter et al. (1995)

	Spinach	Barley	Potato
		µl per mg chl	
Total leaf volume	1177	902	623
Gas space	381	204	100
Aqueous space	810	698	472
Epidermis	36	244	74
Mesophyll	688	379	394
Apoplast	60	41	21
Veins	12	55	30
Mesophyll			
Cell	688	379	394
Vacuole	545	278	300
Chloroplast	113	71.6	70
Stroma	65	34.8	32
Thylakoid	36	29.1	21
Cytosol	24	24.9	20
Mitochondria	3.6	4.1	4.0
Nuclei	2.1	1.3	nd
Epidermis			
Cell	36	244	74
Vacuole	34	243	72
Chloroplast	0.4	1.4	0.10
Cytosol + nuclei	1.8	0.28	4.9
Mitochondria	0.1	nd	0.17

nd, not determined

Table 9.2 Comparison of the percentage of the total volume of the mesophyll cells occupied by subcellular compartments in leaves of spinach, barley and potato. Data obtained from Leidreiter et al. (1995)

	Spinach	Barley (%)	Potato
Mesophyll cell	= 100	= 100	= 100
Vacuole	79.2	73.4	76.1
Chloroplast	16.4	18.9	17.6
Cytoplasma	3.8	6.9	5.2
Mitochondria	0.5	1.1	1.0

18%, the cytosol 5%, and the vacuole 75% of the cell volume (Table 9.2). This uniformity between the three species is unexpected and points to a general design of a mesophyll cell.

9.2.2 Subcellular metabolite distribution

The determination of the distribution of metabolites in leaves *in vivo* requires a method which stops metabolism immediately and avoids alterations in the metabolite pools during the separation procedure. This is achieved by the method of 'non-aqueous fractionation' (Gerhardt and Heldt, 1984; Riens et al., 1991). Leaves are quenched in liquid nitrogen and freeze dried. The dry material is homogenized to a powder and subsequently ultrasonicated in non-aqueous media. In this way the leaf cells are fragmented into small particles. During this procedure, due to lack of water, the metabolites of a certain subcellular compartment remain attached to the dried proteins of this compartment. The cell fragments are separated on a density gradient formed by apolar organic solvents and the distribution of metabolites in the gradient is compared with those of the marker enzymes for the various subcellular compartments. This procedure enables the determination of metabolite distribution between the chloroplastic, cytosolic and vacuolar compartments. One restriction of the method is that it is not possible to separate the compartments of different cell types. But as shown above, in spinach and potato leaves most of the volume is occupied by the mesophyll cells. The volumes of the chloroplasts and cytosol of the epidermis cells are very small. Therefore, the values for the cytosolic and chloroplastic concentrations obtained for these plants represent mainly mesophyll cell concentrations. As shown later the results obtained with barley are similar to those of the other plant species, indicating that this method is also suitable for this plant species.

9.2.3 Concentrations of exported products in the apoplast and the sieve tubes

The determinations of the subcellular volumes described in Section 9.2.1 additionally yield the volumes of the apoplastic space and the sieve tubes (Table 9.1). Both compartments are involved in the export of the products of CO_2 and nitrate fixation. Therefore, knowledge of metabolite concentrations in these compartments is necessary for an understanding of the assimilate export. Additionally, the apoplastic and sieve tube contents may falsify the results obtained by non-aqueous fractionation. The composition of the apoplast can be determined by infiltrating the gas space with an isoosmotic buffer for washing out the water soluble components (Speer and Kaiser, 1991). The following centrifugation allows the extraction of the infiltrated solution and the determination of the apoplastic metabolites. The phloem sap in the sieve tubes is not directly accessible to a researcher. Therefore, the best method is to seek help from a specialist. Aphids are able to introduce their stylet into the sieve tubes and thus to remove the phloem sap for their own nutrition. From such an aphid the stylet can be severed by a laser beam; from the remaining stump the phloem sap runs out and is available for sampling (Barlow and McCully, 1972; Fischer and Frame, 1984).

9.3 Subcellular concentrations of phosphorylated intermediates between chloroplast and cytosol

The fixation of CO_2 takes place in the chloroplast stroma. The main products of CO_2-fixation are the carbohydrates sucrose and starch. All the intermediates of their synthesis pathways are phosphorylated compounds which occur only in the non-vacuolar compartments, as the vacuole contains an active unspecific phosphatase (Matile, 1978). Most of these intermediates occur in the chloroplastic and cytosolic compartments, which can be separated by non-aqueous fractionation. The initial products of chloroplastic CO_2-fixation and reduction are the triosephosphates dihydroxyacetone phosphate and glyceraldehyde phosphate (Figure 9.1). Five of six triosephosphate molecules must remain in the chloroplast for the regeneration of the CO_2-acceptor ribulose-1,5-bisphosphate. The remaining triosephosphate molecule is exported from the chloroplast into the cytosol in a strict counterexchange with inorganic phosphate. This reaction is catalyzed by the chloroplast phosphate translocator (Fliege et al., 1978). 3-Phosphoglycerate is also transported by the phosphate translocator. The determination of the subcellular concentrations of triosephosphates and 3-phosphoglycerate in illuminated leaves show that both are equally distributed between the chloroplasts and the cytosol (Table 9.3). Other phosphorylated intermediates are only poorly transported across the chloroplast envelope membrane. Therefore, the further metabolism of triosephosphates in both compartments is separated. Table 9.3 also shows the distribution of hexose-phosphates. Fructose 1,6-bisphosphate (Fru1,6P_2) is concentrated in the chloroplasts, but glucose 6-phosphate (Glc6P) and fructose 6-phosphate (Fru6P) are concentrated in the cytosol. Of fructose-1,6-bisphosphatase (Fru1,6P_2ase) two isoenzymes occur, one in the cytosol and the other in the chloroplast. Both are the first enzymes after the branching point between sucrose and starch synthesis, catalyzing an irreversible reaction. A number of metabolic signals are known to regulate their activities. The low concentration of Fru1,6P_2 and the high concentration of Glc6P in the cytosol, as compared to the chloroplast, reflect the high rate of sucrose synthesis. One other key enzyme of sucrose synthesis, sucrose phosphate synthase, is known to be activated by Glc6P (Doehlert and Huber, 1983).

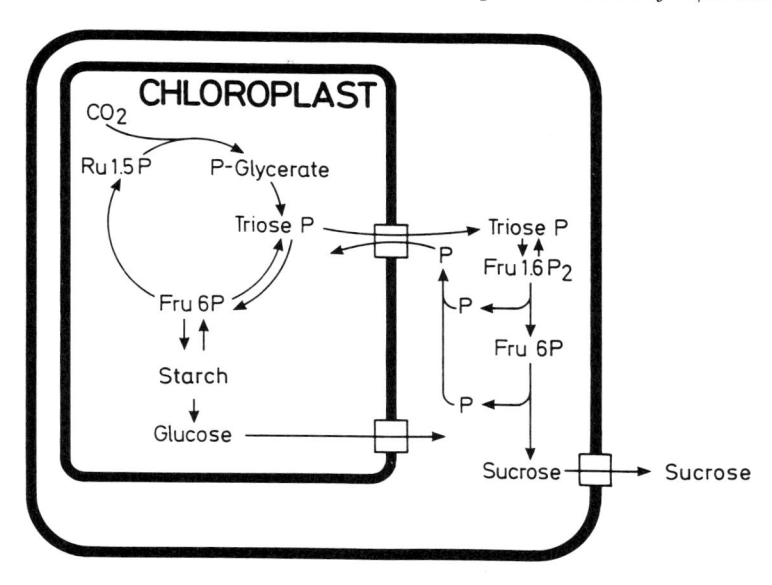

Figure 9.1 Compartmentation of CO_2-fixation; the sucrose and starch synthesis pathways (from Heldt, 1992)

Table 9.3 Subcellular concentrations of phosphorylated intermediates in spinach, barley and potato leaves at the end of the light period. Data obtained from Winter et al. (1993, 1994) and Leidreiter et al. (1995)

	Spinach	Barley (mM)	Potato
PGA			
Stroma	4.3	5.7	2.0
Cytosol	4.2	3.5	1.7
Vacuole	⩽0.01	0.003	⩽0.01
DHAP			
Stroma	0.21	nd	0.32
Cytosol	0.57	nd	0.21
Vacuole	⩽0.001	nd	⩽0.001
Fructose1,6P$_2$			
Stroma	0.55	nd	1.04
Cytosol	0.14	nd	0.08
Vacuole	⩽0.001	nd	⩽0.01
Glucose6P			
Stroma	0.6	nd	0.75
Cytosol	5.9	nd	3.5
Vacuole	⩽0.005	nd	0.06
Fructose6P			
Stroma	0.57	nd	0.79
Cytosol	2.0	nd	2.2
Vacuole	0.003	nd	⩽0.01

nd, not determined

9.4 Subcellular distribution of sugars and malate

Carbohydrates and malate are products of carbon fixation. In most crop species the disaccharide sucrose is the export form of carbohydrates. Its concentration is highest in the compartment of its synthesis, the cytosol (Table 9.4). The vacuolar concentrations are five to ten times lower and normally sucrose is absent in the chloroplast. This observation is consistent with earlier findings that the chloroplast envelope membrane is impermeable to sucrose (Heldt and Sauer, 1971) and that a sucrose carrier activity is found in the tonoplast (Kaiser and Heber, 1984). The K_m value for sucrose uptake into the vacuole has been determined as about 20–30 mM (Martinoia et al., 1987). Hexoses occur almost exclusively in the vacuolar compartments (Table 9.4). The concentrations in the chloroplast and the cytosol are below the limit of detection. Obviously, the uptake of hexoses into the vacuole requires active transport. The origin and the role of hexoses in leaf metabolism is not understood. Huber (1989) found that the ability of leaves to accumulate sucrose and hexoses is correlated with the activity of acid invertases, and Foyer (1987) argued that the apoplastic invertase might play a role in signalling an imbalance in sink/source interactions. In a situation of reduced sink demand sucrose should accumulate in the apoplast and be hydrolyzed by invertase. The resulting hexoses are taken up by the mesophyll cells, where they are phosphorylated and fed back into the sucrose synthesis pathway. This loop would form a futile cycle to dissipate surplus energy. Another possible source of hexoses is the release of glucose during fructan

Table 9.4 Subcellular concentrations of sugars and malate in spinach, barley and potato leaves at the end of the light period. Data obtained from Winter et al. (1993, 1994) and Leidreiter et al. (1995)

	Spinach	Barley (mM)	Potato
Sucrose			
Stroma	⩽0.8	⩽5	⩽3.1
Cytosol	53	232.1	22.9
Vacuole	11	20.3	4.4
Glucose			
Stroma	nd	nd	⩽0.3
Cytosol	nd	nd	⩽0.44
Vacuole	nd	nd	3.1
Fructose			
Stroma	nd	nd	⩽1.1
Cytosol	nd	nd	⩽1.4
Vacuole	nd	nd	9.3
Malate			
Stroma	1.2	nd	⩽2.4
Cytosol	0.8	nd	⩽3.2
Vacuole	6.8	nd	20.6

nd, not determined

synthesis in fructan accumulating tissues (Martinoia et al., 1987). Hexoses occurring in the cytosol are phosphorylated by gluco- and fructokinases with K_m values of about 0.1 mM (Doehlert, 1989; Renz and Stitt, 1993; Schnarrenberger, 1990; Turner et al., 1977).

The observed accumulation in the vacuole described here can be explained by a competition between the gluco- and fructokinases and the active uptake into the vacuole. Earlier studies of hexose uptake into isolated vacuoles were indecisive with respect to the existence of an energy-dependent transport system. Guy et al. (1979) found a slight stimulation of the uptake of glucose and 3-O-methyl-glucose into isolated vacuoles of pea leaves, and Rausch et al. (1987) described similar effects of ATP for *Zea mays* coleoptiles. On the other hand, Martinoia et al. (1987) could not find any stimulating effects of ATP on glucose uptake into barley vacuoles.

A similar subcellular distribution as for hexoses is also found for malate (Table 9.4). The concentration gradient of malate between the cytosol and the vacuole reflects the active uptake system, which has been characterized by Martinoia et al. (1985) from the ATP-dependent accumulation of malate in isolated vacuoles from barley leaves.

9.5 Subcellular distribution of amino acids

In leaves all proteinogenic amino acids are present, many of them at relatively high concentrations. In the three plant species described here, glutamate dominates. Other abundant amino acids are alanine, aspartate, glutamine, glycine and serine. In general, the amino acid concentrations in the chloroplast and the cytosol are very similar, but the vacuolar concentrations are more than ten times lower (Table 9.5). The chloroplast envelope membrane contains dicarboxylate carriers, which catalyze a counterexchange of dicarboxylates such as malate, oxoglutarate, glutamate and aspartate with overlapping specificities, and are involved in the exchange of 2-oxoglutarate and glutamate during nitrogen fixation (Woo et al., 1987). A translocator transporting glutamate and glutamine was characterized by Yu and Woo (1988). For the amino acid transport into the vacuole an ATP-dependent uptake of amino acids has been reported (for Review see Martinoia, 1992). This is in contradiction to the concentration gradients shown in Table 9.5. Winter et al. (1993) argued that these ATP-dependent transport systems may *in vivo* be responsible for the active extrusion of amino acids from the vacuole rather than for the uptake. This conclusion is supported by the observation that in barley the vacuolar amino acid concentration increases during the dark period (Winter et al., 1993).

9.6 Export of assimilates

9.6.1 *Exported assimilates and export pathway*

Source leaves are characterized by the ability to export assimilates. For an understanding of the export process two main questions have to be answered: which are the main export components; and which is the route for their export. One

Table 9.5 Subcellular concentrations of amino acids in spinach, barley and potato leaves at the end of the light period. Data obtained from Winter et al. (1993, 1994) and Leidreiter et al. (1995)

	Spinach	Barley (mM)	Potato
Alanine			
Stroma	5.5	6.5	0.97
Cytosol	3.9	21.4	3.4
Vacuole	0.20	⩽0.10	0.36
Aspartate			
Stroma	14	22.9	3.0
Cytosol	23	32.1	8.9
Vacuole	0.87	⩽0.1	0.36
Glutamate			
Stroma	14	73.6	26.4
Cytosol	21	89.3	41.0
Vacuole	0.55	⩽0.38	1.7
Glutamine			
Stroma	20	17.2	4.8
Cytosol	24	25.7	4.4
Vacuole	0.58	⩽0.12	0.30
Glycine			
Stroma	0.6	7.9	0.5
Cytosol	1.8	25.0	7.3
Vacuole	0.13	0.77	0.27
Serine			
Stroma	4.3	20.7	1.4
Cytosol	7.5	50.0	12.0
Vacuole	0.43	⩽0.09	0.42
Σ Amino acids			
Stroma	58	170	42
Cytosol	86	275	80
Vacuole	2.8	⩽1.7	4.5

way to answer the first question is the analysis of the composition of the phloem sap collected by the aphid method. Two of the three species described here, spinach and barley, are suitable for obtaining sufficient samples of phloem sap. The analysis revealed that sucrose is the only form of carbohydrate which is exported, whereas the export of amino acids is not specific. All the amino acids found in the leaves also occur in the phloem sap. The amino acid patterns in the cytosol and in the phloem sap of spinach and barley leaves are found to be nearly identical (Lohaus et al., 1995).

The mode of export does not seem to be uniform in all plant species. Some plants have a high number of plasmodesmatal connections between mesophyll cells and the companion cell/sieve tube complex. They are regarded as symplastic exporters. In these species often not sucrose but sugars of the raffinose family are the favoured exported carbohydrates (Turgeon and Beebe, 1991). Other plants including spinach, barley and potato have only few plasmodesmatal connections. In these plants an apoplastic step in the phloem loading process has been proposed (Giaquinta, 1983). Evidence for this hypothesis comes from studies with transgenic plants of tobacco (von Schaewen et al., 1990), tomato (Dickinson et al., 1991) and potato (Heineke et al., 1992) expressing a yeast-derived invertase in their apoplast. In all these plants the loading of sucrose was inhibited and the plants were retarded in growth. Recently, Riesmeier et al. (1994), by using antisense technique, described potato plants with a reduced expression of a sucrose carrier gene in the companion cell/sieve tube complex. These plants show a similar phenotype to the invertase plants.

9.6.2 Concentration gradients along the export pathway

When the apoplast is involved in the assimilate export, two membrane passages are required: between the cytosol of mesophyll cells and the apoplastic space and between apoplast and sieve tubes. The characterization of involved transport mechanisms requires the knowledge of the metabolite concentrations in all three compartments. Table 9.6 compares the cytosolic, apoplastic and phloem sap concentrations of sucrose and amino acids in spinach and barley. For both species the apoplastic concentrations are much lower than those in the cytosol and the phloem sap. The ratios between the concentrations of sucrose and of the sum of amino acids are similar in the apoplast as in the cytosol, but much higher in the phloem sap. Apparently the uptake of sucrose into the phloem sap is responsible for the high preference of sucrose export, although sucrose and amino acids seem to be both loaded against a concentration gradient in an energy dependent reaction. All the transporters characterized until now are proton symporters and the driving force is supplied by a proton ATPase in the plasma membrane of the companion cells (for review see Frommer and

Table 9.6 Comparison of metabolite concentrations in the apoplastic compartment with those in the cytosolic compartment and in the sieve tubes. Data obtained from Lohaus et al. (1995)

	Spinach	Barley	
Sucrose			
Cytosol	53	232	mM
Apoplast	1.3	1.5	mM
Phloem	830	1030	mM
Cytosol/apoplast	41	156	
Phloem/apoplast	633	691	
Σ Amino acids			
Cytosol	86	275	mM
Apoplast	3.2	3.1	mM
Phloem	192	186	mM
Cytosol/apoplast	26	89	
Phloem/apoplast	59	60	

Sonnewald, 1995). The transfer of metabolites from the source cells into the apoplast is believed to be catalyzed by an unspecific translocator facilitating a passive efflux. The results shown in Table 9.6 argue against this view, because this concept cannot explain the observed concentration gradients between the cytosol and the apoplast. These gradients continue to exist under conditions when phloem export is inhibited by cold-girdling (Lohaus et al., 1995) or in potato plants with a phloem specific antisense inhibition of the sucrose carrier (Leidreiter, personal communication). As no efflux carrier has been found until now, these data may suggest that the export of sucrose and amino acids is also catalyzed by proton symporting transporters.

9.7 The role of phloem loading for the composition of storage sinks

Sucrose and amino acids of the phloem sap are the sources for sink metabolism. In storage sinks they are stored directly or utilized to build up storage compounds such as starch, proteins or lipids. In Chapter 10 it is pointed out that the step of loading assimilates into the sieve tube/companion cell complex influences the phloem sap composition and defines the supply for the sinks.

Sugar beet stores high amounts of sucrose in its taproots and plant breeding increased the content from 2% to 18% (Burba et al., 1984). Additionally, it has long been known that excessive N-fertilization is counterproductive for sucrose accumulation and increased the N-containing compounds of the taproots. For defining the steps responsible for this altered composition of the taproots, Lohaus et al. (1994) determined the contents of sucrose and amino acids in the leaves, the phloem sap and the taproots of sugar beet plants (Table 9.7). They used three hybrids with differences in their ability to store N-compounds. The sucrose concentrations were identical in all parts of the plants. As in other plant species (see previous section) the phloem loading step is responsible for concentrating sucrose. The sucrose concentration was determined to be 1.3 M, which is higher than that of other species. The difference between the three hybrids could be localized in the phloem loading step. In all hybrids the amino acid contents

Table 9.7 Comparison of metabolite contents in the leaves, the phloem sap and the taproots of three hybrids of sugar beet. Data obtained from Lohaus et al. (1994)

		9EO 106	Hybrid 9EO 205	9AO 131	
Leaves	Sucrose	6.0	6.6	5.8	µmol mg^{-1}chl
	Σ Amino acids	5.0	5.5	5.8	µmol mg^{-1}chl
	Sucrose/Σ AA	1.2	1.2	1.0	
Phloem sap	Sucrose	1360	1290	1340	mM
	Σ Amino acids	62	114	143	mM
	Sucrose/Σ AA	22	11	9	
Taproot	Sucrose	422	403	378	µmol g^{-1}FW
	Σ Amino acids	9.1	13.6	19.4	µmol g^{-1}FW
	Sucrose/Σ AA	46	30	19	

of the leaves were identical. In the phloem sap the amino acid content was lowest in that hybrid with the lowest amount of N-compounds in the taproots. This observation confirms the important role of phloem loading for sink nutrition.

Acknowledgments

The authors thank H. W. Heldt. He initiated all the work described in this chapter, the experiments were carried out in his institute, and he revised the manuscript. The work presented here was supported by the Deutsche Forschungsgemeinschaft.

References

BARLOW, C.A. and McCULLY, M.E. (1972) The ruby laser as an instrument for cutting the stylets of feeding aphids. *Can. J. Zool.* **50**, 1497–1498.

BURBA, M., NITZSCHKE, U. and RITTERBUSCH, R. (1984) Die N-Assimilation der Pflanze unter Berücksichtigung der Zuckerrübe. *Zuckerindustrie* **109**, 613–628.

DICKINSON, C.D., ALTABELLA, T. and CHRISPEELS, M.J. (1991) Slow-growth phenotype of transgenic tomato plants expressing apoplastic invertase. *Plant Physiol.* **95**, 420–425.

DOEHLERT, D.C. (1989) Separation and characterization of four hexose kinases from developing maize kernels. *Plant Physiol.* **89**, 1042–1048.

DOEHLERT, D.C. and HUBER, S.C. (1983) Regulation of spinach leaf sucrose phosphate synthase by glucose-6-phosphate, inorganic phosphate and pH. *Plant Physiol.* **73**, 989–994.

FISCHER, D.B. and FRAME, J.M. (1984) A guide to the use of the exuding-stylet technique in phloem physiology. *Planta* **161**, 385–393.

FLIEGE, R., FLÜGGE, U.-I., WERDAN, K. and HELDT, H.W. (1978) Specific transport of inorganic phosphate, 3-phosphoglycerate and triosephosphates across the inner membrane of the envelope in spinach chloroplasts. *Biochim. Biophys. Acta* **502**, 232–247.

FOYER, C.H. (1987) The basis for source-sink interactions in leaves. *Plant Physiol. Biochem.* **25**, 649–657.

FROMMER, W.B. and SONNEWALD, U. (1995) Molecular analysis of carbon partitioning in solanaceous species. *J. Exp. Bot.* **46**, 587–607.

GERHARDT, R. and HELDT, H.W. (1984) Measurement of subcellular metabolite levels in leaves by fractionation of freeze-stopped material in nonaqueous media. *Plant Physiol.* **75**, 542–547.

GIAQUINTA, R.T. (1983) Phloem loading of sucrose. *Annu. Rev. Plant Physiol.* **34**, 347–387.

GUY, M., REINHOLD, L. and MICHAELI, D. (1979) Direct evidence for a sugar transport mechanism in isolated vacuoles. *Plant Physiol.* **64**, 61–64.

HEINEKE, D., SONNEWALD, U., BÜSSIS, D., GÜNTER, G., LEIDREITER, K., WILKE, I., RASCHKE, K., WILLMITZER, L. and HELDT, H.W. (1992) Apoplastic expression of yeast-derived invertase in potato. Effects on photosynthesis, leaf solute composition, water relations, and tuber composition. *Plant Physiol.* **100**, 301–308.

HELDT, H.W. (1992) Recent advances in photosynthesis. In: *Proceedings of International Symposium on Photochemistry, Photobiology and Photomedicine* (YOON, M. and SONG, P.-S., eds), pp. 158–171.

HELDT, H.W. and SAUER, F. (1971) The inner membrane of the chloroplast envelope as the site of specific metabolite transport. *Biochim. Biophys. Acta* **234**, 83–91.

HO, L.C. (1988) Metabolism and compartmentation of imported sugars in sink organs in relation to sink strength. *Annu. Rev. Plant Physiol. Plant Mol. Biol.* **39**, 355–378.

HUBER, S.C. (1989) Biochemical mechanism for regulation of sucrose accumulation in leaves during photosynthesis. *Plant Physiol.* **91**, 656–662.

KAISER, G. and HEBER, U. (1984) Sucrose transport into vacuoles isolated from barley mesophyll protoplasts. *Planta* **161**, 562–568.

LEIDREITER, K., KRUSE, A., HEINEKE, D., ROBINSON, D.G. and HELDT, H.W. (1995) Subcellular volumes and metabolite concentrations in potato (*Solanum tuberosum* cv. Désirée) leaves. *Bot. Acta* **108**, 439–444.

LOHAUS, G., BURBA, M. and HELDT, H.W. (1994) Comparison of the contents of sucrose and amino acids in the leaves, phloem sap and taproots of high and low sugar-producing hybrids of sugar beet (*Beta vulgaris* L.). *J. Exp. Bot.* **45**, 1097–1101.

LOHAUS, G., WINTER, H., RIENS, B. and HELDT, H.W. (1995) Further studies of the phloem loading process in leaves of barley and spinach. The comparison of metabolite concentrations in the apoplastic compartment with those in the cytosolic compartment and the sieve tubes. *Bot. Acta* **108**, 270–275.

MARTINOIA, E. (1992) Transport processes in vacuoles of higher plants. *Bot. Acta* **105**, 232–245.

MARTINOIA, E., FLÜGGE, U.-I., KAISER, W., HEBER, U. and HELDT, H.W. (1985) Energy dependent uptake of malate into vacuoles isolated from barley mesophyll protoplasts. *Biochim. Biophys. Acta* **806**, 311–319.

MARTINOIA, E., KAISER, G., SCHRAMM, M.J. and HEBER, U. (1987) Sugar transport across the plasmalemma and the tonoplast of barley mesophyll protoplasts. Evidence for different transport systems. *J. Plant Physiol.* **131**, 467–478.

MATILE, P. (1978) Biochemistry and function of vacuoles. *Annu. Rev. Plant Physiol.* **29**, 193–213.

RAUSCH, T., BUTCHER, D.N. and TAIZ, L. (1987) Active transport and proton pumping in tonoplast membrane of *Zea mays* L. coleoptiles are inhibited by Anti-H^+-ATPase antibodies. *Plant Physiol.* **85**, 996–999.

RENZ, A. and STITT, M. (1993) Substrate specificity and product inhibition of different forms of fructokinases and hexokinases in developing potato tubers. *Planta* **190**, 166–175.

RIENS, B., LOHAUS, G., HEINEKE, D. and HELDT, H.W. (1991) Amino acid and sucrose content determined in the cytosolic, chloroplastic and vacuolar compartment and in the phloem sap of spinach leaves. *Plant Physiol.* **97**, 227–233.

RIESMEIER, J.W., FROMMER, W.B. and WILLMITZER, L. (1994) Evidence for an essential role of the sucrose transporter in phloem loading and assimilate partitioning. *EMBO J.* **13**, 1–7.

SCHNARRENBERGER, C. (1990) Characterization and compartmentation, in green leaves, of hexokinases with different specificities for glucose, fructose, and mannose and for nucleoside triphosphates. *Planta* **181**, 244–255.

SPEER, M. and KAISER, W.M. (1991) Ion relations of symplastic and apoplastic space in leaves from *Spinacia oleracea* L. and *Pisum sativum* L. under salinity. *Plant Physiol.* **97**, 990–997.

TURGEON, T. and BEEBE, D.U. (1991) The evidence for symplastic phloem loading. *Plant Physiol.* **96**, 349–354.

TURNER, J.F., CHENSEE, Q.J. and HARRISON, D.D. (1977) Glucokinase of pea seeds. *Biochim. Biophys. Acta* **480**, 367–375.

VON SCHAEWEN, A., STITT, M., SCHMIDT, R., SONNEWALD, U. and

WILLMITZER, L. (1990) Expression of yeast-derived invertase in the cell wall of tobacco and *Arabidopsis* plants leads to accumulation of carbohydrate and inhibition of photosynthesis and strongly influences growth and phenotype of transgenic tobacco plants. *EMBO J.* **9**, 3033–3044.

WINTER, H., ROBINSON, D.G. and HELDT, H.W. (1993) Subcellular volumes and metabolite concentrations in barley leaves. *Planta* **191**, 180–190.

WINTER, H., ROBINSON, D.G. and HELDT, H.W. (1994) Subcellular volumes and metabolites in spinach leaves. *Planta* **193**, 530–535.

WOO, K.C., FLÜGGE, U.-I. and HELDT, H.W. (1987) A two translocator model for the transport of 2-oxoglutarate and glutamate in chloroplasts during ammonia assimilation in the light. *Plant Physiol.* **84**, 624–632.

YU, J. and WOO, K.C. (1988) Glutamine transport and the role of glutamine translocator in chloroplasts. *Plant Physiol.* **88**, 1048–1054.

10

Plasmodesmal-mediated plant communication network: implications for controlling carbon metabolism and resource allocation

SHMUEL WOLF AND WILLIAM J. LUCAS

In the last three decades, the yield of economically important crops has been substantially increased through plant breeding and optimization of growing conditions. The improvement of yield was made possible through some increase in dry-matter production in the leaves, but more importantly, through improvement in the accumulation of dry matter by harvestable organs (Austin *et al.*, 1987; Gifford *et al.*, 1984). For instance, in modern cultivars of potato (*Solanum tuberosum*), tuber dry weight, as a proportion of plant weight (i.e. harvest index), has been increased to 81% as compared to only 7% in the wild species, *Solanum demissum* (Inoue and Tanaka, 1978).

Allocation of carbohydrates towards the various plant organs is governed, in part, by: the rate of photosynthesis; the extent of partitioning of fixed carbon between transport and non-transport carbohydrates (such as sucrose and starch); the rate of sucrose synthesis; intercellular transfer of assimilates to the loading region of the phloem; transport and accumulation of sucrose (the major translocated sugar in crop plants) into the phloem; long-distance transport of the assimilates within the plant vascular system; unloading of the assimilates in the sink regions; and utilization of the sugars within the sink organs (Figure 10.1). This network of biochemical/physiological processes is closely regulated at the cellular level and alteration in any process must necessarily affect the others.

10.1 Cell-to-cell transfer of sucrose in source leaves

It is now generally accepted that photosynthetic carbon assimilation, within mesophyll cells of the mature source leaf, is indirectly controlled by cytosolic levels of certain key intermediates (e.g. inorganic phosphate). Considerable information is available concerning the regulation of the biochemical steps involved in photoassimilate production, and in particular, the synthesis of sucrose within a mesophyll cell (Geiger and Servaites, 1994; Stitt, 1990). These findings have focused on intracellular processes, with the perception of the mesophyll cell as a more or less autonomous entity. However, obviously the

Figure 10.1 Schematic representation of carbohydrate allocation towards various plant organs: (a) rate of photosynthesis; (b) extent of partitioning of fixed carbon between transport and non-transport carbohydrates (such as sucrose and starch); (c) rate of sucrose synthesis; (d) intercellular transfer of assimilates to the loading region of the phloem; (e) transport and accumulation of sucrose (the major translocated sugar in crop plants) into the phloem; (f) long-distance transport of the assimilates within the plant vascular system; (g) unloading of the assimilates in the sink regions; (h) utilization of the sugars within the sink organs

synthesis and flow of sucrose must also be governed by cell-to-cell transport through both the mesophyll and vascular tissues, including the steps associated with sucrose loading into the companion cell–sieve element (CC–SE) complex. Such transport will involve controls exerted at a tissue, rather than the cellular level. Thus, a comprehensive understanding of carbon assimilation will necessarily involve an integration of all processes, from the chloroplast to phloem export at the tissue/organ level.

As an illustration of the problems faced by research in this area, even the simplest component of this process – short-distance transport of sucrose from the mesophyll cells (site of synthesis) to the CC–SE complex – remains equivocal, even after extensive research on this topic which has spanned more than two decades. Based on ultrastructural studies conducted on a wide range of plant species, two mechanisms of sugar transport have been proposed. Symplasmic phloem loading is thought to include intercellular movement, through plasmodesmata, along the entire pathway from the mesophyll cell to the CC–SE complex (Figure 10.2). (For a detailed description of plasmodesmal structure, see Ding *et al*, 1992a; Lucas *et al.*, 1993a). However, in many agronomically important species, the low density, or near absence, of plasmodesmata between bundle sheath/phloem parenchyma (BS/PP) cells and the CC–SE complex suggests that a boundary exists in terms of sucrose transport between these cells. These observations, together with the existence of a higher sucrose concentration in the phloem compared with the mesophyll cells, support the hypothesis that an apoplasmic step is involved in the transfer of sucrose to the CC–SE complex (Giaquinta, 1983; van Bel, 1993). As illustrated in Figure 10.3, sucrose is thought to be released across the PP plasma membrane, into the apoplasm, and from here it is loaded into the CC–SE complex. Recent molecular studies on sucrose transport have provided new insights into this process. For example, transgenic potato plants expressing antisense constructs of a cloned sucrose transporter (Riesmeier *et al.*, 1992) develop a severe phenotype, involving curling and early senescence of leaves, together with the accumulation of carbohydrates within

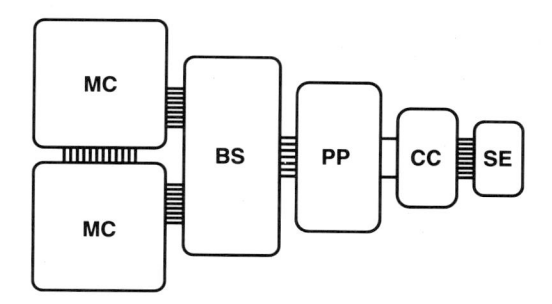

Figure 10.2 Schematic representation of plasmodesmal frequency demonstrating the symplasmic connections between the mesophyll and the CC–SE complex. Number of interconnecting lines indicates the relative frequency of plasmodesmata between the different cell types. Note the low density of plasmodesmata between the phloem parenchyma and companion cells. BS, bundle sheath; CC, companion cells; MC, mesophyll cells; PP, phloem parenchyma cells; SE, sieve elements

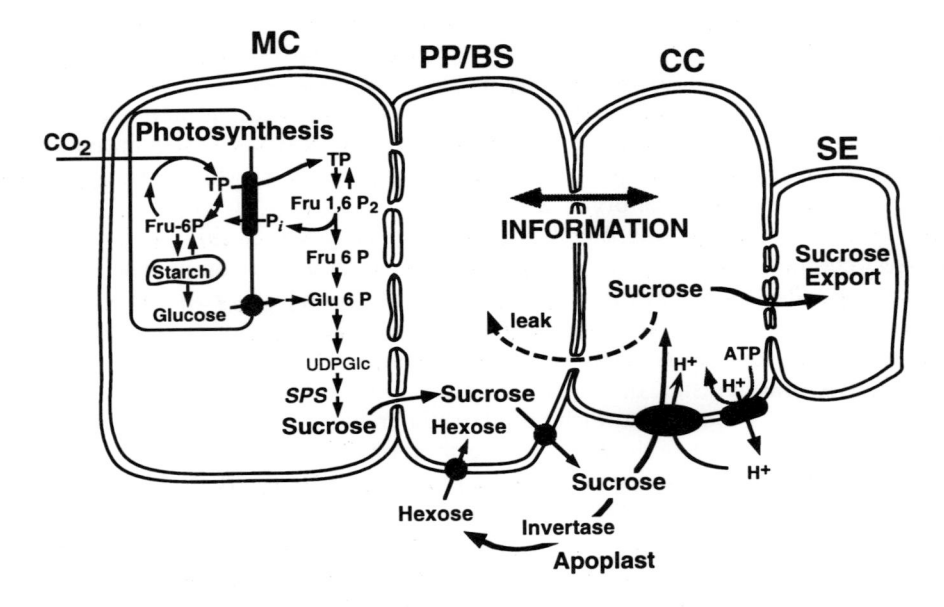

Figure 10.3 Model for the processes involved in carbon assimilation within the mesophyll cell (MC) and translocation of sucrose into the companion cell (CC)–sieve element (SE) complex. Sucrose is released across the phloem parenchyma/bundle sheath (PP/BS) plasma membrane into the apoplasm, and from here it is loaded into the CC–SE complex

these leaves (Riesmeier et al., 1994). These results are consistent with the notion that, in many species, apoplasmic transfer of sucrose is central to the process of phloem loading.

10.2 Conundrum of plasmodesmata connecting PP–CC

Such an apoplasmic step would necessitate the involvement of intercellular co-ordination to achieve efficient delivery of sucrose into the long-distance transport pathway of the phloem. This co-ordination could be achieved either by the exchange of information molecules across the adjoining cell wall, or through the low density of plasmodesmata that interconnect the PP and CC. At first sight, the cell wall route appears more logical, as the presence of open plasmodesmata would act as a leak to the overall process of sucrose loading (Figure 10.3). However, microinjection experiments have indicated that plasmodesmata interconnecting the PP–CC are functional, in terms of allowing the exchange of small membrane-impermeable fluorescent probes (Madore et al., 1986). Additionally, the existence of plasmodesmata at this junction in all species studied suggests that communication through these plasmodesmata may be of critical importance in the regulation of phloem loading and carbon allocation.

That the plasmodesmata between PP and CC retain functional capabilities is also supported by studies on virus movement through plasmodesmata from the mesophyll into the SE (Lucas and Gilbertson, 1994). Furthermore, it has been established that plants have the genetic capacity to truncate plasmodesmata, in situations where symplasmic isolation is essential for physiological function, as is the case for the stomal complex (Palevitz and Hepler, 1985; Wille and Lucas, 1984). Finally, plants are capable of regulating plasmodesmal size exclusion limit (SEL). Thus, it would seem that any back leakage of sucrose that might occur during the passage of putative signalling molecules would be insignificant, when judged against the likely efficacy of this communication route. Direct experimental support for the hypothesis that plasmodesmata are capable of trafficking macromolecules, and indirect evidence that addresses the involvement of such trafficking in the regulation of sucrose delivery to the phloem will shortly be discussed. However, before proceeding to these topics, we will discuss the related conceptual problem of the manner in which carbon allocation is regulated at the whole plant level.

10.3 Long-distance sucrose transport and resource allocation

Production and utilization of dry matter within the plant are interdependent processes, whereas the regulation of dry matter allocation, into different organs, is independent of the production of assimilates. The allocation of photoassimilates towards various sink organs is determined partly by the ability of the organ to acquire assimilates, relative to other sinks; however, the fundamental mechanism involved in orchestrating the integration of photosynthesis and resource allocation remains unresolved.

According to the Münch (1930) hypothesis, translocation between the source leaf and the various sinks is determined by the pressure gradient and the hydraulic conductance associated with the long-distance pathway. It has therefore been suggested that a key factor controlling resource allocation is the efficiency of sucrose metabolism/compartmentation within sink organs (Ho, 1988). Depletion of sucrose in the sink region indirectly increases the pressure gradient and, thus, the translocation rate towards this organ would be enhanced. Conversely, a decrease in sucrose metabolism, which would result in a reduction in the pressure gradient between the source and this sink region, would cause a lower translocation rate and this has been proposed to result in an inhibition of sucrose loading into the phloem at the source. As a consequence of this inhibition of loading, more of the fixed carbon would be retained in the leaf, with a shift in carbon partitioning occurring towards starch accumulation (Geiger and Fondy, 1991). When starch is accumulated, a feedback regulation of photosynthesis takes place (Stitt et al., 1991).

A number of studies appear to challenge the validity of this hypothesis. Many of these studies were performed on transgenic plants in which a specific reaction (or step) in the process of carbon metabolism and/or transport was modulated in order to determine the role of this step in controlling the overall process of carbon partitioning (Heinke et al., 1992, 1994; Riesmeier et al., 1994). For example, expression in transgenic potato plants of an antisense construct of the sucrose transporter cDNA resulted in up to a ninefold increase in the levels of sucrose and starch within source leaves. Even more remarkably, hexose concentrations were found to increase up to 100-fold, yet the photosynthetic rate within such tissue was not significantly different from control leaves (Riesmeier et al.,

1994). Furthermore, cold-girdling-induced rapid changes in steady-state transcript level of the Rubisco small subunit, in spinach source leaves, were evident before any detectable changes occurred in photosynthetic rate or stomatal conductance (Krapp and Stitt, 1995). This latter finding indicates that relatively rapid modulation of gene expression, in response to changes in the source–sink balance, may play a central role in orchestrating the processes of carbon metabolism and resource allocation.

Interestingly, expression of the sucrose transporter antisense construct caused a significant alteration in the allocation of assimilates between the various plant organs (Riesmeier et al., 1994). These plants had only a small root system and tuber development was inhibited; however, around the time of tuber initiation, shoot growth increased significantly, resulting in a dramatic increase in the shoot-to-tuber (+root) ratio. This striking change in resource allocation cannot be explained on the basis of the Münch hypothesis. A more specific explanation is needed in terms of the endogenous mechanism(s) responsible for the time-specific shift in resource allocation towards the shoots of these plants.

Collectively, the above-cited studies indicate that there may well be several hierarchies of control over the processes responsible for resource allocation. At a lower hierarchical level, biochemical controls probably operate, at a cellular level, and would be responsible for sucrose synthesis. At the other end of the spectrum, the highest hierarchical level of control presumably functions at the whole plant level. Here, the control mechanisms must include inputs from all plant parts and involve the integration of both the status of assimilated carbon in all regions as well as the developmental stage of each tissue. Since development and growth must be highly co-ordinated, processing of all input signals would take place before the generation of output signals. These signals must then interact to modulate specific gene expression/enzymatic activity in the various organs.

In this chapter we discuss the likely existence of a novel plant communication network that may provide the molecular signals involved in this putative hierarchical control system. Plasmodesmata appear to play a central role in establishing this network by trafficking information molecules, via the phloem, thereby co-ordinating the processes of carbon partitioning, export from source leaves, and allocation of the photosynthates between the various plant organs.

10.4 Viral infection alters carbohydrate partitioning

Numerous studies have reported that virus infection can alter both photosynthetic capacity and carbohydrate allocation in leaves (Fraser, 1987; Goodman et al., 1986). Photosynthesis is often reduced within virally infected tissues and, whereas soluble sugar levels may also fall, starch often accumulates in the chloroplasts. Recently it has been established that some viruses are capable of inducing complex spatio-temporal alterations in carbon metabolism which clearly transcend simple perturbations to cellular function (Técsi et al., 1995). For example, infection of pea plants with red mottle comovirus strain 'O' (RCMV-O) results in a large increase in the starch content of leaves located between the inoculated leaf and the apex, where virus was absent (Técsi et al., 1992). This interaction between the pea plant and RCMV-O indicates that alteration in starch accumulation is not due to a simple interference in the synthesis, or breakdown, of starch at the infection site, but rather to a process mediated by information perceived by distant tissues.

These observations demonstrate that viral infection may well offer a powerful way to explore the mode by which the plant orchestrates photoassimilate partitioning and allocation. The ability to dissect the role of specific viral genes, by expressing them in transgenic plants, now provides a model experimental system to explore the influence of the viral protein(s) on both leaf/plant physiology and the endogenous mechanism(s) controlling physiological performance (photosynthesis, carbon metabolism, biomass partitioning).

10.5 Viral movement protein alters carbohydrate metabolism and allocation in transgenic tobacco plants

Most plant viruses are known to spread by moving through plasmodesmata (Lucas and Gilbertson, 1994). Molecular studies on plant viruses have established the generality of the concept that viruses encode a protein(s) that is essential for their movement through plasmodesmata from the site of replication to surrounding, uninfected, cells. These proteins have been named movement proteins (MPs). The best characterized viral movement protein is the 30 kDa protein of tobacco mosaic virus (TMV-MP). Immunogold labelling studies have indicated that, in infected plants, this protein becomes localized to mesophyll plasmodesmata (Tomenius et al., 1987). In addition, dye-coupling experiments established that expression of the TMV-MP in transgenic tobacco plants alters the size exclusion limit (SEL) of plasmodesmata from a value of approx. 0.8 kDa to more than 10 kDa (Wolf et al., 1989, 1991).

Transgenic tobacco plants expressing TMV-MP have been used to explore the role of plasmodesmata in terms of the transport of photosynthate. Based on the hypothesis that plasmodesmata provide the cell-to-cell pathway for symplasmic movement of carbohydrates (also amino acids and inorganic ions), it seemed logical to assume that, if diffusion through plasmodesmata were rate limiting for sucrose transport, major changes in plasmodesmal SEL would result in an increase in the rate of sucrose movement from the mesophyll to the site of loading in the phloem. Analysis of diurnal accumulation of carbohydrates indicated that significantly higher levels of starch and sugars accumulated in source leaves of transgenic tobacco plants expressing MP under the control of the 35S CaMV promoter (Lucas et al., 1993b). Moreover, pulse-labelling experiments confirmed that the export rate of ^{14}C-sucrose, during the day, was lower in MP-expressing compared with control tobacco plants (Olesinski et al., 1995).

In addition to the effect of TMV-MP on carbohydrate allocation and export in source leaves, MP-expressing tobacco plants exhibited significant alterations in dry matter partitioning between the various plant organs. Photosynthetic rate as well as total dry matter were only marginally lower in MP-expressing plants, and there was no significant difference in dark respiration as compared to control plants (Balachandran et al., 1995; Olesinski et al., 1995). However, the root mass of TMV-MP-expressing plants was 30–50% of the value measured for control plants, resulting in a twofold decrease in root-to-shoot ratio (Balachandran et al., 1995). These results established that TMV-MP (when constitutively expressed in transgenic tobacco plants) exerts a significant alteration in carbohydrate partitioning and transport in source leaves as well as over the allocation of assimilates between the various plant tissues.

10.6 TMV-MP has pleiotropic effects on plant function

Transgenic plants expressing various mutations in TMV-MP have been employed to determine further the relationship between TMV-MP alteration of plasmodesmal SEL and its effect on carbon metabolism/transport. Studies on plasmodesmal SEL in transgenic plants expressing a temperature-sensitive TMV-MP mutant indicated that, under permissive temperatures (24°C), the plasmodesmal SEL was elevated to levels identical to those in plants expressing wild-type TMV-MP. However, after 6 h at non-permissive temperatures (32°C), SEL was reduced to values similar to those detected in control plants (Wolf et al., 1991). Analysis of carbohydrate levels in transgenic plants exposed to permissive temperatures (24°C) revealed that values were similar to those in plants expressing wild-type TMV-MP. However, pretreatment at non-permissive temperatures (32-34°C) did not cause the sucrose, glucose and fructose to return to control levels. These results unambiguously established that TMV-MP exerts its influence over carbon partitioning in tobacco source leaves via a mechanism that is independent of its capacity to increase plasmodesmal SEL.

Further support for the hypothesis that TMV-MP has pleiotropic effects over plant function was provided by studies performed on transgenic plants expressing various forms of TMV-MP deletion mutants (Wolf et al., 1995). Examination of the role of the TMV-MP C-terminus in mediating plasmodesmal SEL increase revealed that expression of TMV-MPs from which up to 55 amino acids had been deleted still resulted in an increase in plasmodesmal SEL to values greater than 10 kDa (Berna et al., 1991). However, this ability was lost when 77 or 108 amino acids were deleted from the TMV-MP C-terminus. Carbohydrate analyses performed on plants expressing these TMV-MP mutants revealed that diurnal changes in starch and sucrose levels resembled those measured on plants expressing wild-type TMV-MP (Wolf et al., 1995). Similarly, plants expressing truncated MP had root-to-shoot ratios that were similar to those determined on wild-type TMV-MP-expressing plants (Balachandran et al., 1995). These studies further confirmed the hypothesis that the pleiotropic effects of TMV-MP on physiological function can be separated from its ability to increase plasmodesmal SEL.

A further significant result from these TMV-MP mutant studies was the finding that transgenic tobacco plants expressing a 10-amino acid C-terminal deletion exhibited the same diurnal changes in source leaf carbohydrate levels as vector control plants. Moreover, the root-to-shoot ratios were identical in vector control and the TMV-MP C-10 amino-acid-deletion mutant lines. Clearly, TMV-MP contains domains that potentiate its interaction with plasmodesmata to mediate the cell-to-cell transport of its viral RNA (Lucas and Gilbertson, 1994). The domain that potentiates MP-mediated increase in plasmodesmal SEL is presumably associated (required) with viral nucleic acid transport. But TMV-MP must also contain a domain(s) that can interact with endogenous processes, which results in a marked alteration in the regulation of photoassimilate partitioning and translocation (Figure 10.4).

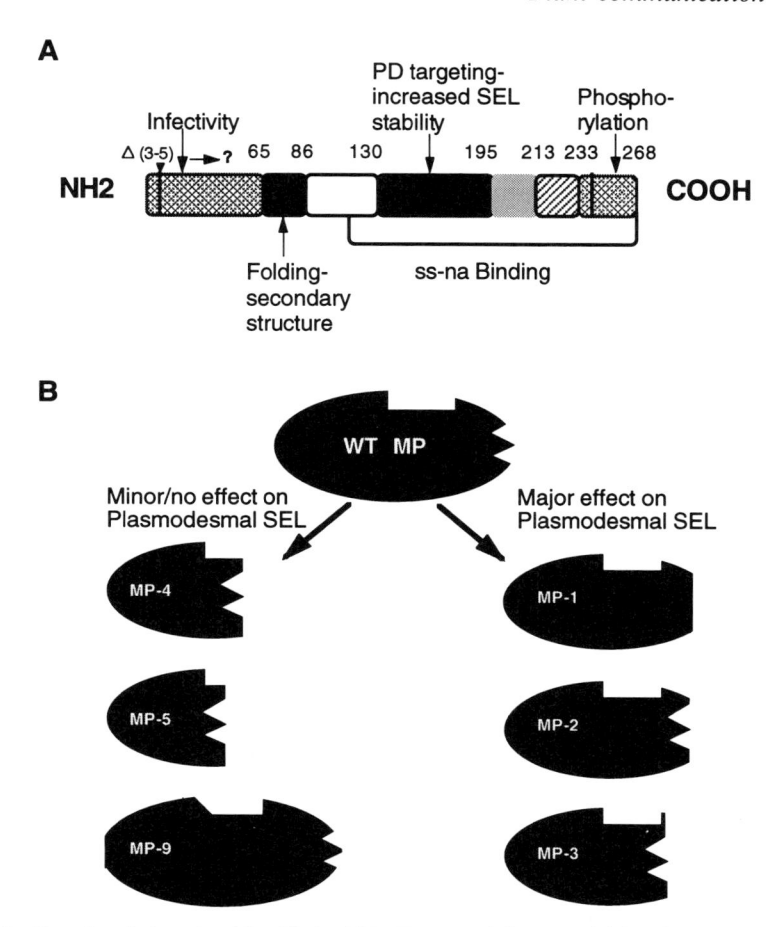

Figure 10.4 Functional domains identified within TMV-MP (A). A model for the pleiotropic effect of TMV-MP (B). Plant lines MP-1, MP-2 and MP-3 express an MP in which the site affecting plasmodesmal SEL is intact, while in plant lines MP-4, MP-5 and MP-9, this site is truncated/modified. The toothed pattern on TMV-MP represents an active site which affects carbohydrate metabolism and resource allocation, while the blunt pattern represents a modification in the structure of this site which blocks the effect of TMV-MP on carbohydrate metabolism (redrawn, with permission, from Wolf et al., 1995)

Strong support for the hypothesis that TMV-MP possesses a unique domain that can alter endogenous processes has been provided by studies performed on transgenic tobacco plants expressing the MP of cucumber mosaic virus (CMV). Considerable similarities exist between CMV-MP and TMV-MP, in terms of predicted structure and the domains required to increase plasmodesmal SEL and transport the viral ribonucleoprotein complex. As expected, the mesophyll plasmodesmal SEL was increased in these CMV-MP-expressing plants, to a value greater than 10 kDa (Ding et al., 1995). However, carbohydrate partitioning within these source leaves as well as resource allocation and plant phenotype remained the same as in control plants (Lucas et al., 1996). Production of CMV-MP chimeras containing various segments of the TMV-MP C-terminal region should permit the identification of the unique motif responsible for the TMV-MP-induced alteration in carbon metabolism and resource allocation.

10.7 Source leaves may be the site of TMV-MP action

All of the above-described studies involved transgenic plants expressing the TMV-MP under the control of the 35S CaMV constitutive promoter. An important step towards understanding the mode by which the TMV-MP exerts its influence both on source leaf physiology and resource allocation would be the identification of the site(s) of action within the plant. Two approaches were taken to ascertain whether TMV-MP is required in both shoots and roots to elicit these effects. Reciprocal grafting between transgenic tobacco expressing TMV-MP and vector control plants indicated that expression of MP is necessary only in leaf tissue to alter the root-to-shoot ratio (Balachandran et al., 1995). When MP expression was restricted to the shoot, carbon allocation was still altered such that root growth was reduced.

Further support for the shoot (leaves) being the site at which TMV-MP exerts its influence over carbon metabolism *and* resource allocation was recently obtained using a different plant system (Olesinski et al., 1996). Potato plants were transformed with the TMV-MP gene under the control of either the nuclear photosynthesis gene ST-LS1 (green tissue) or the class I patatin gene B33 (tubers) tissue-specific promoters. Restricted expression of TMV-MP to tubers did not cause any detectable change in plant phenotype, nor did it cause a detectable alteration in carbohydrate partitioning in source leaves. In contrast, significantly lower levels of starch and sugars were found to accumulate in source leaves when the TMV-MP was expressed in green tissue. As photosynthetic rate and dark respiration were not significantly different between TMV-MP expressing and control plants, the lower levels of carbohydrate accumulation in potato source leaves must result from higher rates of sucrose export. This conclusion was confirmed by direct measurement of sucrose efflux from petioles of detached potato source leaves. Sucrose efflux from leaves of transgenic plants expressing the TMV-MP in green tissue was significantly higher as compared to values obtained using leaves from control plants (Olesinski et al., 1996).

10.8 Developmental control over TMV-MP-mediated alteration in carbohydrate allocation

It could be argued that TMV-MP does not have a specific effect on processes of carbon metabolism and/or transport, but, as a foreign protein, it may act to disrupt normal metabolic events at the cellular level. However, a simple perturbation to cell function, due to expression of TMV-MP in source leaves, would be expected to cause similar effects in both transgenic potato and tobacco plants. Such a disruption would likely be evident at all stages of plant development and should not be affected by alterations in sink–source relationships.

Evidence to the contrary has been provided by the finding that TMV-MP-mediated change in carbohydrate allocation in transgenic potatoes is under developmental control and is closely associated with tuber development. Differences in leaf carbohydrate levels within TMV-MP transgenic and control potato plants were absent under long-day photoperiodic conditions where tuber induction is delayed. Interestingly, alteration in source/sink balance in tobacco plants also eliminated some of the effects exerted by TMV-MP. Here, elevation in temperature caused a striking increase in growth rate and a significantly higher respiration rate in both transgenic tobacco expressing TMV-MP under the control of the 35S CaMV promoter and control plants (Olesinski et al., 1995). Growth under elevated temperatures caused a marked increase in carbon export which eliminated the differences between the two plant lines in net ^{14}C-export and starch accumulation. These results indicate that increased sink demand for photosynthate, in tobacco plants, can override the influence of TMV-MP on starch metabolism, whereas in potato, the effects of the TMV-MP are only detected following tuber development and the onset of enhanced demand for photosynthates.

Resource allocation is closely related to the regulation of growth and development, in as much as growth of different plant parts and organs often requires the import of assimilates from elsewhere in the plant. The different TMV-MP-mediated responses to high temperature in tobacco, compared with tuber development in potato, demonstrate that the interaction of TMV-MP on the regulatory pathway(s) of photosynthate compartmentation and/or transport is highly complex. Clearly, it will be a challenge to elucidate the modes by which TMV-MP exerts its influence on both source leaf carbon metabolism and resource allocation within the whole plant.

10.9 Regulation of sucrose transport by plasmodesmal SEL

It is generally considered that diffusion, down the respective gradient in chemical potential, is responsible for the flux of small molecules through plasmodesmata (Tucker et al., 1989; Tyree, 1970). Based on this assumption, it would be logical to assume that changes in plasmodesmal SEL, *per se*, would have an effect on the rate of sucrose movement from the mesophyll to the BS/PP boundary. Consequently, under the influence of TMV-MP (or CMV-MP), the rate of sucrose export from source leaves should be higher, provided diffusion through plasmodesmata constitutes a rate-limiting step in the overall process of phloem loading. If plasmodesmal SEL within the mesophyll does not constitute the rate-limiting step for this process of sucrose export, carbon metabolism should remain unaffected by TMV-MP expression, and carbohydrate levels should be similar for all plant lines. Characterization of the transgenic tobacco and potato plants, described earlier, clearly established that expression of TMV-MP caused an alteration in both the process(es) of carbon export, via the phloem, and resource allocation. However, as already stressed, the pleiotropic effects of TMV-MP detected in studies on tobacco plants suggest that this influence cannot be simply attributed to changes in plasmodesmal SEL.

The hypothesis that plasmodesmal SEL serves to control the diffusion of sucrose was based on the perception that plasmodesmata are static structures. Regulation of plasmodesmal permeability has been demonstrated in several experimental systems. Callose ($[1\rightarrow3]$-β-D-glucan) synthesis, at the plasmodesmal orifice, has been proposed as a wound-activated sealing mechanism (Robards and Lucas, 1990). Callose synthesis can also be involved in the fine regulation of plasmodesmal function, especially under elevated temperatures (Wolf *et al.*, 1991). More recent studies have demonstrated that plasmodesmal function is affected by the energy status of the cell (Cleland *et al.*, 1994; Tucker, 1993). Reduction in ATP levels, as a result of azide treatment, was accompanied by an increase in plasmodesmal SEL to values ranging from 5 to 10 kDa (Cleland *et al.*, 1994). Ultrastructural studies of pea root tips indicated a loss of plasmodesmal neck constriction and enlargement of the mean diameter after pretreatment of the tissue with 350 mM mannitol. These observations were positively correlated with the transient increase in assimilate import by osmotically stressed cortical cells in the root elongation zone (Schulz, 1995).

These studies demonstrate that plasmodesmata are dynamic entities and are under endogenous control. *De novo* formation of plasmodesmata also occurs under developmental control (Ding *et al.*, 1992b) and this process would obviously be responsive, in a homeostatic manner, to artificially induced changes to symplasmic conductivity. It may well be that such endogenous control over plasmodesmal function can override the TMV-MP-mediated changes in plasmodesmal SEL. It has to be borne in mind that assimilate transport is also a dynamic process, operating under continuous regulation in response to endogenous signals. One facet of these signals may well involve dynamic changes in plasmodesmal function.

10.10 Macromolecular trafficking through plasmodesmata: implications for controlling carbohydrate partitioning and resource allocation

The most significant evidence for characterizing plasmodesmata as dynamic entities was provided by studies which established that these intercellular organelles have the capacity to mediate the trafficking of macromolecules (proteins and nucleic acids). It is now evident that MPs of many different plant viruses can traffic through plasmodesmata (Ding *et al.*, 1995; Fujiwara *et al.*, 1993; Lucas and Gilbertson, 1994; Noueiry *et al.*, 1994). It is logical to assume that viruses utilize the endogenous cellular mechanisms for cell-to-cell trafficking of macromolecules through plasmodesmata. Indeed, microinjection studies have now established that plasmodesmata facilitate the cell-to-cell transport of a 45 kDa plant-encoded transcription factor, KNOTTED1 (Lucas *et al.*, 1995). In addition, this study demonstrated that KNOTTED1 can also mediate the selective trafficking of its own mRNA through mesophyll plasmodesmata. Furthermore, indirect evidence for the extensive movement of endogenous proteins is available from studies on the phloem. Many proteins have been isolated from the enucleate SE that form the long-distance transport system (Fisher *et al.*, 1992; Ishiwatari *et al.*, 1995; Lucas and Wolf, 1993; Lucas *et al.*, 1993a). One such phloem protein has been cloned and the encoded protein produced in *E. coli*, fluorescently-labelled and microinjected into leaf tissue. These studies established that this protein has the capacity to traffic through plasmodesmata (Y. Ishiwatari, T. Fujiwara and W.J. Lucas, unpublished results).

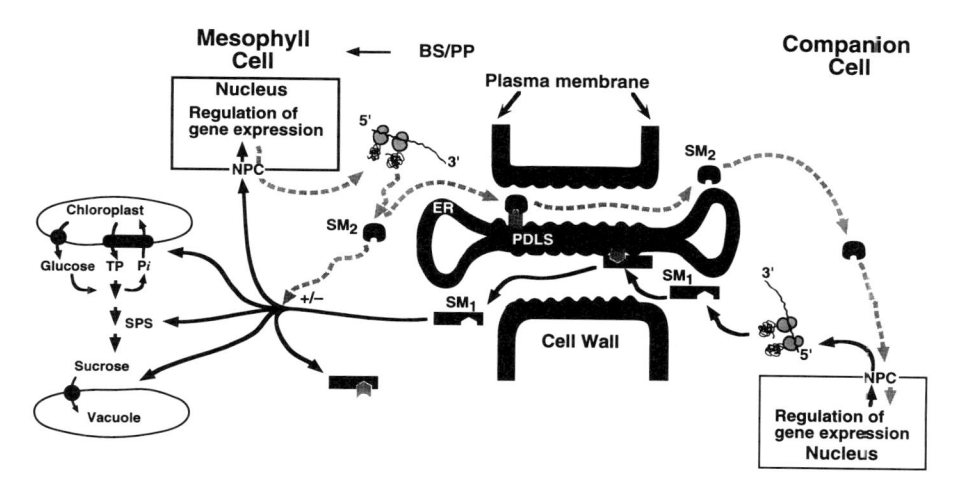

Figure 10.5 Model of a communication network proposed to function between mesophyll and companion cells. In this model, we envisage that selective trafficking of information molecules, via plasmodesmata, forms the basis of the signal molecule (SM) regulatory cascade. Trafficking of signalling molecules (SM_1) from the CC to the mesophyll potentiates the regulation of carbon metabolism, either by a direct interaction with the pathway of sucrose synthesis, or by an activation of a secondary messenger. Alternatively, SM_1 can act as a transcription factor to regulate gene expression within the mesophyll (SM_2). Signalling molecule (SM_2) traffic from the mesophyll to the CC to 'inform' the 'phloem' as to the level of available fixed carbon. Again, SM_2 can induce an alteration in gene expression within the CC to produce two sets of signalling molecules. One set constitutes the feedback component of the 'local' information network and the other set traffics, through plasmodesmata, into the SE for transport to distant plant organs. NPC, nuclear pore complex; PDLS, plasmodesmal localization site

In light of these findings, we advance the hypothesis that trafficking of regulatory (information) molecules, through plasmodesmata, establishes a special supracellular communication network between the CC and the mesophyll which operates to regulate carbon partitioning (Figure 10.5). Given that a plasmodesmal network couples the functional phloem to all tissues of the plant, and that proteins can move through plasmodesmata from the CC to the SE, we propose that an equivalent supracellular communication network also operates to orchestrate resource allocation between all plant organs.

The role of these communication systems in regulating carbon partitioning is founded on three experimental observations: the presence of TMV-MP in transgenic plants alters both carbon metabolism within the source leaf and resource allocation between the various plant organs; the source leaf was identified as the site where TMV-MP exerts this effect on plant function; and the effect of TMV-MP was found to be independent of its influence on plasmodesmal SEL.

According to the model, a continuous exchange of information (molecules), via plasmodesmata, occurs between the CC and the mesophyll. One facet of this information network is the regulation of photosynthesis, within the mesophyll, in terms of the delivery of sucrose to the site of phloem loading (Figure 10.5). Trafficking of signalling molecules (SM_1 in Figure 10.5) from the CC to the mesophyll potentiates the regulation of metabolism via direct interaction with specific enzymes (or their regulatory components; e.g. SPS as discussed by Lucas et al., 1996). Alternatively, SM_1 may act as a transcription factor that can regulate specific steps within this pathway via the synthesis of a second signalling molecule (SM_2 in Figure 10.5). As shown in Figure 10.5, signalling molecules (SM_2) also traffic from the mesophyll to the CC–SE complex. The role of this molecule is to 'inform' the 'phloem' as to the level of available fixed carbon. Again, this arm of the information pathway includes regulation of gene expression, within the CC, to produce signalling molecules. One set of these molecules (SM_n) constitutes the feedback component of this 'local' information network and is transported to the mesophyll to adjust carbon partitioning (e.g. alteration in the timing and/or rate of starch accumulation). A second set of information molecules (hormone, peptide or protein), generated within the CC, is trafficked through plasmodesmata into the SE for transport to distant plant organs. These molecules form a higher-order signalling system that operates to determine resource allocation at the whole-plant level.

TMV-MP may interfere with any of the individual elements involved in this complex information network. It could compete with endogenous proteins for the plasmodesmal localization site(s). However, it could also interfere/compete with endogenous protein function within either the CC or the mesophyll. Such interference would alter the explicit nature of the message(s) generated by the local information network. The outcome from this TMV-MP-induced shift in endogenous signals would cause an alteration in both carbon export and resource allocation. Finally, as photosynthesis is not significantly reduced in transgenic tobacco plants expressing TMV-MP, even though sugar and starch levels are greatly increased, the TMV-MP-induced changes in the endogenous signals must not result in the feedback inhibition of photosynthesis. Thus, once the individual signalling elements have been elucidated, a control model will be available that can accommodate the decoupling of the rate of translocation from photosynthesis.

10.11 Prospects

The hypothesis that higher plants utilize supracellular controls to orchestrate developmental and physiological processes is gaining experimental support. Further, direct evidence has recently been presented that plasmodesmata mediate in the selective cell-to-cell trafficking of macromolecules. That such trafficking has the potential to establish a novel level of regulation over both enzymic function *and* gene expression appears beyond doubt. The exciting challenge is to provide *direct* experimental support for this hypothesis. In order to achieve this goal it will be necessary to develop strategies that will allow the identification of the information molecules that act within the proposed supracellular control network. The first such information molecule identified and characterized will likely open up this field to intense study. Clearly, a full understanding of the underlying mechanisms utilized by the plant to orchestrate resource allocation, on a whole-plant level, will open the way for a wide range of biotechnological applications. Perhaps we are about to witness a whole new revelation in plant biology.

Acknowledgments

This work was supported by United State–Israel Binational Agricultural Research Development Fund No IS-1968-91R (W.J.L. and S.W.), and National Science Foundation grant IBN-94-06974 (W.J.L.). S.W. was on sabbatical leave from the Hebrew University of Jerusalem, Jerusalem, Israel.

References

AUSTIN, R.B., FORD, M.A. and MORGAN, C.L. (1987) Physiological changes associated with genetic improvement in English cereal yields. OECD Workshop: *Genetic Physiology and Photosynthesis of Crop Yield*, Abstract 5. EEC Publications, Brussels.

BALACHANDRAN, S., HULL, R.J., VAADIA, Y., WOLF, S. and LUCAS, W.J. (1995) Alteration in carbon partitioning, induced by the movement protein of tobacco mosaic virus, originates from the mesophyll and is independent of change in plasmodesmal size exclusion limit. *Plant, Cell Environ.* **18**, 1301–1310.

BERNA, A., GAFNY, R., WOLF, S., LUCAS, W.J., HOLT, C.A. and BEACHY, R.N. (1991) The TMV movement protein: role of the C-terminal 73 amino acids in subcellular localization and function. *Virology* **182**, 682–689.

CLELAND, R.E., FUJIWARA, T. and LUCAS, W.J. (1994) Plasmodesmal-mediated cell-to-cell transport in wheat roots is modulated by anaerobic stress. *Protoplasma* **178**, 81–85.

DING, B., TURGEON, R. and PARTHASARATHY, M.V. (1992a) Substructure of freeze substituted plasmodesmata. *Protoplasma* **169**, 28–41.

DING, B., HAUDENSHIELD, J.S., HULL, R.J., WOLF, S., BEACHY, R.N. and LUCAS, W.J. (1992b) Secondary plasmodesmata are specific sites of localization of the tobacco mosaic virus movement protein in transgenic tobacco plants. *Plant Cell* **4**, 915–928.

DING, B., LI, Q., NGUYEN, L., PALUKAITIS, P. and LUCAS, W.J. (1995) Cucumber mosaic virus 3a protein potentiates cell-to-cell trafficking of CMV RNA in tobacco plants. *Virology* **207**, 345–353.

FISHER, D., WU, Y. and KU, M.S.B. (1992) Turnover of soluble proteins in the wheat sieve tube. *Plant Physiol.* **100**, 1433–1441.

FRASER, R.S.S. (1987) *Biochemistry of Virus-infected Plants*. Research Studies Press, Letchworth, Hertfordshire.

FUJIWARA, T., GIESMAN-COOKMEYER, D., DING, B., LOMMEL, S.A. and LUCAS, W.J. (1993) Cell-to-cell trafficking of macromolecules through plasmodesmata, potentiated by the red clover necrotic mosaic virus movement protein. *Plant Cell* **5**, 1783–1794.

GEIGER, D.R. and FONDY, B.R. (1991) Regulation of carbon allocation and partitioning: status and research agenda. In: *Recent Advances in Phloem Transport and Assimilate Compartmentation* (BONNEMAIN, J.L., DELROT, S., LUCAS, W.J. and DAINTY, J., eds), pp. 1–9. Ouest Edition, Nantes.

GEIGER, D.R. and SERVAITES, J.C. (1994) Diurnal regulation of photosynthetic carbon metabolism in C_3 plants. *Annu. Rev. Plant Physiol. Plant Mol. Biol.* **34**, 235–256.

GIAQUINTA, R.T. (1983) Phloem loading of sucrose. *Annu. Rev. Plant Physiol.* **34**, 347–387.

GIFFORD, R.M., THORNE, J.H., HITZ, W.D. and GIAQUINTA, R.T. (1984) Crop productivity and photoassimilate partitioning. *Science* **225**, 801–808.

GOODMAN, R.N., KIRALY, Z. and WOOD, K.R. (1986) *The Biochemistry and Physiology of Plant Disease.* University of Missouri Press, Columbia.

HEINEKE, D., SONNEWALD, U., BÜSSIS, D., GÜNTER, G., LEIDREITER, K., WILKE, I., RASCHKE, K., WILLMITZER, L. and HELDT, H.W. (1992) Apoplastic expression of yeast-derived invertase in potato: effect on photosynthesis, leaf solute composition, water relations, and tuber composition. *Plant Physiol.* **100**, 301–308.

HEINEKE, D., KRUSE, A., FLÜGGE, U.-I., FROMMER, W.B., RIESMEIER, J.W., WILLMITZER, L. and HELDT, H.W. (1994) Effect of antisense repression of the chloroplast triose-phosphate translocator on photosynthetic metabolism in transgenic potato plants. *Planta* **193**, 174–180.

HO, L.C. (1988) Metabolism and compartmentation of imported sugars in sink organs in relation to sink strength. *Annu. Rev. Plant Physiol. Plant Mol. Biol.* **39**, 355–378.

INOUE, H. and TANAKA, A. (1978) Comparison of source and sink potentials between wild and cultivated potatoes. *J. Soil Sci.* **49**, 321–327.

ISHIWATARI, Y., HONDA, C., KAWASHIMA, I., NAKAMURA, S., HIRANO, H., MORI, S., FUJIWARA, T., HAYASHI, H. and CHINO, M. (1995) Thioredoxin h is one of the major proteins in rice phloem sap. *Planta* **195**, 456–463.

KRAPP, A. and STITT, M. (1995) An evaluation of direct and indirect mechanisms for the 'sink-regulation' of photosynthesis in spinach: changes in gas exchange, carbohydrates, metabolites, enzyme activities and steady-state transcript levels after cold-girdling source leaves. *Planta* **195**, 313–323.

LUCAS, W.J. and GILBERTSON, R.L. (1994) Plasmodesmata in relation to viral movement within leaf tissues. *Annu. Rev. Phytopathol.* **32**, 387–411.

LUCAS, W.J. and WOLF, S. (1993) Plasmodesmata: the intercellular organelles of green plants. *Trends Cell Biol.* **3**, 308–315.

LUCAS, W.J., DING, B. and VAN DER SCHOOT, C. (1993a) Plasmodesmata and the supracellular nature of plants. *New Phytol.* **125**, 435–476.

LUCAS, W.J., OLESINSKI, A., HULL, R.J., HAUDENSHIELD, J.S., DEOM, C.M., BEACHY, R.N. and WOLF, S. (1993b) Influence of the tobacco mosaic virus 30-kDa movement protein on carbon metabolism and photosynthate partitioning in transgenic tobacco plants. *Planta* **190**, 88–96.

LUCAS, W.J., BOUCHE-PILLON, S., JACKSON, D.P., NGUYEN, L., BAKER, L., DING, B. and HAKE, S. (1995) Selective trafficking of KNOTTED1 homeodomain protein and its mRNA through plasmodesmata. *Science* **270**, 1980–1983.

LUCAS, W.J., BALACHANDRAN, S., PARK, J. and WOLF, S. (1996) Plasmodesmal companion cell–mesophyll communication in the control over carbon metabolism and phloem transport: insight gained by viral movement protein. *J. Exp. Bot.* **47**, 1119–1128.

MADORE, M.A., OROSS, J.W. and LUCAS, W.J. (1986) Symplastic transport in Ipomoea tricolor source leaves: demonstration of functional symplastic connections from mesophyll to minor veins by a noval dye-tracer method. *Plant Physiol.* **82**, 432–442.

MÜNCH, E. (1930) *Die stoffbewegung in der pflanze.* Gustav Fischer, Jena.

NOUEIRY, A.O., LUCAS, W.J. and GILBERTSON, R.L. (1994) Two proteins of a plant DNA virus coordinate nuclear and plasmodesmal transport. *Cell* **76**, 925–932.

OLESINSKI, A.A., LUCAS, W.J., GALUN, E. and WOLF, S. (1995) Pleiotropic effects of TMV-MP on carbon metabolism and export in transgenic tobacco plants. *Planta* **197**, 118–126.

OLESINSKI, A.A., ALMON, E., NAVOT, N., PERL, A., GALUN, E.,

LUCAS, W.J. and WOLF, S. (1996) Tissue-specific expression of the tobacco mosaic virus movement protein in transgenic potato plants alters plasmodesmal function and carbohydrate partitioning. *Plant Physiol.* **111**, 541–550.

PALEVITZ, B.A. and HEPLER, P.K. (1985) Changes in dye coupling of stomatal cells of *Allium* and *Commelina* demonstrated by microinjection of Lucifer yellow. *Planta* **164**, 473–479.

RIESMEIER, J.W., WILLMITZER, L. and FROMMER, W.B. (1992) Isolation and characterization of a sucrose carrier cDNA from spinach by functional expression in yeast. *EMBO J.* **11**, 4705–4713.

RIESMEIER, J.W., FROMMER, W.B. and WILLMITZER, L. (1994) Evidence for an essential role of the sucrose transporter in phloem loading and assimilate partitioning. *EMBO J.* **13**, 1–7.

ROBARDS, A.W. and LUCAS, W.J. (1990) Plasmodesmata. *Annu. Rev. Plant Physiol. Plant Mol. Biol.* **41**, 369–419.

SCHULZ, A. (1995) Plasmodesmal widening accompanies the short-term increase in symplasmic phloem unloading of pea root tips under osmotic stress. *Protoplasma* **188**, 22–37.

STITT, M. (1990) Fructose-2,6-bisphosphate as a regulatory molecule in plants. *Annu. Rev. Plant Physiol. Plant Mol. Biol.* **41**, 153–185.

STITT, M., BRAUER, M., QUICK, W.P., NEUHAUS, H.E., FICHTNER, K., SCHULZE, E.-D., GUNTHER, L., HEINEKE, D., HELDT, H.W., SONNEWALD, U., VON SCHAEWEN, A. and WILLMITZER, L. (1991) Regulation of metabolism: biochemical and genetic studies. In: *Recent Advances in Phloem Transport and Assimilate Compartmentation* (BONNEMAIN, J.L., DELROT, S., LUCAS, W.J. and DAINTY, J., eds), pp. 10–17. Ouest Edition, Nantes.

TÉCSI, L., WANG, D., SMITH, A.M., LEEGOOD, R.C. and MAULE, A.J. (1992) Red clover mottle virus infection affects sink–source relationships and starch accumulation in pea plants. *J. Exp. Bot.* **43**, 1409–1412.

TÉCSI, L., SMITH, A.M., MAULE, A.J. and LEEGOOD, R.C. (1995) Relationship between viral infection and carbohydrate accumulation and metabolism. In: *Carbon Partitioning and Source–Sink Interactions in Plants* (MADORE, M.A. and LUCAS, W.J., eds), Vol. 13, *Current Topics in Plant Physiology*, pp. 130–140. American Society of Plant Physiologists, Rockville, Maryland.

TOMENIUS, K., CLAPHAN, D. and MESHI, T. (1987) Localization by immunogold cytochemistry of the virus-coded 30K protein in plasmodmesmata of leaves infected with tobacco mosaic virus. *Virology* **160**, 363–371.

TUCKER, E.B. (1993) Azide treatment enhances cell-to-cell diffusion in staminal hairs of *Setcreasea purpurea*. *Protoplasma* **174**, 45–49.

TUCKER, J.E., MAUZERALL, D. and TUCKER, E.B. (1989) Symplastic transport of carboxyfluorescein in staminal hairs of setcreasea purpurea is diffusive and includes loss to the vacuole. *Plant Physiol.* **90**, 1143–1147.

TYREE, M.T. (1970) The symplast concept. A general theory of symplastic transport according to the thermodynamics of irreversible processes. *J. Theoret. Biol.* **26**, 181–214.

VAN BEL, A.J.E. (1993) Strategies of phloem loading. *Annu. Rev. Plant Physiol. Plant Mol. Biol.* **44**, 253–281.

WILLE, A.C. and LUCAS, W.J. (1984) Ultrastructure and histochemical studies on guard cells. *Planta* **160**, 129–142.

WOLF, S., DEOM, C.M., BEACHY, R.N. and LUCAS, W.J. (1989) Movement protein of tobacco mosaic virus modifies plasmodesmatal size exclusion limit. *Science* **246**, 377–379.

WOLF, S., DEOM, C.M., BEACHY, R.N. and LUCAS, W.J. (1991)

Plasmodesmatal functioning is probed using transgenic tobacco plants that express a virus movement protein. *Plant Cell* **3**, 593–604.

WOLF, S., OLESINSKI, A.O., BALACHANDRAN, S. and LUCAS, W.J. (1995) Movement protein expression and carbohydrate partitioning. In: *Carbon Partitioning and Source–Sink Interactions in Plants* (MADORE, M.A. and LUCAS, W.J., eds), Vol. 13, *Current Topics in Plant Physiology*, pp. 117–129. American Society of Plant Physiologists, Rockville, Maryland.

11

Nitrogen uptake and assimilation in roots and root nodules

ALAIN OURRY, ANTHONY J. GORDON AND JAMES H. MACDUFF

11.1 Introduction

Chemical analysis of different soils shows that N occurs predominantly in organic form (1.5–880 mg N g^{-1} dry soil), with amino acids representing 0.45–220 mg N g^{-1} dry soil. The major plant-available forms of mineral N, namely NH_4^+ and NO_3^-, are less abundant, their respective concentrations usually in the range 17–25, and 0.05–36 mg N g^{-1} dry soil (Jones and Darrah, 1993). The higher mobility of NO_3^- compared with NH_4^+ in soil solution gives rise to decreasing availability of N to plant roots in the order NO_3^- > amino acids and NH_4^+, with mean soil solution concentrations ranging, respectively, between 197–1716, 9.9–76.5 and 3.9–14.4 μM. The concentrations and rates of transport to the root surface of these plant-available forms of N depend on numerous biological, chemical and physical processes which can increase or reduce the amount of N available to the roots (Engels and Marchner, 1995; Nye and Tinker, 1977). Consequently, rates of N uptake by plant roots depend both on soil factors, determining the rates of supply of available N, and on plant factors, most immediately the activity and numbers of carriers in epidermal and cortical cells of the roots. At the whole plant level, the uptake of N is regulated by the N demand of the plant which is regarded as a function of plant growth rate (Touraine et al., 1994).

Most of the past improvements to the N economy of agricultural systems have involved the intensification of fertilizer N use in tandem with the breeding of varieties adapted to high rates of N application. Although the historical success of this strategy is evident, in terms of increased yields, it is probable that many of the genotypes selected under these conditions are not the most efficient in terms of nutrient capture or efficiency of N utilization. Furthermore, various negative environmental impacts of intensive fertilization practises have become evident (Clarkson and Hawksford, 1993). New solutions may arise from research targeted towards genetically manipulating the regulation of N uptake by plant growth, the balance between storage and assimilation of NO_3^-, and the coupling between transport and assimilation, amongst other aspects of plant N relations.

11.2 Nitrogen uptake by root systems

11.2.1 Mechanisms of nitrate, ammonium and amino acid uptake

Nitrate

The roots of seedlings never previously exposed to external nitrate are immediately able to absorb nitrate supplied at low concentration (< 250 µM). This constitutive high affinity transport system (CHATS, Figure 11.1), operates at a low rate (< 1 µmol NO_3^- h^{-1} g^{-1} FW) and obeys Michaelis–Menten kinetics. The putative carrier involved may be a component of an NO_3^- sensing mechanism triggering the expression of several enzymes in the assimilatory pathways for NO_3^- (Clarkson, 1986; Larsson, 1994). After a lag phase (between 2–6 h), net uptake rate increases progressively. This transition process is suppressed by inhibitors of protein or RNA synthesis, and its duration corresponds to the time required to induce the synthesis of new proteins and to integrate them into the plasmalemma (Dhugga et al., 1988). This so-called 'inducible high affinity transport system' (IHATS, Figure 11.1) is also substrate saturable. The V_{max} and affinity constant (K_m) vary widely between species, with maximum uptake rates reaching 35 µmol h^{-1} g^{-1} FW attained in *Brassica napus* (Lainé et al., 1993) and with many other external or internal factors (Clarkson, 1986; Engels and Marschner, 1995). The affinity constant of the IHATS is < 40 µM for many species (Lainé et al., 1993; Van de Dijk et al., 1982). Consequently, the potentially limiting parameter of the transport system with respect to NO_3^- uptake by field-grown plants (concentrations of NO_3^- in soil solution commonly in the range 200–1800 µM) is likely to be V_{max} rather than K_m. In view of the electrochemical gradient across the plasma membrane (inside negative) and the sensitivity of the high affinity transport system to temperature and metabolic inhibitors, these carriers are assumed to be active transports. Alkalinization of the external medium and transient depolarization of the plasmalemma occur during NO_3^- uptake, suggesting that IHATS and CHATS involve either $2H^+/1NO_3^-$ symport (Figure 11.1) or a $2OH^-/1NO_3^-$ antiport (Crawford, 1995; Touraine et al., 1994), although the exact stoichiometry remains unclear. The action of H^+ ATPase inhibitors, in reducing uptake and the repolarization of the membrane, has been interpreted as further evidence that CHATS and IHATS activity is linked to proton motive force through the plasma membrane (Ruiz-Cristin and Briskin, 1991). It is not known whether these two transport systems rely on different or identical carriers having a basal constitutive level of expression with the potential for increased expression following upon induction.

At higher external concentrations of NO_3^- (>1 mM), uptake rates increase linearly with substrate concentrations (Siddiqi et al., 1990). This low affinity transport system (LATS, Figure 11.1) has been distinguished from the active high affinity systems by a lower sensitivity to cold temperatures and to several metabolic inhibitors (Glass et al., 1990). It was assumed that the LATS effected a passive influx into the root, but given a probable cytoplasmic concentration of NO_3^- (mM range) and membrane potentials (between -120 and -300 mV), it is likely to be thermodynamically active (King et al., 1992), electrogenic and involve proton co-transport (Crawford, 1995).

Nitrogen uptake and assimilation in roots

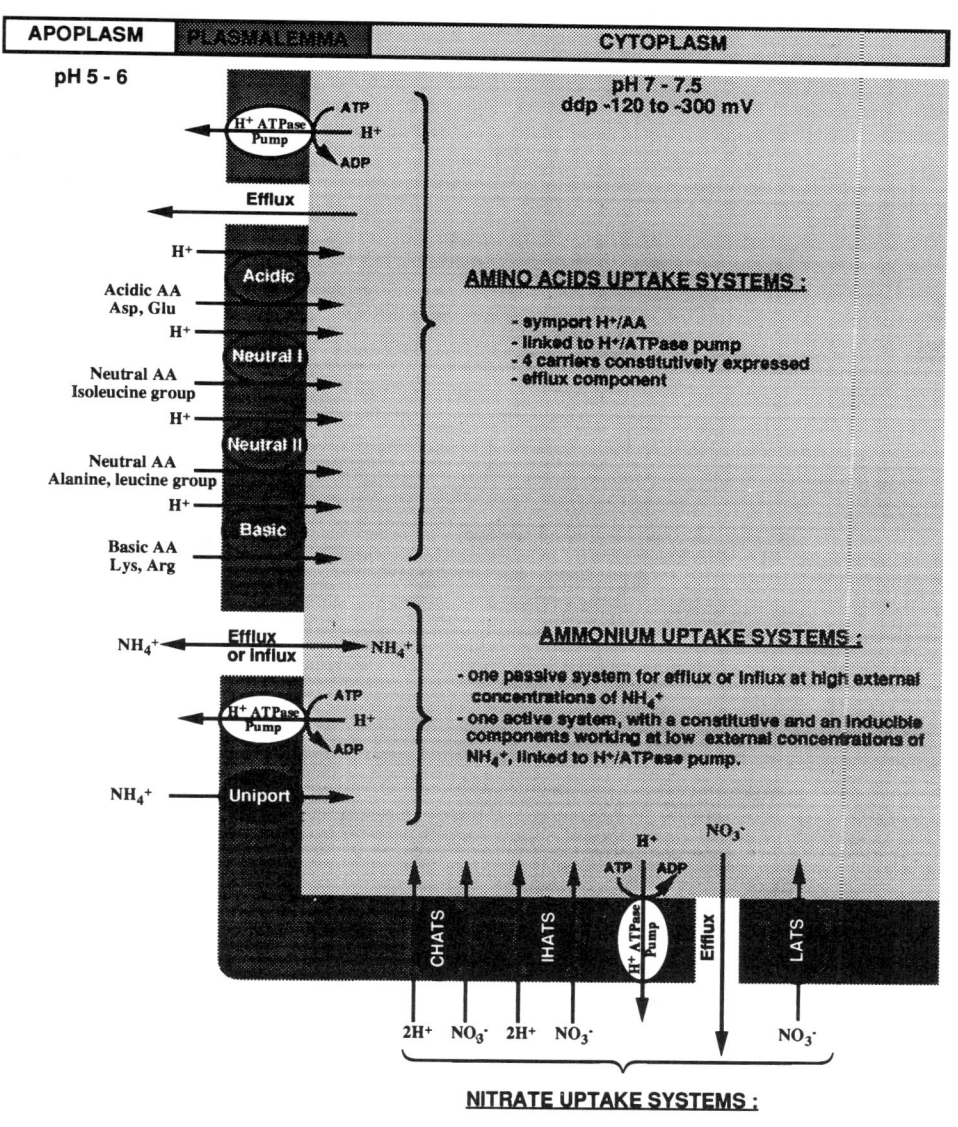

Figure 11.1 A schematic of simplified ammonium, nitrate and amino acid uptake systems in roots of higher plants

Another component of trans-root plasma membrane NO_3^- fluxes is the passive efflux proposed to occur via ion channels (Figure 11.1). Efflux of NO_3^- increases with plant NO_3^- supply and is usually correlated with root NO_3^- content. It has been suggested that efflux may regulate net NO_3^- uptake rate (Deane-

Drummond, 1990). A pump/leak/buffer model for NO_3^- uptake based on this principle has been proposed by Scaife (1989), although the regulatory significance, if any, of efflux is still unclear (Lee, 1993), as is the ecological significance (Macduff and Jackson, 1992). This is partly due to methodological difficulties associated with obtaining accurate measurements of unidirectionnal fluxes over short time periods (<10 min), without physical disturbance of plants.

At the molecular level, a putative inducible NO_3^- transporter has been identified recently in *Arabidopsis thaliana* (Tsay et al., 1993), by using transgenic plants with an insertional mutagenesis and selection for resistance to chlorate (acting as a NO_3^- analogue). This *A. thaliana* CHL1 clone encodes a 65 kDa protein with 12 putative membrane domains, but its membrane localization remains unknown (plasma membrane or tonoplast). Its functional expression in *Xenopus* oocytes has been demonstrated successfully (increased NO_3^- uptake associated with membrane depolarization following CHL1 mRNA injection), and a corresponding mRNA was synthesized predominantly in roots previously induced by NO_3^- supply (Tsay et al., 1993). These results, together with the characterization of a chlorate resistant phenotype (B1 mutant: Doddema and Telkamp, 1979), suggest that CHL1 encodes for an inducible low-affinity NO_3^- transport system, but one that is not compatible with the known physiological behaviour of the constitutive LATS. Despite the CHL1 clone showing no homology with NO_3^- transporters cloned in *Aspergillus nidulans* (Unkles et al., 1991) or *Chlamydomonas reinhardtii* (Quesada et al., 1994), homologous genes have been found in *Arabidopsis* by Northern analysis and in other higher plant species (Tsay et al., 1993), suggesting that further putative NO_3^- carriers will be identified at the molecular level in the near future.

Ammonium

In view of the prevailing electrochemical gradient between apoplasm and symplast, as a cation, the uptake of NH_4^+ has often been regarded as not requiring energy (Figure 11.1). In wheat (Goyal and Huffaker, 1986), maize (Vale et al., 1988), rice (Wang et al., 1993) and barley (Mäck and Tischner, 1994), NH_4^+ uptake exhibits enzyme-like characteristics at low external concentrations; net uptake rates are substrate-saturable, depend only on the external NH_4^+ concentrations, and adjust to these within a few minutes. However, Vale et al. (1988) interpreted NH_4^+ uptake via this saturable mechanism as an energy-dependent process.

Although seedlings grown on N-free media are able to absorb NH_4^+ constitutively at a low rate, higher rates are inducible after a lag of several hours. Based on the observation that the affinity constants (K_m) of the constitutive and the inducible high-affinity transporters are similar in barley, it has been suggested that both rely on the same carrier-mediated NH_4^+ uniport (Figure 11.1), and that during induction the abundance of functional transport proteins is increased (Mäck and Tischner, 1994). Ammonium itself may induce the synthesis of new carriers, as shown by the inhibition of its assimilation by methionine sulphoximide, while its assimilation products may exert a negative feedback control on uptake (Lee and Ayling, 1993).

A linear uptake component (low-affinity system) which does not require protein synthesis, has been reported at high external concentrations of NH_4^+

(>1 mM) in both induced and uninduced seedlings (Mäck and Tischner, 1994; Vale et al., 1988; Wang et al., 1993). It has been assumed that this influx of NH_4^+ occurs via membrane channels, and therefore is an energy-independent process. The same channels could facilitate NH_4^+ efflux when the NH_4^+ cytoplasmic pool is increased (Lee and Ayling, 1993).

At the molecular level, a gene for a high affinity NH_4^+ uptake system has been identified by an *Arabidopsis* cDNA complementation of a yeast mutant deficient for NH_4^+ uptake (Ninneman et al., 1994). This AMT1 cDNA encodes a highly hydrophilic protein with 9–12 putative membrane spanning regions, with an open reading frame of 501 amino acids, giving a coding capacity for a 53 kDa protein. Its sequence shows significant homology with a yeast NH_4^+ transporter and three other proteins with unknown functions, while no similarity was found with NO_3^- (CHL1) or K^+ (AKT1) transporters (Ninneman et al., 1994). In *Arabidopsis*, this clone is mainly expressed in roots and leaves and to a lesser extent in stems. This transport protein expressed in yeast can be inhibited by protonophores, while its sequence predicts the absence of a putative ATP binding site. These results suggest the occurrence of a secondary transport protein, as the energy required for active transport of NH_4^+ is not obtained by direct hydrolysis of ATP, given the evidence from physiological studies (Figure 11.1).

Amino acids

Conventionally, it has been thought that N is absorbed by the roots of higher plants predominantly as NO_3^- and to a lesser extent as NH_4^+. However, amino acids may commonly constitute 15–25% of the N supply (Schobert and Komor, 1987). Under extreme conditions they may account for up to 60% of total N taken up in the field, as in the Arctic species *Eriophorum vaginatum* (Chapin et al., 1993). Amino acids may be transported into root cells by proton-coupled symport (for review, see Bush, 1993). This view is based on evidence that amino acid uptake is linked to membrane depolarization, and to alkalinization of the external medium, similar to that for NO_3^- uptake. Amino acid carrier activity shows saturation kinetics, is strongly reduced by metabolic inhibitors and by compounds that lower the proton gradient across the plasma membrane. Their activity is constitutively expressed, and linked to the functioning of H^+ ATPase pumps (Figure 11.1).

The occurrence of a proton alanine symport at the plasma membrane level has been demonstrated using microsomal membrane vesicles (Bush and Langston-Unkefer, 1988). These authors characterized four different proton/amino acid transporters on the basis of competitive inhibition during measurements of uptake rates of individual amino acids from different mixtures of the 20 common amino acids into isolated membrane vesicles. These four amino acid carriers – one for acidic, two for neutral (isoleucine group and alanine, leucine group) and one for basic amino acids (Figure 11.1) – differ in terms of stereospecificity, arising from the varying positional relationship between the α amino and carboxyl groups, this determining substrate recognition.

Efflux of amino acids from the roots occurs concurrently with uptake, probably by passive diffusion (Jones and Darrah, 1993). Given the dipolar nature of amino acids, these authors speculated that their absorption by the mineral fraction of the soil may release bound solutes making them available to the plant. The net charge of most amino acids is zero between a solution pH of 5 to 7. However, aspartate and glutamate, ubiquitous in roots of many species, hold a negative net charge and may be effective in anion mobilization from the soil. In their work with maize, Jones and Darrah (1993) showed the main site of amino acid exudation to be the root tips, and postulated that amino acid transport activity may also be significant to the C and N budget of the plant, by recapturing lost amino acids from the rhizosphere in competition with the soil microbial population.

Other N transporters at the cellular and plant levels?

Accumulation of NH_4^+, a toxic decouplant, in the root cytoplasm to concentrations beyond the immediate assimilatory capacity, appears to be unusual. This has prompted the view that NH_4^+ carriers other than those found in the plasma membrane, are not required. However, this is contradicted by the molecular data suggesting that an NH_4^+ carrier may be found in leaves and to a lesser extent in stems (Ninneman et al., 1994), although their physiological function in these tissues remains uncertain.

Nitrate and amino acids may be accumulated within the vacuolar compartment of the cell and several tonoplast-bound carriers may be involved in influx and efflux. Recent studies cited by Bush (1993) have provided evidence for ATP-stimulated amino acid transporters in isolated vacuoles. A phenylalanine H^+-coupled antiport system was suggested, while transport of other amino acids across the tonoplast may occur by ATP-regulated, facilitated diffusion. Similarly, a tonoplast carrier for NO_3^- has been suggested, although not yet identified (Redinbaugh and Campbell, 1991). Because xylem and phloem loading and unloading exhibit specificity for amino acids (Schobert and Komor, 1987), and are likely nodes for the control of the redistribution of N within the plant, it is probable that several further specific transporters may be identified. This is further implied by the putative function assigned to the continuous amino acid cycling between shoot and roots (Clarkson, 1986; Imsande and Touraine, 1994) as the means of supplying the root system with information relating to plant N status and effecting the regulation of N uptake to match growth-related demand for N.

Two clones from a cDNA library constructed from above ground parts of *Arabidopsis thaliana* have been identified by functional complementation of yeast mutants deficient for histidine (Hsu et al., 1993) or proline (Frommer et al., 1993) uptake. Further characterization of these NAT2 (53 kDa predicted MW protein) and AAP1 (52.9 kDa predicted MW protein) clones, respectively, suggest that they encode for a putative neutral II amino acid transporter (Hsu et al., 1993) and a neutral I amino acid transporter (Frommer et al., 1993). The corresponding proteins are supposed to be located at the membrane level, as suggested by their hydrophobicity, and to be involved in an H^+/amino acid symport. To our knowledge, the functional expression of these two genes has not yet been reported in plant tissues, so that their occurrence in roots remains uncertain. As with plasma membrane N carriers, the molecular identification of these membrane-bound carriers associated with transport of N into and out of the xylem and phloem, many of which may be expressed at very low levels, will require new strategies to assess both their physiological function and regulatory significance (Clarkson and Hawksford, 1993).

11.2.2 Regulation of N uptake

Evidence of regulation

The literature provides many observations to support the view that N uptake by root systems is a highly regulated process, oriented to match absorption of N with a 'demand' for N that is causally linked to plant growth rate (Touraine *et al.*, 1994). Most have been primarily correlative studies of the regulation of NO_3^- or NH_4^+ uptake; relevant information for amino acid uptake is scarce. For example, when metabolic activity in meristematic tissues of maize is altered by temperature, root NO_3^- uptake is modified accordingly (Engels and Marschner, 1995). Differences in the relative growth rates of different wheat genotypes correlate with rates of NO_3^- uptake (Rodgers and Barneix, 1988). Inter-specific comparisons have shown that differences in maximum NO_3^- uptake rates, expressed per g root FW, relate to differences in partitioning of dry matter between shoots and roots; the higher the shoot/root ratio the higher is the V_{max} for NO_3^- uptake (Lainé *et al.*, 1993). The conclusion drawn from studies comparing species, genotypes or the effects of changing environmental conditions are reinforced by those in which shoot demand for N is held constant. When roots are deprived either locally or entirely, the capacity for uptake is modified. In *Brassica napus* (Lainé *et al.*, 1995), *Hordeum vulgare* (Drew and Saker, 1975), *Triticum aestivum* (Robinson *et al.*, 1994) or *Lactuca sativa* (Burns, 1991), the withdrawal of NO_3^- from part of the root system results in increased absorption by the remaining root, arising from a stimulation of root growth and/or an increased uptake capacity, and compensating for the lack of uptake by NO_3^- deprived roots. Rates of NO_3^- uptake in *Brassica napus* doubled within 24 hours, involving mRNA and protein synthesis and requiring induction by NO_3^-. How such regulation of N uptake is actually co-ordinated with biomass production, at the level of molecular genetics, remains largely unanswered.

Regulation at the cell and plant levels

Within epidermal and cortical cells of the root, it has been suggested that the net uptake rate of NO_3^- and to a lesser extent of NH_4^+, may be regulated under some circumstances by efflux across the plasmalemma into the apoplasm (Deane-Drummond, 1990). The widespread relevance of regulation via efflux has been challenged on energetic and ecological grounds, and on the basis of the narrow range of experimental conditions from which inferences have been drawn (Lee, 1993; Macduff and Jackson, 1992). Regulation of net uptake effected through the influx component is generally regarded as more plausible. Amongst possible effectors, NO_3^- and NH_4^+ are thought to induce the synthesis of their plasma membrane carriers. However, it is unlikely that these ions are involved in demand-driven control of N uptake, as they do not normally occur in the phloem stream, and therefore cannot inform the roots of shoot demand. When supplied together, NH_4^+ inhibits NO_3^- uptake, and it has been shown by using inhibitors of NH_4^+ assimilation that end-products of NH_4^+ assimilation rather than NH_4^+ itself effect the inhibition of NO_3^- uptake (Breteler and Siegerist, 1984).

As phloem- and xylem-borne end-products of the assimilation of mineral N, it is feasible that amino acids could exert an allosteric regulation of NH_4^+ and NO_3^- plasma membrane carriers in root cells (Touraine *et al.*, 1994). Lee *et al.* (1992) employed N starvation, specific inhibitors of GS, GOGAT or aminotransferases (see Figure 11.2) and externally supplied amino acids, with maize to

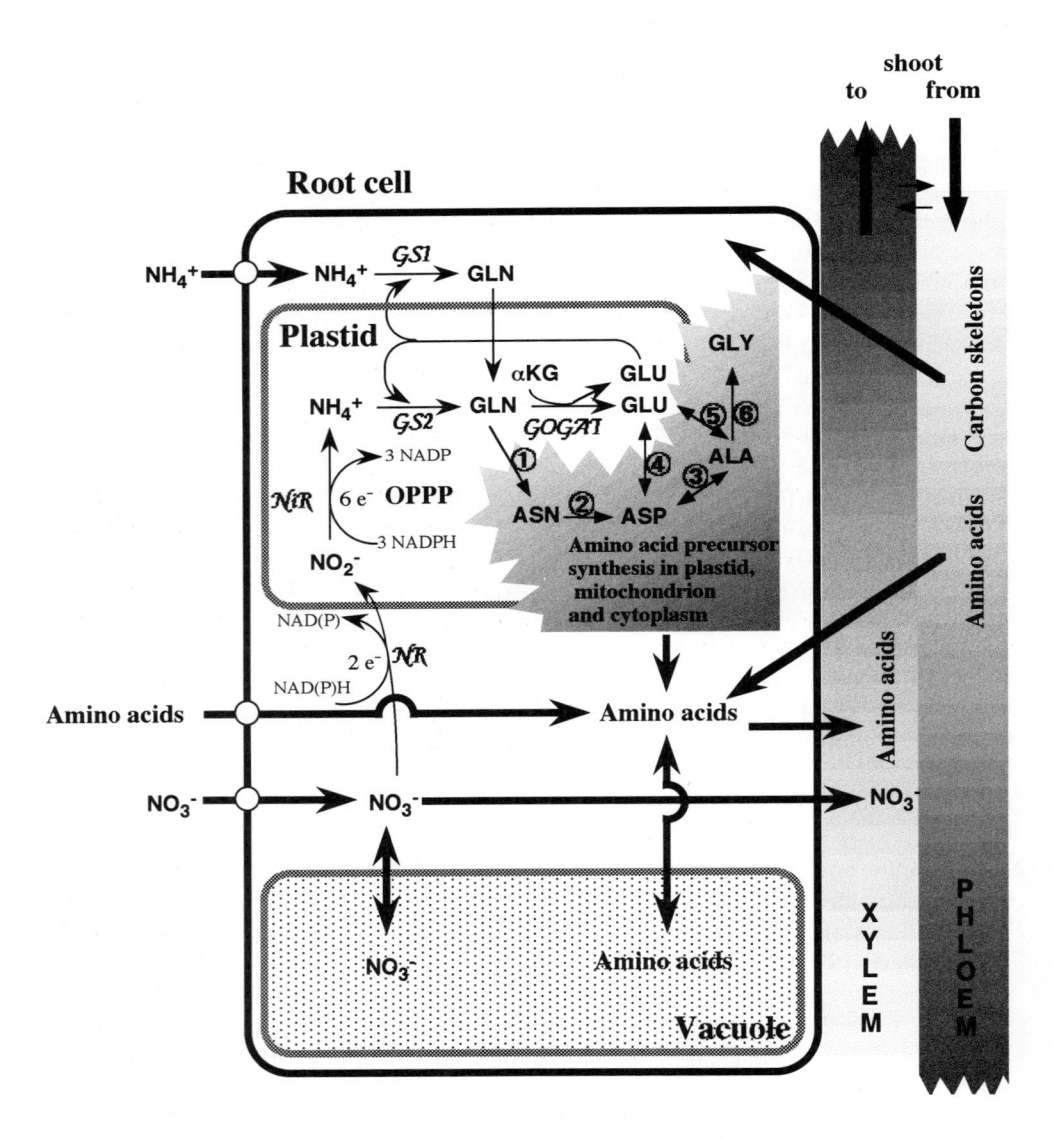

Figure 11.2 Simplified pathways of nitrate and ammonium assimilation in roots of higher plants; NR, nitrate reductase; NiR, nitrite reductase; GS1 and GS2, cytoplasmic and plastidic glutamine synthetase, respectively; GOGAT, glutamate synthase; ①, asparagine synthetase; ②, asparaginase; ③, ④, ⑤ and ⑥, transaminases; OPPP, oxidative/reductive pentose phosphate pathway

demonstrate that treatments raising intracellular concentrations of glutamine and/or asparagine led to the suppression of net uptake of NH_4^+ and NO_3^-. Conversely, conditions which lowered root glutamine and/or asparagine stimulated the net uptake of N. The involvement of phloem-borne amino acids in regulation of N uptake by soybean seedlings was advocated by Muller and Touraine (1992) when they increased the amino acid content of phloem sap by supplying amino acids externally or by immersing the cotyledons in solution containing amino acids and observed an inhibition of NO_3^- uptake. The greatest inhibitory effects were obtained with Asp, Glu, Asn, Arg, Ala, ßAla; less with Ser and Gln and even less with other amino acids. The exact mechanisms, specificities and site of any down-regulation of N transport activity in the roots by phloem-borne amino acids remain to be elucidated. Lee et al. (1992) suggested that cytosolic amino acid pools may be involved in this regulation. However, the occurrence of a lag (> 3 h) between perturbations in amino acid concentrations and in net uptake of NO_3^- has been interpreted as evidence against allosteric interactions (Imsande and Touraine, 1994). The molecular organization of this regulatory system may be elucidated if and when molecular probes for NH_4^+ and NO_3^- carriers become available.

The favoured mechanism for co-ordination of NO_3^- uptake by the roots and NO_3^- assimilation in the shoot is based on the model proposed by Ben-Zioni et al. (1971), developed further and critically discussed by Touraine and co-workers (see Imsande and Touraine, 1994; Touraine et al., 1994). Briefly, excess OH^- equivalents are produced during NO_3^- reduction in the shoot (1 mol OH^- per mol NO_3^- reduced), and strong organic acids such as oxalo-acetate or malate are synthesized to maintain pH homeostasis. K^+-malate is phloem-translocated to the root, where its decarboxylation is regarded as coupled with the excretion of HCO_3^- from the roots in exchange for the influx of NO_3^-. Hence the rate of carboxylate production during NO_3^- assimilation in the shoot may control the rate of NO_3^- uptake by the roots. Potassium is then recycled to the shoot via the xylem together with NO_3^-. This model is supported by the observations that external malate increases NO_3^- uptake, and that stimulation or inhibition of NO_3^- reduction in the shoot affects NO_3^- uptake as predicted (Touraine et al., 1992). Convincing evidence for an NO_3^-/HCO_3^- antiport is lacking, and the application of the model to species in which a considerable proportion of NO_3^- assimilation occurs in the roots has been questioned. Nevertheless, as stoichiometrically synthesized products of NO_3^- assimilation, both amino acids and organic acids constitute convincing candidates as regulatory signals for NO_3^- uptake (Touraine et al., 1994). A mechanism depending on down-regulation by phloem translocated amino acids in tandem with up-regulation mediated by organic acid synthesized in the leaves provides an attractive hypothesis for the demand-driven regulation of NO_3^- uptake, and has also been proposed to describe the change of NO_3^- absorption occurring during transition from vegetative to reproductive growth (Imsande and Touraine, 1994).

The rate of phloem-translocation of carbohydrates from the shoots to the roots may also exert a regulating effect on N uptake, particularly NH_4^+. The rationale behind this hypothesis is first the requirement for downward transport of carbon skeletons to support the mandatory assimilation of absorbed NH_4^+ in the roots. A similar dependence exists for NO_3^- when assimilated in the roots, with a C5-compound (2-oxoglutarate) associated with production of one glutamate in the GS/GOGAT pathway (Gojon et al., 1994). Second, the uptake of NO_3^- against an electrochemical gradient requires metabolic energy, although the specific ATP costs associated with uptake appear to vary widely between species and with level of N supply (Van der Werf et al., 1994). Both the rates of uptake of NH_4^+ and NO_3^-, their assimilation and carbohydrate translocation to the roots vary diurnally, generally decreasing from the beginning of the dark period and recovering during the light period. Simple correlation does not confirm a causal relationship between N uptake and carbon availability in the roots, but it may be significant that sucrose has recently been shown to induce the synthesis of nitrate reductase mRNA (see Hoff et al. (1994) for discussion on sucrose/light effects).

11.3 Pathways of N assimilation in roots and root nodules

11.3.1 Symbiotic N_2 fixation

The genes encoding the components of the enzyme nitrogenase are only found in microorganisms, some of which can fix N_2 in the free living state while others only do so while in association with a plant host. Here we will concentrate on the symbiotic association between (Brady) rhizobium and legumes. In this association chemical signalling, causing the induction of a wide range of rhizobial and legume genes, results in the production of specialized organs called nodules, which, for the most part, form on roots (although species which also form stem nodules under certain conditions also occur).

Examination of a nodule from a typical legume–Rhizobium association would generally reveal a spherical or cylindrical organ composed (in cross-section) of an outer cortical region surrounding the central infected area. The bacteria, initially induced to invade and proliferate in the plant cells, are none the less contained and controlled by the plant. They enlarge into the 'bacteroid' form and are surrounded by a modified plasma membrane termed the peribacteroid membrane. In the low oxygen environment of the nodule interior, partially created by the production in the host plant cell of large concentrations of the haemoglobin-like protein leghaemoglobin, the two component enzyme, nitrogenase, is induced in the bacteroids. Both components are extremely sensitive to inactivation by oxygen.

Thus, the structure and properties of the nodule are such that a low internal O_2 concentration (c. 20 nmolar) is maintained while enabling a rapid flux of O_2 to occur to support the high respiratory demands of N_2 reduction to ammonia. The nodules are also a large sink for photosynthetic products required to fuel the N_2 fixation process (Gordon et al., 1985, 1987). Sucrose is metabolized initially by sucrose synthase (and to a lesser extent alkaline invertase) and thereafter via glycolysis to phosphoenolpyruvate (PEP) (Gordon, 1992; Vance and Heichel, 1991) (Figure 11.2). The major fate of PEP is carboxylation by PEP carboxylase to produce oxaloacetate which is rapidly reduced to malate by malate dehydrogenase. The products of sucrose metabolism have two major functions in nodules. The first is to provide a substrate which can cross the peribacteroid membrane and be oxidized to provide the ATP and reducing power for the fixation of N_2. The second is the involvement in ammonia assimilation, synthesis of N products and their export from the nodule via the xylem.

The dicarboxylic acids malate and succinate are considered to be the most likely compounds to cross the peribacteroid and bacteroid membranes. This has been demonstrated in uptake experiments using isolated symbiosome or peribacteroid units (bacteroids with the plant derived peribacteroid membrane intact) (Day et al., 1989). In addition, nodules formed with *Rhizobium* strains in which the dicarboxylic acid transporter has been mutated are unable to fix N_2 (Streeter, 1995).

The nitrogen product exported from the bacteroids into the plant cytosol is ammonium. Recent findings suggest that NH_4^+ effluxes from the symbiosome space into the plant cytosol via a monovalent-cation-specific, passive, channel-mediated transport system (Tyerman et al., 1995). Their results suggest that NH_4^+ transport could be mediated by Ca^{2+} but that the transport system differs from other transporters permeable to NH_4^+ that have been characterized (Tyerman et al., 1995). Once in the host plant cytosol, ammonium is rapidly assimilated in the GS/GOGAT pathway. Oxaloacetate, produced in the PEP carboxylase reaction, may be transaminated by aspartate amino transferase to produce aspartate and regenerate 2-oxoglutarate for the GOGAT reaction.

These amino acids or the amides, glutamine and asparagine, are either exported directly (e.g. in pea or clover) or are used in the synthesis of ureides which are specifically exported in some species (e.g. soybean and cowpea) (Schubert, 1986).

11.3.2 Regulation of N_2 fixation by N status

The most notable example of the interaction of N compounds with N_2 fixation is that of nitrate. It is well documented that both the nodulation process itself, and N_2 fixation in established nodules, are inhibited in the presence of 5–20 mM nitrate although the presence of 1–2 mM nitrate appears to be beneficial. Two major theories have been advanced to account for inhibition of N_2 fixation in established nodules. The carbohydrate deprivation hypothesis is based on observations that photosynthate supply to nodules is reduced when NO_3^- is supplied to the plant. The nitrite toxicity hypothesis suggests that nitrite, formed in or transported to nodules may interact with components of the N_2-fixing nodules. The reader is directed to a comprehensive and extensive review of these and a number of other possibilities, by Streeter (1988) who noted, however, that despite many hundreds of publications on the subject, 'we do not seem to be closer to understanding or alleviating the inhibitions' than we were at the end of the last century.

Nodulated legume plants grown in the absence of mineral N have, none the less, a limitless supply of atmospheric N_2. The fact that such plants grow in an ordered way without wild fluctuations in the N:C ratio of their organs despite alterations in environmental conditions, strongly suggests that N_2 fixation and assimilation are carefully regulated processes attuned to the needs of the growing plant. A number of experiments have been performed which provide at least circumstantial evidence that a feedback inhibition mechanism occurs which probably involves amino acids delivered to the nodules in the phloem sap (Herdina and Silsbury, 1992; Oti-Boateng and Silsbury, 1993; Silsbury, 1987; Silsbury et al., 1986). This idea was explored in depth in a recent review

(Parsons et al., 1993) where some suggestions about the mechanism of action were made. These include interactions with the oxygen diffusion barrier, and thus the supply of oxygen for bacteroid respiration, control of carbohydrate metabolism or carbon supply to bacteroids, or interruption to the export of N products out of the nodule. However, as yet, there is no direct evidence of phloem delivered N compounds (which might reflect whole plant N status) interacting with any of these processes at any level.

Intuitively one might expect that if amino compounds were not used rapidly in growth, or if amino acids were being synthesized from other sources of N, such as NO_3^- or NH_4^+, the concentrations of amino signal molecules in phloem would increase and thus be sensed in the nodule, and perhaps would feed back into the bacteroid to reduce the amount or activity of nitrogenase. However this has not been demonstrated in (Brady) rhizobia. In contrast, however, complex transcriptional regulation occurs in free living N_2 fixing organisms such as *Klebsiella pneumoniae* and *Rhodobacter capsulatus* (Merrick, 1993).

In addition, post-transcriptional regulation has been described in *Azospirillium brasiliense* and *Rhodospirillum rubrum* where dinitrogenase reductase can be inactivated/activated by addition/removal of ADP-ribose by the enzymes dinitrogenase reductase ADP-ribosyl transferase (DRAT) and dinitrogenase reductase activating glycohydrolase (DRAG) respectively. The enzymes DRAT and DRAG are, themselves, subject to post-transcriptional regulation involving microaerobiosis and NH_4^+ (Zhang et al., 1993, 1994). Thus, in this case, nitrogenase activity can be modified by the concentrations of O_2 and NH_4^+.

11.3.3 Nitrate and ammonium assimilation

Once nitrate has been absorbed from the external medium into the cytosol of root cells it may subsequently be transported across the tonoplast into the vacuole (storage NO_3^- pool, as distinct from the cytoplasmic metabolic NO_3^- pool), translocated via the xylem to the shoot for reduction or storage, or reduced in the roots (Figure 11.2). The biochemistry and molecular genetics of NO_3^- and NH_4^+ assimilation in shoots and roots have been reviewed and discussed extensively elsewhere (see Crawford, 1995; Hoff et al., 1994; and references therein). The substrate-inducible enzyme nitrate reductase (NR) catalyzes the reduction of NO_3^- to NO_2^-, followed by nitrite reductase (NiR) reducing NO_2^- to NH_4^+ (Figure 11.2). How the flux of NO_3^- is regulated between alternative intracellular destinations is poorly understood. Absorbed NH_4^+ is, in contrast, entirely assimilated into amino acids within the roots via the GS/GOGAT pathway. Whereas it is unusual for rates of net uptake and assimilation of NO_3^- in the roots to be equal, with the possible exception of conditions of very low rates of NO_3^- supply in some species (Gojon et al., 1994), in the absence of alternative 'overflow' pathways from the cytosol, the uptake and assimilation of NH_4^+ must be closely matched in order to avoid toxic accumulation.

It is generally presumed that NO_3^- assimilation occurs predominantly in the shoots of most herbaceous species, under most environmental conditions, the energy cost in terms of carbon being much lower than that ensuing from the translocation of photosynthates to the root and their oxidation in assimilation there (Smirnoff and Stewart, 1985). The distribution of assimilatory capacity and activity between shoots and roots varies widely between species and has received considerable attention on energetic and ecological grounds (Andrews, 1986; Gojon et al., 1994). But as Gojon et al. (1994) pointed out, the absence of convenient direct methods (as opposed to in vivo, in vitro or in situ assays of NR, and reduced N/NO_3^- ratios in the xylem) for measuring actual NO_3^- reduction has hindered attempts to identify the factors determining inter-organ distribution of assimilatory activity. These authors have developed the attractive hypothesis that NO_3^- assimilation in roots predominates where either the capacity for NO_3^- uptake or the available supply are insufficient to saturate the assimilatory capacity of the roots. This is compatible with the generally accepted view that the rate of NO_3^- assimilation in the shoot is determined by the flux of NO_3^- into the xylem, upon correction for a fraction destined for storage in the vacuolar compartment of the shoot. The corollary, dependence of NO_3^- uptake rates on rates of NO_3^- assimilation in the shoots was discussed previously.

11.4 Conclusions and perspectives for future engineering

Despite the considerable progress that has been made recently in elucidating the physiological, biochemical and molecular organization of N uptake and assimilatory systems in plant roots, and the increasing contribution of approaches relying on mutant and transgenic plants (Crawford, 1995; Hoff et al., 1994; Touraine et al., 1994), the major gaps remaining in our knowledge inhibit the implementation of fully cogent programmes for genetically engineering these systems. There is also a recurring need to review precise targets and objectives to match the different and evolving, agronomic, industrial and environmental priorities. Frequent and bland statements of the desirabilty of increasing N transport and efficiency in crop species have been made without detailed consideration of either the agronomic, environmental and physiological consequences for the species concerned, or the exact sites/processes that are rate/yield-limiting under the prevailing or most probable ecological conditions.

Continuing areas of uncertainty include:

1. The molecular configuration, in situ distribution and regulation of plasma membrane-bound carriers/channels for NO_3^- and NH_4^+.
2. The extent of environmentally and genetically induced plasticity in the stoichiometry, energetics and respiratory costs of N uptake.
3. Measurement and regulation of in situ rates of NO_3^- assimilation in different plant organs.
4. The regulation of the partitioning of NO_3^- fluxes from the root cytosol between vacuolar accumulation, assimilation and xylem transport.
5. The signal pathway associated with the down-regulation of N uptake by phloem-borne amino acids.
6. Co-regulation/commonality for NO_3^-, NH_4^+ and amino acid uptake.

The uptake, translocation, assimilation, storage and remobilization components of the overall N cycle within plants have each been identified as comprising a range of potential targets for manipulation, both in terms of their structural and molecular organization, and their activity and regulation with respect to N flux and other processes. The high degree of phenotypic and metabolic plasticity inherent to plants, and an ultimate dependence on the external supply of N, make it difficult to predict the consequences of manipulation in one such component on the functioning of others in the cycle, or upon physiological performance at the whole plant level. For example, an apparently straightforward objective of increasing V_{max} for NO_3^- uptake, perhaps by doubling the number of transporters per unit root plasma membrane surface area, might have very different consequences for root growth rates, and hence for uptake of less mobile ions and even harvest-index, depending on the available mineral N. Although genetic manipulation of N carriers towards over-expression seems to be a sensible objective, it must be kept in mind that their activity is usually submitted to a complex regulation in order to match N acquisition and N demand for growth. Alternative sinks to drive uptake may be found, such as N storage which can be growth-uncoupled, and may influence positively N-use efficiency in the long term (Clarkson and Hawkesford, 1993; Imsande and Touraine, 1994).

References

ANDREWS, M. (1986) The partitioning of nitrate assimilation between root and shoot of higher plants, minireview. *Plant Cell Environ.* **9**, 511–519.

BEN-ZIONI, A., VAADIA, Y. and LIPS, S.H. (1971) Nitrate uptake by roots as regulated by nitrate reduction products in the shoot. *Physiol. Plant.* **24**, 288–290.

BRETELER, H. and SIEGERIST, M. (1984) Effect of ammonium on nitrate utilization of dwarf bean. *Plant Physiol.* **75**, 1099–1103.

BURNS, G. (1991) Short- and long-term effects of a change in the spatial distribution of nitrate in the root zone on N uptake, growth and root development of young lettuce plants. *Plant, Cell Environ.* **14**, 21–33.

BUSH, D.R. (1993) Proton-coupled sugar and amino acid transporters in plants. *Annu. Rev. Plant Physiol. Plant Mol. Biol.* **44**, 513–542.

BUSH, D.R. and LANGSTON-UNKEFER, P.J. (1988) Amino acid transport into membrane vesicles isolated from Zucchini: evidence of a proton-amino acid symport in the plasmalemma. *Plant Physiol.* **88**, 487–490.

CHAPIN, F.S. III, MOILANEN, L. and KIELLAND, K. (1993) Preferential use of organic nitrogen for growth by a non-mycorrhizal artic sedge. *Nature* **361**, 150–153.

CLARKSON, D.T. (1986) Regulation of the absorption and release of nitrate by plant cells: a review of current ideas and methodology. In: *Fundamental, Ecological and Agricultural Aspects of Nitrogen Metabolism in Higher Plants* (LAMBERS, H., NEETESON, J.J. and STULEN, I., eds), pp. 3–27. Martinus Nijhoff Publishers, Dordrecht/Boston/Lancaster.

CLARKSON, D.T. and HAWKESFORD, M.J. (1993) Molecular approaches to plant nutrition. *Plant and Soil* **155/156**, 21–31.

CRAWFORD, N.M. (1995) Nitrate: nutrient and signal for plant growth. *Plant Cell* **7**, 859–868.

DAY, D.A., PRICE, G.D. and UDVARDI, M.K. (1989) Membrane interface of the *Bradyrhizobium japonicum* - *Glycine max.* symbiosis: peribacteroid units from soybean nodules. *Austr. J. Plant Physiol.* **16**, 69–84.

DEANE-DRUMMOND, C.E. (1990) Biochemical and biophysical aspects of nitrate uptake and its regulation. In: *Nitrogen in Higher Plants* (ABROL, Y.P., ed.), pp. 1–37. Research Studies Press, Taunton, UK.

DHUGGA, K.S., WAINES, J.G. and LEONARD, R.T. (1988) Correlated

induction of nitrate uptake and membrane polypeptides in corn roots. *Plant Physiol.* **87**, 120–125.
DODDEMA, H. and TELKAMP, G.P. (1979) Uptake of nitrate by mutants of *Arabidopsis thaliana*, disturbed in uptake or reduction of nitrate. II. Kinetics. *Physiol. Plant.* **45**, 332–338.
DREW, M.C. and SAKER, L.R. (1975) Nutrient supply and the growth of the seminal root system in barley. *J. Exp. Bot.* **90**, 79–90.
ENGELS, C. and MARSCHNER, H. (1995) Plant uptake and utilization of nitrogen. In: *Nitrogen Fertilization in the Environment* (BACON, P.E., ed.), pp. 41–81. Marcel Dekker, New York.
FROMMER, B.F., HUMMEL, S. and RIESMEIER, J.W. (1993) Expression cloning in yeast of a cDNA encoding a broad specificity amino acid permease from Arabidopsis thaliana. *Proc. Natl Acad. Sci. USA* **90**, 5944–5948.
GLASS, A.D., RUTH, T.J. and RUFTY, W.R. (1990) Studies of the uptake of nitrate in Barley. II. Energetics. *Plant Physiol.* **93**, 1585–1589.
GOJON, A., PLASSARD, C. and BUSSI, C. (1994) Root/shoot distribution of NO_3^- assimilation in herbaceous and woody species. In: *A Whole Plant Perspective on Carbon-Nitrogen Interactions* (ROY, J. and GARNIER, E., eds), pp. 131–147. SPB Academic Publishing, The Hague.
GORDON, A.J. (1992) Carbon metabolism in the legume nodule. In: *Carbon Partitioning Within and Between Organisms* (POLLOCK, C.J., FARRAR, J.F. and GORDON, A.J., eds), pp. 133–162. BIOS, Oxford.
GORDON, A.J., RYLE, G.J.A., MITCHELL, D.F. and POWELL, C.E. (1985) The flux of ^{14}C-labelled photosynthate through soyabean root nodules during N_2 fixation. *J. Exp. Bot* **36**, 756–769.
GORDON, A.J., MITCHELL, D.F., RYLE, G.J.A. and POWELL, C.E. (1987) Diurnal production and utilization of photosynthate in nodulated white clover. *J. Exp. Bot.* **38**, 84–98.
GOYAL, S.S. and HUFFAKER, R.C. (1986) The uptake of NO_3^-, NO_2^- and NH_4^+ by intact wheat (*Triticum aestivum*) seedlings. *Plant Physiol.* **82**, 1051–1056.
HERDINA and SILSBURY, J.H. (1992) Nodulation and nitrogen fixation of faba bean (*Vicia faba* L) as affected by removal of the cotyledons and nitrate supply. *Ann. Bot.* **69**, 227–230.
HOFF, T., TRUONG, H.-N. and CABOCHE, M. (1994) The use of mutants and transgenic plants to study nitrate assimilation. *Plant Cell Environ.* **17**, 489–506.
HSU, L., CHIOU, T., CHEN, L. and BUSH, D.R. (1993) Cloning a plant amino acid transporter by functional complementation of a yeast amino acid transport mutant. *Proc. Natl Acad. Sci. USA* **90**, 7441–7445.
IMSANDE, J. and TOURAINE, B. (1994) N demand and the regulation of nitrate uptake. *Plant Physiol.* **105**, 3–7.
JONES, D.L. and DARRAH, P.R. (1993) Influx and efflux of amino acids from *Zea mays* L. roots and their implications for N nutrition and the rhizosphere. *Plant and Soil* **155/156**, 87–90.
KING, B.J., SIDDIQI, M.Y. and GLASS, A.D. (1992) Studies of the uptake of nitrate in barley. V. Estimation of root cytoplasmic nitrate concentration using nitrate reductase activity, implications for nitrate influx. *Plant Physiol.* **99**, 1582–1589.
LAINÉ, P., OURRY, A., MACDUFF, J.H., BOUCAUD, J. and SALETTE, J. (1993) Kinetic parameters of nitrate uptake by different catch crop species: effects of low temperatures or previous nitrate starvation. *Physiol. Plant.* **88**, 85–92.
LAINÉ, P., OURRY, A. and BOUCAUD, J. (1995) Shoot control of nitrate uptake rates by roots of *Brassica napus* L.: effects of localized nitrate supply. *Planta* **196**, 77–83.

LARSSON, C.-M. (1994) Responses of the nitrate uptake system to external nitrate availability: a whole plant perspective. In: *A Whole Plant Perspective on Carbon-Nitrogen Interactions* (ROY, J. and GARNIER, E., eds), pp. 31–45. SPB Academic Publishing, The Hague.

LEE, R.B. (1993) Control of net uptake of nutrients by regulation of influx in barley plants recovering from nutrient deficiency. *Ann. Bot.* **72**, 223–230.

LEE, R.B. and AYLING, S.M. (1993) The effect of methionine sulphoximine on the absorption of ammonium by maize and barley roots over short periods. *J. Exp. Bot.* **44**, 53–63.

LEE, R.B., PURVES, J.V., RATCLIFFE, R.G. and SAKER, L.R. (1992) Nitrogen assimilation and the control of ammonium and nitrate absorption by maize roots. *J. Exp. Bot.* **43**, 1385–1396.

MACDUFF, J.H. and JACKSON, S.B. (1992) Influx and efflux of nitrate and ammonium in Italian ryegrass and white clover roots: comparisons between effects of darkness and defoliation. *J. Exp. Bot.* **24**, 525–535.

MÄCK, G. and TISCHNER, R. (1994) Constitutive and inducible net NH_4^+ uptake of barley (*Hordeum vulgare* L.) seedlings. *J. Plant Physiol.* **144**, 351–357.

MERRICK, M.J. (1993) Organization and regulation of nitrogen fixing genes. In: *New Horizons in Nitrogen Fixation* (PALACIOS, R., MORA, J. and NEWTON, W.E., eds), pp. 43–54. Kluwer, Dordrecht/Boston/London.

MULLER, B. and TOURAINE, B. (1992) Inhibition of NO_3^- uptake by various phloem-translocated amino acids in soybean seedlings. *J. Exp. Bot.* **43**, 617–623.

NINNEMAN, O., JAUNIAUX, J.C. and FROMMER, W.B. (1994) Identification of a high affinity NH_4^+ transporter from plants. *EMBO J.* **13**, 3464–3471.

NYE, P.H. and TINKER, P.B. (1977) *Solute Movement in the Soil-Root System*. Blackwell Scientific Publications, Oxford.

OTI-BOATENG, C. and SILSBURY, J.H. (1993) The effects of exogenous amino acids on acetylene reduction activity of *Vicia faba* L. cv. Fiord. *Ann. Bot.* **71**, 71–74.

PARSONS, R., STANFORTH, A., RAVEN, J.A. and SPRENT, J.I. (1993) Nodule growth and activity may be regulated by a feedback mechanism involving phloem nitrogen. *Plant Cell Environ.* **16**, 125–136.

QUESADA, A., GALVAN, A. and FERNANDEZ, E. (1994) Identification of nitrate transporter genes in *Chlamydomonas reinhardtii*. *Plant J.* **5**, 407–419.

REDINBAUGH, M.G. and CAMPBELL, W.H. (1991) Higher plant responses to environmental nitrate. *Physiol. Plant.* **82**, 640–650.

ROBINSON, D., LINEHAM, D.J. and GORDON, D.C. (1994) Capture of nitrate from soil by wheat in relation to root length, nitrogen inflow and availability. *New Phytol.* **128**, 297–305.

RODGERS, C.O. and BARNEIX, A.J. (1988) Cultivar difference in the rate of nitrate uptake by intact wheat plants as related to growth rate. *Physiol. Plant.* **72**, 121–126.

RUIZ-CRISTIN, J. and BRISKIN, D.P. (1991) Characterization of a H^+/NO_3^- symport associated with plasma membrane vesicles of maize roots using $^{36}ClO_3^-$ as a radiotracer analog. *Arch. Biochem. Biophys.* **285**, 74–82.

SCAIFE, A. (1989) A pump/leak/buffer model for plant nitrate uptake. *Plant and Soil* **114**, 139–141.

SCHOBERT, C. and KOMOR, E. (1987) Amino acid uptake by *Ricinus communis* roots: characterization and physiological significance. *Plant Cell Environ.* **10**, 493–500.

SCHUBERT, K.R. (1986) Products of biological nitrogen fixation in higher plants: synthesis, transport and metabolism. *Annu. Rev. Plant Physiol.* **37**, 539–574.

SIDDIQI, M.Y., GLASS, A.D.M., RUTH, T.J. and RUFTY, T.W. JR. (1990) Studies of the uptake of nitrate in barley. I. Kinetics of $^{13}NO_3^-$ influx. *Plant Physiol.* **93**, 1426–1432.

SILSBURY, J.H. (1987) Nitrogenase activity in *Trifolium subterraneum* L. in relation to the uptake of nitrate ions. *Plant Physiol.* **84**, 950–953.

SILSBURY, J.H., CATCHPOOLE, D.W. and WALLACE, W. (1986) Effect of nitrate and ammonium on nitrogenase (C_2H_2 reduction) activity of subterranean clover, *Trifolium subterraneum* L. *Austr. J. Plant Physiol.* **13**, 257–273.

SMIRNOFF, N. and STEWART, G.R. (1985) Nitrate assimilation and translocation by higher plants: comparative physiology and ecological consequences. *Physiol. Plant.* **64**, 133–140.

STREETER, J. (1988) Inhibition of legume nodule formation and N_2 fixation by nitrate. *Crit. Rev. Plant Sci.* **7**, 1–23.

STREETER, J.G. (1995) Integration of plant and bacterial metabolism in nitrogen fixing systems. In: *Nitrogen Fixation: Fundamentals and Applications* (TIKHONOVICH, I.A., PROVOROV, N.A., ROMANOV, V.I. and NEWTON, W.E., eds), pp. 67–76. Kluwer Academic Publishers, Dordrecht.

TOURAINE, B., MULLER, B. and GRIGNON, C. (1992) Effect of phloem-translocated malate on NO_3^- uptake by roots of intact soybean plants. *Plant Physiol.* **93**, 1118–1123.

TOURAINE, B., CLARKSON, D.T. and MULLER, B. (1994) Regulation of nitrate uptake at the whole plant level. In: *A Whole Plant Perspective on Carbon–Nitrogen Interactions* (ROY, J. and GARNIER, E., eds), pp. 11–30. SPB Academic Publishing, The Hague.

TSAY, Y.F., SCHROEDER, J.I., FELDMANN, K.A. and CRAWFORD, N.M. (1993) The herbicide sensitivity gene CHL1 of *Arabidopsis* encodes a nitrate-inducible nitrate transporter. *Cell* **72**, 703–713.

TYERMAN, S.D., WHITEHEAD, L.F. and DAY, D.A. (1995) A channel-like transporter for NH_4^+ on the symbiotic interface of N_2-fixing plants. *Nature* **378**, 629–632.

UNKLES, S.E., HAWKER, K.L., GRIEVE, C., CAMPBELL, E.I., MONTAGUE, P. and KINGHORN, J.R. (1991) *cnra* encodes a nitrate transporter in *Aspergillus nidulans*. *Proc. Natl Acad. Sci. USA* **88**, 204–208.

VALE, F.R., VOLK, R.J. and JACKSON, W.A. (1988) Simultaneous influx of ammonium and potassium into maize roots: kinetics and interactions. *Planta* **173**, 424–431.

VAN DE DIJK, S.J., LANTING, L., LAMBERS, H., POSTHUMUS, F., STULEN, I. and HOHSTRA, R. (1982) Kinetics of nitrate uptake by different species from nutrient-rich and nutrient-poor habitats as affected by the nutrient supply. *Physiol. Plant.* **55**, 103–110.

VAN DER WERF, A., POORTER, H. and LAMBERS, H. (1994) Respiration as dependent on a species' inherent growth rate and on the nitrogen supply to the plant. In: *A Whole Plant Perspective on Carbon–Nitrogen Interactions* (ROY, J. and GARNIER, E., eds), pp. 91–110. SPB Academic Publishing, The Hague.

VANCE, C.P. and HEICHEL, G.H. (1991) Carbon in N_2 fixation: limitation or exquisite adaptation. *Annu. Rev. Plant Physiol. Plant Mol. Biol.* **42**, 373–392.

WANG, M.Y., SIDDIQI, M.Y., RUTH, T.J. and GLASS, A.D.M. (1993) Ammonium uptake by rice roots. I. Fluxes and subcellular distribution of $^{13}NH_4^+$. *Plant Physiol.* **103**, 1249–1258.

ZHANG, Y., BURRIS, R.H., LUDDEN, P.W. and ROBERTS, G.P. (1993) Post-translational regulation of nitrogenase activity by anaerobiosis and ammonium in *Azospirillum brasilense*. *J. Bacteriol.* **175**, 6781–6788.

ZHANG, Y., BURRIS, R.H., LUDDEN, P.W. and ROBERTS, G.P. (1994) Post-translational regulation of nitrogenase activity in *Azospirillum brasilense* htr BC mutants, ammonium and anaerobic switch-off occurs through independent signal transduction pathways. *J. Bacteriol.* **176**, 5780–5787.

12

Probing the carbon and nitrogen interaction: a whole plant perspective

THOMAS W. RUFTY

12.1 Introduction

Carbon and nitrogen are two of the dominant nutrients regulating plant growth. They are integral components in virtually all essential metabolic functions. Where should a discussion of carbon and nitrogen interactions begin? Perhaps the most appropriate conceptual framework comes from 'Interdependence Theory'. Through the work of a number of research groups, it is known that shoot and root growth are co-ordinated (Brouwer, 1962; Brouwer and De Wit, 1969; Ingestad, 1979; Ingestad and Lund, 1979; Raper et al., 1978). For a given growth environment and developmental stage, there will be a characteristic shoot and root dry matter distribution ratio. If the shoot and root dry matter ratio is disrupted, for example by excision of a major portion of either plant part, then disproportionate regrowth occurs until the characteristic shoot and root distribution pattern is reestablished. Further, if the growth environment is altered, such as by altering nutrient availability, the shoot to root growth ratio (S/R) adjusts to a new characteristic set point.

It has been proposed that the physiological basis for shoot and root co-ordination is the 'functional equilibrium' between processes controlling acquisition of carbon and nitrogen (Brouwer, 1963). Carbon is acquired primarily by the shoot, and carbohydrate provision to the root controls the rate of root growth and the capacity for N uptake; N is acquired primarily by the root, and supply of N to the shoot determines carbon fixation by controlling the rate of shoot growth and the associated expansion of photosynthetic capacity. Interdependence theory also assumes that a hierarchy of carbon and nitrogen utilization exists. In low light when photosynthetic rates are low, a greater proportion of the acquired carbon is utilized for sustaining shoot growth, less is supplied to the root, and a higher S/R growth ratio results. When limited amounts of nitrogen are available to the plant, a larger proportion of absorbed nitrogen is retained by the root and less is transported to the shoot. With slower shoot growth, a larger proportion of available carbohydrate 'spills-over' to the root, and the S/R growth ratio declines.

The interdependence between carbon and nitrogen processes is a logical way to explain the empirical observations on shoot and root growth co-ordination. It has become a central component in models characterizing whole plant growth (Thornley, 1976; Wann and Raper, 1979) and, to borrow from Lambers (1983), a virtual paradigm in whole plant physiology. For the purposes here, it provides the conceptual background for an examination of particular aspects of whole plant carbon and nitrogen interactions. Discussion will be centred primarily on one phase of the interdependence cycle, the role of carbon molecules in regulating processes involved in the provision of reduced N to shoot growth centres when nitrate is the primary nitrogen source.

12.2 The whole plant nitrate assimilation pathway

Interdependence Theory dictates that the rate of nitrogen uptake and delivery to the shoot determines the rate of shoot growth. There is much experimental evidence to support such a relationship. Generally, nitrogen acquisition is closely aligned with shoot growth rates (e.g. Raper *et al.*, 1978), and when the external supply of nitrogen is altered, growth of the shoot is the first plant growth response observed (Brouwer and de Wit, 1969; Greenwood, 1976; Morton and Watson, 1948; Tolley-Henry and Raper, 1986). The relationship between the spatially separate events, nitrate uptake into the root system and growth originating primarily at the shoot apical meristem, involves many events that are subject to regulation.

A schematic depicting key events in the nitrate assimilation pathway is shown in Figure 12.1. Mobile ions such as nitrate are apparently absorbed into cells at the epidermis or outer cortex at the root periphery (Bange, 1973; Kochian and Lucas, 1983). Following uptake into the root symplasm, the absorbed nitrate is subject to a number of fates as it moves inward to the vascular tissues. It can be reduced by nitrate reductase (NR), stored in vacuoles, or translocated out of the root symplasm across plasmalemmae of xylem parenchyma cells into the xylem. In most crop plants beyond the seedling stage, the proportion of the nitrate reduced or stored in the root is relatively small ($< 20\%$), so the majority of the absorbed nitrate is translocated. The amino acids generated by nitrate reduction can be incorporated into protein in the root, stored or translocated. In the xylem, the nitrate and amino acids are transported upwards with the transpiration stream, primarily to source leaves. There, the nitrate can be stored or reduced, but not readily translocated further as evidenced by its extremely low concentration in the phloem. The amino acid products of nitrate reduction, instead, provide the source of nitrogen supplied in the phloem to the shoot growth centres, which are the shoot meristem and developing leaves. The end-products of the assimilation process, proteins and nucleic acids, are formed at the growth centres. Portions of the source leaf amino acid pool can be transported back down to the root, supplying reduced nitrogen to root growth centres, or be cycled back from the root to the shoot in the xylem.

This view of the nitrate assimilation process is, of course, simplified. Some of the transpiration stream and the nitrate in it, for example, may flow directly to sink areas and nitrate reduction can occur there. Also, some of the amino acids

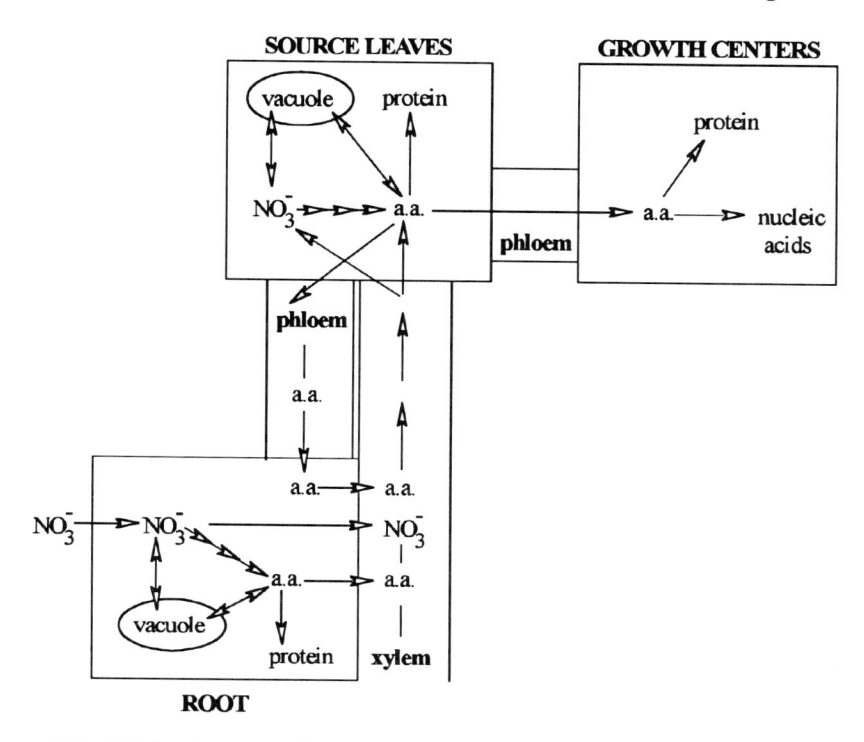

Figure 12.1 Whole plant assimilation of nitrate. The pathway begins with nitrate uptake into cells of the root and ends with protein and nucleic acid formation in cells at growth centres

generated by nitrate reduction in the source leaves can be incorporated into macromolecules in those cells, and some amino acids undoubtedly are released during protein degradation in source leaves and end up in sink leaves. It would be expected, however, that the majority of the reduced nitrogen supplied to the shoot growth regions is the result of flow directly through the series of events shown. The primary evidence for that contention is the limited size of reserves in the plant and the efficiency of the whole plant pathway. In vegetative plants, nitrate and amino acid reserves account for about 15–20% of the whole plant nitrogen content (Rufty *et al.*, 1984a,b). When growth is proceeding at rapid rates (RGRs in the 0.10 to 0.15 range are not unusual), the reserves can meet the plant demand for nitrogen only for 1–2 days. Also, nitrate and amino acids in storage pools throughout the plant are largely bypassed when uptake into the plant is sustained (Rufty *et al.*, 1984a), and they are released relatively slowly when uptake into the plant declines (Chapin, 1991). Furthermore, experiments with ^{15}N nitrate have shown that within 12–24 hours after nitrate is absorbed, the ^{15}N label reflects the spatial distribution pattern characteristic to that of the plant, and 80 to 85% of the ^{15}N is incorporated into the protein fraction, indicating rapid assimilation of nitrate into end-product macromolecules (Rufty *et al.*, 1984a).

Obviously, there are many points of regulation along the whole-plant assimilation pathway where carbohydrate or other carbon molecules can exert control.

From previous reasoning, however, one must conclude that root transport processes are key points of regulation. A sustained supply of reduced-N to shoot growth centres depends on continuous nitrate uptake by the root. In general terms, nitrate uptake is determined by two factors: root growth and the expansion of absorption surface; and uptake per unit of existing root. The provision of carbon to the root could affect uptake of nitrate by altering either process.

12.2.1 Root growth

An important assumption of Interdependence Theory is that growth of the root system is primarily controlled by carbohydrate being supplied from the shoot. This is logical, since carbohydrate pools in the root system are relatively small. Soluble carbohydrate and starch pools together seldom comprise more than 5% of the root dry weight. Furthermore, a large majority of the carbohydrate reserves are relatively unavailable for metabolism, evidently being sequestered in vacuolar storage pools or, in the case of starch, being slowly degraded (Farrar and Williams, 1991; Jackson et al., 1980). Thus, root metabolism and growth are dependent on a sustained flux of carbohydrate from the shoot.

Beyond being a source of energy and carbon skeletons, carbohydrate arriving in the root may directly control growth processes. Specific mechanisms have been considered by Farrar et al. (Farrar, 1992; Farrar and Williams, 1991). They suggest that sucrose could have an immediate impact on cell expansion by acting as an osmoticum. High accumulations of sucrose would be avoided over time, as the sucrose would be released from vacuoles and metabolized. The sucrose would be replaced by nutrient ions being absorbed by the root and organic anions to maintain cell turgor. Farrar et al. also suggest that sucrose could directly regulate the rate of cell division (van't Hof et al., 1973) as well as gene expression of enzymes involved in sucrose degradation (Koch et al., 1992).

One concern with the sucrose control hypothesis has been the presence of non-phosphorylating alternative respiratory pathway activity in the root. In many plants, the alternative pathway is a large portion of total respiration (Lambers et al., 1991). Initially, it was thought that the alternative pathway was engaged only when the phosphorylating cytochrome pathway was saturated, and its presence indicated an energy excess or overflow. Substantial alternative pathway activity at the root tip (Lambers and Posthumus, 1980) would be inconsistent with a carbohydrate limitation or regulation of root growth (Lambers, 1983). However, source–sink experiments have shown that sucrose flux to the root and root growth can increase while substantial activity of the alternative pathway is present and remains unchanged (Farrar and Williams, 1991). Also, recent evidence indicates that electrons can flow through both the cytochrome and the alternative pathways even when the cytochrome pathway is not saturated (Atkin et al., 1995; Wagner and Krab, 1995). In that case, alternative pathway activity is not an indicator of excess energy availability, and its presence would not necessarily be at odds with sucrose control over root growth processes. With the current state of knowledge, the sucrose hypothesis appears credible.

Even if sucrose or one of its metabolites is a direct effector of root extension, however, it should not be assumed that increasing the rate of sucrose delivery to the root, alone, would substantially alter the nutrient ion absorption surface. Experiments with isolated nutritional zones have demonstrated that initiation and growth of lateral roots is strongly influenced by the external presence of nitrogen and phosphate (Burns, 1991; Drew et al., 1973; Granato and Raper, 1989). This is an important observation because fine lateral roots make up the great majority of total root surface area (Dittmer, 1937) and are the primary sites of nutrient and water absorption. While the mechanism(s) controlling the formation of lateral roots has not been resolved at this time, the dependence on external nitrogen and phosphate, and probably their absorption, implies nutritional effects on gene expression. This would explain why lateral root development is minimized in conditions of low N fertility (Drew, 1975), even though low N is often associated with increased carbohydrate accumulation in the root tissue (Rufty et al., 1990; Talouizte et al., 1984a,b).

12.2.2 Nitrogen transport

A substantial amount of evidence indicates that nitrate uptake by existing root is dependent on photosynthesis and carbohydrate supplied from the shoot. Nitrate uptake is lower when plants are exposed to low light or low aerial carbon dioxide concentrations, and during the dark phase of the diurnal cycle (Clement et al., 1978; Hansen, 1980; Pace et al., 1990; Pearson and Steer, 1977; Rideout and Raper, 1994; Rufty et al., 1989). Also, nitrate uptake rapidly decreases following shoot excision, stem ringing, and, in young plants, removal of the endosperm (see references in Jackson et al., 1980), apparently reflecting the limited availability of carbohydrate reserves in the root.

It is logical that nitrate uptake would have a carbohydrate or 'energy' dependence. Nitrate uptake into the root involves an inducible transport system, requiring protein synthesis (Crawford, 1995; Jackson et al., 1973). Following induction, nitrate uptake conforms to Michaelis–Menten kinetics, with a saturable phase at external nitrate concentrations below 1 mM and a linear phase at higher concentrations (Siddiqi et al., 1990). Both phases are associated with H^+ cotransport (Glass et al., 1992; McClure et al., 1990) and dependent on concurrent metabolism, albeit to differing extents (Glass et al., 1990). At least for the saturable phase, nitrate is actively absorbed against a strong electrochemical potential gradient (Clarkson, 1986; Glass, 1988).

In considering control of nitrogen transport into the root, one must be aware of regulation of translocation out of the root symplasm across xylem parenchyma cell membranes. This results in the release of nitrogen molecules into the apoplast of the stele, where molecules move through pits in the walls of the mature xylem vessels and are translocated to the shoot in the transpiration stream (Lauchli, 1976). Translocation into the xylem is regulated separately from uptake into the symplasm, but it too is energy dependent and in most cases more sensitive to metabolic disruption than uptake (de Boer et al., 1983; Pitman, 1977; Touraine and Grignon, 1982). The sensitivity of xylem translocation to carbohydrate availability has potential consequences for the uptake process. Nitrate influx is subject to feedback inhibition, apparently initiated by excess root accumulation of nitrate itself or the amino acid end-products of its assimilation (Clarkson, 1986; Imsande and Touraine, 1994; Siddiqi et al., 1990). Feedback inhibition results in a marked lowering of the capacity (V_{max}) of the inducible, high-affinity uptake system. In situations where xylem translocation is strongly restricted, it could cause a back-up of the regulatory molecules and lead to down-regulation of the uptake process.

Although transport of nitrate into the root is dependent on carbohydrate supplied from the shoot, does carbohydrate actually control the nitrate influx rate? The most likely circumstance is during the dark phase of the diurnal cycle when carbohydrate reserves become depleted. This also is a situation where inhibition of xylem translocation may be the primary regulatory response. In growth chamber experiments with soybean using the ^{15}N-isotope, uptake was decreased by 38% in darkness but translocation was decreased by 73%, which resulted in considerable accumulation of nitrate in the root and possible engagement of the feedback control system (Rufty et al., 1987). There is no reason to suspect the response is limited to soybean. Similar effects on the two transport processes have been seen with other crop plants whenever ^{15}N was used (Ngambi et al., 1980; Pearson and Steer, 1977; Rufty et al., 1989). Furthermore, the possibility exists that the nitrate build-up in darkness may also influence net nitrate uptake by causing higher rates of nitrate efflux out of the root (Pearson et al., 1981).

A degree of caution is needed even in the interpretation of carbohydrate involvement in the restriction of translocation in darkness. As in many situations with carbon and nitrogen interactions, cause and effect are rarely straightforward because of the presence of multiple types of regulation. Nitrogen translocation into the xylem, for example, exhibits characteristics of a circadian rhythm (Rufty et al., 1989). When plants are kept in darkness for extended periods, translocation increases and decreases during the times of normal light and dark periods even though carbohydrate concentrations are stabilized at minimal levels. If translocation rhythms with potassium are analogous, the control mechanisms are located within the root system (Grossenbacher, 1938; Vaadia, 1960). However, during normal light and dark periods, the nitrate translocation rhythms are in phase with, and possibly entrained by, the rhythmic opening and closure of stomata and the movement of water through the plant. Interestingly, stomatal opening and closure themselves are controlled by a circadian rhythm (Heath, 1984; Kerr et al., 1985). The complexity of the changes occurring in the dark phase of the diurnal cycle prevents a clear understanding of the regulatory role for carbohydrate. At this time, it seems reasonable to speculate that the capacity of the nitrate uptake system is set by the carbohydrate supply, perhaps involving the feedback system, and the circadian controls are operative within that boundary.

12.2.3 Nitrate assimilation in the root

Carbohydrate availability clearly can have a strong regulatory influence on nitrate reduction in the root. In experiments where effects on both uptake and reduction are determined, reduction has consistently been found to be more sensitive (Hanisch ten Cate and Breteler, 1981; Jackson et al., 1980). Additions of glucose typically have a large stimulatory effect on reduction, while there is little effect on uptake (Aslam and Oaks, 1975; Rufty et al., 1992). Perhaps the best example of regulation by endogenous carbohydrate availability occurs when plants are nitrogen stressed. The increase in carbohydrate translocation to the root and enhanced root growth are accompanied by an increase in root nitrate reduction (Mattsson et al., 1988; Rufty et al., 1990). The exact nature of the carbohydrate stimulation of reduction is unknown. It could be the result of changes in the availability of reductant or increased synthesis of NR protein.

The energy requirements of processes are not always an indicator of the sensitivity to carbohydrate deficiency. One of the most energetically demanding processes in the nitrate assimilation pathway is the formation of protein. Yet, when carbohydrate availability is limited in the root, protein synthesis appears to be given priority. This can be seen in experiments with tobacco where plants were kept in darkness for an extended period (Rufty et al., 1989). As plants progressively became depleted of carbohydrate reserves, the reduction of ^{15}N-nitrate declined sharply but the proportion of reduced-^{15}N in the insoluble reduced fraction increased by 50%. Thus, once nitrate reduction occurred, the biosynthetic processes involved in protein synthesis operated relatively efficiently.

12.3 Rhythms in nitrate uptake and leaf development

A different context for carbohydrate regulation of root transport activities comes from studies by Raper and colleagues, who examined nitrate and ammonium uptake by vegetative soybean over periods of days (Tolley and Raper, 1985; Tolley-Henry et al., 1988). Their experiments used an automated sampling system to monitor nitrogen depletion from solution continually (Figure 12.2). It was observed that uptake fluctuated rhythmically with maximum rates occurring 3–5 days apart at two root temperatures. Uptake rates varied inversely to the expansion of newly developing trifoliolate leaves on the main stem of the shoot, which were the dominant developmental component in total canopy leaf area expansion at this growth stage. The authors argued that the results were consistent with whole plant interactions between carbohydrate and nitrogen processes. They postulated that rapid leaf expansion would be associated with utilization of a large proportion of available photosynthate, with the low amounts transported to the root supporting only low rates of nitrogen uptake. As the trifoliolate leaf approached full expansion, and the next emerging leaf was still relatively small, greater amounts of photosynthate were transported to the root, increasing the rate of nitrogen uptake. It appeared that nitrogen delivery to the shoot also played a role in the cyclic fluctuations by timing the rate of leaf initiation. This work offers an interesting picture of how carbon and nitrogen interdependence might operate temporally in conjunction with developmental events. The observations are quite similar to those from experiments with woody plants exhibiting episodic growth, where phases of rapid shoot elongation and high 'demand' for carbohydrate are associated with low rates of nitrate uptake (Hershey and Paul, 1983). The extent that other regulatory mechanisms might be involved in such interdependence responses, e.g. endogenous oscillators in the root or feedback control, has not been determined.

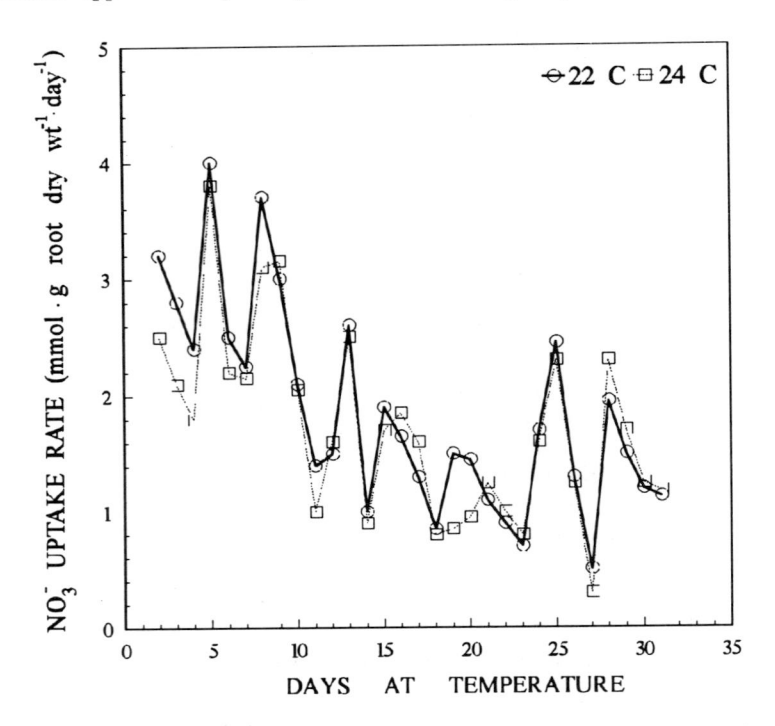

Figure 12.2 Net uptake of nitrate by roots of soybean plants grown at two solution temperatures. Uptake was determined by automated monitoring of nitrate depletion from solution (after Tolley and Raper, 1985)

12.4 Shoot control involving malate and amino acids

Photosynthate is not the only carbon molecule synthesized in the shoot that could exert control over nitrate uptake in the root. From results generated in the early experiments on ion balance in plants (Ben-Zioni et al., 1970; Dijkshoorn, 1962; Dijkshoorn et al., 1968), a theoretical model was assembled by Ben Zioni et al. (1971) which proposed a regulatory role for malate formed in the shoot (Figure 12.3). It was envisioned that, following uptake into the root, nitrate and potassium were transported as counter-ions to the shoot. Nitrate was reduced in the shoot, with the negative charge stoichiometrically transferred to malate. Since the concentration of malate in the shoot was often found to be much lower than the concentration of reduced nitrogen, a portion of the malate evidently was transported to the root with potassium again serving as the counter-ion. The malate or a product of its decarboxylation (HCO_3^- or OH^-) then exchanged with external nitrate. Importantly, this explained how anion uptake could occur independently from and exceed cation uptake. The Dijkshoorn and Ben-Zioni type model has been examined in numerous studies. The results from some appear to support it while results from others do not (e.g. Allen and Raven, 1987; Israel and Jackson, 1982; Kirkby and Armstrong, 1980; Touraine et al., 1988, 1992). An element of uncertainty about its importance remains mainly because of the problems encountered in attempting a precise evaluation in intact plants. It has

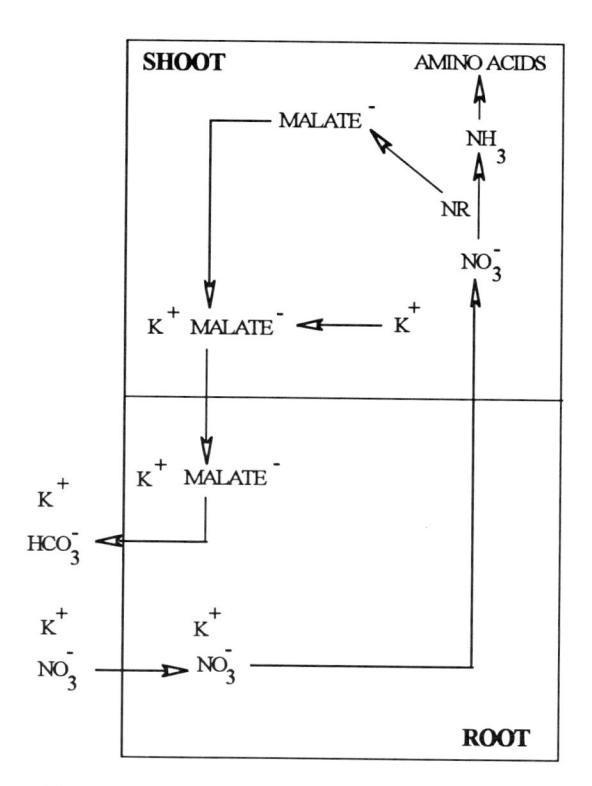

Figure 12.3 A model relating malate synthesis in leaves with nitrate uptake (after Ben-Zioni et al., 1971)

not been possible to estimate accurately the rates of nitrate assimilation in specific locations and subsequent rates of transport in the phloem. A particular complication is the occurrence of nitrate reduction in the root, which was not considered in the initial model. Bicarbonate or hydroxyl ions generated in root nitrate reduction can contribute to the driving force for nitrate uptake without the need for malate formation. Root nitrate reduction is hard to quantify even when using the ^{15}N-label (Rufty et al., 1982).

Uptake of nitrate into the root may also be regulated by amino acids synthesized in the shoot. As intermediates in the nitrate assimilation pathway, amino acids serve as an indicator of the efficiency of formation of the protein and nucleic end-products. Just as importantly, they can be readily translocated throughout the plant. This is in contrast to nitrate, which is not translocated in the phloem, and ammonium, the concentration of which must be limited in the xylem because it is a potential uncoupler of photophosphorylation (Givan, 1979). Amino acids generated by nitrate reduction in the shoot are readily transported in the phloem, and in experiments where phloem sap was enriched with amino acids using a leaf feeding technique, nitrate uptake declined (Muller and Touraine, 1992). Also, supplementing nutrient solutions with amino acids resulted in an uptake inhibition (Breteler and Arnozis, 1985; Doddema and Otten, 1979; Lee et al., 1992). Furthermore, there is a pool of amino acids continually cycling in the phloem and xylem that does not readily mix with storage pools (Cooper and Clarkson, 1989; Vessey and Layzell, 1987). Amino acids account for as much as half of the nitrogen in xylem sap, and the cycling amino acid pool is a significant portion (Cooper and Clarkson, 1989; Rufty and Volk, 1986). This raises the interesting possibility that the cycling amino acids function as a monitoring system for protein synthesis efficiency in the whole plant. It is unclear how amino acids inhibit nitrate uptake, but a likely mechanism would involve feedback controls. An increased concentration of amino acids at the phloem and root symplasm interface, i.e. xylem parenchyma cells, might be sufficient to initiate repression of the nitrate uptake system.

The combination of regulatory effects exerted by malate and amino acids provides a mechanism linking nitrate assimilation in the shoot specifically with the uptake of nitrate into the plant. When resources are optimal and shoot nitrate reduction proceeds at a rapid rate, malate transport to the root and decarboxylation can stimulate the nitrate uptake process. Conversely, when conditions are less favourable and protein synthesis slows, increased amino acid cycling to the root can signal down-regulation of the uptake system. Nitrate assimilation in the shoot is, of course, closely associated with photosynthetic activity, because of the requirements for reductant, ATP, and carbon skeletons. However, the regulatory alternative offered by malate and amino acids would allow carbon based control of root nitrate transport without large scale changes in plant energy status.

12.5 Nitrate assimilation in the shoot

Interdependence Theory, as generally used, does not assign a high degree of regulatory importance to carbon control over nitrate assimilation events in the shoot. Carbon is assumed to regulate root growth and transport, but not activities in the leaf canopy. This probably stems from evidence indicating a close association between nitrogen provision to the shoot and leaf area expansion (e.g. Rufty *et al.*, 1984b; Tolley-Henry and Raper, 1986), which implies that nitrogen is the limiting factor for leaf development and carbon supply is not. There are, nonetheless, intensive demands on the carbon supply and energy systems of leaves during the nitrate assimilation process. How is it that a high degree of assimilation efficiency is maintained?

12.5.1 *Nitrate delivery in the xylem*

A crucial factor in the efficient assimilation of nitrate in leaf tissues is the timing of the delivery of nitrate substrate. Although nitrate can be reduced in leaves in darkness (Aslam and Huffaker, 1982, 1984), conditions are optimal for rapid assimilation only in the light. Perhaps not coincidentally, the rate of nitrate translocation to leaves is markedly different in the light and dark phases of the diurnal cycle. The restriction of transport out of the root in the dark results in temporary compartmentation of nitrate substrate within the root. The nitrate is released in the following light period and transported to leaves (Rufty *et al.*, 1984a, 1987). As indicated in a ^{15}N experiment with soybean (Figure 12.4),

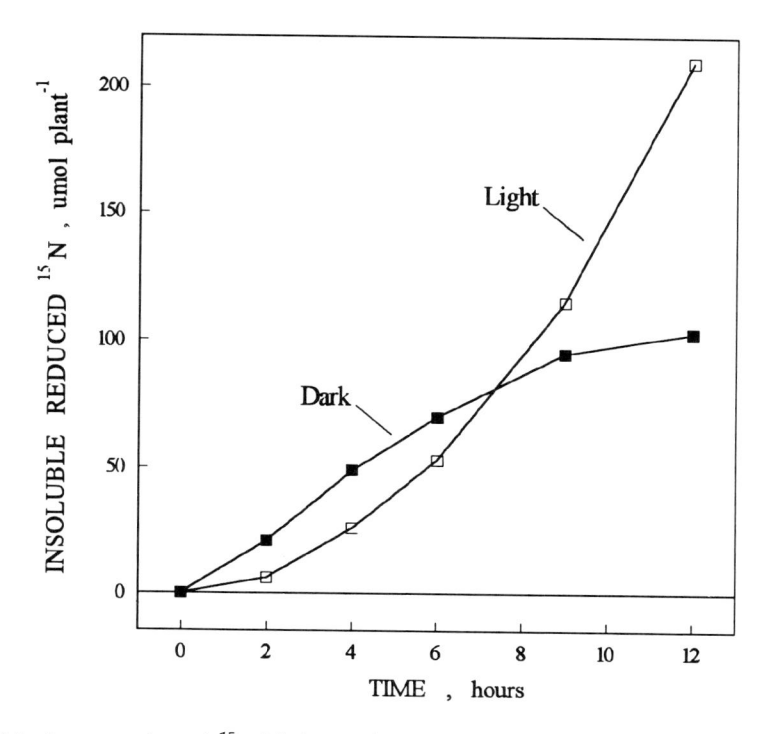

Figure 12.4 Incorporation of ^{15}N label into the protein fraction in soybean plants exposed to ^{15}N-nitrate for 12 hours in the previous dark period or during the sample period in the light

nitrate absorbed in the prior dark period is the main source of N for protein synthesis in the first half of the light period. After that time, nitrate being absorbed by the root in the light becomes the primary source. Because of root compartmentation and the delay in the delivery of nitrate, a large majority of whole plant nitrate assimilation occurs in leaves in the light when the biochemical environment is energetically favourable.

Physiological studies generally support the notion that the main determinant of the nitrate assimilation rate in leaves in the light is the rate of nitrate delivery in the xylem. The metabolic pool of nitrate in leaves is evidently small, because the activity of nitrate reductase, an inducible enzyme, rapidly responds to changes in nitrate flux in the xylem (Kannangara and Woolhouse, 1967; Shaner and Boyer, 1976). Considerable amounts of nitrate can be present in vacuoles, but it is largely bypassed (Rufty *et al.*, 1984a) and only slowly available for reduction when translocation slows (Shaner and Boyer, 1976).

12.5.2 *The role of nitrate reductase*

From a biochemical viewpoint, NR has been considered the rate limiting step in the assimilation of nitrate in leaf tissues (Beavers and Hageman, 1980). This was based primarily on two observations. One is that the extractable activity of NR is much lower than the activity of enzymes further along the pathway, i.e. nitrite reductase, glutamate synthase, and glutamine synthetase. The other is that nitrate is the only substrate in the assimilation sequence to accumulate, as nitrite and ammonium levels in leaf tissues typically are very low. Nonetheless, there was also the recognition that the activity of NR appeared to be greater than the rate of nitrate assimilation occurring *in vivo* in experiments where the two were compared (Beavers and Hageman, 1980; Huffaker and Rains, 1978). The suggestion that NR may be present in large excess has been confirmed in recent studies with mutants and transgenic plants with altered expression of NR. Results clearly show that nitrate assimilation and growth are adversely effected only when NR activity is very low compared to the wild type (Dorbe *et al.*, 1992; Warner and Kleinhofs, 1981). Furthermore, nitrate assimilation and growth are not increased in plants with over-expression of NR (Foyer *et al.*, 1994; Quillere *et al.*, 1994). Based on the evidence available at present, it does not appear that NR activity is a major point of resistance for nitrate assimilation in leaves, at least in the light when most reduction occurs. A small proportion of the nitrate arriving in leaves in the light does accumulate unreduced, but that can be explained by competition between tonoplast membranes and NR for the nitrate entering cell cytosols.

The activity of NR could limit leaf nitrate assimilation in some circumstances. The most obvious possibility is during dark periods. It is known that nitrate reduction can occur in darkness and is strongly influenced by the carbohydrate status of the leaf tissue (Aslam and Huffaker, 1982). However, experiments with ^{15}N show that a larger proportion of the nitrate arriving in leaves in darkness accumulates unreduced compared to that in the light (Rufty *et al.*, 1984a), and NR is down-regulated in darkness. Synthesis of NR appears to be light dependent (Crawford, 1995), and NR catalytic activity can decrease due to phosphorylation (Huber *et al.*, 1994; Kaiser and Huber, 1994). The implication is that low NR activity is a contributing factor in the decreased efficiency of nitrate reduction. It is also conceivable that NR may limit nitrate assimilation during periods of stress when nitrate tends to accumulate in leaves above control levels. Nitrate levels are elevated, for example, during moderate phosphorus stress (Rabe and Lovatt, 1986), and ^{15}N experiments reveal that lowered reduction efficiency occurs even in the light (Rufty *et al.*, 1993). It remains to be determined whether modification of NR activity is responsible for such effects or whether they result from other factors such as decreased availability of reductant.

12.5.3 *Partitioning of carbohydrate*

Interdependence Theory does not specifically address the role of metabolic regulation. One must wonder how partitioning of fixed carbon between nitrogen and carbohydrate pathways is regulated. It is tempting to consider the simplest type of control, that partitioning of fixed carbon into the nitrogen pathway in the light is regulated solely by the availability of nitrogen. Diversion of carbon into the nitrogen pathway could occur whenever there is reduced nitrogen substrate (NH_4^+) present for amino acid synthesis in the chloroplast. The alternative possibility, that partitioning between the pathways might be regulated mainly by co-ordinated interactions between NR and sucrose biosynthesis enzymes, is currently being examined intensively (Huber et al., 1994).

It is also plausible that the availability of reduced nitrogen in the shoot controls partitioning of carbohydrate between the shoot and root. Interdependence Theory has the implicit assumption that sinks closest to the photosynthetic source have priority for carbohydrate, a notion which is supported by evidence in the literature (Cook and Evans, 1976). From previous arguments, it would seem reasonable that the rate of delivery of amino acids to shoot growth centres will determine the capacity for carbohydrate utilization. If a high rate is maintained, the requirements for energy and carbon skeletons at the meristems and in expanding cells will be high. On the other hand, if nitrogen availability at growth centres is lowered, it has been shown experimentally that carbohydrate rapidly accumulates in sink and source leaves (Rufty et al., 1988) and greater amounts can be transported to the root.

12.6 Summary: the integrated plant response

The arguments presented here generally support the basic tenets of Interdependence Theory. It does appear that carbon generated in the shoot and transported to the root can control nitrate uptake by the root. The evidence suggests, however, that it does so primarily by altering root growth and the nutrient absorption surface. Although root growth is driven by carbohydrate flux, morphology (branching) can be influenced by the presence of nitrate, along with phosphate, in the external media. The main control mechanism for nitrate uptake per unit of root evidently is the feedback control system, which does not appear to be regulated directly by carbohydrate availability. However, it is possible that a carbohydrate limitation may initiate feedback effects and indirectly lower uptake rates in darkness, when translocation into the xylem is severely inhibited and nitrate accumulates in the root. An important part of the feedback system may be amino acids accumulating in the root or the amino acid pool cycling between the shoot and root. If this interpretation of the regulatory components is correct, then growth acts as 'coarse' control, determining the capacity for nitrate acquisition, while the feedback system functions as 'fine' control, probably responding to demand signals within the plant.

In accord with Interdependence Theory, the flux of nitrogen to the shoot does appear to control carbon acquisition. But again, regulation appears to involve mainly the growth component, the initiation and expansion of new leaves. The close relationship between nitrogen flux and shoot growth exists because of the high degree of efficiency of nitrate assimilation in leaves in the light and the associated provision of reduced-N to shoot growth centres. Although not dealt with here in depth, the controls for carbon acquisition are somewhat analogous with the controls over nitrate acquisition by the root. There is a coarse regulatory component, growth, and a fine regulatory component, photosynthetic rate (per unit leaf area). Photosynthetic rate also is regulated by feedback controls, which appear responsive to the utilization of carbon end-products (Sharkey et al., 1994). Other than at times of nitrogen stress, nitrogen availability influences photosynthetic rate only indirectly, by affecting the activity of growth centres and the related demand for carbohydrate.

References

ALLEN, S. and RAVEN, J.A. (1987) Intracellular pH regulation in Ricinus communis grown with ammonium or nitrate as N source: the role of long distance transport. *J. Exp. Bot.* **38**, 580–596.

ASLAM, M. and HUFFAKER, R.C. (1982) In vivo nitrate reduction in roots and shoots of barley seedling in light and darkness. *Plant Physiol.* **70**, 1009–1013.

ASLAM, M. and HUFFAKER, R.C. (1984) Dependency of nitrate reduction on soluble carbohydrates in primary leaves of barley under aerobic conditions. *Plant Physiol.* **75**, 623–628.

ASLAM, M. and OAKS, A. (1975) Effect of glucose on the induction of nitrate reductase in corn roots. *Plant Physiol.* **56**, 634–639.

ATKIN, O.K., VILLAR, R. and LAMBERS, H. (1995) Partitioning of electrons beween the cytochrome and alternative pathways in intact roots. *Plant Physiol.* **108**, 1179–1183.

BANGE, G.G.J. (1973) Diffusion and absorption of ions in plant tissue. III. The role of the root cortex cells in ion absorption. *Acta Bot. Neerland.* **22**, 529–542.

BEAVERS, L. and HAGEMAN, R.H. (1980) Nitrate and nitrite reduction. In: *The Biochemistry of Plants*, Vol. 5 (MIFLIN, B.J., ed.), pp. 115–168. Academic Press, New York.

BEN-ZIONI, A., VAADIA, Y. and LIPS, S.H. (1970) Correlations between nitrate reduction, protein synthesis, and malate accumulation. *Physiol. Plant.* **23**, 1039–1047.

BEN-ZIONI, A., VAADIA, Y. and LIPS, S.H. (1971) Nitrate uptake by roots as regulated by nitrate reduction products of the shoot. *Physiol. Plant.* **24**, 288–290.

BRETELER, H. and ARNOZIS, P.A. (1985) Effect of amino compounds on nitrate utilization by roots of dwarf bean. *Phytochemistry* **24**, 653–658.

BROUWER, R. (1962) Nutritive influences on the distribution of dry matter in the plant. *Neth. J. Agric. Sci.* **10**, 399–408.

BROUWER, R. (1963) Some aspects of the equilibrium between overground and underground plant parts. Wageningen, Instituut voor Biologisch en Scheikundig Ondersoek van Landbouwgewassen, Jaarboek, pp. 31–39.

BROUWER, R. and DE WIT, C.T. (1969) A simulation model of plant growth with special attention to root growth and its consequences. In: *Root Growth* (WHITTINGTON, W.J., ed.), pp. 224–242. Butterworths, London.

BURNS, J.G. (1991) Short- and long-term effects of a change in the spatial distribution of nitrate in the root zone on N uptake, growth and root development of young lettuce plants. *Plant, Cell Environ.* **14**, 21–33.

CHAPIN, F.S. III (1991) Integrated responses of plants to stress. *BioScience* **41**, 29–36.

CLARKSON, D.T. (1986) Regulation of the absorption and release of nitrate by plant cells: a review of current ideas and methodology. In: *Fundamental, Ecological and Agricultural Aspects of Nitrogen Metabolism in Higher Plants* (LAMBERS, H., NEETESON, J.J. and STULEN, I., eds), pp. 3–27. Martinus Nijhoff, Dordrecht.

CLEMENT, C.R., HOPPER, M.J., JONES, L.H.P. and LEAFE, E.L. (1978) The uptake of nitrate by Lolium perenne from flowing solution culture. II. Effect of light, defoliation, and relationship to CO_2 flux. *J. Exp. Bot.* **29**, 1173–1183.

COOK, M.G. and EVANS, L.T. (1976) Effect of sink size, geometry and distance from the source on the distribution of assimilates in wheat. In: *Transport and Transfer Processes in Plants* (WARDLAW, I.F. and PASSIOURA, J.B., eds), pp. 393–400. Academic Press, New York.

COOPER, H.D. and CLARKSON, D.T. (1989) Cycling of amino-nitrogen and other nutrients between shoots and roots in cereals. A possible mechanism integrating shoot and root in the regulation of nutrient uptake. *J. Exp. Bot.* **40**, 753–762.

CRAWFORD, N.M. (1995) Nitrate: nutrient and signal for plant growth. *Plant Cell* **7**, 859–868.
DE BOER, A.H., PRINS, H.B.A. and ZANSTRA, P.E. (1983) Bi-phasic composition of transroot electrical potential in roots of Plantago species: involvement of spatially separated electrogenic pumps. *Planta* **157**, 259–266.
DIJKSHOORN, W. (1962) Metabolic regulation of the alkaline effect of nitrate utilization in plants. *Nature* **194**, 165–167.
DIJKSHOORN, W., LATHWELL, D.J. and DE WIT, C.T. (1968) Temporal changes in carboxylate content of ryegrass with stepwise change in nutrition. *Plant and Soil* **29**, 369–390.
DITTMER, H.J. (1937) A quantitative study of the roots and root hairs of a winter rye plant (Secale cereale). *Am. J. Bot.* **24**, 417–420.
DODDEMA, H. and OTTEN, H. (1979) Uptake of nitrate by mutants of Arabidopsis thaliana, disturbed in uptake or reduction of nitrate. III. Regulation. *Physiol. Plant.* **45**, 339–346.
DORBE, M.F., CABOCHE, M. and DANIEL-VEDELE, F. (1992) The tomato nia gene complements a Nicotiana plumbaginifolia nitrate reductase-deficient mutant and is properly regulated. *Plant Mol. Biol.* **18**, 363–375.
DREW, M.C. (1975) Comparison of the effects of a localized supply of phosphate, nitrate, ammonium and potassium on the growth of the seminal root system, and the shoot, in barley. *New Phytol.* **75**, 479–490.
DREW, M.C., SAKER, L.R. and ASHLEY, T.W. (1973) Nutrient supply and the growth of the seminal root system in barley. *J. Exp. Bot.* **24**, 1189–1202.
FARRAR, J.F. (1992) The whole plant: carbon partitioning during development. In: *Carbon Partitioning: Within and Between Organisms* (POLLOCK, C.J., FARRAR, J.F. and GORDON, A.J., eds), pp. 163–179. BIOS, Oxford.
FARRAR, J.F. and WILLIAMS, J.H.H. (1991) Control of the rate of respiration in roots: compartmentation, demand and the supply of substrate. In: *Compartmentation of Plant Metabolism in Non-photosynthetic Tissues* (EMES, M.J., ed.), pp. 167–188. Cambridge University Press, New York.
FOYER, C.H., LESCURE, C., LEFEBVRE, C., VINCENTZ, M. and VAUCHERET, H. (1994) Adaptations of photosynthetic electron transport, carbon assimilation and carbon partitioning in transgenic *Nicotiana plumbaginofolia* plants to changes in nitrate reductase activity. *Plant Physiol.* **104**, 171–178.
GIVAN, C.V. (1979) Metabolic detoxification of ammonia in tissues of higher plants. *Phytochemistry* **18**, 375–382.
GLASS, A.D.M. (1988) Nitrogen uptake by plant roots. *ISI Atlas of Science: Animal and Plant Sciences* **1**, 151–156.
GLASS, A.D.M., SIDDIQI, M.Y., RUTH, T.J. and RUFTY, T.W. JR (1990) Studies of the uptake of nitrate in barley. II. Energetics of $^{13}NO_3^-$ influx. *Plant Physiol.* **92**, 1585–1589.
GLASS, A.D.M., SHAFF, J.E. and KOCHIAN, L.V. (1992) Studies of the uptake of nitrate in barley. IV. Electrophysiology. *Plant Physiol.* **99**, 456–463.
GRANATO, T.C. and RAPER, C.D. JR (1989) Proliferation of maize (Zea mays L.) roots in response to localized supply of nitrate. *J. Exp. Bot.* **40**, 263–275.
GREENWOOD, E.A.N. (1976) Nitrogen stress in plants. *Adv. Agron.* **28**, 1–35.
GROSSENBACHER, K.A. (1938) Diurnal fluctuation in root pressure. *Plant Physiol.* **4**, 669–676.
HANISCH TEN CATE, C.H. and BRETELER, H. (1981) Role of sugars in nitrate utilization by roots of dwarf bean. *Physiol. Plant.* **52**, 129–135.
HANSEN, G.K. (1980) Diurnal variation of root respiration rates and nitrate uptake as influenced by nitrogen supply. *Physiol. Plant.* **48**, 421–427.
HEATH, O.V.S. (1984) Stomatal opening in darkness in the leaves of Commelina

communis, attributed to an endogenous circadian rhythm: control of phase. *Proc. R. Soc. London B* **220**, 399–414.

HERSHEY, D.R. and PAUL, J.L. (1983) Ion absorption by a woody plant with episodic growth. *Hortscience* **18**, 357–359.

HUBER, S.C., HUBER, J.L. and MCMICHAEL, R.W. (1994) Control of plant enzyme activity by reversible protein phosphorylation. *Int. Rev. Cytol.* **149**, 47–98.

HUFFAKER, R.C. and RAINS, D.W. (1978) Factors influencing nitrate acquisition by plants: assimilation and fate of reduced nitrogen. In: *Nitrogen in the Environment*, Vol. 2 (NIELSEN, D.R. and MACDONALD, J.G., eds), pp. 1–43. Academic Press, New York.

IMSANDE, J. and TOURAINE, B. (1994) N demand and the regulation of nitrate uptake. *Plant Physiol.* **105**, 3–7.

INGESTAD, T. (1979) Nitrogen stress in birch seedlings. II. N, K, P, Ca, and Mg nutrition. *Physiol. Plant.* **45**, 149–157.

INGESTAD, T. and LUND, A.B. (1979) Nitrogen stress in birch seedlings. I. Growth technique and growth. *Physiol. Plant.* **45**, 137–148.

ISRAEL, D.W. and JACKSON, W.A. (1982) Ion balance, uptake and transport processes in N_2-fixing and nitrate and urea-dependent soybean plants. *Plant Physiol.* **69**, 171–178.

JACKSON, W.A., FLESHER, D. and HAGEMAN, R.H. (1973) Nitrate uptake by dark-growth corn seedlings. *Plant Physiol.* **51**, 120–127.

JACKSON, W.A., VOLK, R.J. and ISRAEL, D.W. (1980) Enery supply and nitrate assimilation in root systems. In: *Carbon–Nitrogen Interaction in Crop Production* (TANAKA, A., ed.), pp. 25–40. Japan Society for Promotion of Science, Tokyo.

KAISER, W.M. and HUBER, S.C. (1994) Post translational regulation of nitrate reductase in high plants. *Plant Physiol.* **106**, 817–820.

KANNANGARA, C.G. and WOOLHOUSE, H.W. (1967) The role of carbon dioxide, light, and nitrate in the synthesis and degradation of nitrate reductase in leaves of Perilla frutescens. *New Phytol.* **66**, 553–561.

KERR, P.S., RUFTY, T.W. JR and HUBER, S.C. (1985) Endogenous rhythms in photosynthesis, sucrose phosphate synthase activity, and stomatal resistance in leaves of soybean. *Plant Physiol.* **77**, 275–280.

KIRKBY, E.A. and ARMSTRONG, M.J. (1980) Nitrate uptake by roots as regulated by nitrate assimilation in the shoot of caster oil plants. *Plant Physiol.* **65**, 286–290.

KOCH, K.E., NOLTE, K.D., DUKE, E.R., MCCARTY, D.R. and AVIGNE, W.T. (1992) Sugar levels modulate differential expression of maize sucrose synthase genes. *Plant Cell* **4**, 59–69.

KOCHIAN, L.V. and LUCAS, W.J. (1983) Potassium transport in corn roots. II. The significance of the root periphery. *Plant Physiol.* **73**, 208–215.

LAMBERS, H. (1983) 'The functional equilibrium', nibbling on the edges of a paradigm. *Neth. J. Agric. Sci.* **31**, 305–311.

LAMBERS, H. and POSTHUMUS, F. (1980) The effect of light intensity and relative humidity on growth and root respiration of Plantago lanceolata and Zea mays. *J. Exp. Bot.* **31**, 1621–1630.

LAMBERS, H., VAN DER WERF, A. and KONINGS, H. (1991) Respiratory patterns in roots in relation to their functioning. In: *Plant Roots: The Hidden Half* (WAISEL, Y., ESHEL, A. and KAFKAFI, U., eds), pp. 229–263. Marcel Dekker, New York.

LAUCHLI, A. (1976) Symplastic transport and ion release to the xylem. In: *Transport and Transfer Processes in Plants* (WARDLAW, I.F. and PASSIOURA, J.B., eds), pp. 101–112. Academic Press, New York.

LEE, R.B., PURVES, J.V., RATCLIFF, R.G. and SAKER, L.R. (1992) Nitrogen assimilation and the control of ammonium and nitrate absorption by maize roots. *J. Exp. Bot.* **43**, 1385–1396.

MATTSSON, M., LUNDBORG, T. and LARSSON, C.M. (1988) Nitrate utilization in barley: relations to nitrate supply and light/dark cycles. *Physiol. Plant.* **73**, 380–386.

MCCLURE, P.R., KOCHIAN, L.V., SPANSWICK, R.M. and SHAFF, J.E. (1990) Evidence for cotransport of nitrate and protons in maize roots. I. Effects of nitrate on the membrane potential. *Plant Physiol.* **93**, 281–289.

MORTON, A.G. and WATSON, D.J. (1948) A physiological study of leaf growth. *Ann. Bot.* **12**, 281–310.

MULLER, B. and TOURAINE, B. (1992) Inhibition of NO_3^- uptake by various phloem-translocated amino acids in soybean seedlings. *J. Exp. Bot.* **43**, 617–623.

NGAMBI, J.M., CHAMPIGNY, M.L., MARIOTTI, A. and MOYSE, A. (1980) Assimilation des nitrates et photosyntese d'un Mil. Pennisetum americanum 23 DB, an cours d'un nyethemere. *Comptes Rendus de l'Academie des Sciences, Paris*, **291D**, 109–112.

PACE, G.M., VOLK, R.J. and JACKSON, W.A. (1990) Nitrate reduction in response to CO_2-limited photosynthesis. Relationship to carbohydrate supply and nitrate reductase activity in maize seedlings. *Plant Physiol.* **92**, 286–292.

PEARSON, C.J. and STEER, B.T. (1977) Daily changes in nitrate uptake and metabolism in Capsicum annuum. *Planta* **137**, 107–112.

PEARSON, C.J., VOLK, R.J. and JACKSON, W.A. (1981) Daily changes in nitrate influx, efflux and metabolism in maize and pearl millet. *Planta* **152**, 319–324.

PITMAN, M.G. (1977) Ion transport into the xylem. *Annu. Rev. Plant Physiol.* **28**, 71–88.

QUILLÉRÉ, I., DUFOSSÉ, C., ROUX, Y., FOYER, C.H., CABOCHE, M. and MOROT-GAUDRY, J.F. (1994) The effects of deregulation of NR gene expression on growth and nitrogen metabolism of winter-grown Nicotiana plumbaginifolia plants. *J. Exp. Bot.* **45**, 1205–1211.

RABE, E. and LOVATT, C.J. (1986) Increased arginine biosynthesis during phosphorus deficiency. *Plant Physiol.* **81**, 774–779.

RAPER, C.D. JR, OSMOND, D.L. WANN, M. and WEEKS, W.W. (1978) Interdependence of root and shoot activities in determining nitrogen uptake rate of roots. *Bot. Gaz.* **139**, 289–294.

RIDEOUT, J.W. and RAPER, C.D. JR (1994) Diurnal changes in net uptake rate of nitrate are associated with changes in estimated export of carbohydrates to roots. *Int. J. Plant Sci.* **155**, 173–179.

RUFTY, T.W. JR and VOLK, R.J. (1986) Alterations in enrichment of NO_3^- and reduced-N in xylem exudate during and after extended plant exposure to $^{15}NO_3^-$. *Plant and Soil* **91**, 329–332.

RUFTY, T.W. JR, VOLK, R.J., MCCLURE, P.R., ISRAEL, D.W. and RAPER, C.D. (1982) Relative content of nitrate and reduced-N in xylem exudate as an indicator of root reduction of concurrently absorbed ^{15}N-nitrate. *Plant Physiol.* **69**, 166–170.

RUFTY, T.W. JR, ISRAEL, D.W. and VOLK, R.J. (1984a) Assimilation of $^{15}NO_3^-$ taken up by plants in the light and in the dark. *Plant Physiol.* **76**, 769–775.

RUFTY, T.W. JR, RAPER, C.D. and HUBER, S.C. (1984b) Alterations in internal partitioning of carbon in soybean plants in response to nitrogen stress. *Can. J. Bot.* **62**, 501–508.

RUFTY, T.W. JR, VOLK, R.J. and MACKOWN, C.T. (1987) Endogenous NO_3^- in the root as a source of substrate for reduction in the light. *Plant Physiol.* **84**, 1421–1426.

Rufty, T.W. Jr, Huber, S.C. and Volk, R.J. (1988) Alterations in leaf carbohydrate metabolism in response to nitrogen stress. *Plant Physiol.* **88**, 725–730.

Rufty, T.W. Jr, MacKown, C.T. and Volk, R.J. (1989) Effects of altered carbohydrate availability on whole-plant assimiliation of $^{15}NO_3^-$ *Plant Physiol.* **89**, 457–463.

Rufty, T.W. Jr, MacKown, C.T. and Volk, R.J. (1990) Alterations in nitrogen assimilation and partitioning in nitrogen stressed plants. *Physiol. Plant.* **79**, 85–95.

Rufty, T.W. Jr, Volk, R.J. and Glass, A.D.M. (1992) Relationship between carbohydrate availability and assimilation of nitrate. In: *Nitrogen Metabolism of Plants* (Mengel, K. and Pilbeam, D.T., eds), pp. 103–119. Oxford Science Publications, Oxford.

Rufty, T.W. Jr, Israel, D.W., Volk, R.J., Qiu, J. and Sa, T. (1993) Phosphate regulation of nitrate assimilation in soybean. *J. Exp. Bot.* **44**, 879–891.

Shaner, D.L. and Boyer, J.S. (1976) Nitrate reductase activity in maize leaves. I. Regulation by nitrate flux. *Plant Physiol.* **58**, 499–504.

Sharkey, T.D., Socias, X. and Loreto, F. (1994) CO_2 effects on photosynthetic end product synthesis and feedback. In: *Plant Responses to the Gaseous Environment* (Alscher, R.G. and Wellburn, A.R., eds), pp. 55–78. Chapman & Hall, London.

Siddiqi, M.Y., Glass, A.D.M., Ruth, T.J. and Rufty, T.W. Jr (1990) Studies of the uptake of nitrate in barley. I. Kinetics of $^{13}NO_3^-$ influx. *Plant Physiol.* **93**, 1426–1432.

Talouizte, A., Champigny, M.L., Bismuth, E. and Moyse, A. (1984a) Root carbohydrate metabolism associated with nitrate assimilation in wheat previously deprived of nitrate. *Physiol. Veg.* **22**, 19–27.

Talouizte, A., Guiraud, G., Moyse, A., Marol, C. and Champigny, M.L. (1984b) Effect of previous nitrate deprivation on ^{15}N-nitrate absorption and assimilation by wheat seedlings. *J. Plant Physiol.* **116**, 113–122.

Thornley, J.H.M. (1976) *Mathematical Models in Plant Physiology*. Academic Press, New York.

Tolley, L.C. and Raper, C.D. Jr (1985) Cyclic variations in nitrogen uptake rate in soybean plants. *Plant Physiol.* **78**, 320–322.

Tolley-Henry, L. and Raper, C.D. Jr (1986) Expansion and photosynthetic rate of leaves of soybean plants during onset and recovery from nitrogen stress. *Bot. Gaz.* **147**, 400–406.

Tolley-Henry, L., Raper, C.D. and Granato, T.C. (1988) Cyclic variations in nitrogen uptake rate of soybean plants: effects of external nitrate concentration. *J. Exp. Bot.* **39**, 613–622.

Touraine, B. and Grignon, C. (1982) Energetic coupling of nitrate section into the xylem of corn roots. *Physiol. Veg.* **20**, 33–39.

Touraine, B., Grignon, N. and Grignon, C. (1988) Charge balance in NO_3^-–fed soybean. Estimation of K and carboxylate recirculation. *Plant Physiol.* **88**, 605–612.

Touraine, B., Muller, B. and Grignon, C. (1992) Effect of phloem-translocated malate on NO_3^- uptake by roots of intact soybean plants. *Plant Physiol.* **99**, 1118–1123.

Vaadia, Y. (1960) Autonomic diurnal fluctuations in rate of exudation and root pressure of decapitated sunflower plants. *Physiol. Plant.* **13**, 701–717.

Van't Hof, J., Hoppin, D.P. and Yagi, S. (1973) Cell arrest in G1 and G2 of the mitotic cycle of vicia faba root meristems. *Am. J. Bot.* **60**, 889–895.

VESSEY, J.K. and LAYZELL, D.B. (1987) Regulation of assimilate partitioning in soybean. Initial effects following changes in nitrate supply. *Plant Physiol.* **83**, 341–348.

WAGNER, A.M. and KRAB, K. (1995) The alternative respiration pathway in plants: role and regulation. *Physiol. Plant.* **95**, 318–325.

WANN, M. and RAPER, C.D. JR (1979) A dynamic model for plant growth: adaptation for vegetative growth of soybeans. *Crop Sci.* **19**, 461–467.

WARNER, R.L. and KLEINHOFS, A. (1981) Nitrate utilization by nitrate reductase-deficient barley mutants. *Plant Physiol.* **67**, 740–743.

13

The role of mycorrhiza

RÜDIGER HAMPP AND ASTRID WINGLER

13.1 Introduction

The term 'mycorrhiza' describes the symbiotic association of plant roots with hyphal fungi. As about 90% of the land plants belong to families that are commonly mycorrhizal (Molina et al., 1992), mycorrhiza formation is rather the rule than an exception. Therefore, mycorrhizas can be regarded as 'the chief organs involved in nutrient uptake of most land plants' (Harley and Smith, 1983).

According to their morphology and function, mycorrhizas are divided into several classes. In this overview, however, only the two most abundant and economically most important classes, vesicular-arbuscular mycorrhizas (VAM) and ectomycorrhizas, are considered.

Formation of VAM predominates in trees and herbaceous plants of lower latitudes (Read, 1991), such as most crop plants. The fungal partners belong to different genera in the Endogonaceae. Because of the strict dependence on their host, efforts to grow VAM fungi in pure culture have often failed (Bécard and Piché, 1989). Anatomically, VAM are characterized by fungal penetration into root cells (endomycorrhizas) and formation of fungal arbuscules inside cortical cells which, however, are still surrounded by the host plasma membrane. In addition, fungal vesicles are often, but not always, present (Harley and Smith, 1983).

Ectomycorrhiza is the dominant mycorrhizal class in temperate forest ecosystems; for example, trees of the families Betulaceae, Pinaceae, Fagaceae, and Salicaceae are typically ectomycorrhizal (Read, 1991). The fungal partners belong to the basidiomycetes and ascomycetes and many of them have been successfully grown in pure culture. An anatomical feature of ectomycorrhizas is the so-called 'Hartig net' – typical finger-like, branched hyphae within the walls of the cortical cells (Kottke and Oberwinkler, 1986) – and a hyphal mantel enclosing the root (Figure 13.1). The formation of root hairs is suppressed by ectomycorrhiza formation (Harley and Smith, 1983).

Due to extensive fungal growth in the soil, mycorrhiza formation can lead to

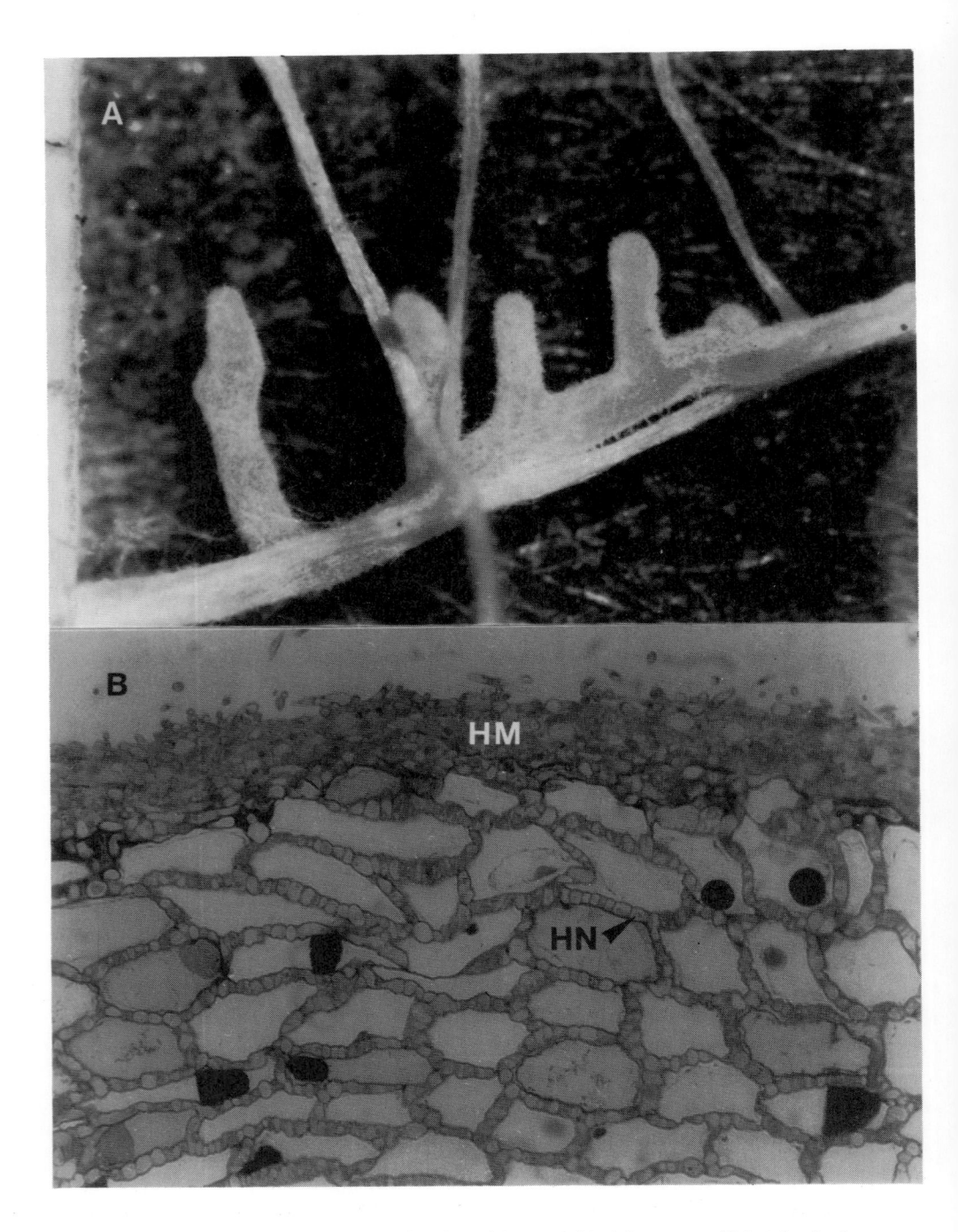

Figure 13.1 Ectomycorrhiza, structural features. (A) Mycorrhizal fine roots; (B) longitudinal cross-section through a mycorrhizal fine root. HM, hyphal mantle; HN, Hartig net (area of solute exchange in the root cortex). Photographs are from studies on the *in vitro* mycorrhization of poplar with fly agaric (Hampp et al., unpublished)

an increased uptake and transfer of nutrients (e.g. nitrogen, inorganic phosphate) to the plant. The plant, on the other hand, supplies the fungus with photoassimilates (Melin and Nilsson, 1957) which can decrease carbon efficiency of the former. In the following sections we will summarize current knowledge on the impact of mycorrhiza formation on carbon and nitrogen metabolism of both partners of symbiosis.

13.2 Carbohydrate metabolism and allocation

13.2.1 Carbon requirement for fungal growth

Carbohydrate dependency of mycorrhizal fungi has been tested by their ability to grow on media containing specific carbon sources. Owing to the lack of cultivation methods for VAM forming fungi, this approach has up to now only been possible with those species involved in ericoid mycorrhizas and ectomycorrhizas (ECM) (for reviews, see Jakobsen, 1991; Wiemken, 1995; Williams, 1992). Major carbon sources are glucose and fructose. An exogenous supply of glucose has been shown to support ectomycorrhizal infection in a range of species (Theodorou and Reddell, 1991). Reports indicating that sucrose, the major form of photoassimilate allocation by a potential host plant, can substantially support mycelial development of ectomycorrhizal fungi have to be regarded with caution. In most cases non-enzymatic hydrolysis could have been responsible for sucrose breakdown due to low pH-values of the cultivation medium. This view is supported by both the lack of detection of fungal invertases, catalyzing sucrose breakdown, and of appreciable rates of sucrose uptake by fungal cells (see below). Apoplast invertase located in the host cell wall appears to be the only physiological way to make sucrose available for these fungi during symbiotic interaction. A histochemical approach towards the radial and longitudinal distribution of acid invertase and sucrose synthase (the latter is a cytosolic enzyme also involved in sucrose breakdown) exhibited a clear dominance of the extractable activity of the former enzyme in the root cortex along with the elongation zone of spruce fine roots while sucrose synthase was confined to the bundle area (Schaeffer, 1995).

Although important for assimilate transfer, the activity of acid invertase in the system *Picea abies*/*Amanita muscaria* was not affected by the degree of fungal infection (Schaeffer *et al.*, 1995) which is different from biotrophic interactions.

13.2.2 Impact of the fungus on host carbohydrate metabolism

Carbohydrate pools

In spite of missing evidence for direct interaction by a certain metabolite (Hampp and Schaeffer, 1995), there is some impact of mycorrhization on carbohydrate pools in the host. *Fraxinus* seedlings inoculated with *Glomus macrocarpum* (VAM) were lower in starch and soluble sugars than non-mycorrhizal seedlings (Borges and Chaney, 1989). Infection of oak seedlings with *Pisolithus tinctorius* (ECM) was significantly correlated with the fructose content of short roots

(Dixon et al., 1981). In mycorrhizal (*A. muscaria*) roots of spruce seedlings, the total amount of sucrose + glucose + fructose was about 30% lower than in non-mycorrhizal fine roots, sucrose being the dominating sugar in both mycorrhizal and non-mycorrhizal fine roots (Rieger et al., 1992). Similarly, Wallander and Nylund (1991) found lower levels of these sugars in mycorrhizal roots of *Pinus sylvestris* when compared to non-mycorrhizal controls.

In contrast, root exudates of a range of plant species infected with *Glomus mosseae* showed no relationship between total sugar content and VAM formation (Azcón and Ocampo, 1984).

In summary, in most cases of VAM, starch levels in leaves/whole plants were reduced while the amount of sugars contained in infected roots was higher in mycorrhizal systems than in non-mycorrhizal controls. The few existing data reported for ECM, however, indicate both decreased pools of starch but also of soluble carbohydrates.

Photosynthesis/leaf metabolism

Gas exchange measurements showed higher rates of net photosynthesis per unit leaf area in VAM plants compared to non-mycorrhizal controls (Allen et al., 1981; Augé et al., 1986; Stahl and Smith, 1984). This is obviously related to an enhanced demand for photoassimilates by ECM as well as VAM (Hampp and Schaeffer, 1995) and is in agreement with many observations that sink strength for photoassimilates can control the rate of photosynthesis (e.g. Herold, 1980; Robbins and Pharr, 1988). The interaction between fungus and plant is, however, dependent on the fungal partner. In ECMs formed between Douglas fir seedlings and *Rhizopogon vinicolor*, *Laccaria laccata*, or *Hebeloma crustuliniforme*, only *Rhizopogon* complied with this concept (Dosskey et al., 1991). The latter symbiosis did not affect plant dry mass but strongly increased the rate of net photosynthesis. In contrast, plants colonized with *Hebeloma* or *Laccaria* showed some suppression of growth but only a marginal enhancement of net photosynthesis. In *Pinus sylvestris* ECMs the fungal partners enhanced rates of photosynthesis, but *L. bicolor* was more effective than *H. crustuliniforme* at reducing host growth and enhancing photosynthesis (Nylund and Wallander, 1989).

The rate of photosynthesis, on the other hand, affects the degree of mycorrhization. Citrus plants grown under long-day photoperiods had greater rates of basipetal transport of photoassimilates, which was paralleled by an improved mycorrhizal infection (Johnson et al., 1982b). Under reduced illumination, additional drain of photoassimilates by the fungal partner can reduce the growth response of mycorrhizal plants compared to that of non-mycorrhizal controls (Tester et al., 1985).

13.2.3 Assimilate allocation in mycorrhizal plants

By using carbohydrates for storage, for the building up of biomass, and for conversion into metabolic energy, mycorrhizal fungi create strong assimilate sinks (e.g. Dosskey et al., 1990, 1991). This should affect carbon allocation in a plant as this process is always regulated by the relative sink strength of competing organs. The mycorrhizal is, however, not necessarily the most prominent sink.

Early flower bud formation, for example, can create a strong sink such that limited carbon availability decreases mycorrhiza formation, as shown for VAM (Johnson et al., 1982a).

Movement of substances between component organisms is a primary feature of symbiosis and has its impact also on growth properties of the host plant. Plants colonized by, for example, VAM fungi, have root systems which acquire a greater percentage of photoassimilates than non-mycorrhizal roots (Clapperton and Reid, 1992; Douds et al., 1988; Koch and Johnson, 1984; Wang et al., 1989). Consequently, they exhibit higher root-to-shoot ratios (Clapperton and Reid, 1990, 1992).

In ECM as well as in VAM, intercellular transport in the apoplast constitutes an important part of the overall exchange of solutes which then have to cross the plasma membranes of both partners. This is also true for VAM where only the host cell wall is penetrated while the host plasma membrane remains intact. A detailed review on this subject is given by Smith and Smith (1990). This infers that photoassimilates are unloaded either from the phloem or from cortical cells into the apoplast where they are open for use by the fungal partner.

Transfer of carbohydrates

Generally, tracer studies indicate rapid translocation of ^{14}C-labelled assimilates to the roots of ectomycorrhizal plants (Cox et al., 1975), especially in young symbiotic interactions (Cairney et al., 1989).

Lewis and Harley (1965) showed that labelled sucrose, the main transport form for photoassimilates in higher plants, applied to the cut axis of excised beech mycorrhizas, was translocated to the tip. Label accumulated particularly in the fungal sheath and was incorporated into the fungal carbohydrates trehalose, mannitol, and glycogen. In seedlings of *Pinus sylvestris* infected with *Suillus variegatus*, label from photosynthetic $^{14}CO_2$ fixation was detected in the cortex, Hartig net, and the hyphal mantle covering the fine roots (Bauer et al., 1991).

An attempt to assay the longitudinal distribution of soluble carbohydrates was made by Rieger et al. (1992). Lyophilized mycorrhizas (*Amanita muscaria*) and fine roots of spruce (length < 2 mm) were dissected into about 0.5 mm thick slices which represented four zones of different physiological functions. A longitudinal distinction of pools of sucrose, glucose, and fructose exhibited a specific response of sucrose. This saccharide showed the lowest amounts in the middle parts of a mycorrhiza, i.e. the area of most intense symbiotic interaction. Correspondingly, levels of fungus-specific compounds such as trehalose or ergosterol were increased in this area (Hampp et al., 1995). Mycorrhization can also be influenced by radial carbohydrate gradients in roots as shown by the preferred location of arbuscules at the inner cortex of some VAM, where carbohydrate concentrations were highest (Koide and Schreiner, 1992).

Growth of, for example, the ectomycorrhizal fungi *Amanita muscaria* and *Hebeloma crustuliniforme* on sucrose was only possible after addition of invertase (Salzer and Hager, 1991; Schaeffer et al., 1995). Root cell walls contain acid invertase (for suspension culture cells and root tissue of spruce, see Salzer and Hager, 1991). It is therefore very probable that sucrose delivered by the host is hydrolyzed in the root apoplast by host enzymes. The products of hydrolysis, glucose and fructose, are obviously not taken up at the same rates. The ECM-forming basidiomycetes *Amanita muscaria* and *Hebeloma crustuliniforme* grew better on glucose than on fructose (Salzer and Hager, 1991). A direct assay of sugar uptake employing ^{14}C-labelled substrates and protoplasts from *Amanita muscaria* confirmed this report. K_m values for the uptake of glucose and fructose were 1.25 and 11.3 mM, respectively. In addition, glucose uptake was only marginally affected by fructose, while glucose was highly competitive with regards to fructose (Chen and Hampp, 1993). As glucose is the substrate giving rise to the formation of fungus-specific trehalose, one might assume that the preferred uptake of this hexose could be specific for basidiomycetes which appear to accumulate trehalose instead of mannitol (the latter is derived from fructose). Similar transport studies with protoplasts of an ascomycete which preferably accumulates mannitol (*Cenococcum geophilum*) (Stülten and Hampp, unpublished) showed, however, a comparable preference for glucose.

Uptake of hexoses by the fungus must not necessarily be energy-dependent as it could be along a concentration gradient due to the conversion into fungal carbohydrates. There is, however, evidence for active transport. Uptake of glucose and fructose by *Amanita* protoplasts was highly dependent on the intracellular ATP/ADP ratio. Depletion of ATP by uncouplers significantly decreased the rates of uptake (Chen and Hampp, 1993).

Regulation of assimilate allocation

Regulation of mycorrhiza under the influence of carbohydrates can be with regard to both formation and function. The involvement of photoassimilates in the initiation and maintenance of a mycorrhiza (Björkman's carbohydrate theory: root sugar concentration must reach a certain threshold value before a mycorrhization can take place) has been thoroughly reviewed by Nylund (1988). This complies with evidence that plants with an ample nitrogen supply have less carbohydrates available for distribution and are thus less attractive for fungal infection (Wallenda et al., 1996).

In an established mycorrhiza, regulation of fungal growth is important for the mutual benefit of the symbiosis. If assimilate extraction from the plant by the fungus becomes too extensive, a decrease in plant performance and fitness could result. For VAM there is evidence that the host plant actively regulates infections in order to optimize its fitness. Limited light, for example, which results in a decrease in photoassimilate production, often reduces the degree of mycorrhizal infection (Koide and Schreiner, 1992). Mechanisms involved in the host-dependent regulation of this symbiosis may be by chemical stimuli, chemical or structural mechanisms to limit the degree of infection, or by the delivery of carbohydrates (Koide and Schreiner, 1992).

With regard to the supply of carbohydrates this could include regulation at the level of sucrose synthesis in source organs, phloem loading of photoassimilates (allocation between source and sink (root) organs), sucrose hydrolysis and assimilate transfer between both partners.

Regulation in source organs starts at the level of assimilate partitioning between starch and sucrose, and the rate of starch synthesis in a source leaf is controlled by the rates of sucrose formation and export. This control is mediated by a complex co-ordination of metabolic processes in two subcellular compartments of the leaf mesophyll, the chloroplast and the cytosol (Stitt, 1990; see also

Chapters 3 and 4). Fru2,6P$_2$, an important cytosolic regulator of glycolysis and present only in micromolar concentrations (Steingraber et al., 1988), affects cytosolic sucrose synthesis by inhibiting the fructose 1,6-bisphosphatase reaction (sucrose synthesis) and activating a pyrophosphate (PPi)-dependent phosphofructokinase, thus possibly enhancing glycolysis. In mycorrhizal spruce seedlings, the levels of Fru2,6P$_2$ were decreased while the extractable activity of sucrose phosphate synthase, a key enzyme of cytosolic sucrose formation, was increased (Hampp et al., 1995). This can be taken as evidence for an increased capacity for sucrose formation in these seedlings.

13.2.4 Mycorrhization and elevated CO_2

Elevated levels of atmospheric CO_2 (global change) are an important topic with regard to improved carbon use efficiency. There is much evidence that the efficiency in taking advantage of elevated CO_2 depends largely on the availability of carbohydrate sinks within a plant (Stitt, 1991). As mycorrhizas are large consumers of photoassimilates, mycorrhizal plants should increase productivity with increasing CO_2 supply. Experimental evidence in support of this hypothesis is controversial. In loblolly pine there was no significant effect on (ecto) mycorrhizal colonization (percentage of fine roots) although elevated CO_2 caused increased root carbohydrate levels (Lewis et al., 1994). In contrast, (VA) mycorrhizal colonization of C_3/C_4 grasses treated with elevated CO_2 was increased (Monz et al., 1994; Morgan et al., 1994), and there is also some indication for similar responses in ECM (Kottke, personal communication).

13.3 Nitrogen acquisition and metabolism

Nitrogen is one of the main factors limiting plant growth. Accordingly, improvement of nitrogen aquisition by mycorrhiza formation should increase plant productivity and competitive ability in N-limited soils.

13.3.1 VA-mycorrhizas

As VAM have mainly evolved in ecosystems with soils of high mineral nitrogen, but low phosphorus availability (Read, 1991), their main function is considered to be to supply phosphorus to the plant (Marschner and Dell, 1994). Thus, emphasis in research was directed to the role of VAM in phosphorus acquisition. Nevertheless, it has now been shown for several mycorrhizal systems that infection with VAM fungi can also improve nitrogen transport to the plant (Marschner, 1995), especially under conditions when nitrate nitrogen is limiting.

Uptake and transfer of nitrogen by fungal hyphae

Ames et al. (1983) first demonstrated the ability of VAM hyphae to take up ammonium nitrogen and to deliver it to celery plants. This could later be confirmed for other VA species. VAM fungi have also been shown to take up and transport nitrate nitrogen (George et al., 1992, Johansen et al., 1993). In view of the high mobility of nitrate and the low mobility of ammonium in soils, an important contribution of fungal hyphae to nitrogen acquisition is expected from ammonium rather than from nitrate uptake (Marschner and Dell, 1994). However, under water-stressed conditions, when mass flow and diffusion of nitrate are negatively affected, hyphal transport of nitrogen from nitrate to the plant can become essential for nitrogen nutrition (Tobar et al., 1994).

In contrast to several ectomycorrhizal fungi (see below), there is no evidence that VAM fungi can utilize organic nitrogen sources (Frey and Schüepp, 1993).

Nitrogen metabolism

Information on the metabolism of inorganic nitrogen taken up by the fungal hyphae is scarce. This is mainly due to the failure to grow VAM fungi in pure culture.

Although Ho and Trappe (1975) were able to detect nitrate reducing capacity in spores of *Glomus mosseae* and *G. macrocarpus*, fungal nitrate reductase is probably not active in mycorrhizas of *T. subterraneum*. Instead, mycorrhiza formation stimulates root nitrate reductase activity indirectly via improved nutrient uptake (Oliver et al., 1983). Thus, in VAM, nitrate is probably transferred to the plant as nitrate before it is reduced to ammonium by plant nitrate and nitrite reductases.

While plants assimilate ammonium via the sequential action of glutamine synthetase (GS) and glutamate synthase (GOGAT), ammonium assimilation in fungi can also occur via NADP-dependent glutamate dehydrogenase (NADP-GDH) (Pateman and Kinghorn, 1976). In mycorrhizas of *T. subterraneum* and *Allium cepa*, however, GDH activity is very low (Smith et al., 1985). In fungal structures of *G. mosseae* separated from the roots via cellulase and pectinase digestion, GS activity could be detected. This activity is high enough to contribute directly to the increased assimilation of ammonium in mycorrhizal compared to non-mycorrhizal roots (Smith et al., 1985). Accordingly, Cliquet and Stewart (1993) found no evidence for ammonium assimilation via GDH in mycorrhizas of maize. Instead, inoculation of the roots with *G. fasciculatum* increased GS activity, accompanied by a higher rate of amino acid synthesis and higher glutamine translocation in the xylem. It can thus be concluded, that formation of VAM leads to a stimulation of ammonium assimilation while the pathway of assimilation, via the GS/GOGAT cycle, remains unchanged.

Interaction with N_2 fixation by root nodules

In legumes growing on phosphorus-deficient soil, increased phosphorus supply due to VA-mycorrhiza formation should enhance nodule formation and N_2 fixation. Accordingly, phosphorus-mediated improvement of symbiotic N_2 fixation and, in addition, nitrogen uptake from the soil, have been shown for *Hedysarum coronarium* (Barea et al., 1987). But, as both symbioses compete for photoassimilates, the beneficial effects of VAM do not apply to phosphorus sufficient conditions (Marschner and Dell, 1994).

13.3.2 Ectomycorrhizas

In many forest ecosystems, rates of nitrogen mineralization of litter are low. Consequently, supply with inorganic nitrogen is often limiting (Read, 1991). In addition, nitrification is usually slow and the poorly mobile ammonium ion predominates as inorganic nitrogen source (Keeney, 1980). Association with ectomycorrhizal fungi can improve nitrogen acquisition by facilitating access to organic nitrogen sources (Read et al., 1989) and by increasing uptake of ammonium (Rygiewicz et al., 1984) via extensive mycelial growth in the soil and circumvention of ammonium depletion zones. Therefore, mycorrhiza formation is an important strategy for improving development and competitive ability of forest trees.

Utilization of organic nitrogen

There is increasing evidence that ectomycorrhizal fungi are important for providing access to organic nitrogen sources which cannot be utilized by the host plants (Read et al., 1989).

Abuzinadah and Read (1988) have shown that ectomycorrhizal fungi can grow with amino acids as sole nitrogen sources. Three of the amino acids readily used by the fungi, glutamate, aspartate and serine, are among the most abundant free amino acids in the soil and should thus be of nutritional significance. The ability of ectomycorrhizal fungi to utilize peptides and proteins is variable: according to their ability to grow with protein as nitrogen source, 'non-protein fungi' (e.g. *Laccaria laccata* and *Lactarius rufus*) can be distinguished from 'protein fungi' (e.g. *Hebeloma crustuliniforme*, *Suillus bovinus*, *Cenococcum geophilum*, *Rhizopogon roseolus* and *Amanita muscaria*) (Abuzinadah and Read, 1986; Finlay et al., 1992). When ectomycorrhizal tree species are inoculated with 'protein fungi', protein as nitrogen source can support high dry weight yields and shoot nitrogen contents, while the respective non-inoculated plants are unable to utilize protein (Abuzinadah et al., 1986).

To make protein accessible, it has to be degraded by extracellular proteases before the amino acids can be taken up. Excretion of protease by the ectomycorrhizal fungi *C. geophilum* (El-Badaoui and Botton, 1989), *H. crustuliniforme* (Zhu et al., 1994) and *A. muscaria* (Guttenberger, Buckenhofer and Spägele, personal communication) is induced by protein and repressed by high ammonium concentrations.

Utilization of inorganic nitrogen

Uptake of inorganic nitrogen and transport to the host plant by ectomycorrhizal fungi was demonstrated by Melin and Nilsson (1952) and Finlay et al. (1988). In accordance with the predominant occurrence of ammonium in the soil, most ectomycorrhizal fungi grow better with ammonium than with nitrate nitrogen in pure culture (Finlay et al., 1992; France and Reid, 1984). Mycorrhizal roots of beech also readily absorb and assimilate ammonium whereas nitrate is a poor nitrogen source (Carrodus, 1966, 1967). In contrast, some ectomycorrhizal fungi such as *Hebeloma cylindrosporum* exhibit higher growth rates with nitrate than with ammonium nitrogen (Scheromm et al.,

1990a) and increase nitrate uptake in association with their host plant (Plassard et al., 1994).

Nitrogen assimilation

When ammonium nitrogen is taken up by the mycorrhizal mycelium of *Paxillus involutus*, it is directly assimilated into amino acids at the uptake site (Ek et al., 1994). The amino acids are then translocated to the roots of the host plants. Rates of uptake and of transfer for nitrate are low, and, in contrast to ammonium, nitrate nitrogen is not assimilated in the mycelium but transferred to the root as nitrate (Ek et al., 1994). This might be different for ectomycorrhizas formed with fungi possessing a high nitrate reductase activity, such as *H. cylindrosporum* (Scheromm et al., 1990b). Assimilation of ammonium in the fungal cells and transfer of amino acids to the roots cells is in accordance with current models of nitrogen metabolism in ectomycorrhizas (e.g. Botton and Chalot, 1995).

Both the GDH pathway and the GS/GOGAT cycle for ammonium assimilation can be active in ectomycorrhizal fungi in pure culture. In *C. geophilum* (Martin et al., 1988a) and *H. cylindrosporum* (Chalot et al., 1991a) NADP-GDH seems to make an important contribution to ammonium assimilation, whereas the GS/GOGAT cycle dominates in *Pisolithus tinctorius* (Kershaw and Stewart, 1992) and *Laccaria bicolor* (Martin et al., 1994). Martin et al. (1986) concluded from ^{15}N-NMR studies with beech mycorrhizas that GDH plays no role in the primary assimilation of ammonium. This is consistent with the strongly supressed fungal NADP-GDH in beech-*Hebeloma* sp. mycorrhizas (Dell et al., 1989). In spruce-*Hebeloma* sp. mycorrhizas, in contrast, the fungal NADP-GDH is obviously involved in ammonium assimilation (Chalot et al., 1991b). These differences demonstrate how variable metabolism in ectomycorrhizas can be. Generalizations should thus be regarded with care.

Carbon/nitrogen interactions and exchange at the symbiotic interface

Amino acids taken up by the fungal mycelium can provide carbon compounds which support respiration or synthesis of other amino acids. On the other hand, assimilation of inorganic nitrogen requires the synthesis of carbon skeletons. In a detailed study, Chalot et al. (1994) investigated metabolism of exogenously supplied ^{14}C-alanine in the ectomycorrhizal fungus *Paxillus involutus*. They demonstrated that alanine can be utilized as a respiratory substrate (conversion to CO_2 and metabolites of the tricarboxylic acid cycle) and provide carbon for the synthesis of amino acids (conversion to glutamate, glutamine and aspartate). Gluconeogenesis from alanine, however, did not occur. This corroborates the view that ectomycorrhizal fungi depend on sugars delivered by their host plant (see above) and have only restricted ability to grow with organic acids as sole carbon source (Palmer and Hacskaylo, 1970).

When ammonium is assimilated in the fungal part of the mycorrhizas by either NADP-GDH or GS/GOGAT, 2-oxoglutarate must be provided. Whether 2-oxoglutarate is delivered by the host plant or whether the fungus itself synthesizes it from imported sugars, was an open question. There is now increasing evidence that 2-oxoglutarate formation takes place in the fungal part of ectomycorrhizas:

1. Ectomycorrhizal fungi in pure culture have all the capabilities needed for the formation of 2-oxoglutarate: they exhibit high growth rates with glucose as sole carbon source (Palmer and Hacskaylo, 1970). Glycolysis (Schaeffer, 1995) and tricarboxylic acid cycle (Martin and Canet, 1986) are functional in ectomycorrhizal fungi, and during ammonium assimilation, oxaloacetate formation by anaplerotic CO_2 fixation takes place in ectomycorrhizal fungi (Martin and Canet, 1986; Martin et al., 1988a). This anaplerotic synthesis of oxaloacetate is necessary for the replenishment of tricarboxylic acid cycle intermediates when 2-oxoglutarate is withdrawn for ammonium assimilation. In fungi, it can theoretically be catalyzed by pyruvate carboxylase or phosphoenolpyruvate carboxykinase (PEPCK), while the plant enzyme phosphoenolpyruvate carboxylase (PEPCase) is generally not active (Casselton, 1976).

2. In mycorrhizas of spruce with *A. muscaria*, anaplerotic CO_2 fixation probably takes place in the fungal part (Wingler et al., 1996). This can be concluded from the fact that mycorrhiza formation drastically reduced activity and protein of the anaplerotic plant enzyme, PEPCase, while CO_2 fixation rates were higher in mycorrhizas compared to non-mycorrhizal short roots (Wingler et al., 1996).

So far, there is no direct evidence as to how nitrogen assimilated by the fungus is transferred to the plant. Since in ectomycorrhizas of beech (Martin et al., 1986) and spruce (Chalot et al., 1991b) nitrogen from ammonium is mainly incorporated into glutamate, glutamine and alanine, it can be speculated that mainly these amino acids are translocated (Botton and Chalot, 1995). This would result in an exchange of carbon in the form of plant-supplied sugars (probably glucose) (Chen and Hampp, 1993) for fungus-derived amino acids (Figure 13.2).

13.4 Conclusions/potentials for engineering

Supply of carbohydrates by the host is the basis for the development of a functional mycorrhiza, and allocation of host photoassimilates to the interactive root areas is triggered in several ways. Most important are obviously establishment and maintenance of gradients between both partners, and between root and shoot. These are achieved by either conversion of, for example, host sucrose into fungus-specific compounds such as trehalose or mannitol, or consumption by fungal growth and respiration, but also for anaplerotic reactions in relation to amino acid synthesis. The latter appears to be an important part in carbon cycling between host and fungus. Carbohydrate needs of mycorrhizal roots obviously decrease sugar availability for the host as shown by reduced starch contents in the latter. The impact on host pools of soluble carbohydrates is, however, more variable and possibly different for both VAM and ECM.

There is now sufficient evidence that the needs for photoassimilates in a developing or functional mycorrhiza induce increased rates of photosynthesis. Such a response could be achieved merely by a steepened sucrose gradient between shoot and root, affecting regulation of carbon partitioning between

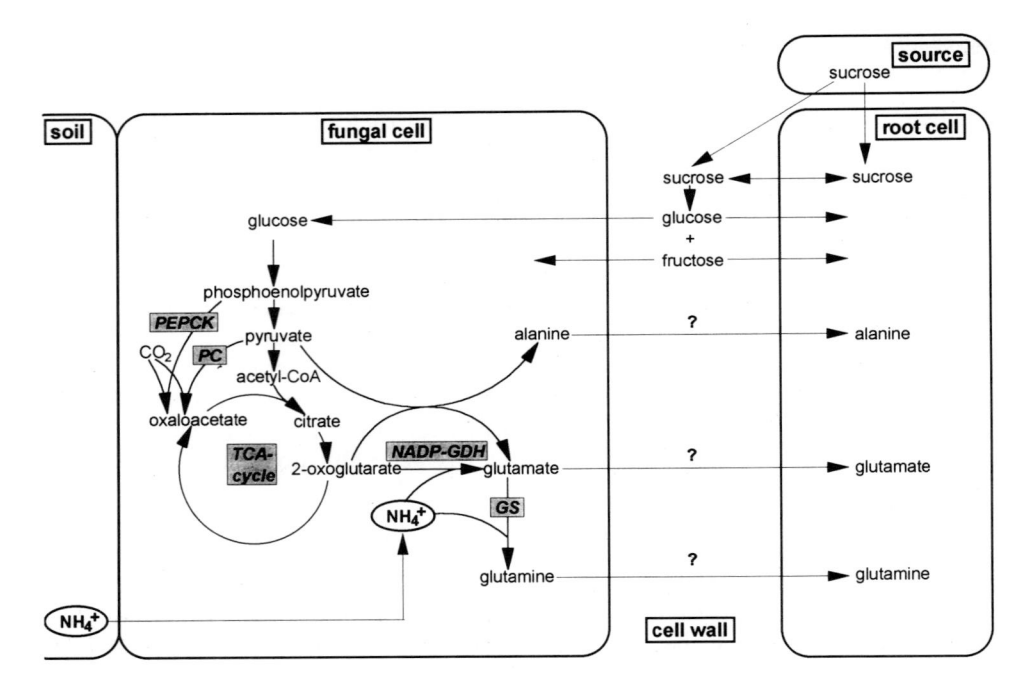

Figure 13.2 Possible ways of interaction of sucrose and amino acid metabolism in ectomycorrhizal roots. Fungal cell (left) and host root cortex cell (right). Host supplied sucrose has to be hydrolyzed by host invertase. Glucose is preferentially taken up by the fungal partner (spruce/fly agaric ectomycorrhiza). There, glucose can support amino acid formation by providing carbon skeletons for anaplerotic reactions (pyruvate carboxylase, PC; phosphoenolpyruvate carboxykinase, PEPCK). Key enzymes for amino acid synthesis are NADP-dependent glutamate dehydrogenase (NADP-GDH) and glutamine synthetase (GS). Part of the amino acids (primarily glutamine) is then shuttled back to the host. Respective transport systems are not yet identified

sucrose and starch and related mesophyll cell metabolite pools (possible regulatory events have been discussed). Genetic manipulation of fungal or host functions could considerably improve our knowledge about the regulation of organismic interaction in such a symbiosis. With regard to ectomycorrhizas, the establishment of *in vitro* mycorrhization of transgenic poplar is very promising (Hampp *et al.*, 1996). Here we aim at both manipulation of fungal/host transport of sugars and amino acids and the supply of glucose for fungal metabolism.

Acknowledgments

The authors gratefully acknowledge financial support from the Deutsche Forschungsgemeinschaft, the Bundesminister für Forschung und Technologie (EUROSILVA), the 'Projekt Europäische Forschungsmaßnahmen zur Luftreinhaltung' (PEF), and the Landesforschungsförderungsprogramm Baden-Württemberg for presentation of their own results.

References and further reading

ABUZINADAH, R.A. and READ, D.J. (1986) The role of proteins in the nitrogen nutrition of ectomycorrhizal plants. I. Utilization of peptides and proteins by ectomycorrhizal fungi. *New Phytol.* **103**, 481–493.

ABUZINADAH, R. A. and READ, D. J. (1988) Amino acids as nitrogen sources for ectomycorrhizal fungi: utilization of individual amino acids. *Trans. Br. Mycol. Soc.* **91**, 473–479.

ABUZINADAH, R.A., FINLAY, R.D. and READ, D.J. (1986) The role of proteins in the nitrogen nutrition of ectomycorrhizal plants. II. Utilization of protein by mycorrhizal plants of *Pinus contorta*. *New Phytol.* **103**, 495–506.

ALLEN, M.F., SMITH, W.K., MOORE, T.S. JR and CHRISTENSEN, M. (1981) Comparative water relations and photosynthesis of mycorrhizal and non-mycorrhizal *Bouteloua gracilis*. *New Phytol.* **88**, 683–693.

AMES, R.N., REID, C.P.P., PORTER, L.K. and CAMBARDELLA, C. (1983) Hyphal uptake and transport of nitrogen from two ^{15}N-labelled sources by *Glomus mosseae*, a vesicular-arbuscular mycorrhizal fungus. *New Phytol.* **95**, 381–396.

AUGÉ, R.M., SCHEKEL, K.A. and WAMPLE, R.L. (1986) Osmotic adjustment in leaves of VA mycorrhizal and non-mycorrhizal rose plants in response to drought stress. *Plant Physiol.* **82**, 765–770.

AZCÓN, R. and OCAMPO, J.A. (1984) Effect of root exudation on vesicular-arbuscular mycorrhizal infection at early stages of plant growth. *Plant and Soil* **82**, 133–138.

BAREA, J.M., AZCÓN-AGUILAR, C. and AZCÓN, R. (1987) Vesicular-arbuscular mycorrhiza improve both symbiotic N_2 fixation and N uptake from soil as assessed with a ^{15}N technique under field conditions. *New Phytol.* **106**, 717–725.

BAUER, T., BLECHSCHMIDT-SCHNEIDER, S. and ESCHRICH, W. (1991) Regulation of photoassimilate allocation in *Pinus sylvestris* seedlings by the nutritional status of the mycorrhizal fungus *Suillus variegatus*. *Trees* **5**, 36–43.

BÉCARD, G. and PICHÉ, Y. (1989) Fungal growth stimulation by CO_2 and root exudates in vesicular-arbuscular mycorrhizal symbiosis. *Appl. Environ. Microbiol.* **55**, 2320–2325.

BORGES, R.G. and CHANEY, W.R. (1989) Root temperature affects mycorrhizal efficacy in *Fraxinus pennsylvanica* marsh. *New Phytol.* **112**, 411–418.

BOTTON, B. and CHALOT, M. (1995) Nitrogen assimilation: enzymology in ectomycorrhizas. In: *Mycorrhiza. Structure, Function, Molecular Biology and Biotechnology* (VARMA, A. and HOCK, B. eds), pp. 325–363. Springer-Verlag, Berlin.

CAIRNEY, J.W., ASHFORD, A.E. and ALLAWAY W.G. (1989) Distribution of photosynthetically fixed carbon within root systems of *Eucalyptus pilularis* plants ectomycorrhizal with *Pisolithus tinctorius*. *New Phytol.* **112**, 495–500.

CARRODUS, B.B. (1966) Absorption of nitrogen by mycorrhizal roots of beech. I. Factors affecting the assimilation of nitrogen. *New Phytol.* **65**, 358–371.

CARRODUS, B.B. (1967) Absorption of nitrogen by mycorrhizal roots of beech. II. Ammonium and nitrate as sources of nitrogen. *New Phytol.* **66**, 1–4.

CASSELTON, P.J. (1976) Anaplerotic pathways. In: *The Filamentous Fungi*, Vol. 2 (SMITH, J.E. and BERRY, D.R., eds), pp. 121–136. Edward Arnold, London.

CHALOT, M., BRUN, A., DEBAUD, J.C. and BOTTON, B. (1991a) Ammonium-assimilating enzymes and their regulation in wild and NADP-glutamate dehydrogenase-deficient strains of the ectomycorrhizal fungus *Hebeloma cylindrosporum*. *Physiol. Plant.* **83**, 122–128.

CHALOT, M., STEWART, G.R., BRUN, A., MARTIN, F. and BOTTON, B. (1991b) Ammonium assimilation by spruce-*Hebeloma* sp. ectomycorrhizas. *New Phytol.* **119**, 541–550.

CHALOT, M., BRUN, A., FINLAY, R.D. and SÖDERSTRÖM, B. (1994) Respiration of [^{14}C]-alanine by the ectomycorrhizal fungus *Paxillus involutus*. *FEMS Microbiol. Lett.* **121**, 87–92.

CHEN, X.-Y. and HAMPP, R. (1993) Sugar uptake by protoplasts of the ectomycorrhizal fungus, *Amanita muscaria* (L. ex Fr.) Hooker. *New Phytol.* **125**, 601–608.

CLAPPERTON, M.J. and REID, D.M. (1990) Effects of sulphur dioxide fumigation on *Phleum pratense* and vesicular-arbuscular mycorrhizal fungi. *New Phytol.* **115**, 465–469.

CLAPPERTON, M.J. and REID, D.M. (1992) Effects of low-concentration sulphur dioxide fumigation and vesicular-arbuscular mycorrhizas on ^{14}C-partitioning in *Phleum pratense* L. *New Phytol.* **120**, 381–387.

CLIQUET, J.-B. and STEWART, G.R. (1993) Ammonia assimilation in *Zea mays* L. infected with a vesicular-arbuscular mycorrhizal fungus *Glomus fasciculatum*. *Plant Physiol.* **101**, 865–871.

COX, G., SANDERS F.E., TINKER, P.B. and WILD, J.A. (1975) Ultrastructural evidence relating to host-endophyte transfer in a vesicular-arbuscular mycorrhiza. In: *Endomycorrhizas* (SANDERS, F.E., MOSSE, B. and TINKER, P.B. eds), pp. 297–312. Academic Press, London.

DELL, B., BOTTON, B., MARTIN, F. and LE TACON, F. (1989) Glutamate dehydrogenases in ectomycorrhizas of spruce (*Picea excelsa* L.) and beech (*Fagus sylvatica* L.). *New Phytol.* **111**, 683–692.

DIXON, R.K., GARRETT, H.E., BIXBY J.A., COX, G.S. and TOMPSON, J.G. (1981) Growth, ectomycorrhizal development and root soluble carbohydrates of black oak *Quercus velutina* seedlings fertilized by 2 methods. *For. Sci.* **27**, 617–624.

DOSSKEY, M.G., LINDERMAN, R.G. and BOERSMA, L. (1990) Carbon-sink stimulation of photosynthesis in Douglas-fir seedlings by some ectomycorrhizae. *New Phytol.* **115**, 269–274.

DOSSKEY, M.G., BOERSMA, L. and LINDERMAN, R.G. (1991) Role for the photosynthate demand of ectomycorrhizas in the response of Douglas fir seedlings to drying soil. *New Phytol.* **117**, 327–334.

DOUDS, D.D. JR, JOHNSON, C.R. and KOCH, K.E. (1988) Carbon cost of the fungal symbiont relative to the net leaf P accumulation in split-root VA mycorrhizal symbiosis. *Plant Physiol.* **86**, 491–496.

EK, H., ANDERSSON, S., ARNEBRANT, K. and SÖDERSTRÖM, B. (1994) Growth and assimilation of NH_4^+ and NO_3^- by *Paxillus involutus* in association with *Betula pendula* and *Picea abies* as affected by substrate pH. *New Phytol.* **128**, 629–637.

EL-BADAOUI, K. and BOTTON, B. (1989) Production and characterization of exocellular proteases in ectomycorrhizal fungi. *Ann. Sci. For.* **46**, 728–730.

FINLAY, R.D., EK, H., ODHAM, G. and SÖDERSTRÖM, B. (1988) Mycelial uptake, translocation and assimilation of nitrogen from ^{15}N-labelled ammonium by *Pinus sylvestris* plants infected with four different ectomycorrhizal fungi. *New Phytol.* **110**, 59–66.

FINLAY, R.D., FROSTEGÅRD, Å. and SONNERFELDT, A.-M. (1992) Utilization of organic and inorganic nitrogen sources by ectomycorrhizal fungi in pure culture and in symbiosis with *Pinus contorta* Dougl. ex Loud. *New Phytol.* **120**, 105–115.

FRANCE, R.C. and REID, C.P.P. (1984) Pure culture growth of ectomycorrhizal fungi on inorganic nitrogen sources. *Microb. Ecol.* **10**, 187–195.

FREY, B. and SCHÜEPP, H. (1992) Transfer of symbiotically fixed nitrogen from berseem (*Trifolium alexandrinum* L.) to maize via vesicular-arbuscular mycorrhizal hyphae. *New Phytol.* **122**, 447–454.

FREY, B. and SCHÜEPP, H. (1993) Acquisition of nitrogen by external hyphae of arbuscular mycorrhizal fungi associated with *Zea mays* L. *New Phytol.* **124**, 221–230.

GEORGE, E., HÄUSSLER, K.-U., VETTERLEIN, D., GORGUS, E. and MARSCHNER, H. (1992) Water and nutrient translocation by hyphae of *Glomus mosseae*. *Can. J. Bot.* **70**, 2130–2137.

HAMPP, R. and SCHAEFFER, C. (1995) Mycorrhiza – carbohydrate and energy metabolism. In: *Mycorrhiza. Structure, Function, Molecular Biology and Biotechnology* (VARMA, A. and HOCK, B., eds.), pp. 267–296. Springer-Verlag, Berlin.

HAMPP, R., SCHAEFFER, C., WALLENDA, T., STÜLTEN, C., JOHANN, P. and EINIG, W. (1995) Changes in carbon partitioning or allocation due to ectomycorrhiza formation: biochemical evidence. *Can. J. Bot.* **73** (Suppl.), S548–S556.

HAMPP, R., ECKE, M., SCHAEFFER, C., WALLENDA, T., WINGLER, A., KOTTKE, I. and SUNDBERG, B. (1996) Axenic synthesis of ectomycorrhiza between hybrid and transgenic aspen and *Amanita muscaria*. *Trees* **11**, 59–64.

HARLEY, J.L. and SMITH, S.E. (1983) *Mycorrhizal Symbiosis*. Academic Press, London.

HEROLD, A. (1980) Regulation of photosynthesis by sink activity – the missing link. *New Phytol.* **86**, 131–144.

HO, I. and TRAPPE, J.M. (1975) Nitrate reducing capacity of two vesicular-arbuscular mycorrhizal fungi. *Mycologia* **67**, 886–888.

JAKOBSEN, I. (1991) Carbon metabolism in mycorrhiza. In: *Methods in Microbiology* **23** (NORRIS, J.R., READ, D.J. and VARMA, A.K., eds.), pp. 149–180. Academic Press, London.

JOHANSEN, A., JAKOBSEN, I. and JENSEN, E.S. (1993) Hyphal transport by a vesicular-arbuscular mycorrhizal fungus of N applied to the soil as ammonium or nitrate. *Biol. Fertil. Soils* **16**, 66–70.

JOHNSON, C.R., GRAHAM, J.H., LEONARD, R.T. and MENGE, J.A. (1982a) Effect of flower bud development in *Chrysanthemum morifolium* on vesicular-arbuscular mycorrhiza formation. *New Phytol.* **90**, 671–676.

JOHNSON, C.R., MENGE, J.A., SCHWAB, S. and TING, I.P. (1982b) Interaction of photoperiod and vesicular-arbuscular mycorrhizae on growth and metabolism of sweet orange *Citrus sinensis*. *New Phytol.* **90**, 665–670.

KEENEY, D.R. (1980) Prediction of soil nitrogen availability in forest ecosystems: a literature review. *For. Sci.* **26**, 159–171.

KERSHAW, J.L. and STEWART, G.R. (1992) Metabolism of ^{15}N-labelled ammonium by the ectomycorrhizal fungus *Pisolithus tinctorius* (Pers.) Coker & Couch. *Mycorrhiza*, **1**, 71–77.

KOCH, K.E. and JOHNSON, C.R. (1984) Photosynthate partitioning in split-root citrus seedlings with mycorrhizal and non-mycorrhizal root systems. *Plant Physiol.* **75**, 26–30.

KOIDE, R.T. and SCHREINER, R.P. (1992) Regulation of the vesicular-arbuscular mycorrhizal symbiosis. *Annu. Rev. Plant Physiol. Plant Mol. Biol.* **43**, 557–581.

KOTTKE, I. and OBERWINKLER, F. (1986) Mycorrhiza of forest trees – structure and function. *Trees* **1**, 1–24.

LEWIS, D.H. and HARLEY, J.L. (1965) Carbohydrate physiology of mycorrhizal roots of beech. III. Movement of sugars between host and fungus. *New Phytol.* **64**, 256–269.

LEWIS, J.D., THOMAS, R.B. and STRAIN, B.R. (1994) Effect of elevated CO_2 on mycorrhizal colonization of loblolly pine (*Pinus taeda* L.) seedlings. *Plant and Soil* **165**, 81–88.

MARSCHNER, H. (1995) *Mineral Nutrition of Higher Plants*, 2nd edn. Academic Press, London.

MARSCHNER, H. and DELL, B. (1994) Nutrient uptake in mycorrhizal symbiosis. *Plant and Soil* **159**, 89–102.

MARTIN, F. and CANET, D. (1986) Biosynthesis of amino acids during [^{13}C]glucose utilization by the ectomycorrhizal ascomycete *Cenococcum geophilum* monitored by ^{13}C nuclear magnetic resonance. *Physiol. Végétale* **24**, 209–218.

MARTIN, F., STEWART, G.R., GENETET, I. and LE TACON, F. (1986) Assimilation of ^{15}NH$_4^+$ by beech (*Fagus sylvatica* L.) ectomycorrhizas *New Phytol.* **102**, 85–94.

MARTIN, F., STEWART, G.R., GENETET, I. and MOUROT, B. (1988a) The involvement of glutamate dehydrogenase and glutamine synthetase in ammonia assimilation by the rapidly growing ectomycorrhizal ascomycete, *Cenococcum geophilum* Fr. *New Phytol.* **110**, 541–550.

MARTIN, F., RAMSTEDT, M., SÖDERHÄLL, K. and CANET, D. (1988b) Carbohydrate and amino acid metabolism in the ectomycorrhizal ascomycete *Sphaerosporella brunnea* during glucose utilization, a ^{13}C NMR study. *Plant Physiol.* **86**, 935–940.

MARTIN, F., CÔTÉ, R. and CANET, D. (1994) NH$_4^+$ assimilation in the ectomycorrhizal basidiomycete *Laccaria bicolor* (Maire) Orton, a ^{15}N-NMR study. *New Phytol.* **128**, 479–485.

MELIN, E. and NILSSON, H. (1952) Transport of labelled nitrogen from an ammonium source to pine seedlings through mycorrhizal mycelium. *Sven. Bot. Tidskr.* **46**, 281–285.

MELIN, E. and NILSSON, H. (1957) Transport of ^{14}C-labelled photosynthate to the fungal associate of pine mycorrhiza. *Sven. Bot. Tidskr.* **51**, 166–186.

MOLINA, R., MASSICOTTE, H. and TRAPPE, J.M. (1992) Specific phenomena in mycorrhizal symbioses: community-ecological consequences and practical implications. In: *Mycorrhizal Functioning* (ALLEN, M.F., ed.), pp. 357–423. Chapman and Hall, New York.

MONZ, C.A., HUNT, H.W., REEVES, F.B. and ELLIOT, E.T. (1994) The response of mycorrhizal colonization to elevated CO_2 and climate change in *Pascopyrum smithii* and *Bouteloua gracilis*. *Plant and Soil* **165**, 75–80.

MORGAN, J.A., KNIGHT, W.G., DUDLEY, L.M. and HUNT, H.W. (1994) Enhanced root system C-sink activity, water relations and aspects of nutrient acquisition in mycotrophic *Bouteloua gracilis* subjected to CO_2 enrichment. *Plant and Soil* **165**, 139–146.

NYLUND, J.E. (1988) The regulation of mycorrhiza formation – carbohydrate and hormone theories reviewed. *Scand. J. For. Res.* **3**, 465–479.

NYLUND, J.E. and WALLANDER, H. (1989) Effects of ectomycorrhiza on host growth and carbon balance in a semi-hydroponic cultivation system. *New Phytol.* **112**, 389–398.

OLIVER, A.J., SMITH, S.E., NICHOLAS, D.J.D. and WALLACE, W. (1983) Activity of nitrate reductase in *Trifolium subterraneum*: effects of mycorrhizal infection and phosphate nutrition. *New Phytol.* **94**, 63–79.

PALMER, J.G. and HACSKAYLO, E. (1970) Ectomycorrhizal fungi in pure culture. I. Growth on single carbon sources. *Physiol. Plant.* **23**, 1187–1197.

PATEMAN, J.A. and KINGHORN, J.R. (1976) Nitrogen metabolism. In: *The Filamentous Fungi*, Vol. 2 (SMITH, J.E. and BERRY, D.R., eds.), pp. 159–237. Edward Arnold, London.

PLASSARD, C., BARRY, D., ELTROP, L. and MOUSAIN, D. (1994) Nitrate uptake in maritime pine (*Pinus pinaster*) and the ectomycorrhizal fungus *Hebeloma cylindrosporum*: effect of ectomycorrhizal symbiosis. *Can. J. Bot.* **72**, 189–197.

READ, D.J. (1991) Mycorrhizas in ecosystems. *Experientia* **47**, 376–391.
READ, D.J., LEAKE, J.R. and LANGDALE, A.R. (1989) The nitrogen nutrition of mycorrhizal fungi and their host plants. In: *Nitrogen, Phosphorus and Sulphur Utilization by Fungi* (BODDY, L., MARCHANT, R. and READ, D.J., eds), pp. 181–204. Cambridge University Press, Cambridge.
RIEGER, A., GUTTENBERGER, M. and HAMPP, R. (1992) Soluble carbohydrates in mycorrhized and non-mycorrhized fine roots of spruce seedlings. *Z. Naturforsch.* **47c**, 201–204.
ROBBINS, N.S. and PHARR, D.M. (1988) Effect of restricted root growth on carbohydrate metabolism and whole plant growth of *Cucumis sativus* L. *Plant Physiol.* **87**, 409–413.
RYGIEWICZ, P.T., BLEDSOE, C.S. and ZASOSKI, R.J. (1984) Effects of ectomycorrhizae and solution pH on [^{15}N]ammonium uptake by coniferous seedlings. *Can. J. For. Res.* **14**, 885–892.
SALZER, P. and HAGER, A. (1991) Sucrose utilization of the ectomycorrhizal fungi *Amanita muscaria* and *Hebeloma crustuliniforme* depends on the cell wall-bound invertase activity of their host *Picea abies*. *Bot. Acta* **104**, 439–445.
SCHAEFFER, C. (1995) Untersuchung des Kohlenhydratstoffwechsels von Ektomykorrhizen (Pilz-Baumwurzel-Symbiosen), Einfluß der Mykorrhizierung von *Picea abies* mit *Amanita muscaria* und *Cenococcum geophilum* auf Enzymaktivitäten und Metabolite des Saccharosestoffwechsels und der Glykolyse. Unpublished PhD thesis, Universität Tübingen.
SCHAEFFER, C., WALLENDA, T., GUTTENBERGER, M. and HAMPP, R. (1995) Acid invertase in mycorrhizal roots of Norway spruce (*Picea abies* [L.] Karst.) seedlings. *New Phytol.* **129**, 417–424.
SCHEROMM, P., PLASSARD, C. and SALSAC, L. (1990a) Effect of nitrate and ammonium nutrition on the metabolism of the ectomycorrhizal basidiomycete, *Hebeloma cylindrosporum* Romagn. *New Phytol.* **114**, 227–234.
SCHEROMM, P., PLASSARD, C. and SALSAC, L. (1990b) Regulation of nitrate reductase in the ectomycorrhizal basidiomycete, *Hebeloma cylindrosporum* Romagn., cultured on nitrate or ammonium. *New Phytol.* **114**, 441–447.
SMITH, S.E. and SMITH, F.A. (1990) Structure and function of the interfaces in biotrophic symbioses as they relate to nutrient transport. *New Phytol.* **114**, 1–38.
SMITH, S.E., ST JOHN, B.J., SMITH, F.A. and NICHOLAS, D.J.D. (1985) Activity of glutamine synthetase and glutamate dehydrogenase in *Trifolium subterraneum* L. and *Allium cepa* L.: effects of mycorrhizal infection and phosphate nutrition. *New Phytol.* **99**, 211–227.
STAHL, P.D. and SMITH, W.K. (1984) Effects of different geographic isolates of *Glomus* on the water relations of *Agropyron smithii*. *Mycologia* **76**, 261–267.
STEINGRABER, M., OUTLAW, W.H. JR and HAMPP, R. (1988) Subcellular compartmentation of fructose 2,6-bisphosphate in oat mesophyll cells. *Planta* **175**, 204–208.
STITT, M. (1990) Fructose-2,6-bisphosphate as a regulatory molecule in plants. *Annu. Rev. Plant Physiol. Plant Mol. Biol.* **41**, 153–185.
STITT, M. (1991) Rising CO_2 levels and their potential significance for carbon flow in photosynthetic cells. *Plant Cell Environ.* **14**, 741–762.
TESTER, M., SMITH, F.A. and SMITH, S.E. (1985) Phosphate inflow into *Trifolium subterraneum*. Effects of photon irradiance and mycorrhizal infection. *Soil Biol. Biochem.* **17**, 807–810.
THEODOROU, C. and REDDELL, P. (1991) *In vitro* synthesis of ectomycorrhizas on Casuarinaceae with a range of mycorrhizal fungi. *New Phytol.* **118**, 279–288.
TOBAR, R., AZCÓN, R. and BAREA, J.M. (1994) Improved nitrogen uptake

and transport from ^{15}N-labelled nitrate by external hyphae of arbuscular mycorrhiza under water-stressed conditions. *New Phytol.* **126**, 119–122.

WALLANDER H. and NYLUND, J.E. (1991) Effects of excess nitrogen on carbohydrate concentration and mycorrhizal development of *Pinus sylvestris* L. seedlings. *New Phytol.* **119**, 405–411.

WALLENDA, J., SCHAEFFER, C., EINIG, W., WINGLER, A., HAMPP, R., SEITH, B., GEORGE, E. and MARSCHNER, H. (1996) Effects of varied soil nitrogen supply on Norway spruce (*Picea abies* [L.] karst.) U. Carbon metabolism in needles and mycorrhizal roots. *Plant and Soil*, **186**, 361–369.

WANG, G.M., COLEMANN, D.C., FRECKMAN, D.W., DYER, M.I., MCNAUGHTON, S.J., ACRA, M.A. and GOESCHL, J.D. (1987) Carbon partitioning patterns of mycorrhizal versus non-mycorrhizal plants: real-time dynamic measurements using $^{11}CO_2$. *New Phytol.* **112**, 489–493.

WIEMKEN, V. (1995) Contributions of studies with in vitro culture systems to the understanding of the ectomycorrhizal symbiosis. In: *Mycorrhiza. Structure, Function, Molecular Biology and Biotechnology* (VARMA, A. and HOCK, B., eds), pp. 411–425. Springer-Verlag, Berlin.

WILLIAMS, P.G. (1992) Axenic culture of arbuscular mycorrhizal fungi. In: *Methods in Microbiology* **24** (NORRIS, J.R., READ, D.J. and VARMA, A.K., eds), pp. 203–220. Academic Press, London.

WINGLER, A., WALLENDA, T. and HAMPP, R. (1996) Mycorrhiza formation on Norway spruce (*Picea abies*) roots affects the pathway of anaplerotic CO_2 fixation. *Physiol. Plant.* **96**, 699–705.

ZHU, H., DANCIK, B.P. and HIGGINBOTHAM, K.O. (1994) Regulation of extracellular proteinase production in an ectomycorrhizal fungus *Hebeloma crustuliniforme*. *Mycologia* **86**, 227–234.

SECTION THREE

Related Metabolism

14

Respiration and the alternative oxidase

HANS LAMBERS

14.1 Introduction

Respiration consumes a major portion of the assimilates fixed in photosynthesis. It proceeds in both non-photosynthetic and photosynthetic organs, in the dark as well as in the light, albeit not necessarily at the same rate (Krömer, 1995). Respiration provides the energy to drive biosynthetic and transport processes in cells. Both NAD(P)H and ATP are end-products of respiration, which are subsequently used in numerous reactions. Most of the ATP derived from respiration is produced in the mitochondria, in a process which involves first the generation of a proton-motive force during electron transport from organic acids to oxygen and, second, the formation of ATP in a reaction which is driven by this proton-motive force. However, plant mitochondria also have respiratory electron transport pathways that allow the oxidation of organic acids to proceed without coupling to oxidative phosphorylation. First there is the rotenone-insensitive NADH dehydrogenase, a bypass of complex I. It catalyzes the transfer of electrons from NADH produced in the TCA cycle to ubiquinone (Q) without concomitant proton extrusion. Second, there is the cyanide-resistant, alternative pathway, which transfers electrons from ubiquinol (Q_r, the reduced form of Q) to oxygen without coupling to proton extrusion.

The phenomenon of cyanide-resistant respiration in higher plants has intrigued plant scientists since the beginning of this century. Much more recently, it was firmly established that the cyanide-resistant path and the cytochrome path differ in the amount of ATP which is produced per oxygen atom which is reduced. This triggered a further interest in this enigmatic pathway. In recent years considerable advances have been made in the characterization of the alternative pathway (Moore and Siedow, 1991). Now, when major progress has been made on aspects of its biochemical and molecular regulation, fascinating new developments are appearing on the horizon, which have shaken established dogmas and forced us to rethink the quantitative significance of the alternative pathway in plant metabolism as well as the physiological role of this pathway (Day *et al.*, 1996; Lambers and Atkin, 1995; Wagner and Krab, 1995).

Until recently, the widely held view was that the alternative pathway acts as an 'overflow' of the cytochrome path, never 'shares' electrons with the cytochrome path and only becomes engaged when the cytochrome path is inhibited or saturated with electrons. Dry et al. (1989) offered a biochemical explanation for this model, as will be detailed in Section 14.2.3. However, recent findings both with isolated mitochondria (Hoefnagel et al., 1995; Ribas-Carbo et al., 1995) and with intact tissues (Atkin et al., 1995), have demonstrated that at least under some circumstances and in some species, the alternative path shares electrons, donated by their common substrate – ubiquinol – with the cytochrome path.

This chapter aims to summarize new developments on the regulation of the two mitochondrial electron transport pathways and to revisit old and discuss new hypotheses on the physiological role of the alternative pathway.

14.2 Structure and regulation of the alternative pathway

Respiratory O_2 consumption of most higher plant tissues as well as that of mitochondria isolated from these tissues is not fully inhibited by inhibitors of the cytochrome path such as KCN, myxothiazol or antimycin, showing that the cyanide-resistant O_2 uptake resides in the mitochondria (Table 14.1; Lambers et al., 1983). This component of respiration is due to a single enzyme, a cyanide-resistant quinol oxidase: the alternative oxidase. The enzyme has a monomer

Table 14.1 A comparison of the cyanide resistance of respiration of intact roots and leaves of a number of species and of oxygen uptake by mitochondria isolated from these tissues. The percentage cyanide resistance of intact tissue respiration was calculated from the rate measured in the presence of 0.2 mM cyanide and that measured in the presence of 0.1 µM FCCP, which uncouples oxidative phosphorylation and electron transport; this was done to obtain a rate of electron transfer through the cytochrome path similar to the state 3 rate. Cyanide resistance of isolated mitochondria was calculated from the rate in the presence and absence of 0.2 mM KCN. Mitochondrial substrates were 10 mM malate plus 10 mM succinate and a saturating amount of ADP. Cyanide-resistant oxygen uptake by isolated mitochondria was fully inhibited by SHAM and disulfiram; in the presence of both KCN and SHAM approximately 10% of the control respiration proceeded in some of the tissues (residual respiration). (Data from Lambers et al., 1983)

		Cyanide resistance (%)	
Species	Tissue	Whole tissue	Mitochondria
Gossypium hirsutum	Roots	36	22
Phaseolus vulgaris	Roots	61	41
Spinacea oleracea	Roots	40	34
Triticum aestivum	Roots	38	35
Zea mays	Roots	47	32
Pisum sativum	Leaves	39	30
Spinacea oleracea	Leaves	40	27

molecular weight of around 36 kD (McIntosh, 1994). Its K_m for oxygen is an order of magnitude higher than that of cytochrome oxidase: 1.7 vs. 0.14 µM (Millar et al., 1994). Using a different methodology, Ribas-Carbo et al. (1994) confirmed the difference in K_m for oxygen and found considerably higher K_m values for the alternative oxidase (10–20 µM).

Since the effect of inhibitors of complex IV (e.g. KCN) and complex III (e.g. antimycin, myxothiazol) on mitochondrial O_2 uptake is the same, the branching point of the alternative path from the cytochrome path must be before complex III. Further work has established that Q is the component common to both pathways and that the alternative oxidase shunts electrons off the cyanide-sensitive, cytochrome pathway. The alternative pathway oxidizes Q and reduces oxygen to water in a reaction which involves no conservation of energy (for recent reviews, see Day et al., 1995; Siedow and Umbach, 1995; Wagner and Krab, 1995). The lack of energy conservation, apart from that associated with complex I, is a major difference in comparison with the cytochrome path.

14.2.1 Molecular biology of the alternative oxidase

The alternative oxidase is encoded in the nucleus (McIntosh, 1994). Sequences of the alternative oxidase cDNA have been reported for a number of plant species and one yeast (Day et al., 1995; McIntosh, 1994). The plant sequences show a high identity, but are much less identical to the yeast sequence (Table 14.2). The plant sequences have a cleavable mitochondrial targeting presequence and some conserved regions which are possibly involved in metal binding in a similar way as observed in superoxide dismutase. Genomic clones have been isolated showing four exons (i.e. coding regions of the gene), separated by introns (i.e. non-coding regions) (Day et al., 1995). For tobacco, two cDNA clones have been identified by different groups from different cultivars, which has led to the suggestion that, at least tobacco plants contain more than one copy of the alternative oxidase gene (Whelan et al., 1995).

The alternative oxidase molecules are embedded in the inner mitochondrial membrane as homodimers, probably linked via cysteine bridges. When the

Table 14.2 A comparison of alternative oxidase cDNA from different species, four higher plant and one yeast species. Two closely related sequences have been reported for Nicotiana tabacum (Vanlerberghe et al., 1994; Whelan et al., 1995). Values shown indicate percentage identity of predicted amino acid sequences (Day et al., 1995)

	Nt 1	Nt 2	Gm	At	Sg	Ha
Nicotiana tabacum 1	100	95	72	74	64	31
Nicotiana tabacum 2		100	79	78	74	32
Glycine max			100	77	69	34
Arabidopsis thaliana				100	69	33
Sauromatum guttatum					100	38
Hansenula anomala						100

monomers are covalently linked via the putative disulphide bonds (oxidized), the oxidase has a low capacity compared to when the monomers are non-covalently associated (reduced) (Umbach and Siedow, 1993). Like many other enzymes, the alternative oxidase is affected allosterically, for example by its substrate (ubiquinone) and by organic acids (pyruvate), as discussed in Section 14.2.3.

Hydropathy analysis of the predicted amino acid sequence of the alternative oxidase indicates three helical structures. The two hydrophobic helices probably span the inner membrane, whereas the hydrophylic helix is likely to be exposed to the intermembrane space, as are the N- and C-termini. The active site is also thought to be on the matrix side (Siedow et al., 1992).

14.2.2 Induction of the alternative oxidase

Induction of the alternative path in higher plants occurs when the activity of the cytochrome path is restricted, for example by application of inhibitors of mitochondrial protein synthesis (Edwards et al., 1974, cited in Day et al., 1995), inhibitors of the cytochrome path (Wagner et al., 1992) or exposure to low temperature (Vanlerberghe and McIntosh, 1992). Interestingly, only those inhibitors of the cytochrome path that enhance superoxide production lead to induction of the alternative oxidase. Moreover, superoxide itself can also induce expression of the alternative oxidase. This has led to the suggestion that active oxygen species, including H_2O_2, are part of the signal(s) communicating cytochrome path restriction in the mitochondria to the nucleus, thus inducing alternative oxidase synthesis (Wagner and Krab, 1995).

In thermogenic tissues, such as the floral appendix of *Sauromatum guttatum*, salicylic acid plays a role in alternative path induction (Raskin, 1992; Raskin et al., 1987; Rhoads and McIntosh, 1992). In other tissues salicylic acid may have a similar function, but this remains to be firmly established (Van der Straeten et al., 1995). Salicylic acid fails to induce transcription when applied in nuclear *in vitro* run-on experiments (Rhoads and McIntosh, 1992). It has therefore been suggested that salicylic acid induces alternative oxidase activity in an indirect manner, through active oxygen species (Wagner and Krab, 1995). This agrees with the finding that salicylic acid application enhances the production of H_2O_2 in cell cultures of parsley (Krauss and Jeblick, 1995). Ethylene also induces the alternative oxidase, for example in ripening fruits (Laties, 1982).

14.2.3 Regulation of the alternative oxidase

The existence of two respiratory pathways, both transporting electrons from ubiquinol to O_2, raises the question if and how electron flow is regulated between the two paths. This is particularly relevant, since the cytochrome path is coupled to proton extrusion, whereas transport of electrons from ubiquinol to O_2 via the alternative path is not coupled to the generation of a proton-motive force. Characterization of the regulation of partitioning of electron flow between the alternative and the cytochrome pathways was first undertaken using isolated mitochondria (Bahr and Bonner, 1973). These authors found that cytochrome pathway inhibitors, such as cyanide or antimycin, diverted electrons from the cytochrome path to the alternative pathway. However, inhibition of the alternative pathway, using compounds such as salicylhydroxamic acid (SHAM), did not switch electrons from the alternative path to the cytochrome pathway. These results suggested that electron flow through the alternative pathway only takes place when electron flow via the cytochrome pathway is either at or near saturation. Bahr and Bonner (1973) were the first to conclude that simple competition for electrons between the alternative path and the cytochrome path cannot explain the experimental data.

Further evidence that the alternative path does not compete for electrons with the cytochrome path has come from titration experiments. This involves using a range of concentrations of an inhibitor of the cytochrome path in the absence and presence of an inhibitor of the alternative path (Bingham and Farrar, 1987; De Visser and Blacquière, 1984; Theologis and Laties, 1978). In such experiments O_2 uptake is measured at a range of concentrations of an inhibitor of the cytochrome path, both in the absence and presence of an inhibitor which fully blocks the alternative path. If inhibition of the alternative path were to increase the flow of electrons through the cytochrome path, inhibitors of the cytochrome path would have a stronger effect under these conditions than in the absence of an inhibitor of the alternative path. Until recently, with one exception, this was never found (Day, 1992). It was therefore widely believed that the alternative path does not compete for electrons with the cytochrome path; it only becomes engaged when the cytochrome path is virtually saturated with electrons. For a long time, a wealth of evidence supported the hypothesis that the alternative path functions as an 'energy overflow' (Lambers, 1985). How could this model be accounted for in biochemical terms?

Application of a procedure that allows simultaneous measurement of oxygen uptake and ubiquinone redox poise (Moore et al., 1988) qualitatively confirmed the Bahr/Bonner model. During succinate oxidation by isolated mitochondria, significant alternative oxidase activity only appeared when the ubiquinone pool was more than 40 to 50% reduced, while the activity of the main pathway increased as a linear function of reduced ubiquinone at levels well below those needed for alternative oxidase activity (Figure 14.1; Dry et al., 1989; Moore et al., 1988). This provided a solid biochemical basis for the observations that the alternative path does not operate until the cytochrome path approaches saturation. However, recent information has cast doubt on the generality of the conclusion that the cytochrome path is invariably saturated before the alternative path contributes to respiratory electron transport.

Addition of pyruvate to isolated mitochondria activates the alternative path (Millar et al., 1993), due to a shift to the left of the curve describing the activity of the alternative path as a function of Q_r/Q_t (Umbach et al., 1994). This is not because pyruvate is used as a respiratory substrate, but due to an allosteric effect (Figure 14.1). Other organic acids have similar allosteric effects (Day et al., 1995; Wagner et al., 1995). If allosteric effectors were to shift the curve describing the activity of the alternative path as a function of Q_r/Q_t very close to that of the cytochrome path, competition between the two pathways might well occur. Indeed, in the presence of pyruvate, but not in its absence, sharing of electrons between the two pathways has been demonstrated in isolated mitochondria (Hoefnagel et al., 1995; Ribas-Carbo et al., 1995). Moreover, evidence for such

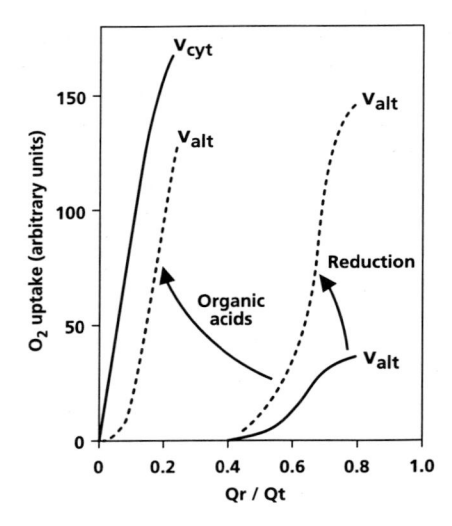

A molecular approach to primary metabolism in higher plants

Figure 14.1 Dependence of the activity of the cytochrome path and of the alternative path on the fraction of ubiquinone that is in its reduced state (Q_r/Q_t). When the alternative oxidase is in its 'reduced' (high-activity configuration), it has a greater capacity to accept electrons. In its reduced state, the alternative oxidase can be affected allosterically by organic acids, which increase its affinity for Q_r (based on information in Dry et al. (1989), Umbach et al. (1994) and Day et al. (1995))

sharing has been found for intact tissues as well (Atkin et al., 1995). The alternative path can therefore, under appropriate conditions, compete with the cytochrome pathway for electrons and switching of electrons from the alternative pathway to the cytochrome pathway can take place following inhibition of the alternative oxidase.

In vitro, the change of the alternative oxidase from the less-active 'oxidized' state to the more-active 'reduced' state can be brought about by artificial reducing agents such as dithiothreitol (Umbach and Siedow, 1993; Table 14.3). Artificial oxidizing agents have the reverse effect (Umbach and Siedow, 1993). There is also evidence that the redox state of the alternative oxidase may shift *in vivo*, for example, during the onset of the respiratory crisis in thermogenic tissues (Umbach and Siedow, 1993) as well as during the normal development of roots of *Glycine max* (Day et al., 1994) and leaves of *Pisum sativum* (Lennon et al., 1995). So far, there is no clear evidence which factor(s) might cause the shift from the 'oxidized' to the 'reduced' state *in vivo*. However, the recent findings that TCA-cycle intermediates affect this configuration in a time-dependent manner suggest that accumulation of citrate, isocitrate and/or malate increases the fraction of the active form (Table 14.3; Vanlerberghe et al., 1995). Addition of oxaloacetate rapidly consumes NADH which is produced during oxidation of (iso)citrate, since the thermodynamic equilibrium of the reaction favours malate formation. Since addition of oxaloacetate favours the 'oxidized' less-active configuration of the alternative oxidase (Table 14.3), it is likely that the shift in configuration involves the net production of NADH.

Table 14.3 The relative amount of 'oxidized' (less-active, covalently bound) alternative oxidase in mitochondria isolated from the leaves of *Nicotiana tabacum*. The mitochondria were supplied with 1 mM ADP and 10 mM of the indicated substrates, except for oxaloacetate, which was supplied at 1 mM (Vanlerberghe et al., 1995)

Substrate	Incubation time (min)	Relative amount of the 'oxidized' alternative oxidase
Citrate	10	0.39
Citrate, oxaloacetate	10	0.81
Isocitrate	10	0.17
Isocitrate, oxaloacetate	10	0.63
Malate (under conditions favouring pyruvate production)	5	0.79
Malate (under conditions favouring oxaloacetate production)	5	0.36
Citrate	2	0.88
	5	0.43
	10	0.19
Succinate	2	1.02
	5	0.88
	10	0.42
Dithiothreitol	10	0.02

14.3 How to assess the activity of the alternative pathway?

The general acceptance of the electron overflow paradigm led to the use of inhibitors of the alternative oxidase to assess alternative pathway activity in isolated mitochondria and intact tissues (Lambers, 1985; Laties, 1982). Inhibition of oxygen uptake following addition of SHAM was taken as a measure of alternative pathway activity in the uninhibited condition, because diversion of electrons onto the saturated cytochrome pathway after inhibitor addition was thought impossible. Numerous caveats attended the application of inhibitors (Møller et al., 1988), but its theoretical underpinnings appeared sound. The recent observations discussed in Section 14.2, cast serious doubt on this inhibitor approach. If organic acid levels *in vivo* are high enough to activate the alternative oxidase allosterically, the alternative pathway can compete with the cytochrome pathway for electrons when neither pathway is near saturation. The addition of SHAM might show little or no inhibition of oxygen uptake, since the electrons were accommodated by the excess capacity available on the cytochrome pathway. Application of specific inhibitors in such a situation would lead to the erroneous conclusion that there was no alternative pathway activity in the uninhibited condition. Moreover, it has also been shown that addition of an inhibitor like SHAM increases Q_r/Q_t, at least in isolated mitochondria (Millar et al., 1995). This implies that the substrate concentration for the alternative path increases upon partial inhibition of the alternative oxidase. Consequently, inhibition is less than expected from the activity of the alternative path, prior to the addition of SHAM. It is still uncertain to what extent this phenomenon occurs *in vivo*, since Wagner and Wagner (1995) found no shift in Q_r/Q_t upon addition of SHAM to intact cells, not even in the presence of uncoupler which did engage the alternative path.

Any observed inhibition of respiration following the addition of an alternative pathway inhibitor to intact tissue or isolated mitochondria only allows the conclusion that some alternative pathway activity was present prior to inhibition. But no quantitative estimate of the activity can be derived from such measurements. Measurements of maximum alternative pathway activity made following the addition of inhibitors of the cytochrome pathway may also be confounding. Such inhibition might lead to perturbation of the reduction state of the pyridine nucleotide pool in the mitochondrial matrix, and lead to activation of the alternative oxidase through reduction of the sulphide bonds (Vanlerberghe *et al.*, 1995). This might result in the measurement of a higher alternative pathway activity than existed prior to inhibition. If inhibitors can no longer be used to assess actual and maximum activity of the alternative path, how should we proceed if we wish to quantify the flux of electrons through the two pathways?

The only promising method for quantitative measurements of alternative pathway activity is the use of oxygen isotope fractionation (Guy *et al.*, 1992; Robinson *et al.*, 1992, 1995), but even this technique is still in the developmental stage. It is based on the observation that cytochrome c oxidase and the alternative oxidase have different fractionation for ^{18}O when reducing oxygen to water. This fractionation can be measured with a mass spectrometer. This allows calculation of the partitioning of electron flow between the two pathways in the absence of added inhibitors, not only in isolated mitochondria but also in intact tissues. Comparative results of oxygen electrode and mass spectrometer studies using isolated mitochondria support the validity of such measurements (Ribas-Carbo *et al.*, 1995).

14.4 The physiological function of the alternative pathway

In trying to understand the physiological role of the cyanide-resistant path, it is important to take into account its non-phosphorylating nature and its operation during high rates of substrate oxidation. In the inflorescence of thermogenic plants, the alternative path is responsible for a major part of the rise in temperature. However, not only the respiration of some reproductive organs, but also that of roots and leaves is (partly) cyanide-resistant. The application of inhibitors of the alternative path has demonstrated that the alternative path also contributes to normal respiration of such tissues. Even though we cannot assess its exact activity, as pointed out in Section 14.3, we do know that in roots the alternative path may be responsible for up to 40% of total root respiration, and sometimes even more. The role of the alternative path in roots and leaves is unlikely to be that of heat production and a number of hypotheses have been put forward to explain the functioning of the alternative path in these organs.

14.4.1 Heat production

In the male reproductive structures of aroids (Meeuse, 1975) and also in a range of other species, including water lilies (Skubatz *et al.*, 1990) and cycads (Skubatz *et al.*, 1993), the alternative path plays a role in thermogenesis. During its 'respiratory crisis' the respiration rate of the ephemeral inflorescence of aroids approaches the rate found in the flying muscles of humming birds, and its temperature may rise to 10°C above ambient. Similar increases in temperature have been reached for water lilies, whereas in cycads the temperature rises only a few degrees. Due to the increase in temperature of the tissue, scents are volatilized and pollinators are attracted. During the respiratory crisis, respiration is largely cyanide-resistant. The amount of alternative oxidase increases dramatically just before the onset of the respiratory crisis (Elthon and McIntosh, 1987). Moreover, the oxidase is largely in its inactive, 'oxidized' state before the onset of the respiratory crisis, and is then transformed into its active state, due to reduction of the putative disulphide bonds (Umbach and Siedow, 1993).

14.4.2 Energy overflow

Until recently, it was generally accepted that the alternative path is only operative when the cytochrome path is inhibited or reaches its saturation. These observations have led to the 'energy overflow hypothesis' (Lambers, 1985), which states that respiration via the alternative path only proceeds in the presence of high concentrations of respiratory substrate which 'flood' the cytochrome path. Now the view has changed, in that the alternative path and the cytochrome path may share electrons from Q_r when pyruvate levels increase. Does this mean that the energy overflow model has to be dismissed completely? Certainly not, but major modifications are required.

Pyruvate, citrate and other organic acids are likely to accumulate when their rate of production is not matched by their oxidation by the mitochondria. Accumulation of TCA-cycle intermediates, like citrate, is likely to affect the reduction of the alternative oxidase (Vanlerberghe *et al.*, 1995). Other acids, for example pyruvate, then enhance the affinity of the alternative oxidase for Q_r. These two effects will feed-forward to activate the alternative oxidase. In combination with the response of the alternative oxidase to Q_r/Q_t, they allow the alternative path to function as an energy overflow, be it in a more subtle way than originally proposed. Thus these recent observations have led to a distinct modification of the energy overflow paradigm (Day *et al*, 1996; Lambers and Atkin, 1995).

Is there any evidence that high endogenous levels of organic acids which reduce the alternative oxidase and/or allosterically activate the alternative oxidase are indeed associated with a large flux of electrons through the alternative path? Such an association occurs during rapid malate decarboxylation in CAM plants. If this decarboxylation occurs in the dark, it is indeed associated with an increased activity of the alternative path (Robinson *et al.*, 1992). Further circumstantial evidence comes from experiments with roots, in which those species which show the highest endogenous pyruvate levels also show the highest activity of the alternative path (Day and Lambers, 1983). Such a situation is likely to arise when the activity of the glycolytic pathway exceeds the flow of electrons that can be accommodated by the cytochrome path. Interestingly, the alternative oxidase is also allosterically activated by glyoxylate, an intermediate of the photorespiratory cycle (Day et al., 1995). This offers the possibility of an enhanced activity of the alternative path under conditions of a high carbon flux through this cycle.

It has been suggested that the continuous oxidation of part of the substrate via a non-phosphorylating electron transport path allows the plant to increase the availability of carbon for sinks which suddenly arise, for example upon a decrease of the water potential in the root environment (Lambers, 1985). However, the significance of an energy overflow might also be that in this manner the alternative oxidase activity prevents the production of harmful levels of superoxide and/or hydrogen peroxide (see Section 14.4.6).

14.4.3 Energy overcharge

An increased demand for ATP in cells in which the cytochrome path operates at maximum capacity may enhance the rate of glycolysis and subsequently lead to increased operation of the alternative path. This was termed the 'energy overcharge model' (De Visser et al., 1986). It may well be relevant under non-steady-state conditions and/or when the activity of the cytochrome path is no longer controlled by adenylates.

14.4.4 Continuation of respiration in the presence of inhibitors, including CO_2

Naturally occurring inhibitors of the cytochrome path, for example sulphide, carbon dioxide and cyanide, may reach such high concentrations in the tissue that respiration via the cytochrome path is partially or fully inhibited (Palet et al., 1991). The presence of an alternative path, which is unaffected by such inhibitors, may allow continued respiration and ATP production, albeit with low efficiency, under such conditions. It has repeatedly been found that leaf respiration in the dark is inhibited at elevated concentrations of CO_2 in the atmosphere, such as are predicted to occur in the next century (e.g. El Kohen et al., 1991; Wullschleger et al., 1994). It is likely that such inhibition is at least partly due to inhibition of the cytochrome path, since in vitro cytochrome oxidase is inhibited by CO_2 concentrations in the range known to inhibit respiration of intact leaves (Gonzalez-Meler, 1995). However, total leaf dark respiration is only expected to decline as a consequence of inhibition of cytochrome oxidase if the partial inhibition of the cytochrome path cannot be accommodated by increased activity of the alternative path, for example because this path was already fully saturated. It is therefore likely that other effects of an elevated atmospheric CO_2 concentration are due to inhibition of other respiratory enzymes, for example succinate dehydrogenase (Gonzalez-Meler, 1995).

14.4.5 NADH oxidation in the presence of a high energy charge

If cells require a large amount of carbon skeletons (e.g. oxoglutarate or succinate) but do not have a high demand for ATP, the operation of the alternative path could prove useful. However, it is hard to envisage such a situation *in vivo*. Whenever the rate of carbon skeleton production is high, there tends to be a great need for ATP to further metabolize and incorporate these skeletons. Also, when the carbon skeletons are used for the synthesis of amino acids, significantly more NADH is required for the reduction of nitrate (if this is the source of N) than generated in the production of carbon skeletons.

Cells from soybean root nodules, which are infected with *Rhizobium* bacteroids and rapidly synthesize organic acids, have less alternative path capacity and activity than any other cells from the same nodules or other tissues of the same plants (Kearns *et al.*, 1992). Hence, rapid synthesis of organic acids is by no means invariably associated with greater activity of the alternative path.

Lance *et al.* (1985) have suggested the need for a non-phosphorylating path to allow rapid oxidation of malate in the absence of a large need for ATP. Such a situation occurs during rapid malate decarboxylation in CAM plants. If this decarboxylation occurs in the dark, it is indeed associated with an increased activity of the alternative path (Robinson *et al.*, 1992). It is likely that the increased pyruvate concentrations, resulting from rapid malate decarboxylation, are responsible for this increase in alternative path activity. However, it has to be kept in mind that malate decarboxylation naturally occurs in the light and it therefore remains to be confirmed that the alternative path plays a vital role in crassulacean acid metabolism.

14.4.6 Avoidance of damage by free radicals charge

It has been proposed that the activity of alternative oxidase prevents the production of superoxide and/or hydrogen peroxide. Superoxide is produced when electron transport through the cytochrome path is impaired, for example due to low temperature or desiccation injury, and this is partly due to a reaction of ubisemiquinone with molecular oxygen (Purvis and Shewfelt, 1993). Superoxide, like other free radicals, may lead to severe metabolic disturbances and a wide range of environmentally induced plant disorders, including chilling damage, are mediated by reactive oxygen species (Scandalios, 1993).

The various interpretations of the physiological function of the alternative oxidase remain speculative in the absence of pertinent results with plants lacking the alternative path. Such plants have recently become available (Vanlerberghe *et al.*, 1994, 1995).

Acknowledgments

The author gratefully acknowledges the constructive comments of Frank Millenaar, Miquel Ribas-Carbo and Bert Simons on the draft of this manuscript.

References

ATKIN, O.K., VILLAR, R. and LAMBERS, H. (1995) Partitioning of electrons between the cytochrome and the alternative pathways in intact roots. *Plant Physiol.* **108**, 1179–1183.

BAHR, J.T. and BONNER, W.D. (1973) Cyanide-insensitive respiration II. Control of the alternative pathway. *J. Biol. Chem.* **248**, 3446–3450.

BINGHAM, I.J. and FARRAR, J.F. (1987) Respiration of barley roots: assessment of the activity of the alternative path using SHAM. *Physiol. Plant.* **70**, 491–498.

DAY, D.A. (1992) Can inhibitors be used to estimate the contribution of the alternative oxidase to respiration in plants? In: *Plant Respiration. Molecular, Biochemical and Physiological Aspects* (LAMBERS, H. and VAN DER PLAS, L.H.W., eds), pp. 37–42. SPB Academic Publishing, The Hague.

DAY, D.A. and LAMBERS, H. (1983) The regulation of glycolysis and electron transport in roots. *Physiol. Plant.* **58**, 155–160.

DAY, D.A., MILLAR, A.H., WISKICH, J.T. and WHELAN, J. (1994) Regulation of alternative oxidase by pyruvate in soybean mitochondria. *Plant Physiol.* **106**, 1421–1427.

DAY, D.A., WHELAN, J., MILLAR, A.H., SIEDOW, J.N. and WISKICH, J.T. (1995) Regulation of the alternative oxidase in plants and fungi. *Austr. J. Plant Physiol.* **22**, 497–509.

DAY, D.A., KRAB, K., LAMBERS, H., MOORE, A.L., SIEDOW, J.N., WAGNER, A.M. and WISKICH, J.T. (1996) The cyanide-resistant oxidase: to inhibit or not to inhibit, that is the question. *Plant Physiol.* **110**, 1–2.

DE VISSER, R. and BLACQUIÈRE, T. (1984) Inhibition and stimulation of root respiration in *Pisum* and *Plantago* by hydroxamate. Its consequences for the assessment of alternative path activity. *Plant Physiol.* **75**, 813–817.

DE VISSER, R., SPREEN BROUWER, K. and POSTHUMUS, F. (1986) Alternative path mediated ATP synthesis in roots of *Pisum sativum* upon nitrogen supply. *Plant Physiol.* **80**, 295–300.

DRY, I.B., MOORE, A.L., DAY, D.A. and WISKICH, J.T. (1989) Regulation of alternative pathway activity in plant mitochondria. Non-linear relationship between electron flux and the redox poise of the quinone pool. *Arch. Biochem. Biophys.* **273**, 148–157.

EL KOHEN, A., PONTAILLER, J.-Y. and MOUSSEAU, M. (1991) Effect of doubling of atmospheric CO_2 concentration on dark respiration in aerial parts of young chestnut trees (*Castanea sativa* Mill.). *CR Acad. Sci. Paris* t. 312, Series III, 477–481.

ELTHON, T.E. and MCINTOSH, L.E. (1987) Identification of the alternative terminal oxidase of higher plant mitochondria. *Proc. Natl Acad. Sci. USA* **84**, 8399–8403.

GONZALEZ-MELER, M.A. (1995) Effect of increasing concentration of atmospheric carbon dioxide on plant respiration. PhD thesis, Universitat de Barcelona.

GUY, R.D., BERRY, J.A., FOGEL, M.L., TURPIN, D.H. and WEGER, H.G. (1992) Fractionation of the stable isotopes of oxygen during respiration by plants – the basis of a new technique to estimate partitioning to the alternative path. In: *Plant Respiration. Molecular, Biochemical and Physiological Aspects* (LAMBERS, H. and VAN DER PLAS, L.H.W., eds), pp. 443–453. SPB Academic Publishing, The Hague.

HOEFNAGEL, M.H.N., MILLAR, A.H., WISKICH, J.T. and DAY, D.A. (1995) Cytochrome and alternative respiratory pathways compete for electrons in the presence of pyruvate in soybean mitochondria. *Arch. Biochem. Biophys.* **318**, 394–400.

KEARNS, A., WHELAN, J., YOUNG, S., ELTHON, T.E. and DAY, D.A. (1992) Tissue-specific expression of the alternative oxidase in soybean and siratro. *Plant Physiol.* **99**, 712–717.

KRAUSS, H. and JEBLICK, W. (1995) Pretreatment of parsley suspension cultures with salicylic acid enhances spontaneous and elicited production of H_2O_2. *Plant Physiol.* **108**, 1171–1178.

KRÖMER, S. (1995) Respiration during photosynthesis. *Annu. Rev. Plant Physiol. Plant Mol. Biol.* **46**, 45–70.

LAMBERS, H. (1985) Respiration in intact plants and tissues: its regulation and dependence on environmental factors, metabolism and invaded organisms. In: *Encyclopedia of Plant Physiology, New Series* (DOUCE, R. and DAY, D.A., eds), pp. 418–473. Springer-Verlag, Berlin.

LAMBERS, H. and ATKIN, O.K. (1995) Regulation of carbon metabolism in roots. In: *Carbon Partitioning and Source-Sink Interactions in Plants* (MADORE, M.A. and LUCAS, W.J, eds), pp. 226–238. American Society of Plant Physiologists, Rockville.

LAMBERS, H., DAY, D.A. and AZCÓN-BIETO, J. (1983) Cyanide-resistant respiration in roots and leaves. Measurements with intact tissues and isolated mitochondria. *Physiol. Plant.* **58**, 148–154.

LANCE, C., CHAUVEAU, M. and DIZENGREMEL, P. (1985) The cyanide-resistant pathway of plant mitochondria. In: *Encyclopedia of Plant Physiology, New Series* (DOUCE, R. and DAY, D.A., eds), pp. 202–247. Springer-Verlag, Berlin.

LATIES, G.G. (1982) The cyanide-resistant, alternative path in plant mitochondria. *Annu. Rev. Plant Physiol. Plant Mol. Biol.* **33**, 519–555.

LENNON, A.M., PRATT, J., LEACH, G. and MOORE, A.L. (1995) Developmental regulation of respiratory activity in pea leaves. *Plant Physiol.* **107**, 925–932.

MCINTOSH, L. (1994) Molecular biology of the alternative oxidase. *Plant Physiol.* **105**, 781–786.

MEEUSE, B.J.D. (1975) Thermogenic respiration in Aroids. *Annu. Rev. Plant Mol. Biol.* **26**, 117–126.

MILLAR, A.H., WISKICH, J.T., WHELAN, J. and DAY, D.A. (1993) Organic activation of the alternative oxidase of plant mitochondria. *FEBS Lett.* **329**, 259–262.

MØLLER, I.M., BÉRCZI, A., VAN DER PLAS, L.H.W. and LAMBERS, H. (1988) Measurement of the activity and capacity of the alternative pathway in intact plant tissues: identification of problems and possible solutions. *Physiol. Plant.* **72**, 642–649.

MOORE, A.L. and SIEDOW, J.N. (1991) The regulation and nature of the cyanide-resistant alternative oxidase of plant mitochondria. *Biochim. Biophys. Acta* **1059**, 121–140.

MOORE, A.L., DRY, I.B. and WISKICH, J.T. (1988) Measurement of the redox state of the ubiquinone pool in plant mitochondria. *FEBS Lett.* **235**, 76–80.

PALET, A., RIBAS-CARBO, M., ARGILES, J.M. and AZCÓN-BIETO, J. (1991) Short-term effects of carbon dioxide on carnation callus cell respiration. *Plant Physiol.* **96**, 467–472.

PURVIS, A.C. and SHEWFELT, R.L. (1993) Does the alternative pathway ameliorate chilling injury in sensitive plant tissues? *Physiol. Plant.* **88**, 712–718.

RASKIN, I. (1992) Role of salicylic acid in plants. *Annu. Rev. Plant Physiol. Plant Mol. Biol.* **43**, 439–463.

RASKIN, I., EHMANN, A., MELANDER, W.R. and MEEUSE, B.J.D. (1987) Salicylic acid: a natural inducer of heat production in *Arum* lilies. *Science* **237**, 1601–1602.

RHOADS, D.M. and MCINTOSH, L. (1992) Salicylic acid regulation of respiration in higher plants. *Plant Cell* **4**, 1131–1139.

RIBAS-CARBO, M., BERRY, J.A., AZCÓN-BIETO, J. and SIEDOW, J.N. (1994) The reaction of the plant mitochondrial cyanide-resistant alternative oxidase with oxygen. *Biochim. Biophys. Acta* **1188**, 205–212.

RIBAS-CARBO, M., BERRY, J.A., YAKIR, D., GILES, L., ROBINSON, S.A., LENNON, A.M. and SIEDOW, J.N. (1995) Electron partitioning between the cytochrome and alternative pathways in plant mitochondria. *Plant Physiol.* **109**, 829–837.

ROBINSON, S.A., YAKIR, D., RIBAS-CARBO, M., GILES, L., OSMOND, C.B., SIEDOW, J.N. and BERRY, J.A. (1992) Measurements of the engagement of cyanide-resistant respiration in the crassulacean acid metabolism plant *Kalanchoe daigremontiana* with the use of on-line oxygen isotope discrimination. *Plant Physiol.* **100**, 1087–1091.

ROBINSON, S.A., RIBAS-CARBO, M., YAKIR, D., GILES, L., REUVENI, Y. and BERRY, J.A. (1995) Beyond SHAM and cyanide: opportunities for studying the alternative oxidase in plant respiration using oxygen isotope discrimination. *Aust. J. Plant Physiol.* **22**, 487–496.

SCANDALIOS, J.G. (1993) Oxygen stress and superoxide dismutases. *Plant Physiol.* **101**, 7–12.

SIEDOW, J.N. and UMBACH, A.L. (1995) Plant mitochondrial electron transfer and molecular biology. *Plant Cell* **7**, 821–831.

SIEDOW, J.N., WHELAN, J., KEARNS, A., WISKICH, J.T. and DAY, D.A. (1992) Topology of the alternative oxidase in soybean mitochondria. In: *Plant Respiration. Molecular, Biochemical and Physiological Aspects* (LAMBERS, H. and VAN DER PLAS, L.H.W., eds), pp. 19–27. SPB Academic Publishing, The Hague.

SKUBATZ, H., WILLIAMSON, P.S., SCHNEIDER, E.L. and MEEUSE, B.J.D. (1990) Cyanide-insensitive respiration in thermogenic flowers of *Victoria* and *Nelumbo. J. Exp. Bot.* **41**, 1335–1339.

SKUBATZ, H., TANG, W. and MEEUSE, B.J.D. (1993) Oscillatory heat-production in the male cones of cycads. *J. Exp. Bot.* **44**, 489–492.

THEOLOGIS, A. and LATIES, G.G. (1978) Relative contribution of cytochrome-mediated and cyanide-resistant electron transport in fresh and aged potato slices. *Plant Physiol.* **62**, 232–237.

UMBACH, A.L. and SIEDOW, J.N. (1993) Covalent and noncovalent dimers of the cyanide-resistant alternative oxidase protein in higher plant mitochondria and their relationship to enzyme activity. *Plant Physiol.* **103**, 845–854.

UMBACH, A.L., WISKICH, J.T. and SIEDOW, J.N. (1994) Regulation of alternative oxidase kinetics by pyruvate and intermolecular disulfide bond redox status in soybean seedling mitochondria. *FEBS Lett.* **348**, 181–184.

VAN DER STRAETEN, D., CHAERLE, L., SHARKOV, G., LAMBERS, H. and VAN MONTAGU, M. (1994) Salicylic acid enhances the activity of the alternative pathway of respiration in tobacco leaves and induces thermogenicity. *Planta* **196**, 412–419.

VANLERBERGHE, G.C. and MCINTOSH, L. (1994) Lower growth temperatures increase alternative oxidase protein in tobacco callus. *Plant Physiol.* **100**, 115–119.

VANLERBERGHE, G.C. VANLERBERGHE, A.E. and MCINTOSH, L. (1994) Molecular genetic alteration of plant respiration: transgenic tobacco expressing sense and antisense alternative oxidase genes. *Plant Physiol.* **106**, 1503–1510.

VANLERBERGHE, G.C., DAY, D.A., WISKICH, J.T., VANLERBERGHE, A.E. and MCINTOSH L. (1995) Alternative oxidase activity in tobacco leaf mitochondria. *Plant Physiol.* **109**, 353–361.

WAGNER, A.M. and KRAB, K. (1995) The alternative respiration pathway in plants: role and regulation. *Physiol. Plant.* **95**, 318–325.

WAGNER, A.M. and WAGNER, M.J. (1995) Measurements of *in vivo* ubiquinone reduction levels in plant cells. *Plant Physiol.* **108**, 277–283.

WAGNER, A.M., VAN EMMERIK, W.A.M., ZWIERS, J.H. and KAAGMAN, H.M.C.M. (1992) Energy metabolism of *Petunia hybrida* cell suspensions growing in the presence of antimycin A. In: *Plant Respiration. Molecular,*

Biochemical and Physiological Aspects (LAMBERS, H. and VAN DER PLAS, L.H.W., eds), pp. 609–614. SPB Academic Publishing, The Hague.

WAGNER, A.M., VAN DEN BERGEN, C.W.M. and WINCENCJUSZ, H. (1995) Stimulation of the alternative pathway by succinate and malate. *Plant Physiol.* **108**, 1035–1042.

WHELAN, J., SMITH, M.K., MEIJER, M., YU, J.-W., BADGER, M.R., PRICE, G.D. and DAY, D.A. (1995) Cloning of an additional cDNA for the alternative oxidase in tobacco. *Plant Physiol.* **107**, 1469–1470.

WULLSCHLEGER, S.D., ZISKA, L.H. and BUNCE, J.A. (1994) Respiratory responses of higher plants to atmospheric CO_2 enrichment. *Physiol. Plant.* **90**, 221–229.

15

Manipulation of oil biosynthesis and lipid composition

MATTHEW J. HILLS AND STEPHEN RAWSTHORNE

15.1 The pathway of lipid biosynthesis

In most plant tissues the lipids are predominantly in the cellular membranes, but in seeds and fruit tissues storage lipids are also found. These storage lipids are valuable sources of a diverse range of oils which have numerous food and non-food applications. It is for this reason that the study of oil biosynthesis has attracted much attention. In particular, the ability to alter specifically the composition of storage oils in existing crop plants towards particular markets, through genetic manipulation with transgenes, has driven much of the research effort into fatty acid and lipid biochemistry over the past ten years. Indeed, genetically engineered oilseed crops are already in the field. Before one is able to carry out these manipulations in a targeted manner it is vital to understand the underlying biochemistry behind the biosynthesis of plant lipids as a whole and of oil in particular. In this chapter we will describe the present state of understanding of the metabolism involved and then illustrate, using examples, how this knowledge has been applied towards manipulation of storage oil composition.

15.1.1 Carbon and energy sources for fatty acid synthesis

The synthesis of plant lipids is a process which involves multiple subcellular compartments and most of the metabolism is common to both membrane and storage lipid synthesis. The initial synthesis of fatty acids in plants occurs in the plastid and the immediate precursor is acetyl-coenzyme A (CoA). The evidence to date suggests that the carbon metabolism leading to the synthesis of acetyl-CoA varies according to the plant tissue and species which have been studied. In chloroplasts of leaf tissues the carbon source may be triose phosphate derived directly from the Calvin cycle although studies also suggest that carbon is imported from the cytosol as pyruvate or acetate. In heterotrophic tissues such as roots and seeds the plastid is entirely dependent on the import of metabolites from the cytosol. This carbon is ultimately derived from carbon translocated from the photosynthetic organs, typically in the form of sucrose. The sucrose is imported into the cells in these heterotrophic tissues and broken down into metabolites which are then imported into the plastid. The metabolite(s) imported by non-photosynthetic plastids for fatty acid biosynthesis are varied. To date glucose 6-phosphate, dihydroxyacetone phosphate, malate, pyruvate, acetate, and acetyl-carnitine have been identified as possible cytosolic precursors for intra-plastidial synthesis of acetyl-CoA. The synthesis of fatty acids requires ATP and reducing power in the form of NADH and NADPH. In the chloroplast these are derived from photosynthetic electron transport in the light. However, in the dark the chloroplast will be dependent on oxidative carbon metabolism in the plastid and/or the cytosol for the generation of reducing power and ATP. Of course in heterotrophic tissues the plastids will be entirely dependent on interaction with the cytosol for the ATP and reducing power needed to drive fatty acid synthesis.

In storage tissues, fatty acid synthesis is not always the major flux of carbon, and starch and storage protein synthesis may well be occurring. In many cases the amounts of these products will be determined by changes in partitioning of carbon which result from temporal effects on the expression of their biosynthetic pathways (Figure 15.1) but in others the products can accumulate throughout the development of the storage organ.

The precise details as to how carbon fluxes to storage products are controlled in order to regulate the carbon partitioning are not well understood but the pathways are likely to compete at the level of the availability of common carbon precursors. Where tissues accumulate storage proteins as well as starch and/or oil, the availability of nitrogen may well have a greater influence than that of carbon. These are important considerations not only for present crops but also for the improvement of alternative crops. In these alternative crops the seeds often contain storage components that are of high added value but the proportion of that component in the seed is small. Improving the yield of the

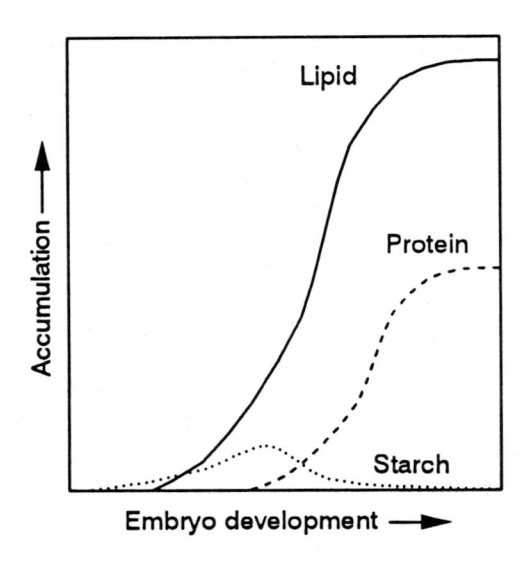

Figure 15.1 Accumulation of storage products during the development of rape embryos

specific storage component in such crops will be important for their adoption into agriculture.

15.1.2 Fatty acid synthesis

A simplified scheme of the pathway of storage triacylglycerol (TAG) synthesis in developing seeds is given in Figure 15.2. In the text we shall refer to the enzymes by the number assigned in the figure to aid explanation. We often use the shorthand notation for fatty acids and the common name and shorthand notation are explained in Table 15.1.

In the main we are describing fatty acid metabolism in seeds since much interest lies in genetically engineering seed oils. However, it must be said that the metabolism of fatty acids in leaves tends to be quite different since so much of the leaf lipid is found in the chloroplast. A much fuller description of the lipid metabolism in plants including the plastidic system of membrane lipid biosynthesis is detailed in the recent review by Ohlrogge and Browse (1995). Fatty acid synthesis commences with the carboxylation of acetyl-CoA to malonyl-CoA by acetyl-CoA carboxylase (ACCase (1)). In other eukaryotic organisms this enzyme is known to be an important point of regulation of fatty acid synthesis and some studies suggest a similar role in plants. Because of the potential importance of this enzyme much research has been focused upon it. This attention has recently revealed an important flaw in the prior understanding of the plant enzyme. In mammals and yeast the enzyme is a multi-functional polypeptide which comprises three distinct domains corresponding to four separate subunits found in the *Escherichia coli* enzyme. Based on evidence from enzyme purification prior to 1994 it was believed widely that the plastidial enzyme in plants was a multi-functional, large polypeptide. It is now clear that this may well be true for Graminaceous plants such as wheat and maize, but in dicotyledons, recent evidence from biochemistry and cloning of genes has revealed that the plastidial ACCase is a multi-subunit type like that in *E. coli* (Sasaki *et al.*, 1995). It is interesting to note that many research groups (including the authors' own!) had expended considerable effort towards the isolation of DNA clones encoding the multifunctional form of ACCase in the expectation that this would enable them to manipulate the activity of this enzyme in the plastid as described below. These genes were isolated but the amino acid sequences predicted by these DNA clones failed to show pre-sequences which would determine plastidial import. This has now been explained on the basis that both dicotyledons and Graminaceous plants also possess a cytosolic ACCase which is a multifunctional polypeptide. It is presumed that this cytosolic ACCase is used for synthesis of malonyl-CoA for the elongation of fatty acids, and for the synthesis of waxes, flavonoids and other diverse compounds.

The malonyl-CoA produced in the plastid is utilized by the fatty acid synthetase (FAS) complex (2). By analogy with the ACCase in dicotyledons the FAS complex in higher plants is of a prokaryotic type. This multi-enzyme complex forms a growing acyl chain which is conjugated to acyl carrier protein (ACP) and produces saturated fatty acids with a chain length of up to 18 carbons (18:0-ACP). In most plants the 18:0-ACP is desaturated by a soluble 18:0-ACP desaturase (3) to 18:1-ACP. The composition of fatty acids made by the plastid

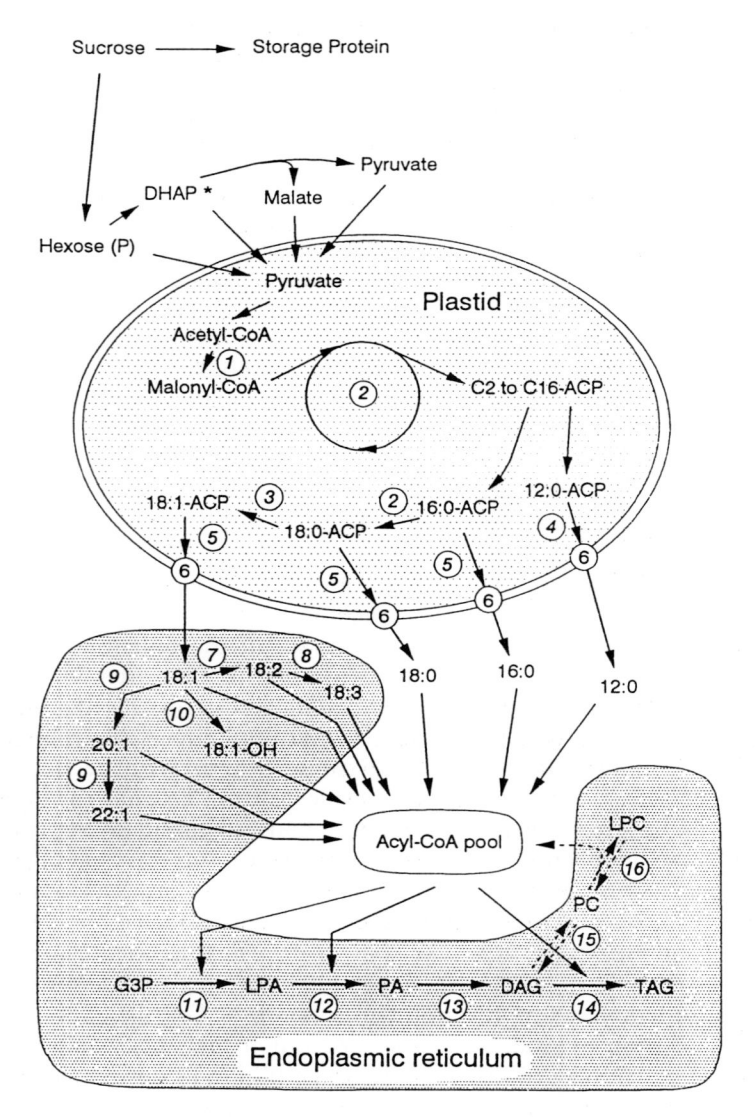

Figure 15.2 Pathway of storage oil synthesis in developing seeds. This presents a simplified overview of the pathways involved, since the substrates for several of the enzymes are complex phospholipids which cannot easily be represented in a diagram of this nature. For clarity, we have omitted several of the plastidic enzymes involved in the fatty acid synthetase system. A more detailed description of lipid synthesis has recently been published (Ohlrogge and Browse, 1995). The medium chain acyl-ACP thioesterase (5) and oleate hydroxylase (10) are not present in developing seeds of rape. We have also omitted, again for clarity, the enzymes of structural lipid synthesis in the plastid. The first part of the pathway mirrors that of the ER pathway from G3P to DAG. ①, acetyl-CoA carboxylase; ②, fatty acid synthetase; ③, stearoyl-ACP desaturase; ④, medium chain acyl-ACP thioesterase; ⑤, acyl-ACP thioesterase; ⑥, acyl-CoA synthetase; ⑦, oleate desaturase; ⑧, linoleate desaturase; ⑨, fatty acid elongase; ⑩, oleate hydroxylase; ⑪, glycerol 3-phosphate acyltransferase; ⑫, lysophosphatidic acid acyltransferase; ⑬, phosphatidate phosphatase; ⑭, diacylglycerol acyltransferase; ⑮, DAG:cholinephosphotransferase; ⑯, lysophosphatidylcholine acyltransferase; * DHAP, dihydroxyacetone phosphate

Table 15.1 Fatty acids are often referred to by a shorthand notation in the diagrams and occasionally in the text and except where relevant the position of the double bond will not be included. The carbon number is given before the colon and the number of double bonds after it. The double bonds are always in the *cis* conformation unless mentioned in the text. The Δ symbol shows the double bond is counted as carbon number from the carboxyl end

Fatty acid	Shorthand
Lauric	12:0
Palmitic	16:0
Stearic	18:0
Petroselinic	18:1-Δ^6
Oleic	18:1-Δ^9
Linoleic	18:2-Δ^9,Δ^{12}
Linolenic	18:3-Δ^9,Δ^{12},Δ^{15}
Erucic	22:1-Δ^{13}

can be seen as the outcome of the competition for acyl-ACP substrates between the enzymes which catalyze fatty acid synthesis and acyl-ACP thioesterase (4,5) and acyl-ACP acyltransferase (important in chloroplasts), which terminate synthesis. In most oilseed crop species, the enzymes which terminate synthesis act very slowly on short and medium chain acyl-ACPs in comparison to the enzymes involved in chain elongation and desaturation. However, at 16 carbons, the chain termination reactions begin to compete and once 18:1-ACP is synthesized the situation is reversed and the enzymes which catalyze chain termination win out.

15.1.3 Glycerolipid synthesis

The synthesis of membrane lipids and storage oils occurs through the Kennedy pathway on the endoplasmic reticulum so oleic and other fatty acids must therefore be exported from the plastid. Furthermore, the modification of the acyl chains through hydroxylation (10), and further desaturation (7,8) and/or elongation (9) also occurs outside the plastid through the action of membrane-bound enzymes. These modifications are important determinants of the end uses of storage oils in crop plant species. The export of 18:1 from the plastid is not fully understood but it is believed to involve acyl-CoA synthetase (6) which is located on the outer membrane of the plastid envelope. The acyl-CoAs formed as a result of the export and modification form a pool which is utilized by the acyltransferases of the Kennedy pathway (11,12,13,14). These enzymes acylate glycerol 3-phosphate sequentially at the *sn*-1 and *sn*-2 positions to form phosphatidic acid. The phosphate is removed from the phosphatidic acid to produce diacylglycerol (DAG) which is the branch point between the production of phosphatidylcholine (PC) for membranes and of triacylglycerol (TAG) for storage oils. TAG is formed by acylation of the *sn*-3 position by diacylglycerol acyltransferase (DGAT,14). Phosphatidylcholine, the major extra-plastidial membrane lipid, is formed by the addition of a phosphocholine group to DAG in a reversible reaction catalyzed by DAG:cholinephosphotransferase (CPT, 15). The enzyme LPC-AT (16) also catalyzes the synthesis of PC from lyso-PC, which again is a reversible reaction. The desaturases of the ER (7,8) act on 18:1 moieties of PC to form 18:2 and 18:3. The polyunsaturated fatty acids can then be returned to the acyl-CoA pool through the reverse reaction of LPC-AT (16) which is important in many plants or to DAG through the action of CPT (15). The acyl-CoAs and DAG containing 18:2 and 18:3 are then available for incorporation into TAG. In many species the acyltransferases of the Kennedy pathway are highly selective for the acyl groups they will transfer. This results in a non-random distribution of fatty acids in the TAG which constitutes the storage oil, and so plays a major role in determining the quality of the product for its use in the food and non-food markets.

15.2 Genetic manipulation of lipid biosynthesis

Having described the basic pathway of lipid biosynthesis we will now illustrate, using examples drawn from the current literature, how our knowledge of the pathways involved is being applied to manipulate lipid synthesis towards commercial objectives. In the course of this it will become clear that our knowledge was, in many cases, less complete than was thought and that simple objectives can demand complex manipulations.

15.2.1 *Acetyl-CoA carboxylase manipulation*

As described earlier, the role of ACCase in determining flux of carbon into fatty acids is an important research objective. The current understanding is certainly not clear. Previous work has shown that in developing oilseed rape embryos the activity of ACCase increases prior to the increase in storage oil content suggesting that the increase in the former is not causal of the latter. However, it is now clear that ACCase is present in both the plastid and the cytosol and that assays of whole embryo extracts will not reveal how the activity of the plastidial form changes during development. In leaf tissues other evidence based upon studies of substrate and product pools in fatty acid synthesis has shown that ACCase does exert control over the pathway in chloroplasts. The technique of control analysis is now being applied to study this problem. For chloroplasts isolated from cereals, the use of graminicide inhibitors, which specifically affect the plastidial ACCase, to alter the enzyme activity have also shown that this enzyme exerts a strong control over fatty acid synthesis. In dicotyledonous plants the use of transgenic plants expressing antisense constructs of the nuclear-encoded subunits of the ACCase is also being used to try to decrease the activity of the enzyme *in vivo* and measure the consequences of this for fatty acid synthesis.

The reduction of plastidial ACCase activity using expression of antisense RNA has wider application than simply in the study of flux control over fatty acid biosynthesis. Reducing the activity of this enzyme may be desirable if the acetyl-CoA was needed for a pathway other than fatty acid synthesis, i.e. a change in the partitioning of carbon between pathways. This is a major commercial objective for the production of the biopolymer polyhydroxybutyric acid (PHB) in plants. This biodegradable polymer with many direct commercial applications is synthesized from acetyl-CoA by certain bacterial species. Commercial production of PHB is currently through fermentation technology. This is a costly process which prohibits the wide application of PHB-derived plastics. The advances made in plant biotechnology now raise the possibility that PHB could be produced in a crop plant with a lower cost than through fermentation, thereby facilitating wider application of an environmentally friendly product.

The current aim is to express the three bacterial enzymes required for PHB synthesis in the plastids of an oil storing seed such as oilseed rape. By downregulating flux of acetyl-CoA into fatty acids and providing an alternative sink in the form of PHB synthesis it is hoped to produce a crop which will yield PHB on an economically viable scale. The potential of this approach has already been demonstrated by expressing the PHB synthetic enzymes in the chloroplasts of *Arabidopsis thaliana* leaves. These transgenic *Arabidopsis* accumulate PHB to a level of 14% in their leaves without any applied perturbation to fatty acid synthesis (Nawrath *et al.*, 1994). Manipulation of an oil crop species in this way is not a trivial exercise and requires at least four new genes (three biosynthetic genes and one antisense gene) to be introduced into the transgenic crop plant. Even then this is not simple since the bacterial biosynthetic enzymes must be targeted to the plastidial compartment, the expression of all of the transgenes must be confined to the embryo tissue, and their expression must not be mutually inhibitory. One approach to this is to produce individual transgenic lines containing one of the transgenes driven by an embryo-specific promoter sequence, and then to intercross these lines to produce a final line containing all four transgenes. This example provides a very clear picture of the complexity required to achieve what seems a straightforward objective.

15.3 Modification of fatty acid chain length

15.3.1 *Medium chain fatty acids*

Synthesis of fatty acids by the FAS complex continues until the acyl chain is long enough to be efficiently cleaved from ACP by acyl-ACP thioesterase (5). Once removed from the ACP, the acyl chain cannot be further elongated by the fatty acid synthetase. In plastids of developing rape seeds and most edible oil crops, this enzyme uses 18:1-ACP most efficiently as its substrate but it will also catalyze the cleavage of 16:0 and 18:0 from ACP as well. This leads to the production of oils which contain mainly 18 carbon fatty acids with a little 16:0. Some crop plants, such as coconut and oil palm and members of the Lauraceae and Lythraceae families, synthesize storage oils which contain large amounts of medium chain fatty acids. It was predicted that the developing seeds of these species contained an acyl-ACP thioesterase (4) which had a specificity for acyl chains of 12 carbons which effectively terminated fatty acid synthesis at 12:0. In the late 1980s Pollard and co-workers at Calgene in California purified a 12:0

selective acyl-ACP thioesterase (4) from the California Bay tree and a cDNA encoding it was cloned from a cDNA library. Expression of the cDNA in *E. coli* confirmed its selectivity for medium chain fatty acids.

The group at Calgene expressed the coding sequence of the medium chain thioesterase in developing seeds of oilseed rape using a strong seed specific promoter. The seed oil of some of the transformants contained significant amounts of 12:0 (Voelker et al., 1992). The transformants were then taken through several rounds of crossing and selection and a 12:0 content of 40–50% in the oil has been achieved. In 1995 the high 12:0 rape was grown on a large scale in the USA making it the first transgenic industrial oilseed crop with an engineered oil composition to be grown commercially. Although a 12:0 content of about 50% of total fatty acids is a significant level, a proportion of at least 70% would be preferred since it would increase the yield of 12:0/hectare in the crop and so increase the economic viability of 12:0 production in oilseed rape. However, in oilseed rape there was another biochemical barrier which prevented much higher levels from being achieved that needed to be overcome. The acyl composition of the outer (*sn*-1 and 3) and middle (*sn*-2) fatty acids of the triacylglycerol was analyzed in the high-12:0 transgenic seed. It was found that 12:0 was essentially excluded from the middle (*sn*-2) position of the oil. This was not entirely unexpected since it had been shown previously that *sn*-1-acyl-glycerol-3-phosphate acyltransferase (also known as *lyso*phosphatidate acyltransferase; LPAT, 12), which catalyzes the transfer of fatty acids to the *sn*-2 position of the TAG selects very strongly against medium chain fatty acyl groups in oilseed rape. The reaction is shown in Figure 15.3 where the LPAT is supplied with a mixture of acyl-CoA substrates but selects the 18:1 very strongly, thus excluding the 12:0 (and also 22:1 which is described in the section on very long chain fatty acids).

The next aim of the Calgene group was therefore to clone a gene encoding an LPAT which can use medium chain acyl groups as substrates and to put this gene into their plants with elevated 12:0 content in the seed oil. They purified the LPAT from coconut which catalyzes this reaction efficiently with 12:0-CoA and cloned the cDNA encoding this protein. The cDNA was expressed in *E. coli* and its identity confirmed by measuring the increase in the transfer of medium chain fatty acids into the *sn*-2 position of the *E. coli* lipids (Knutzon et al., 1995). It now remains to transform the coconut LPAT cDNA into the plants

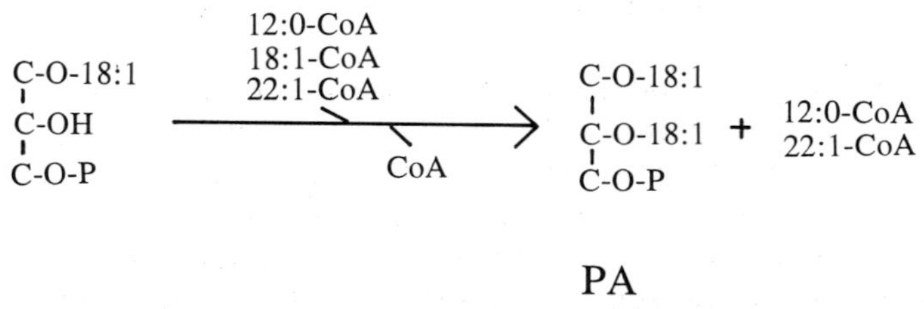

$$\begin{array}{c} \text{C-O-18:1} \\ | \\ \text{C-OH} \\ | \\ \text{C-O-P} \end{array} \quad \xrightarrow{\substack{\text{12:0-CoA} \\ \text{18:1-CoA} \\ \text{22:1-CoA}}} \quad \begin{array}{c} \text{C-O-18:1} \\ | \\ \text{C-O-18:1} \\ | \\ \text{C-O-P} \end{array} + \begin{array}{c} \text{12:0-CoA} \\ \text{22:1-CoA} \end{array}$$

PA

Figure 15.3 The second acylation step catalyzed by lysophosphatidate acyltransferase (LPAT) in the endoplasmic reticulum of rape seeds. Given a mixture of 12:0-CoA, 18:1-CoA and 22:1-CoA, the rape microsomal LPAT will only use the 18:1-CoA, thus excluding 12:0 and 22:1 from the *sn*-2 position of the oil

expressing the 12:0-ACP thioesterase gene to determine whether the increase that is required can be achieved.

The success in engineering rape to produce medium chain fatty acids has led to a great deal of detailed research on acyl-ACP thioesterases from a range of plants and they appear to fall into two types which act on either saturated acyl-ACPs of various chain lengths depending on the source plant, or on 18:1-ACP (Jones et al., 1995). Some species from the genus *Cuphea* synthesize oils containing mainly C8 fatty acids, others mainly C10 and so on. In these species the specificity of the acyl-ACP thioesterase defines the chain length of the fatty acids in the oil. It can be envisaged that it will be possible to produce oils containing fatty acids with defined chain length from 8 to 18 in transgenic rape by using the appropriate thioesterase genes from the various *Cuphea* species. While working towards this goal the Calgene team have shown that it is possible to alter the acyl chain specificity of an acyl-ACP thioesterase by domain swapping and even site directed mutagenesis of selected amino acid residues. By altering just three key amino acids of the C12 specific thioesterase from seeds of the California Bay tree described above they were able to convert it to a predominately C14 specific enzyme (Yuan et al., 1995). In the future it should be possible to tailor such enzymes to enable production of oils to the exact requirements of an end user.

Studies of the *E. coli* strain which is expressing the acyl-ACP thioesterase have led to some very interesting observations on the regulation of fatty acid synthesis in *E. coli* and these have a bearing on our understanding of fatty acid synthesis in plants. The plant FAS is a type II multi-enzyme system very similar to that in bacteria, whereas the animal FAS is comprised of a single protein which contains all the catalytic activities required to synthesize a fatty acid. In the plant/bacterial system it is not clear whether all of the enzymes are associated as a large complex with the acyl-ACP being handed on directly from one enzyme to the next in a co-ordinated fashion or whether the acyl-ACPs are free in solution and diffuse between the various enzymes. By measuring the content of 12:0 in *E. coli* expressing the plant medium chain acyl-ACP thioesterase gene at defined stages as the bacteria move between the log and stationary phases of growth, it can be determined that the thioesterase is acting on free acyl-ACP pools rather than acyl-ACPs which are bound to a multi-enzyme complex. Whether this is also true in plants remains to be determined.

15.3.2 Very long chain fatty acids

Originally rapeseed oil contained large amounts of erucic acid (22:1) but since this was shown to have potential health risks the pathway of its synthesis was blocked through a mutagenesis programme yielding the low erucic, high oleic (18:1) oil produced today. However, 22:1 rape oil is still used in the production of a range of oleochemicals because several of the physico-chemical properties of 22:1 and its derivatives are superior to those of 18:1. Its use could be extended considerably into the production of other chemicals if the high 22:1 rape oil from which it is obtained contained 22:1 in a purer form. The problem of tight chain length specificity of the LPAT of oilseed rape described in the section on medium chain fatty acids extends to very long chain fatty acids. The 22:1

content of high erucic rape oil is currently at about 50% and the oil must be processed to remove contaminating fatty acids, adding to the cost of production. It has been calculated that the market for 22:1 would open up considerably if a level of 22:1 of 85% or better could be achieved. Since the LPAT (12) almost totally excludes 22:1 from the sn-2 position of the oil, the theoretical maximum level of 22:1 achievable is presently 66.7% (see Figure 15.3). To increase this to 100% requires the expression of an LPAT which can use erucic acid at least as efficiently as other fatty acids. The seeds of meadowfoam and poached egg plant (both *Limnanthes* species) have high levels of very long chain fatty acids which are similar in structure to 22:1. The LPAT (12) from *Limnanthes* can use 22:1 more efficiently than other fatty acids and a great deal of effort has been expended in several labs to clone a cDNA encoding it. A temperature sensitive mutant of *E. coli*, created by a mutation in the *plsC* gene which encodes the bacterial LPAT, was used to screen cDNA expression libraries made from developing seeds of *Limnanthes* and other plants. The approach was based on a complementation of the mutation by an expressed plant LPAT which restored growth at non-permissive temperatures for the *plsC* mutant strain. It was successfully applied at about the same time by laboratories in Germany and the UK who showed that LPAT, when expressed in *E. coli*, could catalyze the transfer of erucic acid to the sn-2 position at very high rates (Brown et al., 1995; Hanke et al., 1995). It remains to be seen whether expression of the *Limnanthes* genes in high erucic rape will cause the required increase in the proportion of 22:1 in the oil but the *E. coli* expression work gives cause for optimism.

15.4 Modification of double bonds in the fatty acids of seed oils

15.4.1 *Increasing the content of 18:0 in oils*

There has been much interest in trying to alter the content and composition of unsaturated fatty acids, not only in seed oils but also in membrane lipids which we will describe later in the chapter. In animals and plants, double bonds are created in fatty acids by the action of desaturases. In many plants the first desaturase (3) forms a *cis* Δ^9 double bond in 18:0-ACP. This enzyme is plastidial and uses ferredoxin in transferring electrons. In rape, most 18:0-ACP is converted to 18:1-ACP, which is then cleaved by the acyl-ACP thioesterase described in the previous section. By contrast, plastids in cocoa seeds produce larger amounts of 18:0 than 18:1. Incorporation of 18:0 into storage TAG gives a solid fat known as cocoa butter. The supply and hence price of cocoa butter have varied considerably over the past twenty years which has created a lot of interest in producing cocoa butter substitutes. Some chemical and biotechnological processes have been developed by the food industry to do this by transesterification of 18:0 into vegetable oil. However, the advent of plant molecular biology has allowed the possibility of engineering rape to make cocoa type butter as its seed oil. This was achieved once the Δ^9 desaturase was purified from rape seeds and the cDNA cloned. The cDNA was expressed in an antisense orientation in the developing seeds of rape which reduced the levels of the Δ^9 desaturase and caused an increase in the 18:0 content of the seeds to 40% in some cases (Knutzon et al., 1992).

15.4.2 Reducing levels of polyunsaturated fatty acids

Another target of the seed breeding companies is to increase the level of 18:1 in the seed oil at the expense of 18:2 and 18:3 to improve stability to oxidation. This has been achieved by mutagenesis to knock out the 18:1 desaturase (8) in a number of crops such as sunflower and maize with an 18:1 content of greater than 80% being achieved. However, one mutant of rape which contained 87% 18:1 in the seed oil also had very high levels of 18:1 in the cell membranes of the rest of the plant. This led to severe developmental effects especially at lower temperatures, with the root system being particularly affected (Kinney, 1994). These negative effects on the plant as a whole can be sidestepped if down-regulation of the 18:1 desaturase is achieved by expression of antisense RNA to the 18:1 desaturase driven by a seed-specific promoter, since membrane lipid synthesis in the rest of the plant will be unaffected. However, physiological studies of the 18:1 desaturase mutant in *Arabidopsis* have shown that this may not be enough since the seed membranes still contain reduced levels of 18:2 and 18:3 leading to developmental problems during germination at lower temperatures once the seeds are planted (Miquel and Browse, 1994).

15.4.3 Unusual monounsaturated fatty acids

The vegetable oils produced for human consumption all contain mainly unsaturated fatty acids with the first double bond being at the ninth carbon from the carboxyl end. In the cases of rape and olive the major fatty acid is the monounsaturated oleic acid (18:1). The seeds of some wild plants contain predominantly a monounsaturated fatty acid where the double bond is in another position. Since these fatty acids can be cleaved to yield precursors that are useful in the production of polymers and detergents, there is interest in studying the enzymes which make them. One such plant is coriander whose seed oil contains over 80% petroselinic acid which is an isomer of oleic acid with a *cis* Δ^6 rather than a *cis* Δ^9 double bond. Petroselinic acid can be cleaved oxidatively to form lauric acid (12:0) which is used in the manufacture of detergents as mentioned in the section on medium chain fatty acids, and adipic acid which is a C6 dicarboxylic acid used in nylon synthesis. A cDNA encoding the enzyme was cloned using a PCR approach based on the sequence of the Δ^9 desaturase and this has been expressed in oilseed rape to produce petroselinic acid at low levels (Murphy et al., 1992). The story, however, proved to be more complicated than at first imagined and detailed work has shown that instead of acting at the Δ^6 position of 18:0-ACP, the desaturase from coriander actually acts at the Δ^4 position of 16:0-ACP. A further round of elongation produces the 18-carbon monounsaturated fatty acid thereby shifting the double bond to the six position to produce petroselinic acid (Ohlrogge, 1994). In addition, a separate acyl-ACP thioesterase with a specificity for petroselinic acid is also present in the seeds of coriander. Therefore, in order to obtain petroselinic acid in large amounts in transgenic oilseed rape it is likely that a minimum of three genes will be required, namely those encoding a desaturase, an elongation enzyme, and the thioesterase.

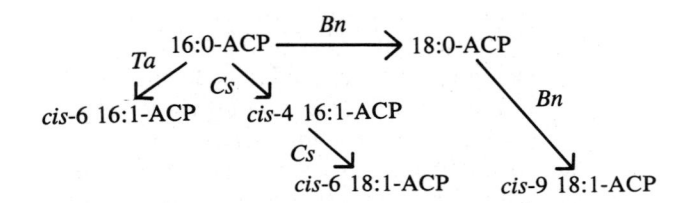

```
                          Bn
         Ta   16:0-ACP ─────────→ 18:0-ACP
           ↙      Cs ↘                  ↘ Bn
  cis-6 16:1-ACP   cis-4 16:1-ACP
                       Cs ↘                  ↘
                        cis-6 18:1-ACP    cis-9 18:1-ACP
```

Figure 15.4 Synthesis of monounsaturated fatty acids with the double bond at different carbon numbers from the carboxy terminus. The synthesis of only three monounsaturated fatty acids has been examined thus far, each showing differences in the desaturation and elongation steps. In rape, the cis-9 18:1 (oleic acid) can be further extended to cis-13 22:1 (erucic acid) by the fatty acid elongase of the ER. Ta, *Thunbergia alata*; Cs, *Coriandrum sativum*; Bn, *Brassica napus*

A related desaturase which creates a double bond in the Δ^6 position of 16:0-ACP has been found in the seeds of *Thunbergia alata* which is in the Acanthaceae family. The gene encoding this enzyme has been cloned and sequenced. Comparison of the amino acid sequences of the Δ^4, Δ^6, and Δ^9 desaturases reveals that they are all very similar (Cahoon et al., 1994). A scheme showing the synthesis of monounsaturated fatty acids in the plants studied so far is given in Figure 15.4. The Δ^9 desaturase has been crystallized and the 3-D structure determined. Analysis of the structures of the other desaturases will be very useful since they might show the mechanism by which soluble desaturases place their double bonds in the hydrocarbon chain. Once understood, it may then be possible to alter the positional and chain length specificity of desaturases by mutagenesis and so determine the reaction product and ultimately the type of oil produced in a plant.

15.4.4 *Fatty acid hydroxylase*

There are other enzymes which are thought to catalyze partial or modified desaturation reactions such as epoxidases or hydroxylases. Fatty acids containing such functional groups have many industrial uses in the production of polymers, paints and lubricants amongst others. Ricinoleic acid, or 12-hydroxyoleic acid is produced in large amounts by seeds of the castor oil plant (*Ricinus communis*). A gene encoding the 12-hydroxylase (10) which catalyzes synthesis of ricinoleic acid was cloned by random sequencing of seed specific cDNA clones. It was assumed that since the reaction catalyzed by the 12-hydroxylase was likely to be similar to that by the 12-desaturase, it would be possible to recognize the clone by sequence homology. This proved to be the case and expression of the hydroxylase in tobacco led to the synthesis of ricinoleic acid in the membrane lipids, albeit at low levels.

15.5 Effects of manipulation of fatty acid composition on plant physiology

When altering the fatty acid composition of seed oils, the expression of the

Manipulation of oil biosynthesis

Figure 15.5 The *fab2* mutant (right) is a severe dwarf at 22°C compared to the wild type (left) as a result of an increase in (18:0) in its membrane lipids (with permission from Ohlrogge and Browse, 1995)

various cDNA constructs is usually driven by a seed specific promoter. Many membranes are adversely affected by the incorporation of unusual fatty acids and an example of the effects of removing the 18:2 and 18:3 from membranes was described above. A very good example (Figure 15.5) of the adverse effect of altering the fatty acid composition of membrane lipids was described recently where even a relatively small increase in the level of 18:0 in the membranes caused an extreme dwarf phenotype (Lightner *et al.*, 1994).

In this case the change was brought about by a mutation of the Δ^9 18:0-ACP desaturase gene in *Arabidopsis* which caused an increase in the level of 18:0 from 3% to about 10%, though this varied between tissues. The plants with increased 18:0 were less than about 5% of the size of wild type and this was found to be due to a decrease in the volume of the cells rather than a reduction in cell number. If plants were grown at elevated temperatures the phenotype could be reversed to some extent suggesting that membrane structure is important in causing the unusual development.

15.5.1 Chilling sensitivity

Genetically engineered plant lipids have also been used to investigate another important physiological phenomenon: chilling sensitivity. Membrane lipids have long been suspected to play an important role in chilling sensitivity in tropical and subtropical plants, though a number of factors are thought to contribute to the phenomenon. It is thought that membrane fluidity is critical and that lateral separation of non-fluid components causes disruption of membrane structure causing function to be impaired. It has been suggested that the fatty acid composition of phosphatidylglycerol (PG) of the plastids is an important component in the temperature response of plants. Plants containing large amounts of 16:0, *trans*-3 16:1 and 18:0 fatty acids in PG (so-called high melting temperature (htm) PG) tend to be chilling sensitive whereas those with low levels of htm PG (i.e. higher levels of unsaturated fatty acids) are chilling insensitive. Although it is not shown in Figure 15.2, chloroplasts contain the enzymes of glycerolipid synthesis and convert glycerol 3-phosphate to DAG and thence to structural membrane lipids such as the galactolipids (Ohlrogge and Browse, 1995). The plastidial enzyme glycerol 3-phosphate acyltransferase (GPAT) appears to be important in regulating the degree of unsaturation of fatty acids in PG. The GPAT from squash (a chilling sensitive plant) transfers saturated fatty acids to the phospholipid whereas the *Arabidopsis* GPAT transfers unsaturated fatty acids. The role of GPAT in chilling sensitivity was addressed by expressing the squash and *Arabidopsis* enzymes in tobacco. Tobacco plants expressing the squash GPAT became chilling sensitive whereas those expressing the *Arabidopsis* enzyme did not (Murata *et al.*, 1992). Likewise, the expression of the *E. coli* GPAT in *Arabidopsis* increased the proportion of saturated fatty acid in the PG caused chilling sensitivity in a normally chilling tolerant plant (Wolter *et al.*, 1992). It could be envisaged that normally chilling sensitive, subtropical crops could be grown in temperate latitudes by expression of the appropriate GPAT cDNA. A fly in the ointment which calls into question this explanation of the role of the degree of fatty acid unsaturation in PG was reported more recently (Wu and Browse, 1995). A mutant of *Arabidopsis* which has greatly elevated levels of 16:0 in PG is *not* chilling sensitive except at prolonged exposure to very low temperatures (2°C), thus reopening this long-running debate.

15.6 Conclusion

Through a continued increase in the understanding of the diverse biochemistry of plant lipid synthesis, the ability to manipulate, in a targeted way, oil biosynthesis in plants has improved greatly. However, much remains to be understood. We still do not know how the rate of oil synthesis is controlled and whether greater rates of oil synthesis are possible in current oilseed crops. Increasing the oil yield of a crop which contains fatty acids which are of high added value in non-food applications is clearly an important objective. To implement this, considerable work is required to understand the biochemistry of fatty acid modification and subsequent incorporation into oil, although progress with desaturases and hydroxylases has been encouraging. Such studies have shown that the presence of a particular fatty acid in storage TAG is often determined by several genes. This suggests that we should be cautious in setting out towards modifying seed oils in existing crops such as oilseed rape. Notwithstanding this, the presence of transgenic oilseed rape crops in the field producing high added value oils represents a major step in an area with considerable potential for further development.

References

BROWN, A.P., BROUGH, C.L., KROON, J.T.M. and SLABAS, A.R. (1995) Identification of a cDNA that encodes a 1-acyl-sn-glycerol-3-phosphate acyltransferase from *Limnanthes douglasii*. *Plant Mol. Biol.* **29**, 267–278.

CAHOON, E.G., CRANMER, A.M., SHANKLIN, J. and OHLROGGE, J.B. (1994) Δ^6 hexadecanoic acid is synthesized by the activity of a soluble Δ^6 palmitoyl-acyl carrier protein desaturase in *Thunbergia alata* endosperm. *J. Biol. Chem.* **269**, 27519–27526.

HANKE, C., WOLTER, F.P., COLEMAN, J., PETEREK, G. and FRENTZEN, M. (1995) A plant acyltransferase involved in triacylglycerol biosynthesis complements an *Escherichia coli* sn-1-acylglycerol-3-phosphate acyltransferase mutant. *Eur. J. Biochem.* **232**, 806–810.

JONES, A., DAVIES, H.M. and VOELKER, T.A. (1995) Palmitoyl-acyl carrier protein (ACP) thioesterase and the evolutionary origin of plant acyl-ACP thioesterases. *Plant Cell* **7**, 359–371.

KINNEY, A.J. (1994) Genetic modification of the storage lipids of plants. *Curr. Opin. Biotechnol.* **5**, 144–151.

KNUTZON, D.S., THOMPSON, G.A., RADKE, S.E., JOHNSON, W.B., KNAUF, V.C. and KRIDL, J.C. (1992) Modification of *Brassica* seed oil by antisense expression of a stearoyl-acyl carrier protein desaturase gene. *Proc. Natl Acad. Sci. USA* **89**, 2624–2628.

KNUTZON, D.S., LARDIZABEL, K.D., NELSON, J.S., BLEIBAUM, J.L., DAVIES, H.M. and METZ, J.G. (1995) Cloning of a coconut endosperm cDNA encoding a 1-acyl-snglycerol-3-phosphate acyltransferase that accepts medium-chain-length substrates. *Plant Physiol.* **109**, 999–1006.

LIGHTNER, J., JAMES, D.W., DOONER, H.K. and BROWSE, J. (1994) Altered body morphology is caused by increased stearate levels in a mutant of *Arabidopsis*. *Plant J.* **6**, 401–412.

MIQUEL, M.F. and BROWSE, J.A. (1994) High-oleate oilseeds fail to develop at low temperature. *Plant Physiol.* **106**, 421–427.

MURATA, N., ISHIZAKI-NISHIZAWA, O., HIGASHI, S., HAYASHI, H., TASAKA, Y. and NISHIDA, I. (1992) Genetically engineered alteration in the chilling sensitivity of plants. *Nature* **356**, 710–712.

MURPHY, D.J., FAIRBAIRN, D.J. and SLOCOMBE, S.P. (1992) Genes for altering plant metabolism, Patent. WO 94/01565.

NAWRATH, C., POIRIER, Y. and SOMERVILLE, C. (1994) Targeting of the polyhydroxybutyrate biosynthesis pathway to the plastids of *Arabidopsis thaliana* results in high levels of polymer accumulation. *Proc. Natl Acad. Sci. USA* **91**, 12760–12764.

OHLROGGE, J.B. (1994) Design of new plant products: engineering of fatty acid metabolism. *Plant Physiol.* **104**, 821–826.

OHLROGGE, J. and BROWSE, J. (1995) Lipid biosynthesis. *Plant Cell* **7**, 957–970.

SASAKI, Y., KONISHI, T. and NAGANO, Y. (1995) The compartmentation of acetyl-coenzyme A carboxylase in plants. *Plant Physiol.* **108**, 445–449.

VOELKER, T.A., WORRELL, A.C., ANDERSON, L., BLEIBAUM, J., FAN, C., HAWKINS, D.J., RADKE, S.H. and DAVIES, H.M. (1992) Fatty acid biosynthesis redirected to medium chains in transgenic oilseed plants. *Science* **257**, 72–73.

WOLTER, F.P., SCHMIDT, R. and HEINZ, E. (1992) Chilling sensitivity of *Arabidopsis thaliana* with genetically engineered membrane lipids. *EMBO J.* **11**, 4685–4692.

WU, J. and BROWSE, J. (1995) Elevated levels of high-melting-point phosphatidylglycerols do not induce chilling sensitivity in an *Arabidopsis* mutant. *Plant Cell* **7**, 17–27.

YUAN, L. VOELKER, T.A. and HAWKINS, D.H. (1995) Modification of the substrate specificity of an acyl-acyl carrier protein thioesterase by protein engineering. *Proc. Natl Acad. Sci. USA* **92**, 10639–10643.

16

Resource use efficiency and crop performance: what is the link?

CHRISTOPHER J. POLLOCK

16.1 Introduction

The plant species present in natural ecosystems are competing for resources, and the patterns of allocation of these resources between the different components are what defines the system. Resource-poor ecosystems tend also to be species-poor, but often demonstrate high levels of specific adaptations to sub-optimal conditions or extremely high efficiencies of capture and utilization of scarce resources. By contrast, domesticated species and varieties are subject to radically different constraints. Effective cultivation and management maximizes 'added value' (the difference in the societal costs of the inputs and outputs) and allocates resources accordingly. If those societal costs equate (as they often do) to the difference between the simple monetary costs of production and the rewards of sale, then individual producers will tend to maximize production until demand is saturated. Such systems will tend towards increased inputs and towards the production of plant varieties which respond to such inputs. For example, in the period between 1900 and 1980, application of nitrogenous fertilizer to UK crops rose eightfold. Over the same period, average cereal yields doubled (Leach, 1976). The cultivars developed in the 1970s responded better to high available soil N than did earlier varieties, but the nitrogen efficiency of production (amount of grain produced per unit of added N) actually fell. Concerns over environmental pollution and a temporary local abundance of food in Western Europe has increased the perceived significance of input efficiency when compared with production efficiency, but an effective methodological basis for improving the efficiency of resource capture and allocation in cultivated plants does not yet exist.

It is instructive to consider why this gap exists. Put simply, classical genetics, the plant breeding which stemmed from it and the agronomy which underpinned it, were not unduly concerned with mechanism. The only requirement for any trait to be manipulated was that it should be selectable, be under some appropriate level of genetic (rather than environmental) control and exhibit usable diversity within an accessible gene pool. This somewhat empirical process was reinforced by the fact that many key agronomic characteristics which were the targets for selection were themselves highly integrated indices of plant performance which combined multiple activities separated in time and space. Take, for example, cereal grain yield. This is the product of yield per grain, number of grains per ear and number of ears per plant. The availability of resources for grain filling depends upon the totality of resource capture in the time from germination, upon competition for resources within the plant and upon the contribution of temporary storage sinks. Selection for increased yield can and has produced changes in all of these parameters, despite the fact that there is not a comprehensive, detailed biochemical and molecular characterization of any of them. Physiology and biochemistry have provided elegant explanations for previous breeding advances (Woolhouse, 1981), and molecular biology has helped to generate high-resolution genetic maps, with a huge number of molecular markers which can be used to direct selection (Arús and Moreno-González, 1993), but breeders have still tended to generate advances empirically through increasing the duration of resource capture rather than its efficiency and through increasing harvest index rather than growth rate.

By contrast, it is much more difficult to define effective selection criteria for input efficiency without understanding the processes involved. The initial processes cover the capture of a range of dilute inputs, of which carbon dioxide and inorganic nitrogen are two of the most significant. These uptake processes are linked to the movement of water and to the capture of light, and are subject to a complex range of interlocking controls which balance the subsequent metabolism of carbon and nitrogen. As the other chapters in this book indicate, there is an increased understanding of the nature of these controls, and transgenic approaches are being used to assess the significance of individual steps. In this chapter, I will try to identify the opportunities for improving the efficiency of input capture, the potential limitations of such an approach and the possible impacts on agriculture. I will concentrate upon the metabolism of carbon, not because nitrogen metabolism is unimportant but because I would argue that there is a broader understanding at the physiological level of the interactions between photosynthesis, growth and storage than there is of the uptake, transport, and metabolism of nitrogen.

16.2 Increased photosynthesis gives increased yield (sometimes!)

The demonstration that there could be a positive correlation between increased carbon assimilation per unit leaf area (input efficiency) and usable yield under conditions of elevated CO_2 concentrations (e.g. Hardy *et al.*, 1978) drove an entire generation of biologists to look for the limiting factors in carbon fixation and to search for practical ways of overcoming them in the field. It is, however, instructive to look more carefully at Hardy's data (Table 16.1). In particular the variation between species in the response to elevated CO_2 is enormous, suggesting considerable differences in the ability of different plants to utilize additional fixed carbon. Later work, associated with the study of plant responses to anthropogenic increases in atmospheric CO_2, has confirmed variability of response and demonstrated the delicate interactions between carbon assimilation and water or nutrient status (Eamus, 1991; Nie *et al.*, 1995).

Table 16.1 Yield of various field-grown crops enriched with CO_2 (from Hardy et al., 1978)

	CO_2 enriched yield as percentage of air control	
	Pre-anthesis enrichment	Post-anthesis enrichment
Grain legumes		
Soybeans		198
Peanuts		130
Peas		153
Phaseolus		159
Cereals		
Wheat	116	114
Rice	131	118
Barley	150	128
Oats	104	
Fibre crops		
Cotton	126	

On the one hand, this can be seen as a subtle illustration by natural systems of the shortcomings of the concept of efficiency in the strictly agronomic sense. In a nitrate-limited, droughted sward, enhanced potential for photosynthesis under elevated CO_2 may serve to reduce the investment of scarce resources in protein or to reduce water use rather than to increase net carbon assimilation. Nitrogen use efficiency or water use efficiency increases at the expense of the expected increase in net assimilation rate. On the other hand, it suggests that carbon dioxide is not always a limiting resource. Corner, in his superlative book, *The Life of Plants*, suggests to those of us with parochial views of plant diversity that the transition to the land would have led to a generalized excess of carbon, and he cites the massive cellulosic structures of trees, the profuse sugary secretions of nectaries, and the exudations of roots as examples of such abundance (Corner, 1964). Only in crop plants has man's needs threatened to shift excess into scarcity.

16.3 Integration links supply and demand (via assimilate abundance?)

If increased availability of inputs does not always lead to increased productivity, what is the evidence that this decline in efficiency affects both resource capture and utilization? Perhaps the most obvious demonstration is the acclimation of photosynthesis which occurs after prolonged transfer of plants to elevated CO_2. The initial increase is often followed by a slow decline back to a rate only slightly above that in ambient air. This is generally accompanied by an accumulation within the leaves of one or other of the stable, neutral products of photosynthesis: sucrose or starch (Pollock and Farrar, 1996). It has long been argued that increased assimilate abundance could inhibit photosynthesis (Neales and Incoll, 1968) but it has become clear recently that high concentrations of sugars in leaves actually reduce the level of expression of key genes coding for enzymes of carbon fixation (Jang and Sheen, 1994).

It is also a truism that reductions in demand for fixed carbon in distant sinks generally lead to accumulation of sugars in mature source leaves. The process can be rapid. Chilling seedlings of *Lolium temulentum* reduces growth almost immediately. Within 30 minutes there is a detectable increase in the sucrose content of mature source leaves when compared to control plants maintained at 20°C (Pollock *et al.*, 1995). The process is freely reversible, indicating that chilling these plants is not associated with damage to either sources or sinks. From observations like this, one can erect a hypothesis which links changes in demand in sink tissues to an increase in sucrose or starch concentration in source leaves and to the subsequent down-regulation of photosynthesis at the level of gene expression. Once more, such a response could be considered as a reduction in the efficiency of resource capture (less carbon dioxide which is fixed by the same leaf area) or as a balancing of competing processes which maximizes the effective use of other resources (mineral nutrients, water, etc.).

16.4 Flexible allocation strategies do occur: the distinctiveness of temperate grasses

If down-regulation of photosynthesis is considered as a reduction in efficiency of carbon fixation, what are the opportunities to modify this process? Although specific modifications in metabolism via transgenic technology have been shown to alter partitioning of fixed carbon in leaves (Stitt *et al.*, 1990), it is equally instructive to consider the range of allocation strategies found *in vivo*. Most plants accumulate carbohydrates in their leaves during the day, but the magnitude and chemical nature of these storage sugars vary markedly (Lewis, 1984; Pollock, 1986). Under conditions of slow growth, the majority of the carbon fixed in photosynthesis in some species remains within the leaves. In temperate grasses and cereals, this accumulation is initially as sucrose, and then as fructan polymers stored in the vacuole (Pollock and Cairns, 1991). If export of sucrose is blocked completely by excising the leaves and then illuminating them, the process of sequential accumulation of sucrose, low molecular weight fructan oligosaccharides and high molecular weight fructan is speeded up and increased greatly in magnitude (Cairns and Pollock, 1988; Housley and Pollock, 1985; Wagner *et al.*, 1983). Under such conditions there is close correspondence between increase in carbohydrate and the net assimilation rate measured by gas exchange (Natr, 1967). These marked increases in total sugars (up to 65% of the total dry weight) do not appear to be accompanied by an initial decline in CO_2-saturated photosynthesis (Table 16.2), suggesting that the chemical and physical sequestration of sugars as fructans into the large volume of the central vacuole represents a strategy for minimizing feedback control (Pollock *et al.*, 1995).

The ability to convert sucrose into fructan in the leaves of temperate C_3 grasses is not constitutive. Induction of this process occurs under a range of conditions where supply of fixed carbon is in excess of demand (Pollock and Cairns, 1991). Net carbon fixation can be increased by increasing irradiance, extending the photoperiod (Pollock *et al.*, 1988), elevating the carbon dioxide content of the surrounding air (Housley and Pollock, 1985) or by feeding

Table 16.2 Photosynthetic oxygen evolution, chlorophyll content and amount of soluble carbohydrate in mature leaves of *Lolium temulentum* following excision (from Housley and Pollock, 1985)

Time after excision (days)	Photosynthetic rate (μmol O_2 m^{-2} h^{-1})	Chlorophyll (mg m^{-2})	Carbohydrate (g m^{-2})
0	6.1	36.5	0.2
1	6.2	38.2	4.0
2	6.6	39.3	8.0
3	4.9	29.7	13.4
4	4.1	18.9	13.1
5	1.6	9.2	12.0

exogenous sugars (Wagner *et al.*, 1983). Sink activity can be reduced by chilling (Simpson *et al.*, 1991), pruning (Wagner *et al.*, 1983) or N-depletion (Archbold, 1940). In all cases, an increase in sucrose content is followed by the progressive appearance of higher molecular mass fructans.

Administration of inhibitors of gene expression to detached leaves of *Lolium temulentum* prior to the induction of increased sucrose contents blocks the subsequent appearance of fructan without affecting the net accumulation of carbon (Table 16.3). Under these conditions, all the photosynthate which, under normal circumstances, would be metabolized into fructan, remains as sucrose. Delaying the administration of inhibitors for more than six hours after sucrose accumulation has commenced prevents them from having an effect, suggesting that increased assimilate abundance affects gene expression within six hours in such a way as to induce the synthesis of fructan (Winters *et al.*, 1994). A number of gene products have been characterized from this tissue whose abundance changes in response to changes in assimilate status (Winters *et al.*, 1995); the extractable activity of enzymes capable of the synthesis of fructans

Table 16.3 Effect of inhibitors of gene expression on the accumulation of fructan and total water-soluble carbohydrate in excised leaves of *Lolium temulentum* illuminated for 24 h (from Cairns and Pollock, 1988)

Inhibitor	Concentration (μM)	Total water-soluble carbohydrate (mg g^{-1} fresh mass)	Percentage as fructan
Water control	–	43.3	58.4
L-MDMP[a]	10	44.5	53.0
D-MDMP[a]	10	40.8	1.0
Cycloheximide	100	40.5	2.5
Cordycepin	1000	44.9	2.8
α-amanitin	1000	48.6	4.2

[a]MDMP: 2-(4-methyl-2,6-dinitroanilino)-N-methyl propionamide. The D-isomer is active as an inhibitor of protein synthesis. The L-isomer serves as an inactive control

from sucrose *in vitro* also rises (Cairns and Ashton, 1994). The hypothesis as currently proposed (Pollock *et al.*, 1996) is that a wide range of environmental stimuli can be transduced into a change in assimilate status in source leaves. In contrast to many plants, such changes, in leaves of temperate grasses and cereals, induce a massive change in resource allocation within the leaves, favouring vacuolar accumulation of fructans (Pollock and Farrar, 1996). The exact nature of the signal transduction pathway is not understood, although the down-regulation of photosynthetic gene expression by assimilate is thought to be mediated via hexokinase (Jang and Sheen, 1994; Sheen, 1994).

The contrasting patterns of resource allocation described above have been established on the basis of experiments on a very restricted range of higher plants. More extreme strategies do, of course, exist (CAM, C_4, etc.) but it should be remembered that there has always been a strong experimental bias towards crop plants in the study of plant physiology, biochemistry and molecular biology. It seems feasible, therefore, that there may be greater genetic variation in the mechanisms used by plants to balance resource allocation than has hitherto been characterized. Such a proposition is consistent with Grime's delineation of plant species into functional types (Grime, 1993) based upon the strategies they use to exploit resources. The historical domestication of plants and their subsequent improvement would strongly favour types which exhibit opportunistic rather than conservative strategies for resource acquisition and allocation.

16.5 Whole plant resource allocation: how efficiently are sources and sinks integrated?

Obviously, the combined growth rate of all the sink organs within a plant cannot exceed the capacity of sources to supply their needs. What is, perhaps, a more significant question is the extent to which the total growth *capacity* of sinks is integrated with source output. There are intuitive reasons for believing this is probably so. If it were not, large amounts of metabolic machinery could be present but be inactive, with a concomitant cost in both 'capital' and 'maintenance' functions. The *size* of root systems is undoubtedly regulated relative to source output (Farrar, 1992) and there is increasing evidence for coarse control of sink metabolism in line with assimilate abundance (Pollock and Farrar, 1996).

At the physiological level, reduction in assimilate abundance in roots via defoliation is associated with a decline in growth rate (Evans, 1972), in respiratory activity and in the activity of key enzymes (Williams *et al.*, 1992). These changes can be reversed by the administration of respirable carbohydrate (sucrose or hexose), suggesting that alterations in sugar supply to individual sinks may represent an effective mechanism for integrating both activity and capacity. In roots as well as in leaves, genes have been isolated whose expression is sensitive to carbohydrate concentration (Koch *et al.*, 1995), and sucrose supply can also modulate the expression of genes concerned with regulation of the cell cycle (Hemerly *et al.*, 1993). One can, therefore, postulate an overall model (Figure 16.1) in which resource allocation can be balanced under fluctuating environments based upon the sensing of changes in assimilate abundance. What such a model does not address is the integration of developmental changes into

	Elevated Sucrose	By acting at the level of:		
		Gene expression	Fine control	Substrate concentration
Source Leaf	Depresses Photosynthesis	+		
	Stimulated Carbohydrate Storage	+	+	+
	Accelerates Senescence?	+		
Transport Pathway	Alters Turgor Gradient			+
	Alters Vascular Differentiation?	+		
Storage Sink	Stimulates Carbohydrate Storage	+	+	+
Growing Sink	Stimulates Sucrose Catabolism	+	+	+
	Stimulates Growth	+		
	Stimulates Respiration	+	+	?

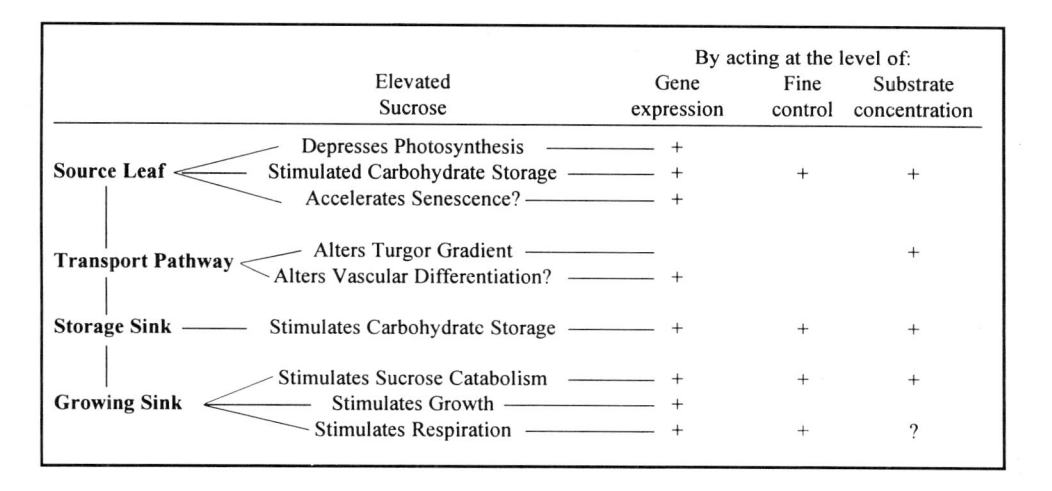

Figure 16.1 Diagrammatic representation of the ways in which sucrose or a related product of its metabolism may act to integrate activity in higher plants. The approach is based upon that presented by Farrar (1992)

the patterns of resource allocation. How, for example, do small sinks, such as recently fertilized cereal grains, establish themselves? Do key gene products in some sinks show lower sensitivity to assimilate abundance as a method of maintaining priority? What effect does such a strategy have on overall efficiency expressed across the lifetime of the plant? I do, however, consider that the evidence for 'active management' of these processes is strong and that carbohydrate metabolism is likely to be an important regulatory component.

16.6 Concluding remarks: the scope for improvement in the efficiency of resource use

According to the approach of Grime and co-workers (Grime, 1993), most crop plants would be derived from opportunist genera which would show high rates of resource acquisition, competitive foraging strategies and relatively poor tolerance to edaphic or biotic stress. By the same token, such plants would allocate significant resources to reproductive structures as a mechanism for avoiding such stress. Is this an 'efficient' strategy in the terms outlined above? It obviously has high fitness, since it has persisted successfully across a wide range of habitats. Competitively it is efficient, since aggressive foraging strategies not only improve performance but also reduce the resources available to competitors. The range of regulatory mechanisms outlined above and in the other chapters will tend to maximize the value of captured resources in terms of investment in new growth etc. but I suggest that they may not always maximize efficiency in the narrow sense of the 'effectiveness of capture' of one or more potentially limiting inputs. Perhaps an economic analogy is appropriate. If the rate of return is sufficiently high, then managing one's current investment portfolio may be a more effective use of time than seeking additional external capital in order to buy more shares.

It follows, therefore, that crop plants and their antecedents may have inherent advantages as the source of genetic variation in both maximum resource collection and in allocation. Other species, particularly from resource-limited habitats or climax vegetation types, may, however, be much more effective at sequestering available resources and at tolerating fluctuations in availability. Comparative biochemistry is much less fashionable than it was ten to fifteen years ago, but the advances in experimental ecology have, I think, served to identify good targets for such studies. In addition, increased knowledge of the syntenic relationships amongst genomes from different species (Moore et al., 1995) has facilitated the isolation of related genes from a wide range of species. A greater understanding of the nature of resource acquisition and allocation in different functional types is, in my opinion, a necessary prerequisite for the improvement of the efficiency and sustainability of crop production. This latter goal will become of increasing importance as more and more marginal land is incorporated into production systems in order to meet increasing food needs.

Basic knowledge on the mechanisms employed by higher plants to regulate both the magnitude and rate of key metabolic processes has increased dramatically as a result of the development of molecular biology. Applying this knowledge to the full range of plant diversity will not only continue to increase our understanding of plant metabolism and development but will generate increasingly effective techniques to manipulate them.

Acknowledgment

Financial assistance for some of the research described above was provided by the Biotechnology and Biological Sciences Research Council.

References and further reading

ARCHBOLD, H.K. (1940) Fructosans in the monocotyledons, a review. *New Phytol.* **39**, 185–219.

ARÚS, P. and MORENO-GONZÁLEZ, J. (1993) Marker-assisted selection. In: *Plant Breeding – Principles and Prospects* (HAYWARD, M.D., BOSEMARK, N.O. and ROMAGOSA, I., eds), pp. 314–331. Chapman and Hall, London.

CAIRNS, A.J. and ASHTON, J.E. (1994) Fructan biosynthesis in excised leaves of *Lolium temulentum* L. VI. Optimisation and stability of enzymatic fructan synthesis. *New Phytol.* **126**, 3–10.

CAIRNS, A.J. and POLLOCK, C.J. (1988) Fructan biosynthesis in excised leaves of *Lolium temulentum* L. I. Chromatographic characterisation of oligofructans and their labelling patterns following $^{14}CO_2$. *New Phytol.* **109**, 399–405.

CORNER, E.J.H. (1964) *The Life of Plants*. Weidenfeld and Nicholson, London.

EAMUS, D. (1991) The interaction of rising CO_2 and temperature with water-use efficiency. *Plant Cell Environ.* **14**, 843–852.

EVANS, P.S. (1972) Root growth of *Lolium perenne* III. Investigation of the mechanisms of defoliation-induced suppression of elongation. *NZ J. Agric. Res.* **15**, 347–355.

FARRAR, J.F. (1992) The whole plant: carbon partitioning during development. In: *Carbon Partitioning Within and Between Organisms* (POLLOCK, C.J., FARRAR, J.F. and GORDON, A.J., eds). Bios Scientific Publishers, Oxford.

GRIME, J.P. (1993) Stress, competition, resource dynamics and vegetation processes. In: *Plant Adaptation to Environmental Stress* (FOWDEN, L., MANSFIELD, T. and STODDART, J., eds), pp. 45–63. Chapman and Hall, London.

HARDY, R.W.F., HAVELKA, U.D. and QUEBEDEAUX, B. (1978) Increasing crop productivity: the problem, strategies, approach and selected rate limitations related to photosynthesis. In: *Proceedings of the Fourth International Congress on Photosynthesis, 1977* (HALL, D.O., COOMBS, J. and GOODWIN, T.W., eds), pp. 695–719. The Biochemical Society, London.

HEMERLY, A.S., FERREIRA, P., ENGLER, J. DE A., VAN MONTAGU, M., ENGLER, G. and INZE, D. (1993) cdc2a expression in *Arabidopsis* is linked with competence for cell division. *Plant Cell* **5**, 1711–1723.

HOUSLEY, T.L. and POLLOCK, C.J. (1985) Photosynthesis and carbohydrate metabolism in detached leaves of *Lolium temulentum* L. *New Phytol.* **99**, 499–502.

JANG, J.-C. and SHEEN, J. (1994) Sugar sensing in higher plants. *Plant Cell* **6**, 1665–1679.

KOCH, K.E., XU, J., DUKE, E.R., MCCARTY, D.R., YUAN, C.-X., TAN, B.-C. and AVIGNE, W.T. (1995) Sucrose provides a long distance signal for coarse control of genes affecting its metabolism. In: *Sucrose Metabolism, Biochemistry, Physiology and Molecular Biology* vol. 14 (PONTIS, H.G., SALERNO, G.L. and ECHEVERRIA, E.J., eds), pp. 266–277. American Society of Plant Physiologists, Rockville.

LEACH, G. (1976) *Energy and Food Production.* ch. 3, pp. 14–20. IPC Science and Technology Press, Oxford.

LEWIS, D.H. (1984) Occurrence and distribution of storage carbohydrates in vascular plants. In: *Storage Carbohydrates in Vascular Plants* (LEWIS, D.H., ed.), pp. 1–52. Cambridge University Press, Cambridge.

MOORE, G., DEVOS, K.M., WONG, Z. and GALE, M.D. (1995) Grasses line up and form a circle. *Curr. Biol.* **5**, 737–739.

NATR, L. (1967) Time-course for photosynthesis and maximum figures for the accumulation of assimilates in barley leaf segments. *Photosynthetica* **1**, 29–36.

NEALES, T.F. and INCOLL, L.D. (1968) The control of leaf photosynthesis rate by the level of assimilate concentration in the leaf: a review of the hypothesis. *Bot. Rev.* **34**, 431–454.

NIE, G.Y., LONG, S.P., GARCIA, R.L., KIMBALL, B.A., LAMORTE, R.L., PINTER, P.H., WALL, G.W. and WEBBER, A.N. (1995) Effects of free-air CO_2 enrichment on the development of the photosynthetic apparatus in wheat, as indicated by changes in leaf proteins. *Plant Cell Environ.* **18**, 855–864.

POLLOCK, C.J. (1986) Fructans and the metabolism of sucrose in higher plants. *New Phytol.* **104**, 1–14.

POLLOCK, C.J. and CAIRNS, A.J. (1991) Fructan metabolism in grasses and cereals. *Annu. Rev. Plant Physiol. Plant Mol. Biol.* **42**, 77–101.

POLLOCK, C.J. and FARRAR, J.F. (1996) Source: sink relations – the role of sucrose. In: *Photosynthesis and the Environment* (BAKER, N.R., ed.), 261–279. Kluwer, The Netherlands.

POLLOCK, C.J., EAGLES, C.F. and SIMS, I.M. (1988) Effect of photoperiod and irradiance changes upon development of freezing tolerance and accumulation of soluble carbohydrate in seedlings of *Lolium perenne* grown at 2°C. *Ann. Bot.* **62**, 95–100.

POLLOCK, C.J., EAGLES, C.F., HOWARTH, C.J., SCHUNMANN, P.H.D. and STODDART, J.L. (1993) Temperature Stress. In: *Plant Adaptation* to *Environmental Stress* (FOWDEN, L., MANSFIELD, T. and STODDART, J., eds), pp. 109–132. Chapman and Hall, London.

POLLOCK, C.J., WINTERS, A.L., GALLAGHER, J. and CAIRNS, A.J. (1995) Sucrose and the regulation of fructan metabolism in leaves of temperate gramineae. In: *Sucrose Metabolism, Biochemistry, Physiology and Molecular Biology* (PONTIS, H.G., SALERNO, G.L. and ECHEVERRIA, E.J., eds), pp. 156–166. American Society of Plant Physiologists, Rockville.

POLLOCK, C.J., CAIRNS, A.J., SIMS, I.M. and HOUSLEY, T.L. (1996) Fructans as reserve carbohydrates in crop plants. In: *Photoassimilate Distribution in Plants and Crops* (ZAMSKI, E. and SCHAFFER, A.A., eds), pp. 97–113. Marcel Dekker, New York.

SHEEN, J. (1994) Feedback – control of gene-expression. *Photosyn. Res.* **39**, 427–438.

SIMPSON, R.J., WALKER, R.P. and POLLOCK, C.J. (1991) Fructan exohydrolase activity in leaves of *Lolium temulentum* L. *New Phytol.* **119**, 527–536.

STITT, M., VON SCHAEWAEN, A. and WILLMITZER, L. (1990) 'Sink' regulation of photosynthetic metabolism in transgenic tobacco plants expressing yeast invertase in their cell wall involves a down-regulation of the Calvin cycle and an up-regulation of glycolysis. *Planta* **174**, 217–230.

WAGNER, W., KELLER, F. and WIEMKEN, A. (1983) Fructan metabolism in cereals: induction in leaves and compartmentation in protoplasts and vacuoles. *Zeitschr. Pflanzenphysiol.* **112**, 359–372.

WILLIAMS, J.H.H., WINTERS, A.L. and FARRAR, J.F. (1992) Sucrose, a novel plant growth regulator. In: *Molecular, Biochemical and Physiological Aspects of Plant Respiration* (LAMBERS, H. and VAN DER PLAS, L.H.W., eds), pp. 463–469. SBP, The Hague.

WINTERS, A.L., WILLIAMS, J.H.H., THOMAS, D.S. and POLLOCK, C.H. (1994) Changes in gene expression in response to sucrose accumulation in leaf tissue of *Lolium temulentum* L. *New Phytol.* **128**, 591–600.

WINTERS, A.L., GALLAGHER, J.A., POLLOCK, C.J. and FARRAR, J.F. (1995) Isolation of a gene expressed during sucrose accumulation in leaves of *Lolium temulentum* L. *J. Exp. Bot.* **46**, 1345–1350.

WOOLHOUSE, H.W. (1981) Crop physiology in relation to agricultural production: the genetic link. In: *Physiological Processes Limiting Plant Productivity* (JOHNSON, C.B., ed.), pp. 1–21. Butterworths, London.

Index

abscisic acid 106, 185
acetate 311, 312
acetohydroxyacid synthase 169
acetolactate synthase 169
acetylcarnitine 312
acetylcoenzyme A (CoA) 188, 311–313
acetylcoenzyme A carboxylase (ACCase) 313, 316
active oxygen species (AOS) 4, 16, 18, 19, 298
  detoxification 16
acyl carrier protein (ACP) 313, 315, 320
  16:0 ACP 322
Δ^9 18:0-ACP desaturase 313, 320, 322, 323
18:1 desaturase 321
acylACP thioesterase 319, 321
acylCoA 316, 318
acylCoA synthetase 315
sn-1-Acyl-glycerol-3-phosphate acyltransferase 318
acyltransferases 315
adenosine diphosphate-glucose (ADPGlc) 84
  synthesis 88, 96
adenosine diphosphate-glucose pyrophosphorylase (ADPGlc PPase) 52, 82, 83–85, 87, 88, 90, 91, 94–96, 109, 113, 138
  bacterial ADPGlcPPases 90
  brittle-2 mutant 83
  glg C16 gene 95
  lys residue 195 of E. coli enzyme 90
S-adenosyl methionine (SAM) 112
adipic acid 321
Agrobacterium rhizogenes 70
alanine 211, 241, 284, 285
alanine dehydrogenases 162
aldolase 51, 52
alfalfa 21
alternative oxidase 296–305
  gene (cDNA) 297
alternative respiratory pathway 258, 295, 296, 298–300, 302
Amanita muscaria 279, 280
amino acids 56, 71, 131–133, 135, 138, 155, 237, 257, 305
  aromatic 166, 168
  biosynthesis 113, 128, 155, 188, 195, 266, 285
  carrier 241
  cycling 264, 267
  efflux 241
  essential 166
  exudation 242
  neutral I transporters 242
  neutral II transporters 242
  proteinogenic 211
  transfer of amino acids (mycorrhizal) 284
  transport (H^+/amino acid symport) 242
  uptake 241
ammonia 116, 133–135, 155, 188, 237, 241, 244–246, 248
  assimilation 56, 282
  carrier-mediated NH_4^+ uniport 240, 245
  constitutive high affinity transport system (CHATS) 238
  effluxes 247
  re-assimilation 135
  transport 247
  uptake 240, 243, 261
ammonium transporters 241
  AKT1 241
  AMT1 241
  high affinity systems 241
  low affinity transport systems (LATS) 238, 240
α-Amylases 53
amylolytic starch degradation 53
amylopectin 81–83, 98, 99
  structure 97
amyloplast 81, 96
amylose 81, 83, 93, 98, 99
  biosynthesis 99

337

Index

Anabaena 90, 92
anapleurotic pathway 125, 128, 136, 285
anoxia 189, 194, 195
anthranilate 169
　derived alkaloids 168
antimycin 296, 297, 299
antioxidant 19
antisense 15, 46, 52, 66, 69
aphid 208, 212
apoplasm 206–213, 221
　invertase 210, 277
　loading 70
　space 207, 213
Arabidopsis thaliana 25, 83, 90, 95, 112, 114, 132, 135, 168, 171, 185, 240–242, 297, 321, 323, 324
　AAP1 mutants 242
　ABI1 mutants 185
　NAT2 mutants 242
　soz 1 mutants 25
ascomycetes 275
ascorbate (AsA) 20, 25
ascorbate peroxidase (APX) 8, 20, 21
asparagine 111, 163, 244
　transport 164
asparagine synthetase (AS) 108, 111, 156, 162, 163
aspartate 156, 163, 164, 166, 185, 211, 242, 283, 284
Aspartate amino transferase 156, 162, 247
aspartate dehydrogenases 162
aspartate kinase (AK) 164, 165
　bacterial 165
　feedback-insensitive activity 166
　isozymes 164
3-aspartic semialdehyde 164, 166
3-aspartyl phosphate 164
3-aspartyl phoshpate, 2-dihydropicolinate synthase (DHPS) 165
　feedback-insensitive activity 166
Aspergillus nidulans 240
assimilate (see also sucrose)
　accumulation 71, 333, 334
　acquisition 327, 328, 330, 333, 334
　export 67, 208
　allocation 223, 224, 227–229, 231, 332–334
　partitioning 51, 107, 280
　sinks 278
assimilatory power 7
ATP 11, 84, 182, 295, 312
　plasma membrane 183
　synthesis 7
　synthetase (ATPase) 12, 15, 45, 70
　synthetase inhibitors 238
　ATP/ADP ratio 43, 71, 107, 116
　ATP/NADPH ratio 12, 13
auxin 106, 168

bacteroid 246
barley 92, 206, 207, 209, 210, 212, 213, 329
basidiomycetes 275, 280
Benson-Calvin cycle (see photosynthesis and carbon assimilation) 9, 105, 125, 187
benzoic acid 19
betulaceae 275

biomass accumulation 179
　dry matter allocation 223
　dry weight 141
biopolymer polyhydroxybutyric acid (PHB) 317
blue-green algae 184
Bryophyllum fedtschenkoi 192
bundle sheath/phloem parenchyma (BS/PP) boundary 229
bark storage protein (BSP) 109

C_3 plants 7
C_3-C_4 intermediate 66
C_4 plants 7, 11, 128, 187
calcium 116
　binding motif 185
　calmodulin dependent PP2B 184
　dependent CDPK 116
　phospholipid-derived signalling 182
callose ([1→3]-β-D-glucan) synthesis 230
calmodulin-domain protein kinase (CDPK) 183
Camissonia claviformis 13
cAMP 182
carbohydrate 54
　accumulation 210, 228
　availability 111
　Björkman's carbohydrate theory 280
　depletion stress 109
　deprivation hypothesis 247
　gene modulation 108, 109, 116
　partitioning 226, 227
carbon 227
　allocation 223, 278
　assimilation 125
　export 67, 110, 229
　^{14}C-export 229
　$^{14}CO_2$ feeding 50
　importing cells 110
　limited conditions 110
　metabolism 105
　modulated genes 108, 109
　partitioning 63, 69, 70, 125, 128, 223, 224, 231, 285
　reomobilization genes 108
　supply 280
carbon dioxide 50
　assimilation 8, 11–13, 16, 128, 208, 209
　dark reactions 43
　enrichment 131, 281, 329
carbon-nitrogen
　balance 105, 108–110, 112
　ratio 131, 247
　regulation 189
2-carboxy-arabinitol-1-phosphate (CA1P) 45, 48
1-(O-carboxyphenylamino)-1-deoxyribulose-5-phosphate 168
catalase 19, 24
　isoforms 24
　photoinactivation 24
cell
　division 258
　expansion 258
　membranes 311

turgor 258
volume 207
cellulase 282
Cenococcum geophilum 280
chaperones 160
  Gro EL 160
  GS-chaperone complex 160
chemotaxis 185
chilling (see stress) 18, 66, 330, 331
  sensitivity 324
  temperatures 18
chitinase (acidic promoter) 161
Chlamydomonas 97
  reinhardtii 240
chlorate 240
Chlorella 7, 112, 114
chlorophyll
  degradation 58
  fluorescence 17, 18
  singlet state 16
  triplet formation 17
chloroplasts 41, 53, 206
  ACCase 313
  envelope membranes 57, 58
  Fru1,6P$_2$ase 50
  genes 110
  NAD(P)H dehydrogenase 24
  phosphate translocator 208
  stroma 206–213
chlororespiration 24
chlorsulphuron 169, 170
chorismate 167, 168
  derived aromatic amino acids 168
chromoplasts 54, 58
circadian controls 260
circadian oscillator 128, 181
circadian rhythm 189, 192, 260
citrate 300, 303
climax vegetation types 334
cold girdling 214, 224
companion cell 70, 231
  sieve element (CC-SE) complex 221
  sieve tube complex 212, 213
control analysis/theory 46, 51, 52, 86, 316
control coefficient 15, 51, 70, 87, 94
control strength 50
cotton 329
cotyledons 110
coupling factor (CF$_0$-CF$_1$; see also ATP synthetase) 45
crassulacean acid metabolism (CAM) 54, 128, 192, 305
crop yield 8, 55, 63, 70
cucumber mosaic virus (CMV) 227
cucurbita 110
cyanobacteria 10, 205
cyclic electron transport 6, 10, 12
cyclohexamide 193
cyclophilin 184
cyclosporin 184
cysteine 22
cytochrome c oxidase (see also respiration) 302, 304
  complex I 297
  complex III 297

complex IV (KCN) 297
cytochrome b_6/f complex 5, 6, 12, 14, 15, 17, 187
cytokinins 10c
cytosol 22, 54, 65, 71, 207, 313

debranching enzymes 53, 82, 83
dehydroascorbate reductase 20
3-deoxy-D-arabino-heptulosonate-7-phosphate (DAHP) synthase 167
desaturases 320, 322
diacylglycerol (DAG) 315, 324
diacylglycerol acyltransferase 316
diacylglycerol:cholinephosphotransferase 316
diarrhetic shellfish poisoning 184
dicarboxylic acid transporter 211, 247
2,3-dihydrodipicolinate 164
dihydrodipicolinate synthase (DHPS) 164
dihydroxyacetone phosphate (DHAP) 53, 129, 167, 208, 312
dinitrogenase reductase 248
  activating glycohydrolase 248
dinitrogenase reductase ADP-ribosyl transferase (DRAT) 248
dithiothreitol 300
drought stress 106
Dunaliella 15

ectomycorrhizal fungi (ECM) 275, 277, 279, 281–285
embryo-specific promoter 317
endoglycosidases 111
endogonaceae 275
endomycorrhizas 275
endoribonuclease 162
endosperm 72, 110
5-*Enol*pyruvyl-shikimate-3-phosphate synthase (EPSPS) 168
environmental pollution 327
epidermis 205, 206
episodic growth 261
epoxidases 322
ergosterol 279
Eriophorum vaginatum 241
erucic acid 319, 320
erythrose 4-phosphate (E4P) 43, 50, 167
Eschericia coli
  ADPGlcPPase gene 83
  3-aspartyl phoshpate, 2-dihydropicolinate synthase (DHPS) 165
  glg C gene 94
  gor deletion mutant 23
  mutants 168
  pyrophosphatase 65
ethylene 106, 298
  sensing proteins 185
excitation energy 17
  distribution 10
exciton 4, 17

fagaceae 275
fatty acids 56, 71, 311–313, 315, 316, 320, 322–324
  polyunsaturated 316
  unsaturated 320

Index

fatty acid synthetase (FAS) 317, 319
Fenton reaction 20
ferredoxin 5, 15, 18, 45, 320
ferredoxin-dependent GOGAT (see also Glutamine:2-oxoglutarate aminotransferase) 112, 134, 135, 163
ferredoxin-NADP oxidoreductase (FNR) 6, 18
ferredoxin-quinone reductase (FQR) 6
ferredoxin/thioredoxin reductase 45
flavine adenine dinucleotide (FAD) 132
flavonoids 313
flower induction 109
flux control 8, 186, 316
   coefficient 46, 93
free radicals (see active oxygen species, AOS) 305
freezing stress 20
fructan 210, 330–332
fructokinase 211, 281
fructose 70, 81, 226, 277, 279,
fructose 1,6 bisphosphate (Fru1,6P$_2$) 43, 50, 64–67, 129, 139, 208
fructose 1,6-bisphosphatase (Fru1,6P$_2$ase) 3, 45, 64, 208, 281,
   antisense 66, 69
fructose 2,6-bisphosphate (Fru2,6P$_2$) 65, 107, 116, 190, 281
fructose 6-phosphate (Fru6P) 43, 50, 64, 68, 69, 208
fructose 6-phosphate 2-kinase/fructose 2,6-phosphate phosphatase (6PF2K/Fru2,6P$_2$ase) 179, 182, 188–190
fruit
   chloroplasts 54, 55
   photosynthesis 55
   ripening 298
   set 109
   olive 56
fungi 275
fusicoccin 193
galactolipids 324
gerontoplasts 58
gibberellic acid 68
Glomus mosseae 278
glucagon 186
α-(1,4)-glucan 82
gluconeogenesis 65, 284
glucopyranose (α-D-glucopyranosyl residues) 81, 82
glucose 53, 81, 226, 277, 279
   responsive genes 106, 116
   mediator complex (SNF1 to SNF6) 117
   mutants *snf* family (*1–4*) 183
   transporter 53, 54
glucose 1-phosphate (Glc1P) 53, 65, 70, 84, 90
glucose 6-phosphate (Glc6P) 50, 54, 58, 67–69, 129, 139, 190, 312
   dependent starch synthesis 57
   transport 58
glucose-6-phosphate dehydrogenase 45
glucosinolates 168, 171
glutamate 107, 130, 133, 138, 157, 164, 170, 195, 211, 242, 246, 283–285
   biosynthesis 156
glutamate dehydrogenase (GDH) 108, 111, 112, 156, 162, 164
   assimilatory function 164
glutamate semialdehyde 171
glutamate synthase (see also glutamine:2-oxoglutarate aminotransferase, GOGAT) 109, 111, 112, 134, 135, 156, 162, 163, 244, 266, 282
glutamine 107, 115, 129–131, 133, 134, 138, 156, 157, 163, 164, 195, 211, 244, 284, 285
   translocation 282
glutamine:2-oxoglutarate aminotransferase (Fd-dependent, GOGAT; see also glutamate synthase) 112, 135, 162, 163, 244
glutamine synthetase (GS) 111, 112, 131, 133, 155, 244, 266, 282
   assembly related factor 160
   chaperone complex 160
   chloroplastic (GS2) 133, 134, 136, 157, 158, 160
   chloroplastic (GS2) isozymes 105
   cytoplasmic (GS1) 133, 134, 136, 157–161
   cytoplasmic (GS1) gene expression 162
   mutants 135
   polypeptides 159
   ribozymes 162
glutamine synthetase (GS/GOGAT) cycle 134, 164, 246, 247, 282, 284
   vascular localized isoenzymes 161
γ-glutamyl cysteine 22
γ-glutamyl cysteine synthetase (γ-ECS) 22
glutathione 19, 20, 22
   biosynthesis 22
glutathione reductase (GR) 20, 22
   activity 22
   gor gene 22
glutathione synthetase (GSH-S) 22
glyceraldehyde phosphate (G3P) 208
glyceraldehyde-3-phosphate dehydrogenase (GAPDH) 3, 45, 39
   pathway 284
   subunit 49
glycerate 136
glycerolipid synthesis 56, 324
glycerol-3-phosphate 324
glycerol-3-phosphate acyltransferase 324
glycine 133, 135, 138, 188, 211
   oxidation 188
glycine decarboxylase 135
Glycine max 297, 300
glycogen 279
glycogen synthase 93
glycolysis 65, 66, 70
glyoxylate 304
glyoxysomes 24
glyphosate 168
   resistant plants 168
cGMP 182
grain yield 328
green pepper 57
GTP 182
guard cells 193
Gossypium hirsutum 296

Haber-Weiss reaction 20
Hansenula anomala 297

Hartig net 275
Harvest index 219, 328
Hebeloma crustuliniforme 278, 279
herbicides 135, 170, 179
   chlorsulphuron 170
   paraquat (methylviologen) 20–23
   Sceptor 170
heterotrophic plastid 57, 58
hexokinase 73, 116, 332
hexose (see glucose, fructose) 81, 132, 210
   concentrations 223
   /sucrose ratio 72, 73
   uptake 211
hexose-phosphate 208
   translocator 57, 67
histidine 242
histidine kinase 185
homoserine dehydrogenase 166
hormones 189
hydraulic conductance 223
hydrogen peroxide (H_2O_2) 19, 20, 298, 304, 305
   induced signal 23
   scavenging enzymes (see also ascorbate peroxidase) 23
hydroxylases 322
hydroxyl radical 19
3-hydroxy-3-methyl glutaryl CoA reductase (HMGR) 188
3-hydroxy-3-methyl glutaryl CoA reductase (HMGR) kinase 181, 183
hydroxypyruvate reductase 136, 160
hygromycin phosphotransferase (HPT) 161
hypersensitive response 19, 61

indole 168
   glucosinolates 171
inorganic phosphate (Pi) 44, 64, 65, 67, 69, 84, 85, 139, 189, 190, 208, 219
inorganic pyrophosphate (PPi) 65, 70
insulin 186
interdependence theory 255, 256, 258, 264, 266, 267
invertase 70–73, 109, 115, 139, 279
   acid 71, 210, 277,
   alkaline 71, 246
   apoplastic 210, 277
   cytosolic 71
   fungal 277
   genes 105
   seed coat-associated 72
ion channels 180
iron-sulphur centre (FeS) antisense-Reiske 15
irradiance (see light) 16
isoamylase 83, 99
isocitrate 301
isoleucine 169, 170, 241
isomerase 43

Juanulloa aurantiaca 9
juvenility 109

Kennedy pathway 315
3-keto-arabinitol-1,5bisphosphate 44
α-ketoglutarate (2-oxoglutarate) 156, 211, 246, 284, 285
kinase (see protein kinase)
kinase kinases 181
kinase-associate protein phosphatase (KAPP) 185
KNOTTED1 230

Laccaria laccata 278
lac operon 106
lauric acid 321
leaves 110
   development 264
   senescence 134
leghaemoglobin 246
leucine 169, 170, 241
light 16
   /dark regulation 45
   harvesting 4, 16
   harvesting chlorophyll *a/b* protein (LHCII) 10, 187
   harvesting pigments 4
   harvesting reactions 110
   incident 13
   reactions 43
   saturation 8
   transition to dark 48
Limnanthes 320
lipid 56, 311
   biochemistry 311
   peroxidation 19
   reserves 110
   synthesis 56, 108
Lolium temulentum 330, 331
low temperature (see chilling, stress) 66
low-affinity NH transport systems (LATS) 238, 240
lysine 16–166
   biosynthesis 165
lysine-ketoglutarate reductase 166
lysophosphatidate acyltransferase (LPAT) 318, 320

maize 92, 114, 191, 296
   endosperm 90, 94
   leaf SPS 69
   leaf SPS (expression in tomato) 139–146
maize mutants
   brittle-2 83
   miniature-1 72
   shrunken-1 72
   shrunken-2 83
   Su 1 82
   sugary 1
malate 128, 192, 210, 211, 245, 247, 262, 264, 300, 312
mind1 > decarboxylation 305
malate dehydrogenase 45, 246
maltose 53
mannitol 279, 280, 285
malonyl-CoA 313
Marchantia 10
Mehler-peroxidase cycle 8
Mehler reaction 7, 12, 16, 18
membrane
   fluidity 324

Index

impermeable fluorescent probes 222
lipid biosynthesis 313
lipids 315, 320
Mesembryanthemum crystallinum 53, 54
metabolism
    control 48
    engineering 63
    partitioning 63
    sinks 71
    subcellular metabolite concentrations 205, 211–215
methionine 164
methionine sulphoximine 240
methylanthranilate 169
methyl jasmonate 106
microbial challenge 167
microcystin 184, 188, 189
mitochondria 134, 295, 298, 299
mitogen-activated protein (MAP) kinases 181, 182
monodehydroascorbate (MDHA) 8, 20
monodehydroascorbate reductase (MDHR) 20
movement proteins (MPs) 225–232
    tobacco mosaic virus movement protein (TMV-MP) 225, 226
Münch (1930) hypothesis 223
mutants (see *Arabidopsis thaliana*; see maize)
mycorrhiza 275, 277, 279, 281, 283
    ectomycorrhiza (ECM) 275, 277, 279, 281–285
    endomycorrhizas 275
    ericoid 277
    infection 280

NADH 295, 312
NADH dehydrogenase (chloroplast) 24
    ndh genes 24
NADH dehydrogenase (mitochondrial) complex 1 24, 295
NADH-dependent glutamine:2-oxoglutarate aminotransferase (GOGAT) 134, 163
NADP-dependent glutamate dehydrogenase (NADP-GDH) 282, 284
NADPH 5, 295, 312
NADPH/NADP ratio 43, 56
Nicotiana plumbaginifolia 21, 129–132
    35S-NR transformants 129,132
Nicotiana tabacum (see tobacco) 297, 301
    ΔNR transformants 132
nitrate 155, 237, 242, 244, 245, 248, 257, 261, 305
    anion/cation effects 262
    assimilation 116, 127, 181, 188, 195, 248, 256, 264, 266
    carriers 240–245
    carrier, $2H^+$ / $1NO_3^-$ symport 238
    carrier, HCO_3^-/NO_3^- antiport 245
    carrier, $2OH^-/1NO_3^-$ antiport 238
    carrier, CHL1 241
    counter-ions 262
    delivery 265
    efflux 239
    inducible high affinity transport system (IHATS) 238
    reduction 7, 56, 256, 260, 282

    sensing mechanism 238
    storage pool 248
    translocation 264
    uptake 240, 243, 145, 259, 261–264, 267
nitrate reductase (NR) 109, 112, 115, 125, 133, 140, 162, 179, 181, 182, 188, 189, 191, 193, 194, 196, 246, 248, 256, 261, 265, 266, 267, 282, 284
    ΔNR transformants 132
    activation 132
    cofactors 132, 187
    constitutive expression 129, 130–132
    fungal enzyme 282
    gene (*nia*) 125
    inhibitor protein (NIP) 132, 180, 191, 192 197
    mRNA stability 131, 133
    multisite phosphorylation 181
    protein 130
    protein kinase 189, 191, 193, 194
    serine-543 192
nitrification 283
nitrite 112, 133, 155, 266, 282
    nitrite toxicity hypothesis 247
nitrite reducatase (NiR) 112, 133, 155, 266, 282
nitrogen 155, 328
    assimilation 105, 111, 125, 128, 161
    assimilation genes 108
    availability 48, 312
    deficiency 66, 130, 261
    export genes 108
    fertilization 214
    fixation 155, 211, 246, 247
    import/metabolism 108
    metabolism 277
    organic sources 283
    regulation 185
    remobilization genes 108
    responsive genes 116
    starvation 244
    status 189
    uptake 243, 256
    use efficiency 329
    utilization 130, 205, 237
nitrogen fixation 155, 211, 246, 247
    inhibition 247
nitrogenase 246, 248
nodules 134
non-aqueous fractionation 207, 208
non-cyclic electron transport 6, 8, 10, 12, 13, 20
non-photochemical quenching (of chlorophyll *a* fluorescence; q_N) 17, 18
non-protein fungi 283
nuclear genes 110
nucleic acid transport (viral) 226
nutrients 54

oxygen
    evolution 9, 11, 12, 16
    isotope fractionation 302
    uptake 299
    sensitivity 6
oats 329

oil biosynthesis 311
okadaic acid 184, 187–189
oleic acid 321
olive (see also fruit) 56
organic acids 156, 284, 295, 300, 303, 305
   synthesis 189
ornithine 170
osmotic stress 194
oxaloacetate (OAA) 56, 128
oxidative pentose phosphate pathway 45, 53, 54, 56
oxidative phosphorylation 295
oxidative stress 20, 21
oxoglutarate (α-ketoglutarate) 156, 211, 246, 284, 285
ozone 20, 21, 23

P680 5, 17
P680$^+$ 16
P700 5
paraquat 20, 21, 23
patatin 73, 108, 111
   class I patatin gene 228
pathogen responses 19, 106
peanuts 329
peas (*Pisum sativum*) 296, 300, 329
pectinase 282
pedicel 72
peribacteroid membrane 246, 247
perivascular parenchyma 146
peroxidases (PX; see also ascorbate peroxidase) 19
peroxisomes 24
petroselinic acid 321
Phaseolus vulgaris 296, 329
phenylalanine 166
   H$^+$-coupled antiport system 242
phloem 219
   complex 70
   export 221
   function 71
   loading 70, 146, 213, 214, 222, 229, 231, 280
   parenchyma (BS/PP) boundary 229
   protein 230
   sap 208, 212–214
   translocation 245
phosphatase (see protein phosphatase) 181, 185, 186, 208
phosphatidic acid 315
phosphatidylcholine (PC) 315
phosphatidylglycerol (PG) 324
phosphinothricin 157
phospho*enol*pyruvate (PEP) 7, 56, 66, 128, 167, 168, 187
phospho*enol*pyruvate carboxylase (PEPCase) 54, 108, 110, 127, 128, 130, 131, 136, 156, 162, 179, 181, 182, 188, 189, 192, 193, 195, 196, 285
   anapleurotic role 193
   kinase 128, 180, 192, 193
phospho*enol*pyruvate carboxykinase (PEPCK) 195, 285
phosphoesters 182
phosphofructokinase (6-phosphofructo-2-kinase (PFK) 64, 65, 190

3-phosphoglycerate (PGA) 7, 42, 49, 53, 64, 66, 67, 84, 85, 87, 91
   Pi ratio 52
phosphoglycerate kinase (PGK) 49
2-phosphoglycolate 42
phosphohistidine 185
phospholipids 56
5-phosphoribosylanthranilate 168
phosphoribosylanthranilate isomerase (PAI) 168
   genes 168
phosphoribulokinase 3, 51
phosphorolysis 187
photochemistry 13,
photoinhibition 16–18, 21
photorespiration 7, 8, 11–13, 42, 55, 116, 133, 135, 155
   flux 138
   nitrogen recycling 112
photosynthate (see assimilate) 51
photosynthesis (C_3) 4, 49, 110, 111, 208, 209, 278, 328
   C_4 193
   absorbance spectra 10
   acclimation 14, 16, 69, 329
   capacity 13, 14
   charge seperation 4, 10
   electron transport 4, 10, 13, 16
   genes 16, 108
   H$^+$/ATP ratio 12
   phosphate limitation 66
   quantum efficiency 8, 12, 17
   quantum efficiency (PSI) 14–16
   quantum efficiency (PSII) 14–16
   quantum yield 8, 9, 11–13
   quantum yield (PSII) 135
   reaction centres (rc_I; rc_{II}) 15
   steady state 49
photosynthetic control 18
photosystem I (PSI) 5, 9, 18, 187
photosystem II (PSII) 5, 9, 18, 21, 187
phylogenetic trees 182
phytoalexins 168
phytoglycogen 82, 99
pinaceae 275
Pinus sylvestris 278, 279
planimetric analysis 206
plant development 107
plasmodesmata 70, 221–223, 225, 231,
   function 230
   neck constriction 230
   size exclusion limit (SEL) 226, 230, 229
plastocyanin 5, 15
plastoquinone/plastoquinol 6, 10, 14, 15, 17
   pool 18
polysaccharide 81
potassium channels 184
potassium cyanide 296
potato 73, 83, 90, 92, 170, 191, 206, 207, 209, 210, 212, 213
pre-amylopectin 99
pressure gradient 223
prokaryotes 90
proline 170, 242

Index

promoters
  rolC promoter 70
  β-phaseolin gene promoter 166
  Rubisco SSU (Rbcs) 69, 141
  35S CaMV 69, 73, 130, 132, 140, 170, 225
proteases 111
  extracellular 283
protein
  degradation 257
  fungi 283
  import 7
  investment 109
  lysine content 165
  phosphorylation 15, 179, 180–182, 186
  remobilization 111, 116
  stability 133
  synthesis 56, 180, 261, 264
proteinase inhibitors 73
protein kinases (PSK) 68, 107, 179, 181, 182, 185, 186
  autoinhibitory domains 182
  autophosphorylated 185
  classification 183
  histidine/aspartate kinases 188
  intracellular localisation motifs 182
  ligand binding 180
  nodulin 26 protein 183
protein (serine/threonine) kinases 182, 184
  AGC group 182, 183
  AMPK family (*snf*) 183
  CaMK group 182, 183, 195
  CMGC group 182, 183
  receptor RLK5 185
protein phosphatases 180, 184
  kinase-associated (KAPP) 185
  PP1 184, 188
  PP1/2A/2B gene family 185
  PP2A 184, 188–190, 194, 195
  PP2B 184
  PP2C 185, 187, 188, 194
  regulatory subunits 184
  Type 1 / Type 2A 187
proteolysis 85
proton-motive force 6, 298
pseudo-cyclic electron transport 7, 8
pullulanase 99
pyridine nucleotides (see NADH, NADPH)
  pool 302
pyridoxal phosphate 91
pyrophosphate 65
pyrophosphatase 65
pyrophosphate:fructose-6-phosphate-1-phosphotransferase (PFP) 64
  phosphate limitation 66
pyrophosphate (Pi-dependent) phosphofructokinase 281
Δ-pyrroline-5-carboxylate 170
Δ-pyrroline-5-carboxylate reductase (P5CR) 170
Δ-pyrroline-5-carboxylate synthetase (P5CS) 171
pyruvate 164, 188, 299, 303, 311, 312
pyruvate carboxylase (PC) 286
pyruvate decarboxylase kinase 185

pyruvate dehydrogenase multienzyme complex (PDC) 116, 188
  kinase 188
  phosphatase 188
pyruvate Pi dikinase (PPdK) 187
  kinase 180
  regulatory protein (PPRP) 187

Q-cycle 6, 12
Q_A 10, 17
Q_B 10
quantum efficiency (φ; see photosynthesis) 8, 12, 17
  ϕ_{PSI} 14–16
  ϕ_{PSII} 14–16
quantum yield 8, 9, 11–13
quantum yield of PSII 135
quenching (of chlorophyll *a* fluorescence) 17

R enzyme (see also starch) 99
raffinose 212
reaction centres (rc_I; rc_{II}) 4, 15
red mottle comovirus 224
reducing power (see also NADPH, ferredoxin) 11
reducing sugars (see also hexoses) 73
resource acquisition 333, 334
  allocation 223, 224, 227–229, 231, 332–334
  capture 327, 328, 330
  partitioning 107
respiration 256, 295
  alternative respiratory pathway 258, 295, 296, 298–300, 302
  carbon dioxide 55, 56
  crisis 303
  cyanide-resistant quinol oxidase 296
  cyanide-resistant respiration 295, 299, 302–304
  cytochrome c oxidase pathway 296, 299–301, 304
  electron overflow 301
  electron transport 295
  energy conservation 297
  energy overcharge model 304
  energy overflow hypothesis 303
  metabolism 71
  rate 229
relative growth rates (RGR) 141–145
Rhizobia / Rhizobium 205, 246, 305
Rhizopogon vinicolor 278
ribose 5-phosphate 42, 43, 51
ribozymes 162
ribulose 1,5-bisphosphate (Ru1,5P$_2$) 7, 8, 42–44, 50, 51
ribulose-1,5-bisphosphate carboxylase-oxygenase (Rubisco) 7, 8, 41–45, 48, 55, 95, 108, 110, 135, 138, 192
  activation 45
  activity 187
  antisense 46
  carbamylation 44, 48
  carboxylation reaction 41
  fall-over 48
  nocturnal inhibitor 45
  protein 41, 48

ribulose-1,5-bisphosphate carboxylase-oxygenase (Rubisco) activase 3, 45, 48, 49, 187
ribulose 5-phosphate (R5P) 42, 43
ribulose 5-phosphate kinase 42, 45, 51
rice 90, 329
ricinoleic acid 322
Ricinus communis 322
root 238
  cortex 110
  extension 259
  growth 267
  hairs 275
  nodules 134
  respiration 302
  symplasm 259
  /shoot ratio 225, 226, 279

salicaceae 275
salicylhydroxamic acid (SHAM) 299, 301, 302
salicylic acid 19, 298
Salmonella typhimurium 87, 90
Sauromatum guttatum 297, 298
scenedesmus 7
second messengers 180
sedoheptulose 1,7-bisphosphate (S1,7P$_2$) 42, 43
sedoheptulose 1,7-bisphosphatase (S1,7P$_2$ase) 45, 50, 52
  antisense 52
seed coat 72
serine 133, 135, 138, 180, 192, 211, 283, 543
shikimate pathway 166–168
  synthesis 7
shoot
  apical meristem 256
  excision 259
  /root ratio 243, 255
sieve
  element (SE) 221, 222
  tube 205, 207, 208, 214
signal transduction 109, 113, 180
  supracellular communication network 231
  trafficking macromolecules cell-to-cell 230
singlet oxygen 16
sink
  capacity 54
  development 71
  leaves 66
  strength 71, 73, 278
  tissues 14
  /source interactions 210
size exclusion limit (SEL) 223, 225
SNF1 to SNF6 117
snf family (*1–4*) 183
source leaves 70
source-sink balance 224
source-sink relationship 139
soybeans 329
Spinacea oleracea (spinach) 90, 92, 206, 207, 209, 210, 212, 213, 296
sporamin 108, 111, 112
sporulation 185
starch 43, 50, 69, 71, 81, 188, 208, 277, 312
  accumulation 73, 223, 224, 229
  biosynthesis 52, 64, 83, 94
  branching enzyme (SBE) 93, 97, 109, 113

breakdown (degradation) 53, 67, 83
cluster model 97
D-enzymes 53
granule 82
granule-bound starch synthase 97
isoamylase 99
metabolism 52
mutants 81
phosphorylase 53, 108, 113
phosphorylytic pathway of starch breakdown 53
pullulanase 99
R enzyme 99
transitory 52
starch synthase (SS) 82, 93, 94, 96, 97, 113
  soluble (SSSII) 97
  UDP-glucose-specific 83
starvation 109
  responsive genes 109
state transitions (I,II) 10
stem girdling 259
storage
  carbohydrate 44
  lipids 311
  oil 315, 316
  proteins 73
  protein synthesis 312
  sinks 71
  sugars 330
stress 21
  chilling 18, 330, 331
  cold girdling 214, 224
  drought 170
  low temperature 66
  osmotic stress 194
  oxidative stress 20, 21
  ozone 20, 21, 23
  salinity 170
  starvation 109
  starvation responsive genes 109
  stem girdling 259
  tolerance 21
  water 131, 132
succinate 247
  oxidation 299
succinate dehydrogenase 304
sucrose 43, 69, 70, 81, 188, 208, 226, 279, 330
  accumulation 69
  active transport system 70, 210
  biosynthesis 64, 67, 267
  carrier gene 213, 223
  carrier/transporter 70, 210, 221
  control hypothesis 258
  efflux 228
  export 213, 228
  /glutamine ratio 130
  /hexose ratio 72, 73
  import/cleavage 108
  loading 221
  metabolism 63
  suc2 gene 71, 72
  hydrolysis 280
  uptake 277

utilization 70, 71
sucrose 6-phosphate (Suc6P) 64
sucrose phosphate synthase (SPS) 64, 67, 69, 70, 108, 128, 130, 131, 139, 179, 181, 188–191, 193, 194, 196, 281
   activity 68, 140, 144
   antisense repression 69
   co-suppression 69
   covalent modulation 139
   kinase 188, 191
   multisite phosphorylation 181
   /NR kinase 181
   protein 68
   serine-158 68, 191
   /starch partitioning 190
   /starch ratio 140
   synthesis 50, 52, 65, 67, 70, 128, 139, 140, 181, 189, 190, 191, 195, 219, 224, 280
sucrose synthase 70–73, 108, 195, 246, 277
   genes 73, 109
   shrunken-1 72
   shrunken-2 83
sucrose-phosphate phosphatase (SPP) 64
sugars (see sucrose, hexose 71
   beet 214
   enhanced senescence 112
   responsiveness 113
   uptake 280
sulphate reduction 7
sulphur dioxide 22
sulphide 304
superoxide (O_2^-) 7, 16, 19, 20, 298, 304, 305
superoxide dismutase (SOD) 7, 19, 20, 297
   Cu/Zn SOD 20, 21
   Fe SOD 20, 21
   Mn SOD 21
   over-expression 23
symbiosis 279
symplasm
   movement 225
   phloem loading 221, 212
   loading 70
Synechocystis 10, 90, 92
systemic acquired resistance (SAR) 19

taproots 214
tricarboxylic acid (TCA) cycle 128, 188, 295, 300, 303
thioesterase (C12 specific) 319
thermogensis 303
   plants 302
   tissues 298
thioesterase 318
thiol groups
   reduction 45, 46, 51
   regulation 46, 49
thioredoxin 15, 45
   system 187
threonine 164–166, 180
Thunbergia alata 322
thylakoid membrane 5
   ATPase 7
   charge seperation 4,10
   development 10

membrane proteins 15, 187
protein phosphatase 187
transmembrane electrical potential ($\Delta\Psi$) 6
transmembrane pH gradient (ΔpH) 6
trans-thylakoid eletrochemical potential difference ($\Delta\mu_H^+$) 7
tobacco (see also *Nicotiana plumbaginifolia*) 10, 21, 213, 301, 324
tobacco mosaic virus movement protein (TMV-MP) 225, 226
tomato 213
   SPS over-expression 139–146, 191
trafficking (macromolecules) 230–231
transcription factors 180
   KNOTTED1 230
transesterification 320
transit peptide 95
translocation 223, 259
   glutamine 161
transport proteins 57
   proton alanine symport 241
   proton co-transport 238
   proton sucrose cotransport 70
   proton symporters 213
   proton symporting transporters 214
   proton/amino acid transporters 241
trehalose 279, 285
triacylglycerol (TAG) 56, 315, 318
   synthesis 313
triose phosphate (TP) 43, 50, 64, 67, 208
   translocator 64, 67
triplet chlorophylls 16
Triticum aestivum 296
tryptamine 171
tryptophan 166, 167, 168
   biosynthetic pathway 168
   decarboxylase 171
tubers 110
   development 229
tyrosine 166, 167, 180
   kinases 182
   phosphatases 184

ubiquinone (Q) / ubiquinol (Q_r) 295–303
   Q_r/ total ubiquinone pool (Q_t) ratio 299, 301–303
ubisemiquinone 305
UDP-glucose (UDPGlc) 64, 65, 70, 83, 126
UDP-glucose pyrophosphorylase (UDPGlcPase) 65, 70, 126
ureides 247
uronic acids 25

vacuole 206–213
   fructan 110
valine 169, 170
vascular system 157, 221, 222
vegetative storage protein 109, 111
vesicular-arbuscular mycorrhizas (VAM) 275–282, 285
Vicia faba 72
virus
   infection 224
   movement 222

volume (subcellular) 206, 207

water
  availability 43
  status 189
  stress 131, 132
  use efficiency 329
waxes 313
wheat 329
  endosperm 90
  seed 92
wound respiration 65
wounding 106, 167
wrinkled pea 93

xylem 259
sap 264
  translocation 260
xylulose 1,5-bisphosphate 44
xylulose 5-phosphate 43

yeast invertase 140
yield (see biomass accumulation) 328, 329
  limiting factors 328

Zea mays (see maize) 296
zeaxanthin 17
Z-scheme 6, 9, 11, 12